Cryosphere and Environmental Change

Cryosphere and Environmental Change

Editor: Nicolas Judd

RCALLISTO
REFERENCE
www.callistoreference.com

Callisto Reference,
118-35 Queens Blvd., Suite 400,
Forest Hills, NY 11375, USA

Visit us on the World Wide Web at:
www.callistoreference.com

ISBN: 978-1-64116-105-3 (Hardback)

Cataloging-in-Publication Data

Cryosphere and environmental change / edited by Nicolas Judd.
    p. cm.
Includes bibliographical references and index.
ISBN 978-1-64116-105-3
1. Cryosphere. 2. Cold regions. 3. Climatic changes. 4. Global environmental change. I. Judd, Nicolas.
QC880.4.C79 C79 2019
551.31--dc23

Table of Contents

Preface

The cryosphere is that portion of the earth's surface where water is in solid form. This includes sea ice, river ice, glaciers, permafrost, etc. The cryosphere is an integral part of the climate system with feedbacks and linkages with clouds, moisture fluxes, atmospheric and oceanic circulation. Various physical properties of snow and ice affect the energy exchanges between the surface and the atmosphere. Some of these properties include surface reflectance, thermal diffusivity, latent heat, etc. The environmental change observed in terms of global warming results in major changes to the partitioning of snow and rainfall. It also affects the timing of snowmelt. The field of study dedicated to the study of cryosphere is known as cryology. This book strives to provide a fair idea about this discipline and to help develop a better understanding of the latest advances within this field. Coherent flow of topics, student-friendly language and extensive use of examples make this book an invaluable source of knowledge.

The researches compiled throughout the book are authentic and of high quality, combining several disciplines and from very diverse regions from around the world. Drawing on the contributions of many researchers from diverse countries, the book's objective is to provide the readers with the latest achievements in the area of research. This book will surely be a source of knowledge to all interested and researching the field.

In the end, I would like to express my deep sense of gratitude to all the authors for meeting the set deadlines in completing and submitting their research chapters. I would also like to thank the publisher for the support offered to us throughout the course of the book. Finally, I extend my sincere thanks to my family for being a constant source of inspiration and encouragement.

Editor

A revised calibration of the interferometric mode of the CryoSat-2 radar altimeter improves ice height and height change measurements in western Greenland

Laurence Gray[1], David Burgess[2], Luke Copland[1], Thorben Dunse[3], Kirsty Langley[4], and Geir Moholdt[5]

[1]Department of Geography, Environment and Geomatics, University of Ottawa, Ottawa, ON K1N 6N5, Canada
[2]Geological Survey of Canada, Natural Resources Canada, Ottawa, ON K1A 0E8, Canada
[3]Department of Geosciences, University of Oslo, 0316 Oslo, Norway
[4]Asiaq, Greenland Survey, 3900 Nuuk, Greenland
[5]Norwegian Polar Institute, 9296 Tromso, Norway

Correspondence to: Laurence Gray (laurence.gray@sympatico.ca)

Abstract. We compare geocoded heights derived from the interferometric mode (SARIn) of CryoSat to surface heights from calibration–validation sites on Devon Ice Cap and western Greenland. Comparisons are included for both the heights derived from the first return (the "point-of-closest-approach" or POCA) and heights derived from delayed waveform returns ("swath" processing). While swath-processed heights are normally less precise than edited POCA heights, e.g. standard deviations of ~ 3 and ~ 1.5 m respectively for the western Greenland site, the increased coverage possible with swath data complements the POCA data and provides useful information for both system calibration and improving digital elevation models (DEMs). We show that the pre-launch interferometric baseline coupled with an additional roll correction ($\sim 0.0075° \pm 0.0025°$), or equivalent phase correction ($\sim 0.0435 \pm 0.0145$ radians), provides an improved calibration of the interferometric SARIn mode.

We extend the potential use of SARIn data by showing the influence of surface conditions, especially melt, on the return waveforms and that it is possible to detect and measure the height of summer supraglacial lakes in western Greenland. A supraglacial lake can provide a strong radar target in the waveform, stronger than the initial POCA return, if viewed at near-normal incidence. This provides an ideal situation for swath processing and we demonstrate a height precision of ~ 0.5 m for two lake sites, one in the accumulation zone and one in the ablation zone, which were measured every year from 2010 or 2011 to 2016. Each year the lake in the ablation zone was viewed in June by ascending passes and then 5.5 days later by descending passes, which allows an approximate estimate of the filling rate. The results suggest that CryoSat waveform data and measurements of supraglacial lake height change could complement the use of optical satellite imagery and be helpful as proxy indicators for surface melt around Greenland.

1 Introduction

Temporal change in ice sheet surface elevation derived from satellite altimeters has been used in mass balance estimates and the associated contribution to sea level rise (e.g. Davis and Ferguson, 2004; Rémy and Parouty, 2009; Shepherd et al., 2012; Hurkmans et al., 2014; Zwally et al., 2015). Satellite radar altimeters have traditionally operated at Ku band (~ 13 GHz) and used parabolic transmit–receive dish antennas with a diameter of ~ 1 m so that the main beam illuminates an area beneath the satellite with a diameter of ~ 15 km and area of ~ 180 km^2. With a typical bandwidth of ~ 300 MHz the range resolution is ~ 50 cm and, as delay time increases beyond the point at which the first surface returns are received, an increasing area contributes to the received signal. These returns are termed "pulse limited", with the initial signal originating from the area within the main beam closest to the satellite, often referred to as the "point-of-closest-approach" (POCA). With these parameters, the diameter of the initially sampled POCA area over the ocean is $\sim$ 1.2–1.5 km, but this is not necessarily the case over glacial ice. The initial area contributing to the leading edge of the waveform (the delay time variation in received power) over an ice cap or ice sheet depends on the topography. All we know is that it must originate from somewhere within the area illuminated by the main antenna beam and that part of the POCA surface area must be orthogonal to the incident wave. Considering the large variability in ice cap topography and surface conditions, it is not unexpected that the waveforms from glacial ice will vary significantly in shape and power. The fact that the geographic position of the POCA is, a priori, unknown is one of the major problems in traditional radar altimetry and methods to get around this limitation have been studied extensively (Brenner et al., 1983; Bamber, 1994; Brenner et al., 2007; Hurkmans et al., 2012; Levinsen et al., 2016).

The European Space Agency (ESA) launched CryoSat as the first in their Earth Explorer series of satellites, which are designed to explore and demonstrate new techniques and methods in Earth observation. As such, CryoSat was designed to include a new mode of operation to address some of the limitations of traditional radar altimetry when used over sea ice, ice caps, and ice sheet margins. The new approach uses bursts of pulses in which the frequency of the pulses within each burst is high enough that coherent Doppler processing can be used to focus the energy in the along-track direction and ultimately create a footprint for which the along-track position is known, but the footprint centre can still be displaced from the sub-satellite track dependent on the cross-track slope. The along-track processing approach is referred to as "delay-Doppler" and was pioneered by Raney (1998). The suggestion that cross-track interferometry could solve the cross-track footprint position problem in radar altimetry is due to Jensen (1999). For glacial terrain the new SARIn mode of operation provides a relatively small geocoded footprint which allows, for the first time, a systematic comparison of satellite radar altimeter elevations with surface heights from surface and airborne campaigns.

The first CryoSat satellite equipped with the Synthetic Aperture Interferometric Radar Altimeter (SIRAL) was launched in 2005 but failed to enter orbit. A replacement satellite was launched in 2010 and, as of March 2017, is still operating satisfactorily, almost 4 years beyond its design life. CryoSat operates in three modes: a conventional low-resolution mode (LRM) which is used over oceans and the interior of Antarctica and Greenland, a synthetic aperture mode (SAR) for use over sea ice, and the interferometric SARIn mode over all the other glacial ice areas on Earth. A comprehensive description of CryoSat is given by Wingham et al. (2006). Here we are concerned primarily with SARIn mode calibration and with demonstrating some unique capa-

bilities of this new mode of satellite radar altimetry. These depend primarily on the ability to geocode the position of the relatively small footprint.

After the initial commissioning phase of the satellite in spring and summer 2010, intermediate and final products were available from ESA. For glacial ice the ESA level 2 (L2) product contains the position and height of the geocoded POCA positions. An additional L2i product is available, which contains the same geocoded height solution as the L2 product as well as information on the waveform which can be used to help eliminate poor data and solutions. An intermediate product (L1b) has also been made available which includes the waveform power, phase, coherence, satellite position and velocity, etc., and all the corrections and timing information necessary to calculate the position and height of the POCA footprint. This has been useful to those users wishing to study processing techniques; for example, by having access to the intermediate L1b product it has been possible to demonstrate that the returns which are time delayed beyond the initial POCA position can be used in areas with suitable cross-track slopes to create "swath-processed" elevations (Gray et al., 2013). Initially, airborne data had been used to demonstrate the possibility of swath-mode processing of delay-Doppler data (Hawley et al., 2009). The L1b products have also been used in several studies of change in Antarctica and Greenland (e.g. Helm et al., 2014; Nilsson et al., 2016; Christie et al., 2016; Smith et al., 2017), smaller Arctic ice caps (Gray et al., 2015; Foresta et al., 2016), and lake height (Kleinherenbrink et al., 2014). In these studies, the authors claim improvements in the results over the standard level 2 product due to the specialized processing.

Three versions of the various CryoSat products have been distributed by ESA since commissioning; these are the so-called baseline A, B, and C products. Details of the improvements can be found through the ESA Earth Online website devoted to the CryoSat mission (https://earth.esa.int/web/guest/missions/esa-operational-eo-missions/cryosat). Here we have used only the latest baseline C products, particularly because the waveforms in these products span a range window distance of $\sim$ 240 m, twice the distance available in the baseline B products. Some comparisons are also made between results derived from the baseline C L1b files and those provided in the L2 products.

In this study we use CryoSat and surface height data from two well-studied sites in the Canadian Arctic and Greenland to improve the calibration of the SARIn mode. Further, we show that the waveforms do change significantly with surface melt and that it is possible to detect the formation of supraglacial lakes. By using a modified swath processing scheme, we also show that it is possible to measure lake height and height change.

2 Methods

Our processing methods were described in Gray et al. (2013, 2015). The current Matlab processing provides both POCA and swath-mode results, and here we note any changes since the earlier work. The method to generate POCA heights is comparable to those described in Helm et al. (2014), Nilsson et al. (2016), and Smith et al. (2017) and were motivated by similar concerns, particularly the performance of the L2 "retracker": this is the algorithm designed to find the position of the POCA return in each waveform.

The delay-Doppler processing (Raney, 1998) for the SARIn mode of CryoSat is described in Wingham et al. (2006) and Kleinherenbrink et al. (2014). In this method 64 pulses are used in each transmitted burst and fast Fourier transform processing is used to create 64 unfocussed beams so that, with appropriate superposition of results from a sequence of bursts, multiple "looks" can be averaged for each ground footprint. In practice there are less than 64 looks contributing to each waveform in the L1b file, normally ~ 57. In the along-track direction the footprints are separated by ~ 280–300 m and the resolution is ~ 380 m (Bouzinac, 2012). In the cross-track direction the footprint size is dictated by the cross-track slopes and by any smoothing of the waveform in the processing. The position of the POCA footprint derived from each waveform will be in the plane, including the satellite position, and the lines defined by the cross-track and nadir directions. The POCA area will be centred on the closest point in the intersection of this plane with the terrain surface so that when ascending and descending orbits cross the two POCA footprints will not be the same when there is a cross-track slope. Consequently, it is not appropriate to compare results from the interpolated orbital cross-over point. The L1b files contain two echo-scaling parameters for each waveform, which allow a calibration of the waveform power to watts. The logarithmic (dB) values used in the results then represent logarithmic ratios scaled with respect to 1 W.

2.1 Selecting the POCA position from the SARIn waveform

If the altimeter response from terrain were "predictable" it would be beneficial to use the complete waveform in the estimation of the position in delay time of the surface, and this is the basis of the ESA L2 processing. However, our experience with the L1b SARIn waveforms over glacial ice shows that the shape and magnitude of the waveform can vary significantly, even in one area at one time (see examples in Sect. 4). The average return power as a function of delay time from the first surface sample will vary with the illuminated surface area, the reflectivity of the surface, and any near-surface layering on the ice cap. The cross-track slope and fixed sampling in delay time (3.125 ns) defines the basic cross-track footprint size so that the waveform shape beyond the POCA

depends primarily on the variation in topography in the cross-track direction. This is essentially independent of the position of the POCA, resulting in our decision to estimate the POCA position based on the first significant leading edge in the waveform. Our approach (Gray et al., 2015) uses the point of inflexion (maximum slope) on the first significant waveform increase and is similar to that adopted by Nilsson et al. (2016) and Smith et al. (2017). Helm et al. (2014) used a threshold level of the first significant leading edge for their work in Greenland and Antarctica, following the work of Davis (1997), who advocated a threshold retracker to minimize the dependency on varying microwave penetration into, and backscattering from, various snow–firn–ice layers. The importance of the cross-track footprint size in dictating the shape of the waveform has been demonstrated by the success of the straightforward waveform simulation based primarily on topography shown in Gray et al. (2013).

Although the L1b waveforms already represent averaged values, some additional smoothing has been done on the complex waveform data. The low-pass filter uses a 3 dB width of ~ 4 samples and is designed to avoid introducing any bias in the waveform phase. Smoothing the SARIn waveform data is performed only in the range direction with a relatively small impact on the cross-track footprint size (Gray et al., 2015) and none on the along-track resolution. The resulting reduction in phase noise improves the POCA footprint geocoding, as the phase provides the cross-track look angle. It is not appropriate to average any of the L1b waveform data in the azimuth direction because there can be jumps in the delay time to the first waveform sample. The processing steps to generate geocoded heights are described in Gray et al. (2015) using the results of the calibration described in Sect. 3 below. Solutions are derived for the phase at the estimated POCA position in the waveform and for this phase are $+2\pi$ and -2π. Comparison with the height of the reference digital elevation model (DEM) is used to select the most likely of the three solutions. Some waveforms are not used for POCA generation. This can occur for various reasons: the coherence at the POCA point is less than 0.7, the power for the average of the first five waveform values is too high (> -150 dB, for baseline C), the ratio of the maximum waveform power to the average of the first five values is too low (< 6 dB), or there is not a clear leading edge in the waveform. These criteria are rather arbitrary and may be changed for different sites, depending on the results. For example, we found that using a more stringent POCA coherence requirement improved the overall results for the western Greenland site.

2.2 Swath-mode processing

The techniques used to process the returns delayed beyond the POCA position are essentially as described in Gray et al. (2013). In that work the bias errors associated with the uncertainty in the baseline roll angle (Galin et al., 2012) were reduced by comparing the derived east–west slope on the western flank of Devon Ice Cap with the reference data slope and changing the baseline roll angle to minimize this error. This step has not been undertaken here as it presumes a good-quality reference DEM which is not necessarily available.

Waveform smoothing can lead to a situation in which results may be oversampled in the cross-track direction. The swath-processed results from any one waveform will form a straight line in the cross-track direction and the final samples in cross-track are generated by binning and averaging the results in segments of the cross-track line. The separation between ground-range cross-track samples is nominally ~ 100 m. Criteria for minimum values of the filtered coherence and returned power are set and are usually ~ 0.84 and -150 dB respectively for baseline C data. The phase unwrapping and ambiguity checking method is similar to that described by Smith et al. (2017).

The swath processing of the summer CryoSat data for supraglacial lake height (Sect. 4.2) omitted the cross-track binning stage and produced an elevation for each sample in the waveform. Only heights derived from waveform samples with phase values equivalent to small look angles ($< \sim 0.2°$), high power ($> \sim -140$ dB), and high coherence (> 0.95) were used. These minimum values virtually guarantee that there will be a small contribution from the range ambiguous zone and that phase unwrapping or ambiguity checking is unnecessary. The resulting geographic positions were compared to the best-available visible imagery, usually Landsat 8 images, and north, south, east, and west boundaries around the lake feature were set. The resulting height estimate was then obtained by averaging all estimates within the lake boundary.

2.3 Measuring the height difference between the reference surface and CryoSat heights

We used two methods to compare the derived CryoSat heights with the surface reference data. For Devon Ice Cap the reference data included intercalibrated snowmobile-based differential GPS transects and airborne scanning laser altimeter data from both the NASA Airborne Terrain Mapper (ATM; Krabill et al., 2002; Krabill, 2014) and the TUD ALS (https://earth.esa.int/documents/10174/134665/ESA-CryoVEx-ASIRAS-2014-report) systems. For the Greenland site, we have relied on the ATM data collected on NASA IceBridge flights. The first method stepped through all the CryoSat results and searched for reference heights within 400 m of the centre of the CryoSat footprint. The height differences between the CryoSat and reference heights were corrected for the slope between the centres of the two footprints using interpolation with the reference DEM. If there were many reference values, as can be the case for the western Greenland site, then a second simpler method was used: a search was made for reference points within 50 m and the height differences were tabulated and

averaged without the slope correction stage. Virtually all the reference height data for both sites were obtained under cold conditions in April or early May and we assumed that any accumulation or change in the backscatter conditions between January and mid-May would lead to a relatively small change in the CryoSat height. This provided the rationale for comparing all the CryoSat results from the January to May passes with the April or May reference height data.

2.4 Estimating height errors in the CryoSat data

Ku band radar waves can penetrate the surface and the CryoSat-to-surface height bias will vary depending on the conditions of the surface and near surface (Gray et al., 2015; Nilsson et al., 2015). Consequently, we use the standard deviation of the height differences about the mean height difference as the primary measure of the quality of the CryoSat measurements. The relatively small error in the ATM or ALS laser surface heights (~ 20 cm; Krabill et al., 2002) is ignored, and any impact due to the difference in the footprint size is not considered.

When estimating the height errors for the supraglacial lakes it is not appropriate to quote the standard error (standard deviation divided by the square root of the number of samples averaged), because the samples will not be independent and there is the possibility of small bias error in the result. The errors were therefore estimated on a case-by-case basis by looking at any cross-track slope across a lake feature, using the standard deviation itself, and checking independent estimates from ascending and descending passes over the same feature. The standard deviation about the mean was typically ~ 0.5 m, and the mean difference between the ascending and descending passes over the same accumulation zone lake feature in August was ~ 0.25 m. Table 1 includes the error estimates from two lakes and shows that relatively good precision can be achieved for these strong targets, better than the potential error for individual POCA estimates.

3 Results: SARIn mode calibration

The key parameters for SARIn mode geocoding are the range to the surface and the satellite look angle between the normal to the WGS84 ellipsoid and the footprint centre in the cross-track plane. The former involves consideration of timing and the retracker algorithm for the POCA results, but it is the latter which requires careful calibration for both POCA and swath-mode results.

The satellite look angle, α, is related to two other angles through

$$\alpha = \beta - \delta, \tag{1}$$

where β is the interferometric angle defined below and δ is the roll angle of the interferometric baseline, all defined in the cross-track plane containing the line normal to the

Table 1. Information on the conditions and results of the analysis of the CryoSat data for the two lake features L1 (70.275° N, 48.56° W) and L2 (70.178° N, 48.55° W) shown in Figs. 13 and 14. The "Min dB" column reflects the lower limit of the sample power used in the averaging of the height estimates contained within the window around the surface depression.

Year L1	Date	Pass direction	Local time	Min dB	Standard deviation, m (no. of samples)	Look angle (deg.)	Height (m)	Height error (estimated, ±m)
2010	4 Aug	Ascending	01:03	−130	0.52 (61)	0.08 ± 0.006°	1606.5	0.5
	9 Aug	Descending	13:33	−135	0.58 (61)	−0.18 ± 0.006°	1606.3	0.6
2011	7 Aug	Ascending	06:48	−130	0.48 (51)	−0.05 ± 0.006°	1609.1	0.5
	12 Aug	Descending	19:18	−134	0.41 (57)	−0.22 ± 0.006°	1609.5	0.5
2012	9 Aug	Ascending	12:33	−125	0.63 (52)	−0.05 ± 0.006°	1613.8	0.6
	15 Aug	Descending	01:03	−127	0.29 (59)	−0.2 ± 0.006°	1614.3	0.4
2013	12 Aug	Ascending	18:16	−135	0.59 (73)	−0.1 ± 0.006°	1612.5	0.6
	18 Aug	Descending	06:46	−139	0.33 (48)	−0.2 ± 0.006°	1612.8	0.4
2014	16 Aug	Ascending	00:01	−138	0.56 (75)	0.05 ± 0.006°	1610.8	0.6
	21 Aug	Descending	12:32	−130	0.46 (74)	−0.18 ± 0.006°	1611.2	0.5
2015	19 Aug	Ascending	05:46	−138	0.64 (48)	−0.04 ± 0.006°	1610.2	0.6
	24 Aug	Descending	18:16	−139	0.45 (34)	−0.15 ± 0.006°	1610.3	0.5
2016	21 Aug	Ascending	11:31	−140	0.49 (34)	−0.02 ± 0.006°	1606.1	0.5
	27 Aug	Descending	00:01	−140	0.51 (18)	−0.17 ± 0.006°	1606.7	0.6
L2								
2010	4 Aug	Ascending	01:03	−120	0.75 (41)	0.01 ± 0.006°	1571.5	0.7
2011	7 Aug	Ascending	06:48	−134	0.42 (57)	−0.14 ± 0.006°	1573.1	0.5
2012	9 Aug	Ascending	12:33	−130	0.35 (54)	−0.1 ± 0.006°	1576.0	0.4
2013	12 Aug	Ascending	18:16	−135	0.32 (16)	−0.16 ± 0.006°	1573.2	0.5
2014	15 Aug	Ascending	00:01	−135	0.37 (30)	−0.03 ± 0.006°	1572.1	0.4
2015	19 Aug	Ascending	05:46	−137	0.43 (32)	−0.13 ± 0.006°	1572.5	0.5
2016	21 Aug	Ascending	11:31	−133	0.47 (27)	−0.10 ± 0.006°	1573.1	0.5

WGS84 ellipsoid. The angle β is related to the interferometric phase through (Galin et al., 2012)

$$\beta = -\mathrm{asin}(\chi / kB), \qquad (2)$$

where χ is the phase provided in the L1b file, k is the wavenumber, B is the length of the interferometric baseline, and the CryoSat altimeter transmits through the left antenna and receives from both. The sense of the look and interferometric angle is as follows. For zero roll an observer siting on the CryoSat satellite facing in the direction of motion with their feet pointing towards the Earth will "see" a footprint to the right of the sub-satellite track when the look angle α is positive. The roll angle δ is also provided in L1b files. For the same observer configuration, a positive roll angle corresponds to the left antenna being higher than the right-hand one.

Any bias in the look angle, $\Delta\alpha$, can then be related to biases in the baseline (ΔB), phase ($\Delta\chi$), and roll angle ($\Delta\delta$) through

$$(\alpha + \Delta\alpha) = -\mathrm{asin}\left(\frac{(\chi + \Delta\chi)}{k(B + \Delta B)}\right) - (\delta + \Delta\delta). \qquad (3)$$

Using the approximations that $\sin(x) = x$ for small x and $B \gg \Delta B$ leads to an expression for the bias in roll angle as

$$\Delta\alpha = -\frac{\Delta\chi}{kB} + \frac{\chi}{kB}\left(\frac{\Delta B}{B}\right) - \Delta\delta. \qquad (4)$$

The CryoSat satellite and processing chain contains careful controls, which should minimize any extraneous inter-channel phase shift $\Delta\chi$ on the satellite (Bouzinac, 2012). Even if a residual phase bias exists, due perhaps to an uncompensated path length difference between the two receivers, it can be expressed in the same form as the roll-angle correction $\Delta\delta$ and the two can be considered together. The second term in Eq. (4) reflects the possibility of a bias between the actual and pre-launch measurement of the interferometric baseline – the distance between the two antenna phase centres. This was part of the post-launch SARIn mode calibration carried out by Galin et al. (2012). This work used results from satellite roll manoeuvres over mid-latitude ocean tracks to show that the interferometric angle should be scaled by a factor of 0.973 ± 0.002, which is equivalent to scaling the baseline by a factor of 1.0277. The third term in Eq. (4), the uncertainty in the baseline roll angle $\Delta\delta$, is important because the baseline roll angle is derived from one of three star trackers mounted on a support bench on the satellite. Galin et al. (2012) identified a problem with the reported roll angle and suggested that this was due to bending of the support bench under a changing thermal environment. However, recent work by ESA (Scagliola et al., 2017) showed that the roll-angle problem arose, at least partly, because of an error in processing the star-tracker data. Consequently, there is currently an unknown bias in the reported value of the base-

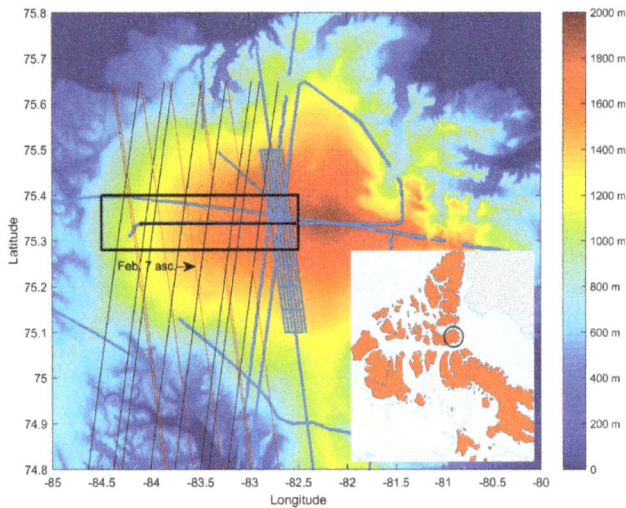

Figure 1. Location of the test area on the western slopes of the Devon Ice Cap (black rectangle). The sub-satellite positions of the spring 2011 ascending and descending passes crossing the test area are shown by the red and black lines respectively. The positions of the reference surface height data are shown in blue, the elevation profile in Fig. 3c is a black line, and the sub-satellite track for the waveform power in Fig. 6a is labelled. The insert shows the position of Devon Ice Cap (circled) in the Canadian Arctic Archipelago.

line roll angle which can vary pass to pass. In the following sections, we use SARIn data over well-documented glacial ice to investigate any residual bias in the roll angle provided in the L1b files and to study the influence of changing the baseline length in processing L1b files.

3.1 Calibration test sites

We used data from two sites, the western flank of Devon Ice Cap (Fig. 1) and an area in western Greenland including the Jakobshavn Glacier (Fig. 2), as both have excellent reference surface height data. Our calibration approach depends on the presence of a predominantly east–west slope, which is why the test area in Fig. 1 is limited in the north–south direction. By using terrain with an east–west slope we obviate the necessity for roll tilting the satellite. Figure 3 illustrates the difference in the slopes for the two test sites. The significant increase in slope variation in the western Greenland site represents a more challenging situation for satellite radar altimetry than the more modest slope variation on the western flank of Devon Ice Cap, and this is the reason we have concentrated on comparing the results from these two test sites.

3.2 Calibration based on data from Devon Ice Cap

The western portion of Devon Ice Cap has suitable cross-track slopes for swath-mode height estimation for both ascending and descending passes, and this area was used in the demonstration of swath-mode processing (Gray et al., 2013).

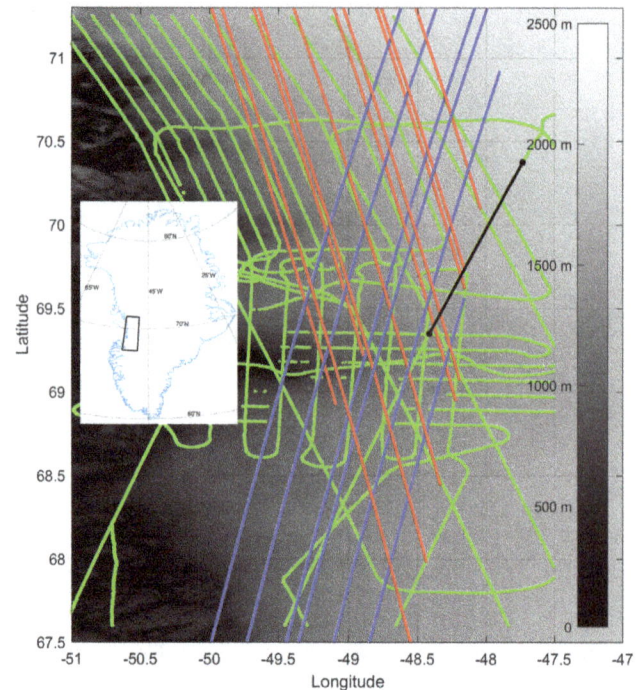

Figure 2. The positions of the reference ATM surface elevations flown by NASA IceBridge missions over the western Greenland site in spring 2011 are shown in green. Sub-satellite CryoSat tracks for the period 20 January to 16 May 2011 are shown by red (ascending) and blue (descending) lines. The inset map shows the position of the test area in Greenland and the background image is a black–white representation of the GIMP reference DEM (Howat et al., 2014). The position of the height profile in Fig. 3a is shown by the black line.

While the possible range of average cross-track slopes can be ~ 0.5 to $\sim 2°$, here we have restricted the use of results to east–west slopes of ~ 0.7–$1.5°$ over a distance of > 5 km as this range generally provides a better suppression of the ambiguous range contribution. Figure 1 shows the positions of the spring 2011 surface height reference data obtained from NASA and ESA supported overflights and from surface snowmobile dGPS transects, all superimposed on a colour representation of the reference DEM. The sub-satellite tracks of 15 CryoSat passes are also shown. Results from all the passes in this time period were compared to the reference surface heights as conditions on Devon Ice Cap change little between January and May, and we assume that any change in surface height or change in the bias between the surface and CryoSat height was small with respect to the error in the CryoSat heights.

The histogram of the difference between the reference and CryoSat swath-mode heights obtained with the pre-launch baseline estimate (1.1676 m; Bouzinac, 2012) showed a bimodal distribution and the average bias changed between ascending and descending passes, ~ -0.5 and ~ 2.5 m respectively. As we could find no reasonable geophysical expla-

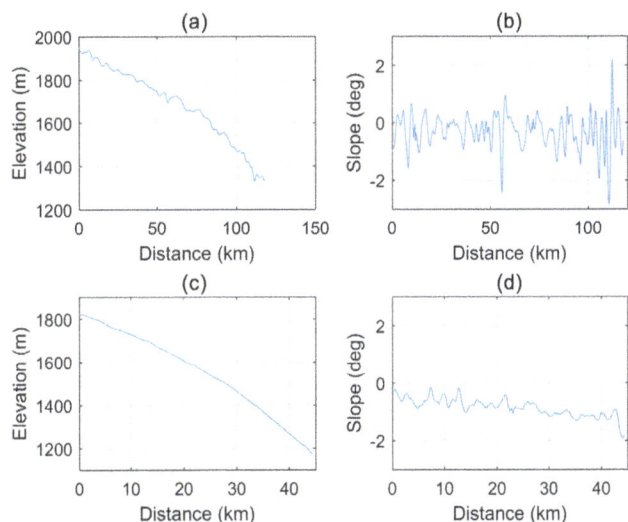

Figure 3. Illustration of the difference in slopes for a typical western Greenland transect (70.37° N, −47.73° W to 69.35° N, −48.42° W; black line in Fig. 2) derived from an ATM flight line from 6 April 2011 (**a**; elevation and **b**; slope) and the east–west transect (black line in Fig. 1) from western Devon Ice Cap (**c**; elevation and **d** slope, Fig. 1).

Figure 4. Illustration of the changing bias between the reference and CryoSat (CS) swath-mode heights for the Devon test site as an additional roll bias is subtracted from the roll figure given in the L1b file. Results for seven ascending and eight descending passes in the winter–spring of 2011 are shown in red and black respectively. The reference–POCA height variation with the added roll-angle bias is shown with the dashed lines.

nation for this difference, the possibility of a roll-angle bias was investigated. If there were a roll-angle bias on an ascending pass the swath-processed height estimates would be displaced either up- or down-slope depending on the sense of the bias. However, with a descending pass and the same roll-angle bias, the results will be displaced in the opposite direction and the height bias will have the opposite sign from that obtained with the ascending pass. To investigate this effect further, all the data in this time period were reprocessed with an additional roll-angle bias added to the value provided in the L1b file. Figure 4 illustrates the results of an experiment in which the 15 2011 passes (seven ascending and eight descending) were each reprocessed nine times with an additional roll correction varying from −0.02 to +0.02°. The results were then compared to the reference height data collected in early May 2011. As expected, the sense of the height difference changes between ascending and descending passes but the curves do not overlap well. While the results from the 8 2011 descending passes do cluster nicely, this was not the case for the 2012 data (Gray et al., 2016), and neither year shows consistent results for the ascending pass results.

Consequently, it appears that the roll angle provided in the L1b file has a time variable bias, apparently due to a problem in processing the star-tracker data (Scagliola et al., 2017). The uncertainty in the roll angle in this example appears to be of the order of 0.006° or ∼ 100 µ radians, not inconsistent with the observations in Galin (2012). While there will be a contribution from the range ambiguous zone in swath-mode processing, which could introduce a small bias, this does not appear to be the primary source of these differences. The roll-

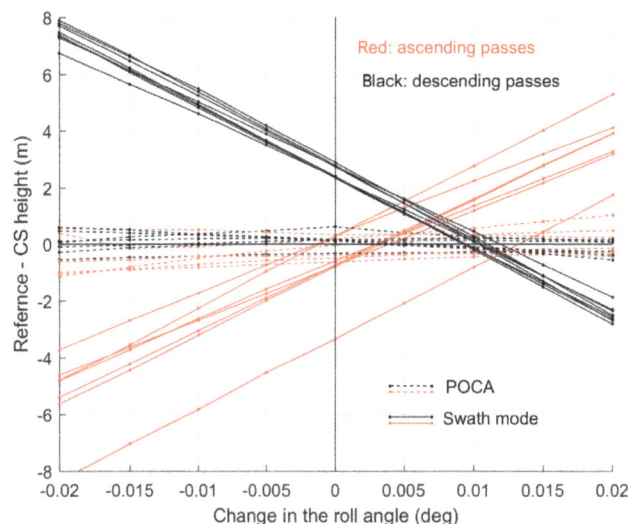

angle uncertainty, and resulting unknown bias in the baseline roll angle, appears to be a limitation to the use of swath-mode heights. Note that in Fig. 4 there is essentially no slope to the plots of the height difference versus roll-angle bias for the POCA height estimates. This is a direct consequence of the fact that while the POCA estimates are mapped incorrectly when there is a roll-angle error, the derived height can still be appropriate for the wrong position because the incident wave may still be essentially perpendicular to the surface (Gray et al., 2013).

The variable east–west cross-track slope also provides a suitable test area to check the phase to cross-track angle conversion dictated by the baseline (Eq. 2 above). Figure 5 illustrates the results of an experiment in which the results obtained with a phase-to-angle conversion based on the pre-launch baseline are compared to the calibration given by Galin et al. (2012). The two histograms on the left used the pre-launch baseline while the histograms on the right used the angle scaling from Galin et al. (2012). Figure 5c shows the bimodal distribution referred to earlier, and Fig. 5a shows the improved results with a significantly narrower error distribution when a bias of 0.0075° is subtracted from the roll angle provided in the L1b file. The uncertainty in this additional roll bias has been estimated as ±0.0025°. When the phase-to-angle conversion is scaled by 0.973 (Fig. 5b and d), the results show a broader distribution and poorer results.

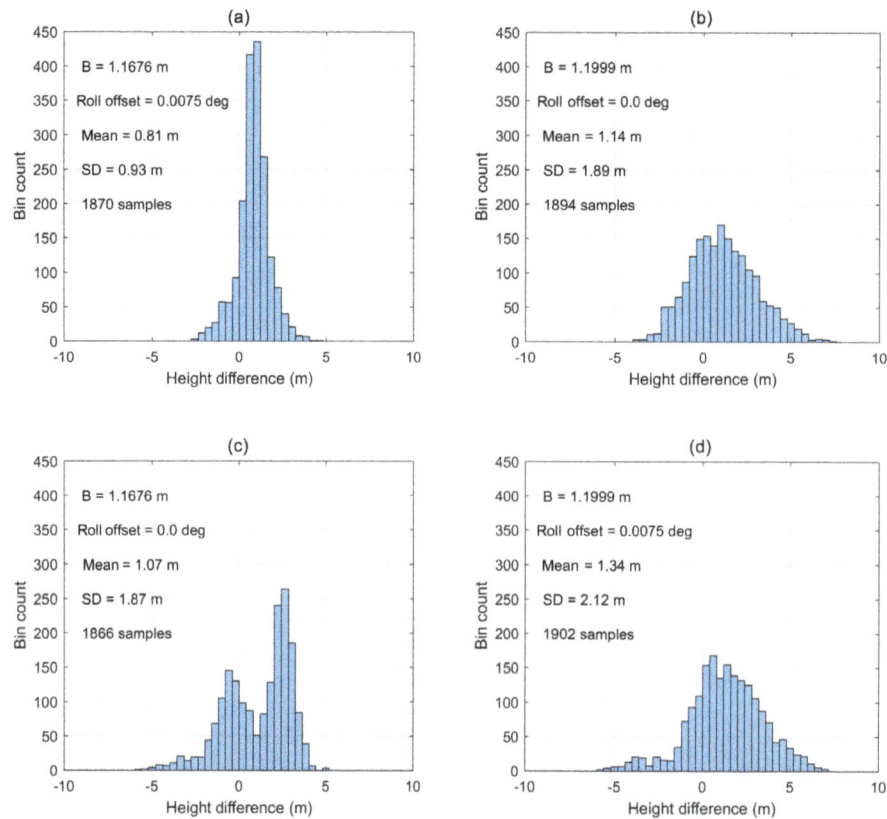

Figure 5. Histograms of the reference minus CryoSat swath heights for Devon Ice Cap: **(a)** pre-launch baseline and a roll-angle offset of 0.0075°; **(b)** modified baseline with zero roll offset; **(c)** pre-launch baseline with zero roll offset; and **(d)** modified baseline with a roll offset of 0.0075°.

3.3 Calibration based on data from western Greenland

We use IceBridge data from an area in central western Greenland (Fig. 2, insert), including the Jakobshavn Glacier, that has shown significant surface height loss in recent years due to change in both output flux and surface mass balance (Joughin et al., 2008; Qi and Braun, 2013) and has excellent reference surface height data (Krabill et al., 2002; Krabill, 2014).

Figure 2 illustrates the positions of the reference surface height data obtained from the four NASA IceBridge flights flown on 31 March and 6, 7, and 23 April 2011 superimposed on a black and white representation of the GIMP DEM (Howat et al., 2014). This DEM was used as the reference DEM for all the CryoSat processing in this area. Data from the ATM L2 files have been used for this work and compared with height results from all the CryoSat passes between 16 February and 23 April.

It is important to recognize the differences in this test site in relation to that on Devon Ice Cap. The two profiles in Fig. 3 show that even in the accumulation area of this part of western Greenland the slope variation is much larger than on the east–west profile interpolated from the airborne laser altimeter flown over of Devon Ice Cap. The difference is

also very apparent in the CryoSat results: Fig. 6 compares two image representations of the waveform power for 22 km segments of the 7 February 2011 ascending pass over Devon Ice Cap and the 21 April 2011 descending pass over the western Greenland test site. For the ascending pass over Devon Ice Cap the POCA position will be on the left, close to the beginning of the 240 m range window, as indicated by the stronger return signals in red. However, for the western Greenland site the peak return is often in the middle of the waveform. The difference in the signals may be influenced by the different conditions but it is clear that the dominant reason for the differences in waveforms are the differences in the cross-track slopes. The larger slope variation in western Greenland clearly influence the CryoSat returns, and the waveform shape is now much more variable than those from the Devon test site. This situation favours a retracker which looks for the first significant leading edge, rather than one that assumes a particular model for the waveform and then fits the waveform to that model, as is the case for the ESA L2 SARIn product. Some details of the retrackers used in the baseline C L2 products are given in Buffard (2015).

Figure 7 compares the results obtained with our geocoding and that obtained with CryoSat L2 retracker. Our processor picks out the POCA position satisfactorily (black dots

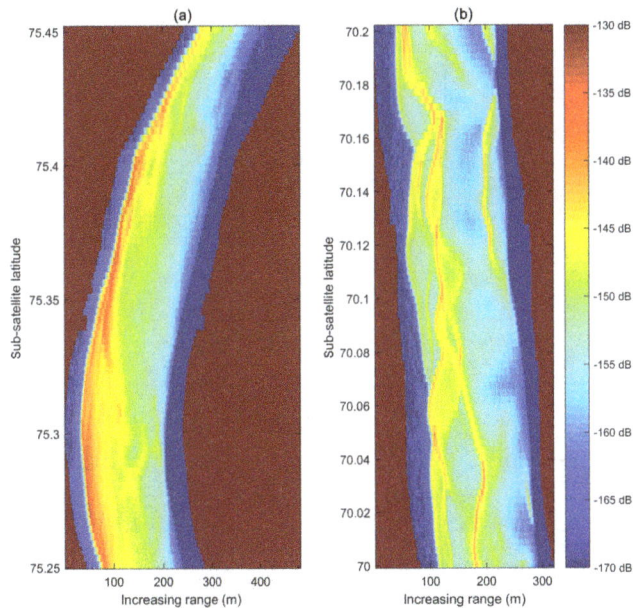

Figure 6. Waveform power for 22 km segments of **(a)** the 7 February 2011 ascending pass over Devon Ice Cap (Fig. 1) and **(b)** the 21 April 2011 descending pass (Fig. 12a) over the western Greenland test site. The return power in dB is represented in colour and the individual waveforms have been shifted in the *x* direction depending on the time delay to the first sample and the satellite elevation above the WGS84 ellipsoid.

Figure 7. (a) Waveform power without any *x* axis shifts using the same dB colour scale as in Fig. 6b. The detected POCA positions are shown in **(a)** for each waveform with black dots, and they clearly correspond to the leading edge of the waveforms. The purple dots are the estimated positions in the waveforms of the L2 POCA solution. **(b)** Geographic positions of the geocoded footprints (black dots) are compared to the positions of the ESA L2 solutions (red dots). The solid black line is the sub-satellite track.

on Fig. 7a) and leads to the mapping solution shown in Fig. 7b. The positions of the CryoSat L2 solutions are shown in Fig. 7b as purple dots and are often different by many kilometres. The solutions are close only when the waveforms show a clear maximum close to start of the waveform (e.g. at ∼ 70.05° N). Using the position of the L2 solution, the off-nadir look angle and equivalent phase can be calculated. Then the position in the waveform with that phase is identified and marked as purple dots in Fig. 7a. This shows that the L2 retracker normally does not identify the point-of-closest-approach correctly, primarily because of the strong peaks in middle of the waveform.

In comparing our CryoSat POCA height results with the ATM surface height results we found that the results here were not as precise as those obtained over the Devon test site. However, when slightly more stringent editing was used, in particular by increasing the minimum POCA coherence requirement to 0.8 from 0.7, then the results were improved. The histograms of the ATM minus CryoSat heights for the 2011 spring data are shown in Fig. 8. Again the poor results from the baseline C CryoSat L2 files are apparent (Fig. 8a), particularly the much larger number of height errors greater than 20 m. Results from exactly the same waveforms have been used in this comparison, as the L2 results were removed for those waveforms already removed through the L1b editing. While it is unfair to compare results from an operational

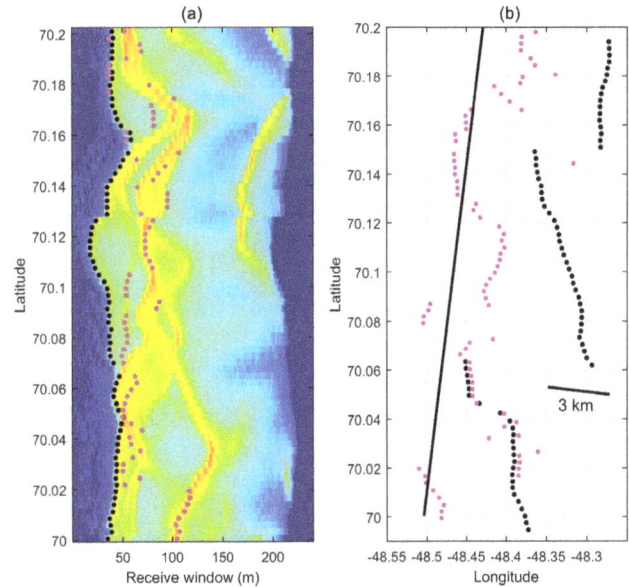

algorithm which must work everywhere to one which can be tuned for different areas and includes editing based on the coherence and the return power, it is fair to say that the current L2 retracker is inherently unsuitable for the western Greenland site. The L2 results are better in other areas, such as the ridges on Austfonna, ice rises and ice shelves in Antarctica, and parts of the Devon Ice Cap. In these areas the waveforms show a more consistent shape and the dominant return is close to the start of the waveform.

The comparison between results obtained with the angle scaling factor from the Galin et al. (2012) calibration (Fig. 9a and c) and without (Fig. 9b and d) mirrors the results discussed in the previous section for Devon Ice Cap. The results imply that the pre-launch baseline coupled with an additional roll-angle offset (or equivalent phase shift) improves the results for both western Greenland and Devon Ice Cap.

There is an important difference in the results for this test site in relation to Devon. For Devon, the ATM–POCA height difference was essentially independent of the roll-angle offset between −0.02 and 0.02° (Fig. 4), but this was not the case for the western Greenland site. A comparison of the average ATM–POCA height difference over 16 passes as a function of the additional roll-angle bias (Fig. 10a) shows that the CryoSat POCA height is not independent of the roll-angle bias but increases for both positive and negative bias errors from a value of ∼ 0.0075° ± 0.0025°. As the CryoSat

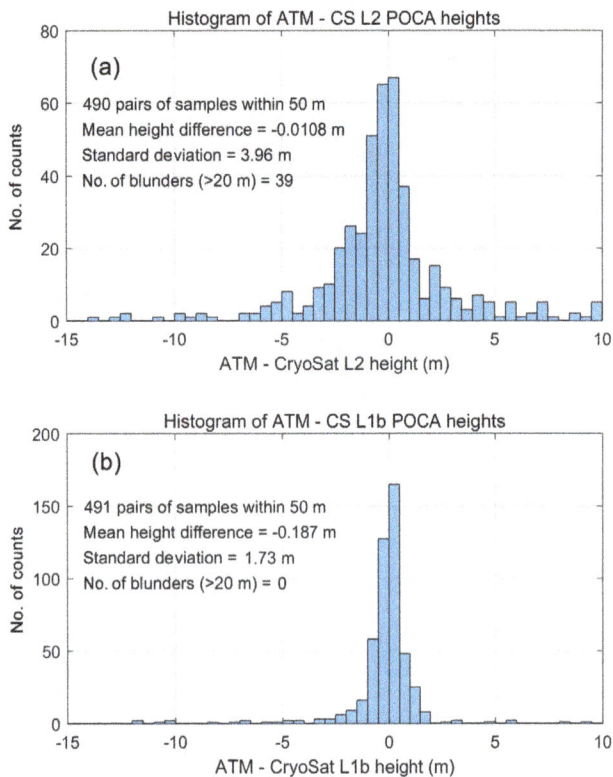

Figure 8. Comparisons of the ATM–CryoSat POCA height difference histograms for the western Greenland test site. **(a)** The ESA L2 solution. **(b)** Results from the current maximum slope leading edge retracker. The mean and standard deviation in **(a)** have been calculated after removal of the 39 blunders. CryoSat data from all the passes between 16 February and 23 April 2011 have been used in this comparison, and results from the same waveforms used in both cases.

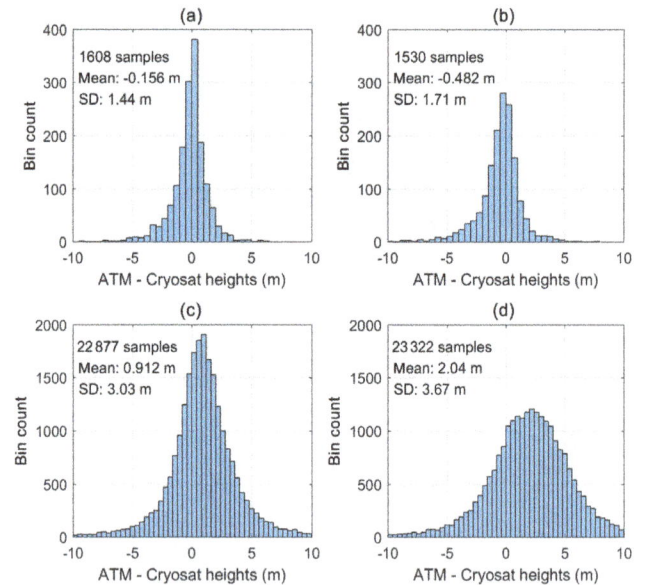

Figure 9. Comparisons of the western Greenland ATM–CryoSat height difference histograms for the solution using the pre-launch baseline coupled with an additional roll offset (**a**; POCA, **c**; swath-mode solutions). **(b)** and **(d)** use the Galin et al. (2012) calibration for the POCA and swath solutions respectively. Results from the CryoSat passes for the period 20 January to 16 May 2011 have been used.

results are mapped incorrectly in the cross-track direction, the larger cross-track slopes imply that the distance in the cross-track direction which is essentially orthogonal to the incident wave is smaller in western Greenland than for the relatively smooth surface of western Devon Ice Cap. Consequently, this will lead to a CryoSat POCA height error as the mapping process takes the centre of the footprint outside the region which is orthogonal to the incident wave. Figure 10b shows the variation in the standard deviation of the swath-mode ATM–CryoSat heights for each pass (dotted lines) and the average over all 16 passes (black line). The offset in the position of the minimum from zero roll-angle bias also supports the contention that on average there is a difference between the actual baseline roll angle and the value reported in the L1b file based on one of the three star trackers or that there is an equivalent phase shift. For batch processing, we have used the L1b roll angle minus 0.0075°, but this may change with more experience with the bias.

There is another discrepancy in these results that warrants explanation. From Fig. 9a we see that the average ATM–POCA height difference is −0.16 m, but with the same waveform data the height difference from swath-mode processing is +0.91 m (Fig. 9c), so that the two processing methods are giving average heights different by 1.07 m. With the Galin et al. (2012) calibration the discrepancy is even worse: 2.52 m. Further, there is an apparent discrepancy with the results from Devon Ice Cap where previously (Gray et al., 2015), and now, we see the CryoSat height as being somewhat below the physical surface. The explanation for the anomalous average ATM–POCA result for western Greenland, where the average CryoSat POCA height is slightly above the surface, appears to be related to the results in Fig. 10a. If there is an error in the roll angle this will lead to an increase in detected height irrespective of the sign of the roll-angle error. This will lead to an asymmetric distribution and the mean height will be biased high. Note that the distribution in Fig. 9a is somewhat asymmetric, more so than that in Fig. 9c for the swath-processed data where the sign of any roll-angle error would dictate the sign of the height error. For areas like the western Greenland test site this implies that the roll-angle bias error will tend to bias the average POCA height high with respect to the surface.

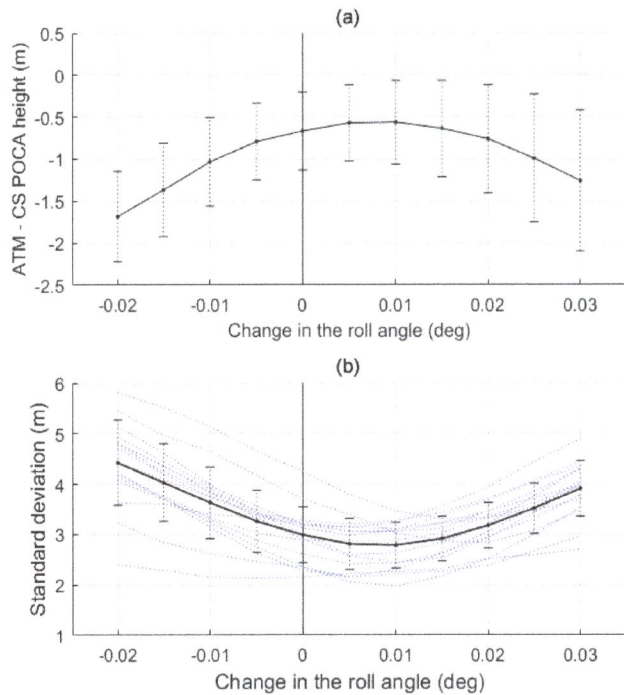

Figure 10. Illustration of the average of the ATM–CryoSat height differences for **(a)** 16 passes plotted against the additional roll-angle bias used in the processing. The error bars are ± 1 standard deviation about the mean. **(b)** Variation in the standard deviation (SD) of the ATM–CryoSat height difference for the individual passes (blue dotted lines) and the average over all the passes (solid black line).

4   Unique capabilities of the SARIn mode

In this section we use our methodology and revised calibration to demonstrate some unique capabilities of the SARIn mode, first by illustrating signature change with surface conditions in western Greenland and secondly by showing that it is possible to detect supraglacial lakes in the waveform data and estimate the surface height and height change with relatively good precision.

4.1   The effect of surface melt on SARIn waveforms

The influence of melt on SARIn signatures should be considered when presenting temporal height change for any region which may have undergone surface melt (Nilsson et al., 2015; Gray et al., 2015). Figure 11 illustrates one example of the influence of melt on the strength of the SARIn waveform data. The position of this 14 July 2011 descending pass is shown in Fig. 12a and begins at ~ 2200 m elevation and crosses the Jakobshavn Glacier at ~ 1000 m; then the elevation increases slightly before ending at ~ 1100 m. At high elevations, the returns are comparable to those obtained under cold winter–spring conditions, but at lower elevations, ~ 1700–1900 m, there is a decrease of ~ 15–20 dB in average waveform power. It is well known that the in-

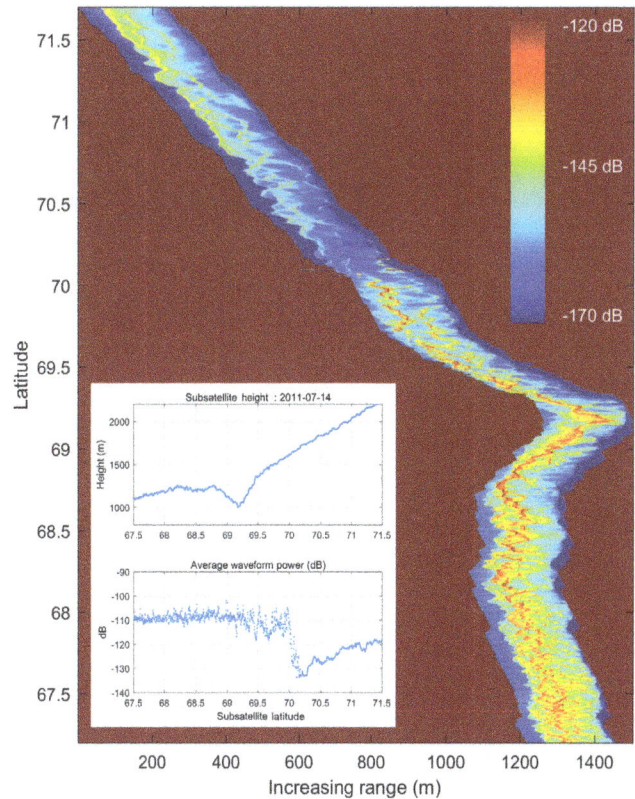

Figure 11. The background image illustrates the swath waveform power in colour with a dB scale for the 14 July 2011 descending pass over the western Greenland test area (Fig. 12a). The waveforms making up this pseudo-image have been shifted in the x direction to account for the changing delay time to the first sample and the varying satellite height above the WGS84 ellipsoid. The insert shows the sub-satellite terrain elevation and the waveform average power both plotted against latitude.

troduction of even a small amount of liquid water in snow dramatically alters the emissivity and backscatter (Ulaby et al., 1986). For example, a significant drop in QuikSCAT 13.3 GHz backscatter was shown to be linked to melting from weather station data (Nghiem et al., 2001). The presence of water droplets in snow increases absorption and reduces the penetration depth, which in turn leads to an increase in brightness temperature and decrease in radar backscatter (Wang et al., 2016). Consequently, we associate the relatively low reflectivity at these elevations to a damp snow layer. At lower elevations (< 1600 m) not only is the average return larger but also the waveform-to-waveform variability is much higher, indicative of occasional specular reflection from a wet surface facing the radar. Also, the strongest returns in most of the waveforms in this area are not from the leading edge but vary in position across the waveform so that a retracker that uses all of the waveform will not accurately measure the position and height of the POCA.

Figure 12 illustrates the average waveform power plotted against elevation for five descending passes (Fig. 12a)

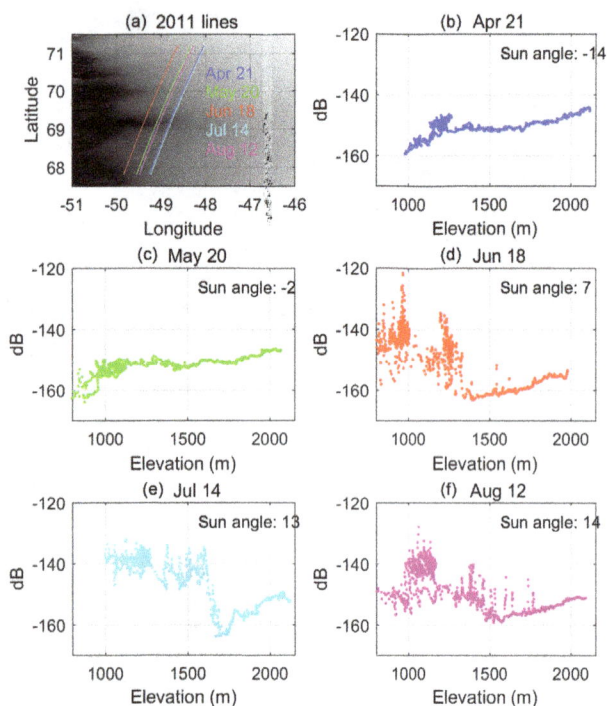

Figure 12. Plots of the average waveform power for the five 2011 descending passes shown in **(a)**. The five plots are from descending passes on **(b)** 21 April, **(c)** 20 May, **(d)** 18 June, **(e)** 14 July, and **(f)** 12 August and illustrate average waveform power as a function of elevation.

acquired during the summer of 2011. At elevations up to $\sim 1300\,\text{m}$ the 18 June pass (Fig. 12d) shows the high waveform-to-waveform variability that we suggest is due to occasional specular reflection, but this was not observed in the earlier passes in April (Fig. 12b) and May (Fig. 12c). By 14 July (Fig. 12e) the region with strong and variable power includes elevations up to $\sim 1600\,\text{m}$ and the August pass (Fig. 12f) shows some strong waveform returns at even higher elevations. Comparable results were obtained from the five repeat passes 369 days later in 2012, but the descending pass on 20 July 2013 showed the wet snow signature at lower elevations ($\sim 1500\,\text{m}$) without any indication of occasional specular reflections. This is consistent with the relatively colder conditions at that time in 2013 with respect to both 2011 and 2012 (see e.g. Fettweis, 2016, http://climato.be/melt-2016). The Supplement includes figures equivalent to Fig. 12 for all the years from 2012 to 2016.

4.2 Supraglacial lakes

During summer melt around the periphery of the Greenland Ice Sheet water pools in surface depressions as supraglacial lakes (Echelmeyer et al., 1991), forming first at lower elevations and then to higher elevations as melt progresses. With increasing positive air temperatures, surface meltwater will infiltrate to lower elevations so that the snow at the edges of

the depression will tend to become saturated and melt before the snow in the centre of the depression. In many cases, small supraglacial streams form, which will add energy to melt snow or ice where they enter the surface depression. Optical satellite imagery and DEM data have been used to study the distribution, extent, depth, and drainage of these features when there is an open water surface (Box and Ski, 2007; McMillan et al., 2007; Sneed and Hamilton, 2007; Liang et al., 2012; Fitzpatrick et al., 2014; Leeson et al., 2015; Pope et al., 2016; Ignéczi et al., 2016). While Landsat and MODIS imagery has been used to estimate total lake volume of relatively large areas (e.g. Pope et al., 2016), the limitations due to clouds and atmospheric conditions hamper routine use for quantitative melt estimates. Here we demonstrate that CryoSat SARIn data can provide complementary information to that available from visible satellites by showing that measurements of surface height and height change can be derived from SARIn data over individual supraglacial lakes. SARIn data can be obtained reliably day or night and in all weather conditions but are very limited in surface coverage.

If CryoSat passes directly over a typical unfrozen supraglacial lake one would expect a strong specular reflection which would not be at the leading edge of the waveform, as it must be surrounded by ice at higher elevations. Even if the lake has some snow cover or a partially unfrozen surface, the flat surface will still enhance the return and could lead to a strong peak in the waveform. Figure 13 illustrates some strong signals in the middle of the waveforms of a 50 km section of the 7 August 2011 ascending pass over the test area in western Greenland. These may originate from extended surfaces orthogonal, or nearly orthogonal, to the incident wave. We have selected one such strong signal, labelled as "L1" in Fig. 13, which is detected in results from ascending and descending passes from all the summers from 2010 to 2016. The Supplement contains a sequence of 14 summer MODIS images from 2012 to 2016 which show that the L1 and L2 features are above the snow line for all 5 years and that the surface of these depressions did not become totally ice free. Figure 14 shows the positions of the sub-satellite tracks superimposed on a summer 2016 Landsat 8 image and that there were dark regions, presumably wet snow, at the positions of the topographic lows marked as L1 and L2. The relative strength of the CryoSat return signals for the seven ascending passes for both features are shown in Fig. 15 and the year-to-year derived height in Fig. 16 with details provided in Table 1. The sequence of dates for the repeat ascending passes are 4 August 2010, 7 August 2011, 9 August 2012, 12 August 2013, 16 August 2014, 19 August 2015, and 21 August 2016 reflecting the 369.25-day repeat orbit cycle. The repeat descending passes are 5.5 days after the ascending passes.

Our interpretation of the strengths of the lake signatures and the surface elevation is as follows: considering the low surface velocity ($\sim 3.5\,\text{m}\,\text{yr}^{-1}$; Joughin et al., 2010, 2016)

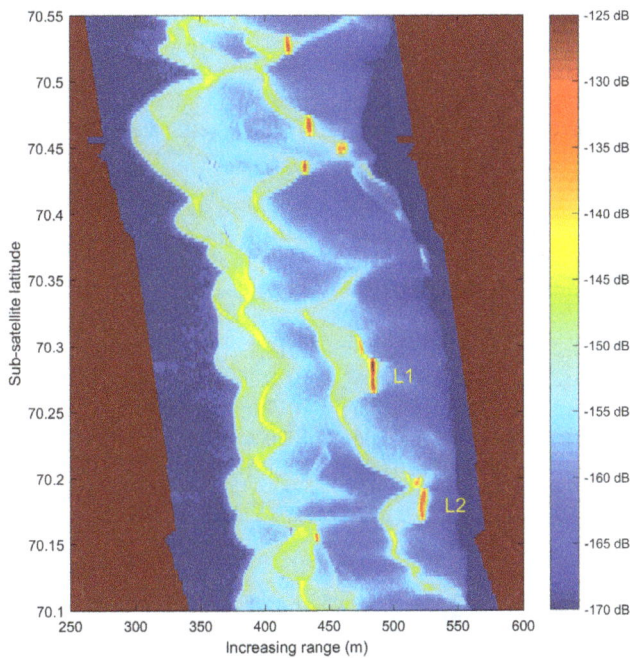

Figure 13. Illustration of part of the waveform power from an ascending pass over western Greenland on 7 August 2011 (Fig. 14). The bright returns labelled as L1 and L2 are at elevations ∼ 1609 m and 1573 m respectively and represent topographic lows where water could collect.

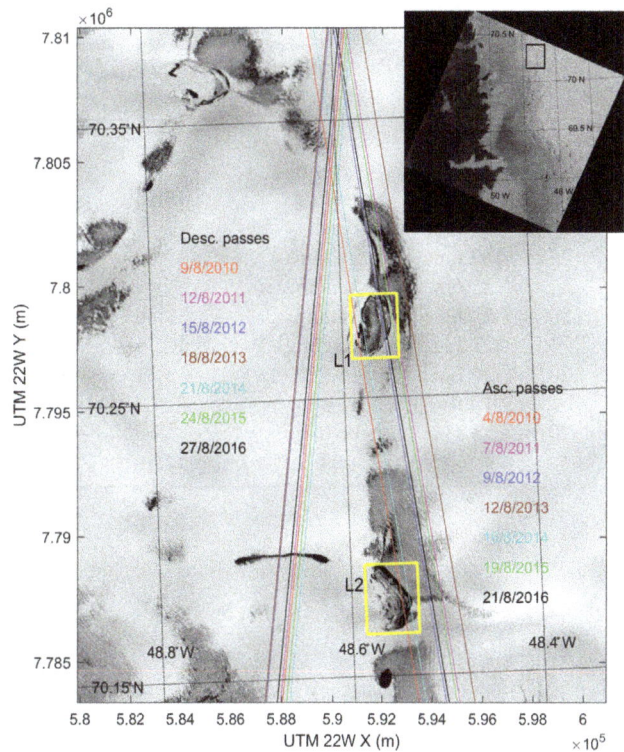

Figure 14. Ascending and descending sub-satellite repeat tracks over, or close to, the L1 and L2 features for all the years from 2010 to 2016 superimposed on part of the Landsat 8 image of 9 August 2016 (inset image).

and elevation (∼ 1600 m) at this position, it is unlikely that either of these depressions drained in the manner of the lakes in the ablation zone in any of the summers. The increase in height from the summer 2010 to 2012 (Fig. 16) may reflect the melt at these positions, which was particularly strong in 2012 (see e.g. Fettweis, 2016, http://climato.be/melt-2016). However, the decrease in elevation in subsequent years is then a problem. The discovery that water can persist for years in firn aquifers (Koenig et al., 2014; Forster et al., 2014) suggests that the decrease in elevation after 2012 may reflect a slow percolation of the meltwater into the firn. Clearly, the specific causes of the decrease in elevation of L1 and L2 after 2012, and the difference between the L1 and L2 height change, are not known.

The Landsat 8 image from 6 July 2016 (Fig. 17) includes one 2.4×1 km lake at 70.37° N, 49.79° W, and ∼ 1020 m in elevation, which was detected in the CryoSat waveforms from all the ascending and descending repeat passes listed on Fig. 17 between 2011 and 2016. By the time of the Landsat 8 image in 2016 most of the snow had melted and we surmise that melt had been on-going during June and early July for the years 2011–2016 at this position and at the times of the CryoSat overpasses (Fig. 17 and Table 2). Figure 18 illustrates the lake height for all passes except for the 2013 descending pass, which was too far to the west of the lake for reliable results. In contrast to the high elevation, low melt "lake" described above, now there is a clear height in-

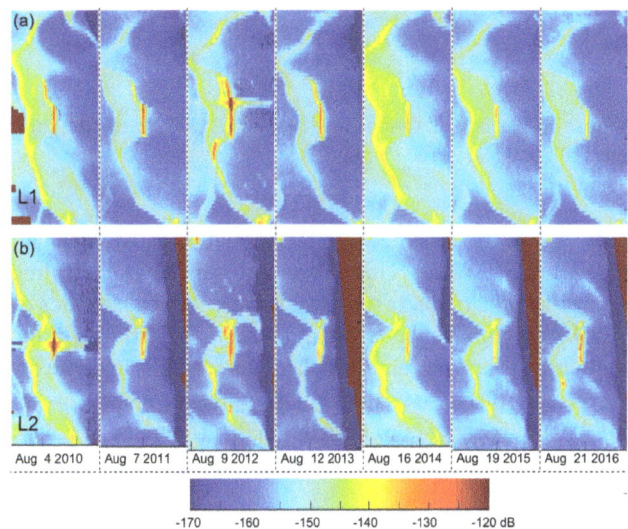

Figure 15. "Images" of part of the CryoSat waveforms for the areas including "L1" **(a)** and "L2" **(b)** in western Greenland for the August dates in each year from 2010 to 2016. The x and y axes of each "image" are increasing range and increasing along-track position (north up).

Table 2. Information on the conditions and results of the analysis of the CryoSat data for the lake shown in Fig. 17 and overflown by CryoSat on the dates shown.

Year	Date	Pass direction	Local time	Min dB	SD height (no of samples)	Look angle (deg.)	Height (m)	Height error (estimated ± m)
2011	14 June	Ascending	9:33	−125	0.74 (38)	$0.03 \pm 0.006°$	1014.7	0.7
	20 June	Descending	22:03	−130	0.59 (41)	$-0.07 \pm 0.006°$	1016.8	0.6
2012	16 June	Ascending	15:18	−115	0.77 (56)	$0.003 \pm 0.006°$	1020.4	0.8
	22 June	Descending	3:48	−120	0.80 (45)	$-0.01 \pm 0.006°$	1022.3	0.8
2013	19 June	Ascending	21:02	−130	0.57 (48)	$-0.02 \pm 0.006°$	1017.1	0.6
	25 June	Descending						
2014	23 June	Ascending	2:46	−130	0.37 (23)	$0.06 \pm 0.006°$	1018.7	0.4
	28 June	Descending	15:16	−135	0.53 (22)	$0.03 \pm 0.006°$	1020.2	0.5
2015	26 June	Ascending	8:31	−130	0.67 (39)	$-0.001 \pm 0.006°$	1017.6	0.7
	1 July	Descending	21:01	−130	0.87 (29)	$-0.07 \pm 0.006°$	1020.1	0.9
2016	28 June	Ascending	14:16	−130	0.74 (62)	$0.03 \pm 0.006°$	1021.4	0.7
	4 July	Descending	2;46	−130	0.64 (29)	$0.02 \pm 0.006°$	1022.1	0.6

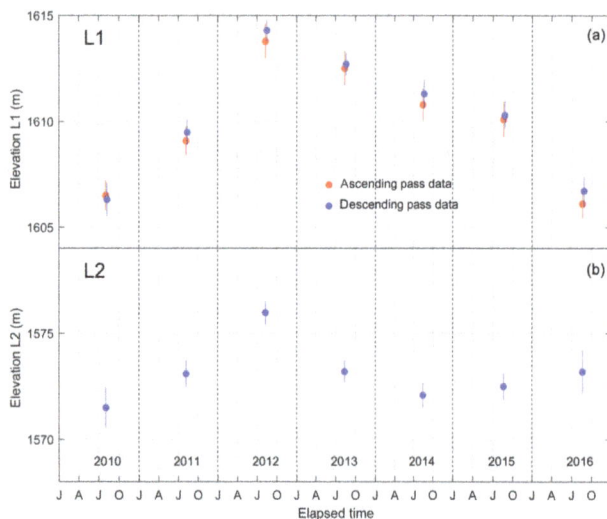

Figure 16. Surface elevation of L1 (**a**) and L2 (**b**) between the summers of 2010 and 2016.

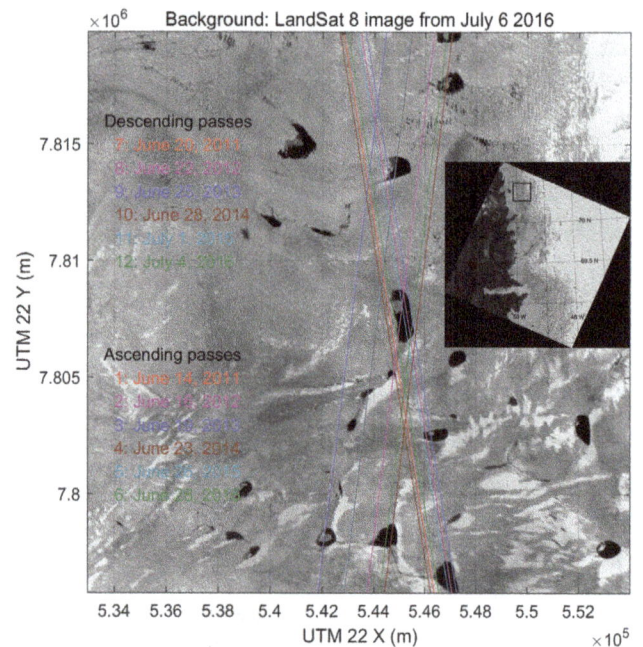

Figure 17. Landsat 8 image from 6 July 2016 of an area in the ablation zone of the western Greenland test site which includes a lake viewed by CryoSat on all the repeat ascending and descending passes listed on the image. The insert image shows the magnified position in the full Landsat 8 frame.

crease in the 5.5 days between the ascending and descending passes over the lake. This allows an estimate of the filling rate at the time of the two passes. If we assume a lake area of $2 \pm 0.5\,\mathrm{km}^2$ this implies a filling rate of $\sim 0.2.\,10^6$–$2.10^6\,\mathrm{m}^3$ meltwater added per day. This lake does drain sometime after the start of July (see the MODIS sequence in the Supplement) but appears not to have drained at the times of any of the CryoSat overpasses.

5   Discussion

In this section we discuss the two SARIn processing approaches and the limitations and successes of the current CryoSat SARIn products for glacial ice.

There are two important advantages with swath processing: firstly, there is no need for a retracker and, secondly, the swath data are obtained predominantly from the region directly beneath the satellite and the look angles for the swath footprints can be less than those for the POCA (for those areas with cross-track slopes appropriate for swath processing). With the small look angles, the footprint illumination cross-track is essentially uniform. Consequently, assuming a small contribution from the range ambiguous area, the phase

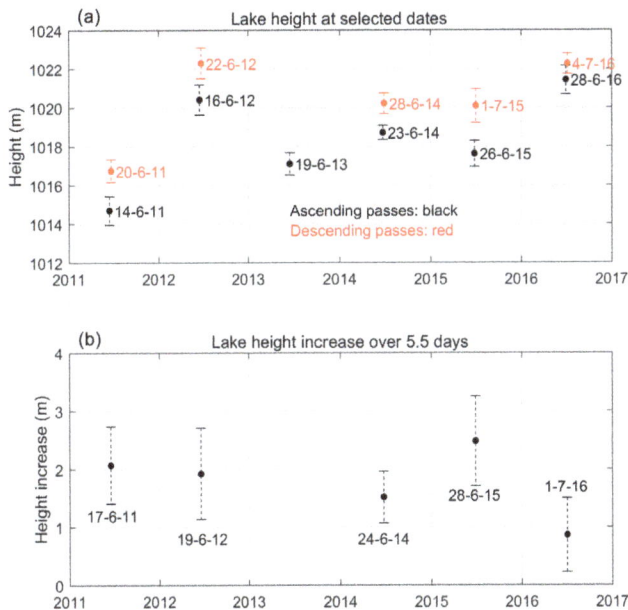

Figure 18. (a) Surface height of the lake in Fig. 17 at the times of the overpasses; **(b)** height increase during the 5.5 days between the ascending and descending passes. The dates listed in the lower plot are at the middle of the 5.5-day period between the ascending and descending passes.

should represent the geometric centre of the footprint so that the range, satellite state vectors, and the various angles lead to reliable heights. Unfortunately, the roll-angle problem discussed earlier compromises the swath-mode results as the resulting cross-track mis-mapping will normally lead to a height error (Gray et al., 2013).

POCA processing requires a retracker and the look angle can extend into the range in which the illumination cross-track is affected by the antenna pattern variation so that the phase may not reflect the geometric centre of the footprint; instead, it may be displaced towards the sub-satellite track. With interferometric swath processing, precise knowledge of the baseline and baseline angles is important (Rosen et al., 2000), and with the CryoSat roll-angle problem individual POCA heights are normally more precise than swath-mode heights. For height change estimates, however, both POCA and swath-mode results can be combined as long as any bias is accounted for. Foresta et al. (2016) used primarily swath-mode results in a study of elevation change of Icelandic ice caps, showing the improved surface coverage of swath mode and that height change information could be derived from these results. Also, Smith et al. (2017) combined swath and POCA data to document surface height change on the Thwaites Glacier. However, in this case "metre-scale biases", correlated over tens of kilometres but independent orbit-to-orbit, were partially corrected by combining with the POCA data.

The known problem in processing the star-tracker data (Scagliola et al., 2017), and the resulting varying error in the reported value of the baseline roll angle, can have an impact on the precision of the CryoSat height results. Any roll-angle error translates directly into a cross-track mapping error so that the resulting height error then depends on the angle between the incident wave and the tangent to the cross-track surface. If this angle is 90° and the surface slope changes slowly over a few hundred metres, then the error is small as the geocoding algorithm produces the correct elevation for the mis-mapped footprint. Although we show that the roll-angle problem had essentially no impact on the Devon POCA results, it did have an impact on the POCA results from the western Greenland test site. In this case the cross-track slopes varied more rapidly than for Devon and lead to the situation where an incorrect roll angle could lead to an increase in the CryoSat height with respect to the surface irrespective of the sense of the roll-angle error. This we suggest is the origin of the unrealistic result that the average POCA height was slightly above the physical surface for the western Greenland site.

POCA heights originate from ridges and peaks and, when the cross-track slope is appropriate for swath processing, the swath-mode results will normally originate from the area beneath the satellite so the two approaches are complementary in surface coverage. As discussed above, there can be a bias between POCA and swath heights which needs to be considered if the results are merged. The potential height error for individual estimates is normally less for POCA data than for swath-mode heights but the exception is the precision with which one can estimate the height of relatively large supraglacial lakes when the lake is beneath the satellite and viewed at close to normal incidence. In this case, we have a very strong signal in the middle of the waveform, any range ambiguous contribution should be small, and no retracker is required for the geocoding solution. Further, with this viewing geometry the problem of an incorrect roll angle leads to a small error in the lake surface height and a precision of ∼ 0.5 m is possible for the surface height of a large lake. Work is underway to better evaluate the extent to which CryoSat data can help in quantifying the time and extent of melt around Greenland.

The ability to geocode the relatively small footprint possible with the SARIn mode over glacial ice creates a huge advantage for this mode over the traditional low-resolution radar altimetry. Future radar altimeters employing coherent along-track processing, either fully focussed or delay-Doppler, coupled with cross-track interferometry could play a very important role in monitoring change on many ice caps and glaciers.

6   Conclusions

Here we list the specific conclusions arising from our analysis of the SARIn data over Devon Ice Cap and western Greenland.

A more consistent fit can be obtained between CryoSat and surface heights using the pre-launch baseline coupled with an additional roll-angle bias of $\sim 0.0075° \pm 0.0025°$. Although the additional bias may originate with the angle measurement, it could equally well be an equivalent additional phase correction of $\sim 0.0435 \pm 0.0145$ radians to the value of 0.612 radians currently used in the baseline C product (Bouzinac, 2012).

A retracker which uses the first significant leading edge of the waveform normally leads to more reliable elevations than a retracker that uses the whole waveform; this appears to be particularly true for areas like western Greenland in which the shape of the waveform is very variable and the peak signal is often in the middle of the waveform.

Swath-mode results complement the POCA results but are normally less precise. The exception is the precision with which the heights of supraglacial lakes can be obtained when the satellite flies almost directly over the lake.

The uncertainty in the CryoSat baseline roll angle affects primarily swath-mode results but can also impact the precision of POCA results when the surface topography is comparable to that in the western Greenland test site.

While more work is required to establish to what extent CryoSat SARIn waveforms and heights can improve our knowledge of melt in the ablation zone of the Greenland Ice Sheet, these initial results indicate that CryoSat SARIn data can help provide useful information on the variation of year-to-year melt.

Competing interests. The authors declare that they have no conflict of interest.

Acknowledgements. This work was supported by the European Space Agency through the provision of CryoSat-2 data and the support for the CRYOVEX airborne field campaigns in both the Canadian Arctic and Greenland. NASA supported the IceBridge flights over the Canadian Arctic and Greenland, while NSIDC facilitated provision of the airborne laser data. The Technical University of Denmark (TUD) managed the ESA supported flights over Devon. The IceBridge and TUD teams are gratefully acknowledged for the acquisition and provision of the airborne data used in this work. The Polar Continental Shelf Project (Natural Resources Canada) provided logistic support for field work in the Canadian Arctic, and the Nunavut Research Institute and the community of Resolute Bay gave permission to conduct research on the Devon Ice Cap. Support for D. Burgess was provided through the Climate Change Geoscience Program, Earth Sciences Sector, Natural Resources Canada and the GRIP programme of the Canadian Space Agency. Support for K. Langley was provided by ESA project Glaciers-CCI (4000109873/14/I-NB) and GlacioBasis Nuuk of the Greenland Ecosystem Monitoring programme. T. Dunse and G. Moholdt were supported by ESA-Prodex project 4000 110 725/724 "CRYOVEX" and T. Dunse was supported by the Nordforsk-funded project Green Growth Based on Marine Resources: Ecological and Sociological Economic Constraints (GreenMAR). Wesley Van Wychen and Tyler de Jong helped with the 2011 kinematic GPS survey on Devon. NSERC funding to L. Copland is gratefully acknowledged. We also acknowledge NASA and NSIDC for the provision of the Landsat 8 and MODIS imagery.

Tommaso Parrinello, ESA, and Michele Scagliola, Aresys, provided information on the problem in processing the star-tracker data. We also appreciate the work of the editor, Ian Howat, and the four anonymous reviewers who provided thoughtful and helpful reviews.

Edited by: I. M. Howat

References

Bamber, J. L.: Ice sheet altimeter processing scheme, Int. J. Remote Sens., 15, 925–938, doi:10.1080/01431169408954125, 1994.

Bouzinac, C.: CryoSat-2 Product Handbook, Tech. Report, European Space Agency, available at: http://emits.sso.esa.int/ emits-doc/ESRIN/7158/CryoSat-PHB-17apr2012.pdf (last access: 27 April 2017), 2012.

Box, J. E. and Ski, K.: Remote sounding of Greenland supraglacial melt lakes: implications for subglacial hydraulics, J. Glaciol., 53, 257–265, doi:10.3189/172756507782202883, 2007.

Brenner, A. C., Bindschadler, R. A., Thomas, R. H., and Zwally, H. J.: Slope-induced errors in radar altimetry over continental ice sheets, J. Geophys. Res., 88, 1617–1623, 1983.

Brenner, A. C., DiMarzio, J. P., and Zwally, H. J.: Precision and accuracy of satellite radar and laser altimeter data over the continental ice sheets, IEEE T. Geosci. Remote. Sens., 45, 321–331, doi:10.1109/TGRS.2006.887172, 2007.

Buffard, J.: CryoSat-2 Level 2 product evolutions and quality improvements in Baseline C, ESA technical report XCRY-GSEG-EOPG-TN-15-00004, available at: https://earth.esa.int/ web/guest/document-library (last access: April 2017), 2015.

Christie, F. D. W., Bingham, R. G., Gourmelen, N., Tett, S. F. B., and Muto, A.: Four-decade record of pervasive grounding line retreat along the Bellingshausen margin of West Antarctica, Geophys. Res. Lett., 43, 5741–5749, doi:10.1002/2016GL068972, 2016.

Davis, C. H.: A robust threshold retracking algorithm for measuring ice-sheet surface elevation change from satellite radar altimeters, IEEE T. Geosci. Remote, 35, 974–979, doi:10.1109/36.602540, 1997.

Davis, C. H. and Ferguson, A. C.: Elevation change of the Antarctic ice sheet, 1995-2000, from ERS-2 satellite radar altimetry, IEEE Trans. Geosci. Remote Sens., 42, 2437–2445, doi:10.1109/TGRS.2004.836789, 2004.

Echelmeyer, K., Clarke, T. S., and Harrison, W. D.: Surficial glaciology of Jakobshavns Isbræ, West Greenland: Part I. Surface morphology, J. Glaciol., 37, 368–382, 1991.

ESA: CryoSat Ground Segment, Level 1 data, baseline C release, available at: https://earth.esa.int/web/guest/-/ cryosat-user-tool-7386, last access: 1 May 2017.

Fettweis, X.: The 2015–2016 Greenland ice sheet season as simulated by MARv3.5.2, available at: http://climato.be/melt-2016 (last access: 1 May 2017), 2016.

Fitzpatrick, A. A. W., Hubbard, A. L., Box, J. E., Quincey, D. J., van As, D., Mikkelsen, A. P. B., Doyle, S. H., Dow, C. F., Hasholt, B., and Jones, G. A.: A decade (2002–2012) of supraglacial lake volume estimates across Russell Glacier, West Greenland, The Cryosphere, 8, 107–121, doi:10.5194/tc-8-107-2014, 2014.

Foresta, L., Gourmelen, N., Pálsson, F., Nienow, P., Björnsson, H., and Shepherd, A.: Surface Elevation Change and Mass Balance of Icelandic Ice Caps Derived from Swath Mode CryoSat-2 Altimetry, Geophys. Res. Lett., 43, 12138–12145, doi:10.1002/2016GL071485, 2016.

Forster, R. R., Box, J. E., van der Broeke, M. R., Miege, C., Burgess, E. W., van Angelen, J. H., Lenaerts, J. T. M., Koenig, L. S., Paden, J, Lewis, C., Gogenini, S. P., Leuschen, C., and McConnell, J. R.: Extensive liquid meltwater storage in firn within the Greenland ice sheet, Nature Geosc., 7, 95–98, doi:10.1038/ngeo2043, 2014.

Galin, N., Wingham, D. J., Cullen, R., Fornari, M., Smith, W. H. F., and Abdall, S.: Calibration of the CryoSat-2 Interferometer and Measurement of Across-track Ocean Slope, IEEE T. Geosci. Remote., 51, 57–72, 2012.

Gray, L., Burgess, D., Copland, L., Cullen, R., Galin, N., Hawley, R., and Helm, V.: Interferometric swath processing of Cryosat data for glacial ice topography, The Cryosphere, 7, 1857–1867, doi:10.5194/tc-7-1857-2013, 2013.

Gray, L., Burgess, D., Copland, L., Demuth, M. N., Dunse, T., Langley, K., and Schuler, T. V.: CryoSat-2 delivers monthly and inter-annual surface elevation change for Arctic ice caps, The Cryosphere, 9, 1895–1913, doi:10.5194/tc-9-1895-2015, 2015.

Gray L., Burgess, D., Copland L., Dunse, T., Hagen, J.O., Langley, K., Moholdt, G., Schuler, T., and Van Wychen, W.: On the bias between ice cap surface elevation and Cryosat results. Proc. 'Living Planet Symposium 2016', Prague, Czech Republic, 9–13 May 2016, ESA SP-740, August 2016.

Hawley, R. L., Shepherd, A., Cullen, R., and Wingham, D. J.: Ice-sheet elevations from across-track processing of airborne interferometric radar altimetry, Geophys. Res. Lett., 36, L25501, doi:10.1029/2009GL040416, 2009.

Helm, V., Humbert, A., and Miller, H.: Elevation and elevation change of Greenland and Antarctica derived from CryoSat-2, The Cryosphere, 8, 1539–1559, doi:10.5194/tc-8-1539-2014, 2014.

Howat, I. M., Negrete, A., and Smith, B. E.: The Greenland Ice Mapping Project (GIMP) land classification and surface elevation data sets, The Cryosphere, 8, 1509–1518, doi:10.5194/tc-8-1509-2014, 2014.

Hurkmans, R. T. W. L., Bamber, J. L., and Griggs, J. A.: Brief communication "Importance of slope-induced error correction in volume change estimates from radar altimetry", The Cryosphere, 6, 447–451, doi:10.5194/tc-6-447-2012, 2012.

Hurkmans, R. T. W. L., Bamber, J. L., Davis, C. H., Joughin, I. R., Khvorostovsky, K. S., Smith, B. S., and Schoen, N.: Time-evolving mass loss of the Greenland Ice Sheet from satellite altimetry, The Cryosphere, 8, 1725–1740, doi:10.5194/tc-8-1725-2014, 2014.

Ignéczi, Á., Sole A. J., Livingstone S. J., Leeson A. A., Fettweis X.,

Selmes N., Gourmelen N., and Briggs, K.: Northeast sector of the Greenland Ice Sheet to undergo the greatest inland expansion of supraglacial lakes during the 21st century, Geophys. Res. Lett., 43, 9729–9738, doi:10.1002/2016GL070338, 2016.

Jensen, J. R.: Angle measurement with a phase monopulse radar altimeter, IEEE T. Antenn. Propag., 47, 715–724, 1999.

Joughin, I., Howat, I. M., Fahnestock, M., Smith, B., Krabill, W., Alley, R. B., Stern, H., and Truffer, M.: Continued evolution of Jakobshavn Isbræ following its rapid speedup, J. Geophys. Res., 113, F04006, doi:10.1029/2008JF001023, 2008.

Joughin, I., Smith, B. E., Howat, I., Scambos, T., and Moon, T.: Greenland Flow Variability from Ice-Sheet-Wide Velocity Mapping, J. Glaciol., 56, 415–430, doi:10.3189/002214310792447734, 2010.

Joughin, I., Smith, B. E., Howat, I., and Scambos, T.: MEaSUREs Multi-year Greenland Ice Sheet Velocity Mosaic, Version 1, Boulder, Colorado USA, NASA National Snow and Ice Data Center Distributed Active Archive Center, doi:10.5067/QUA5Q9SVMSJG (last access: June 2016), 2016.

Kleinherenbrink, M., Ditmar, P. G., and Lindenbergh, R. C.: Retracking Cryosat data in the SARIn mode and robust lake level extraction, Remote Sens. Environ., 152, 38–50, doi:10.1016/j.rse.2014.05.014, 2014.

Koenig, L. S., Miege, C., Forster, R. R., and Brucker, L.: Initial in situ measurements of perennial meltwater storage in the Greenland firn acquifer, Geophys. Res. Lett., 41, 81–85, doi:10.1002/2013GL058083, 2014.

Krabill, W. B.: IceBridge ATM L2 Icessn Elevation, Slope, and Roughness, Version 2. [ILATM2.002], Boulder, Colorado USA, NASA National Snow and Ice Data Center Distributed Active Archive Center, doi:10.5067/CPRXXK3F39RV, (last access: October 2016), 2014, updated 2016 (data available at: https://nsidc.org/icebridge/portal/map).

Krabill, W. B., Abdalati, W., Frederick, E. B., Manizade, S. S., Martin, C. F., Sonntag, J. G., Swift, R. N., Thomas, R. H., and Yungel, J. G.: Aircraft laser altimetry measurement of elevation changes of the Greenland ice sheet: technique and accuracy assessment, J. Geodyn., 34, 357–376, 2002.

Leeson, A. A., Shepherd, A., Briggs, K., Howat, I., Fettweis, X., Morlighem, M., and Rignot, E.: Supraglacial lakes on the Greenland ice sheet advance inland under warming climate, Nature Climate Change, 5, 51–55, doi:10.1038/nclimate2463, 2015.

Levinsen, J. F., Simonsen, S. B., Sorensen, L. S., and Forsberg, R.: The Impact of DEM Resolution on relocating Radar Altimetry Data over Ice Sheets, IEEE J. Sel. Topics in Appl. Earth Obs. and Rem. Sens., 9, 3158–3163, doi:10.1109/JSTARS.2016.2587684, 2016.

Liang, Y. L., Colgan, W., Lv, Q., Steffen, K., Abdalati, W., Stroeve, J., Gallaher, D., and Bayou, N.: A decadal investigation of supraglacial lakes in West Greenland using a fully automatic detection and tracking algorithm, Remote Sens. Environ., 123, 127–138, 2012.

McMillan, M., Nienow, P., Shepherd, A., Benham, T., and Sole, A.: Seasonal evolution of supraglacial lakes on Greenland Ice Sheet, Earth Planet Sc. Lett., 262, 484–492, 2007.

Nghiem, S. V., Steffen, K., Kwok, R., and Tsai, W.-Y.: Detection of snowmelt regions on the Greenland ice sheet using diurnal backscatter change, J. Glaciol., 47, 539–547, doi:10.3189/172756501781831738, 2001.

Nilsson, J., Vallelonga, P., Simonsen, S. B., Sørensen, L. S., Forsberg, R., Dahl-Jensen, D., Hirabayashi, M., Goto-Azuma, K., Hvidberg, C. S., Kjaer, H. A., and Satow, K.: Greenland 2012 melt event effects on CryoSat-2 radar altimetry, Geophys. Res. Lett., 42, 3919–3926, doi:10.1002/2015GL063296, 2015.

Nilsson, J., Gardner, A., Sandberg Sørensen, L., and Forsberg, R.: Improved retrieval of land ice topography from CryoSat-2 data and its impact for volume-change estimation of the Greenland Ice Sheet, The Cryosphere, 10, 2953–2969, doi:10.5194/tc-10-2953-2016, 2016.

Pope, A., Scambos, T. A., Moussavi, M., Tedesco, M., Willis, M., Shean, D., and Grigsby, S.: Estimating supraglacial lake depth in West Greenland using Landsat 8 and comparison with other multispectral methods, The Cryosphere, 10, 15–27, doi:10.5194/tc-10-15-2016, 2016.

Qi, W. and Braun, A.: Accelerated Elevation Change of Greenland's Jakobshavn Glacier Observed by ICESat and IceBridge, IEEE GRS Letters, 10, 1133–1137, doi:10.1109/LGRS.2012.2231954, 2013.

Raney, R. K.: The delay/Doppler radar altimeter, IEEE T. Geosci. Remote, 36, 1578–1588, 1998.

Rémy, F. and Parouty, S.: Antarctic Ice Sheet and Radar Altimetry: A Review, Remote Sens., 4, 1212–1239, doi:10.3390/rs1041212, 2009.

Rosen, P., Hensley, S., Joughin, I., Li, F., Madsen, S., Rodriguez, E., and Goldstein, R.: Synthetic Aperture Radar Interferometry, Proc. IEEE, 88, 333–382, 2000.

Scagliola, M., Fornari, M., Bouffard, J., and Parrinello, T.: The CryoSat interferometer: end-to-end calibration and achievable performance for Adv. Space Res., submitted, 2017.

Shepherd, A., Ivins, E. R., A., G., Barletta, V. R., Bentley, M. J., Bettadpur, S., Briggs, K. H., Bromwich, D. H., Forsberg, R., Galin, N., Horwath, M., Jacobs, S., Joughin, I., King, M. A., Lenaerts, J. T. M., Li, J., Ligtenberg, S. R. M., Luckman, A., Luthcke, S. B., McMillan, M., Meister, R., Milne, G., Mouginot, J., Muir, A., Nicolas, J. P., Paden, J., Payne, A. J., Pritchard, H., Rignot, E., Rott, H., Sandberg Sørensen, L., Scambos, T. A., Scheuchl, B., Schrama, E. J. O., Smith, B., Sundal, A. V., van Angelen, J. H., van de Berg, W. J., van den Broeke, M. R., Vaughan, D. G., Velicogna, I., Wahr, J. D., Whitehouse, P. L., Wingham, D. J., Yi, D., Young, D., and Zwally, H. J.: A Reconciled Estimate of Ice-Sheet Mass Balance, Science, 338, 1183–1189, doi:10.1126/science.1228102, 2012.

Sneed, W. and Hamilton, G.: Evolution of melt pond volume of the surface of the Greenland Ice Sheet, Geophys. Res. Lett., 34, L03501, doi:10.1029/2006GL028697, 2007.

Smith, B. E., Gourmelen, N., Huth, A., and Joughin, I.: Connected subglacial lake drainage beneath Thwaites Glacier, West Antarctica, The Cryosphere, 11, 451–467, doi:10.5194/tc-11-451-2017, 2017.

Ulaby, F., Moore, R., and Fung, A.: Microwave Remote Sensing: Active and Passive, Vol. 2, Norwood, Massachusetts, Artech House, 816–920, 1986.

Wang, L., Toose, P., Brown, R., and Derksen, C.: Frequency and distribution of winter melt events from passive microwave satellite data in the pan-Arctic, 1988–2013, The Cryosphere, 10, 2589–2602, doi:10.5194/tc-10-2589-2016, 2016.

Wingham, D., Francis, C. R., Baker, S., Bouzinac, C., Cullen, R., de Chateau-Thierry, P., Laxon, S. W., Mallow, U., Mavrocordatos, C., Phalippou, L., Ratier, G., Rey, L., Rostan, F., Viau, P., and Wallis, D.: CryoSat-2: a mission to determine the fluctuations in earth's land and marine ice fields, Adv. Space Res., 37, 841–871, 2006.

Zwally, H. J., Li, J., Robbins, J. W., Saba, J. L., Yi, D., and Brenner, A. C.: Mass gains of the Antarctic ice sheet exceed losses, J. Glaciol., 61, 1019–1036, doi:10.3189/2015JoG15J071, 2015.

Experimental observation of transient δ^{18}O interaction between snow and advective airflow under various temperature gradient conditions

Pirmin Philipp Ebner[1], Hans Christian Steen-Larsen[2,3], Barbara Stenni[4], Martin Schneebeli[1], and Aldo Steinfeld[5]

[1]WSL Institute for Snow and Avalanche Research SLF, 7260 Davos Dorf, Switzerland
[2]LSCE Laboratoire des Sciences du Climat et de l'Environnement, Gif-Sur-Yvette CEDEX, France
[3]Center for Ice and Climate, Niels Bohr Institute, University of Copenhagen, Copenhagen, Denmark
[4]Department of Environmental Sciences, Informatics and Statistics, University Ca' Foscari of Venice, Venice, Italy
[5]Department of Mechanical and Process Engineering, ETH Zurich, 8092 Zurich, Switzerland

Correspondence to: Martin Schneebeli (schneebeli@slf.ch)

Abstract. Stable water isotopes (δ^{18}O) obtained from snow and ice samples of polar regions are used to reconstruct past climate variability, but heat and mass transport processes can affect the isotopic composition. Here we present an experimental study on the effect of airflow on the snow isotopic composition through a snow pack in controlled laboratory conditions. The influence of isothermal and controlled temperature gradient conditions on the δ^{18}O content in the snow and interstitial water vapour is elucidated. The observed disequilibrium between snow and vapour isotopes led to the exchange of isotopes between snow and vapour under non-equilibrium processes, significantly changing the δ^{18}O content of the snow. The type of metamorphism of the snow had a significant influence on this process. These findings are pertinent to the interpretation of the records of stable isotopes of water from ice cores. These laboratory measurements suggest that a highly resolved climate history is relevant for the interpretation of the snow isotopic composition in the field.

1 Introduction

Water stable isotopes in polar snow and ice have been used for several decades as proxies for global and local temperatures (e.g. Dansgaard, 1964; Lorius et al., 1979; Grootes et al., 1994; Petit et al., 1999; Johnsen et al., 2001; EPICA Members, 2004). However, the processes that influence the isotopic composition of precipitation at high latitudes are complex, making direct inference of palaeotemperatures from the isotopic record difficult (Cuffey et al., 1994; Jouzel et al., 1997, 2003; Hendricks et al., 2000). Several factors affect the vapour and snow isotopic composition, which give rise to ice-core isotopic composition, starting from the process of evaporation in the source region, transportation of the air mass to the top of the ice sheet and post-depositional processes (Craig and Gordon, 1964; Merlivat and Jouzel, 1979; Johnsen et al., 2001; Ciais and Jouzel, 1994; Jouzel and Merlivat, 1984; Jouzel et al., 2003; Helsen et al., 2005, 2006, 2007; Cuffey and Steig, 1998; Krinner and Werner, 2003). Mechanical processes such as mixing, seasonal scouring or spatial redistribution of snow can alter seasonal and annual records (Fisher et al., 1983; Hoshina et al., 2014). Post-depositional processes associated with wind scouring and snow redistribution are known to introduce a "post-depositional noise" in the surface snow. Comparisons of isotopic records obtained from closely located shallow ice cores have allowed for estimations of a signal-to-noise ratio and a common climate signal (Fisher and Koerner, 1988, 1994; White et al., 1997; Steen-Larsen et al., 2011; Sjolte et al., 2011; Masson-Delmotte et al., 2015). After deposition, interstitial diffusion in the firn and ice affects the water-isotopic signal but back-diffusion or deconvolution techniques have been used to establish the original isotope signal (Johnsen, 1977; Johnsen et al., 2000).

Snow is a bicontinuous material consisting of fully connected ice crystals and pore space (air) (Löwe et al., 2011). Because of the proximity to the melting point, the high vapour pressure causes a continuous recrystallization of the snow microstructure known as snow metamorphism, even under moderate temperature gradients (Pinzer et al., 2012). The whole ice matrix is continuously recrystallizing by sublimation and deposition, with vapour diffusion as the dominant transport process. Pinzer et al. (2012) showed that a typical half-life of the ice matrix is a few days. The intensity of the recrystallization is dictated by the temperature gradient and this can occur under midlatitude or polar conditions. Temperature and geometrical factors (porosity and specific surface area) also play a significant role (Pinzer and Schneebeli, 2009; Pinzer et al., 2012).

The interpretation of ice-core data and the comparison with atmospheric model results implicitly rely on the assumption that the snowfall precipitation signal is preserved in the snow–ice matrix (Werner et al., 2011). Classically, ice-core stable-isotope records are interpreted as reflecting precipitation-weighted signals and compared to observations and atmospheric model results for precipitation, ignoring exchanges between surface snow and atmospheric water vapour (e.g. Persson et al., 2011). However, recent studies carried out on top of the Greenland and Antarctic ice sheets combining continuous atmospheric water-vapour-isotope observations with daily snow surface sampling document a clear day-to-day variation of isotopic composition of surface snow between precipitation events as well as diurnal change in the snow isotopes (Steen-Larsen et al., 2014a; Ritter et al., 2016; Casado et al., 2016). This effect was interpreted as being caused by the uptake of the synoptic-driven atmospheric water-vapour-isotope signal by individual snow crystals undergoing snow metamorphism (Steen-Larsen et al., 2014a) and the diurnal variation in moisture flux (Ritter et al., 2016). However, the impact of this process on the isotope-temperature reconstruction is not yet sufficiently understood, but crucial to constrain. This process, compared to interstitial diffusion (Johnsen, 1977; Johnsen et al., 2000), will alter the isotope mean value. The field observations challenge the previous assumption that sublimation occurred layer-by-layer with no resulting isotopic fractionation (Dansgaard, 1964; Friedman et al., 1991; Town et al., 2008; Neumann and Waddington, 2004). It is assumed that the solid undergoing sublimation would not be unduly enriched in the heavier isotope species due to the preferential loss of lighter isotopic species to the vapour (Dansgaard, 1964; Friedman et al., 1991). Because self-diffusion in the ice is about 3 orders of magnitude slower than molecular diffusion in the vapour, the amount of isotopic separation in snow is assumed to be negligible.

Snow has a high permeability (Calonne et al., 2012; Zermatten et al., 2014), which facilitates diffusion of gases and, under appropriate conditions, airflow (Gjessing, 1977; Colbeck, 1989; Sturm and Johnson, 1991; Waddington et al.,

1996). In a typical Antarctic and Greenland snow profile, strong interactions between the atmosphere and snow occur, especially in the first 2 m (Neumann and Waddington, 2004; Town et al., 2008), called the convective zone. In the convective zone, air can move relatively freely and therefore exchange occurs between snow and the atmospheric air. Air flowing into the snow reaches saturation vapour pressure nearly instantly through sublimation (Neumann et al., 2008; Ebner et al., 2015a). Models of the influence of the so-called "wind pumping" effect (Fisher et al., 1983; Neumann and Waddington, 2004), in which the interstitial water vapour is replaced by atmospheric air pushed through the upper metres of the snow pack by small-scale high and low pressure areas caused by irregular grooves or ridges formed on the snow surface (dunes and sastrugi), have assumed that the snow grains would equilibrate with the interstitial water vapour on timescales governed by ice self-diffusion. However, no experimental data are available to support this assumption. With this in mind the experimental study presented here is specifically developed to investigate the effect of ventilation inside the snow pack on the isotopic composition. Only conditions deeper than 1 cm inside a snowpack are considered. Previous work showed that (1) under isothermal conditions, the Kelvin effect leads to a saturation of the pore space in the snow but does not affect the structural change (Ebner et al., 2015a), (2) applying a negative temperature gradient along the flow direction leads to a change in the microstructure due to deposition of water molecules on the ice matrix (Ebner et al., 2015b), and (3) a positive temperature gradient along the flow had a negligible total mass change of the ice but a strong reposition effect of water molecules on the ice grains (Ebner et al., 2016). Here, we continuously measured the isotopic composition of an airflow containing water vapour through a snow sample under both isothermal and temperature gradient conditions. Microcomputed-tomography (μCT) was applied to obtain the 3-D microstructure and morphological properties of snow.

2 Experimental set-up

Isothermal and temperature gradient experiments with fully saturated airflow and defined isotopic composition were performed in a cold laboratory at around $T_{lab} \approx -15\,°C$ with small fluctuations of $\pm 0.8\,°C$ (Ebner et al., 2014). Snow produced from de-ionized tap water in a cold laboratory (water temperature: $30\,°C$; air temperature: $-20\,°C$) was used for the snow sample preparation (Schleef et al., 2014). The snow was sieved with a mesh size of 1.4 mm into a box and isothermally sintered for 27 days at $-5\,°C$ to increase the strength in order to prevent destruction of the snow sample due to the airflow and to evaluate the effect of metamorphism of snow. The morphological properties of the snow are listed in Table 1. The sample holder (diameter 53 mm, height 30 mm, 0.066 L) was filled by a cylinder cut out from the sintered

Table 1. Morphological properties and flow characteristics of the experimental runs: µCT measured snow density (ρ), porosity (ε), specific surface area per unit mass (SSA), mean pore space diameter (d_{mean}), superficial velocity in snow (u_D), corresponding Reynolds number ($Re = d_{mean} \cdot u_D / \nu_{air}$), average inlet temperature of the humidifier and at the inlet ($T_{in, mean}$), average outlet temperature at the outlet ($T_{out, mean}$) and average temperature gradient (∇T_{ave}). Experiment (1) corresponds to the isothermal conditions, experiment (2) to air warming and experiment (3) to air cooling in the snow sample.

	ρ kg m^{-3}	ε –	SSA m^2 kg^{-1}	d_{mean} mm	u_D m s^{-1}	Re –	$T_{in, mean}$ °C	$T_{out, mean}$ °C	∇T_{ave} K m^{-1}
Experiment (1)	202	0.78	28	0.39	0.03	0.76	−15.5	−15.5	–
Experiment (2)	202	0.78	30	0.36	0.03	0.70	−15.4	−14.0	+47
Experiment (3)	220	0.76	27	0.37	0.031	0.74	−12.3	−14.1	−60

snow. To prevent airflow between the snow sample and the sample holder walls, the undisturbed snow disk was filled in at a higher temperature (about −5 °C) and sintering was allowed for about 1 h before cooling down at the start of the experiment. The set-up of Ebner et al. (2014) was modified by additionally inserting a water-vapour-isotope analyser (model: L1102-I Picarro, Inc., Santa Clara, CA, USA) to measure the isotopic ratio $\delta^{18}O$ of the water vapour contained in the airflow at the inlet and outlet of the sample holder. The experimental set-up consisted of three main components (humidifier, sample holder and the Picarro analyser) connected with insulated copper tubing and Swagelok fitting (Fig. 1). The tubes to the Picarro analyser were heated to prevent deposition of water vapour and thereby fractionation. The temperature was monitored with thermistors inside the humidifier and at the inlet and outlet of the snow sample. A dry air pressure tank controlled by a mass flow controller (EL-Flow, Bronkhorst) generated the airflow. A humidifier, consisting of a tube (diameter 60 mm, height 150 mm, 0.424 L volume) filled with crushed ice particles (snow from Antarctica with low $\delta^{18}O$ composition), was used to saturate the dry air entering the humidifier with water vapour at an almost constant isotopic composition. The air temperature in the humidifier and at the inlet of the snow sample was maintained at the same value (accuracy ±0.2 K) to limit the influence of variability in absolute vapour pressure and isotopic composition. We measured the $\delta^{18}O$ of the water vapour produced by the humidifier before and after each experimental run ($\delta^{18}O_{hum}$). The outlet flow ($\delta^{18}O_a$) of the sample holder was continuously measured during the experiment to analyse the temporal evolution of the isotopic signal. All data from the Picarro analyser were corrected to the humidity reference level using the established instrument humidity-isotope response (Steen-Larsen et al., 2013, 2014b). In addition, VSMOW-SLAP correction and drift correction were performed. We followed the calibration protocol and used the calibration system described in detail by Steen-Larsen et al. (2013, 2014b).

The sample holder described by Ebner et al. (2014) was used to analyse the snow by µCT. Tomography measurements were performed with a modified µ-CT80 (Scanco Medical). The equipment incorporated a microfocus X-ray

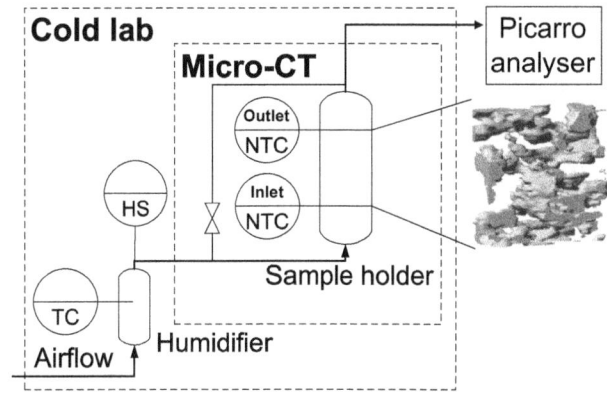

Figure 1. Schematic of the experimental set-up. A thermocouple (TC) and a humidity sensor (HS) inside the humidifier measured the mean temperature and humidity of the airflow. Two thermistors (NTC) close to the snow surface measured the inlet and outlet temperature of the airflow (Ebner et al., 2014). The Picarro analyser measured the isotopic composition $\delta^{18}O$ of the outlet flow. Inset: 3-D structure of $110 \times 42 \times 110$ voxels ($2 \times 0.75 \times 2$ mm^3) obtained by the µCT.

source, operated at 70 kV acceleration voltage with a nominal resolution of 18 µm. The samples were scanned with 1000 projections per 180° in high-resolution setting, with typical adjustable integration time of 200 ms per projection. The field of view of the scan area was 36.9 mm of the total 53 mm diameter and subsamples with a dimension of $7.2 \times 7.2 \times 7.2$ mm^3 were extracted for further processing. The reconstructed µCT images were filtered using a $3 \times 3 \times 3$ median filter followed by a Gaussian filter ($\sigma = 1.4$, support = 3). The Otsu method (Otsu, 1979) was used to automatically perform clustering-based image thresholding to segment the grey-level images into ice and void phase. Morphological properties in the two-phase system were determined based on the exact geometry obtained by the µCT. Tetrahedrons corresponding to the enclosed volume of the triangulated ice matrix surface were applied to the segmented data to determine porosity (ε) and specific surface area (SSA). The mean pore size distribution was estimated using the opening-size-distribution operation. This operation can be imagined as

Table 2. $\delta^{18}O$ is the vapour in the humidifier ($\delta^{18}O_{hum}$) and of the snow in the sample holder ($\delta^{18}O_s$) at the beginning ($t = 0$) and end ($t =$ end) of each experiment and the final $\delta^{18}O$ content of the snow in the sample holder at the inlet ($z = 0$ mm) and outlet ($z = 30$ mm). Experiment (1) corresponds to the isothermal conditions, experiment (2) to air warming and experiment (3) to air cooling in the snow sample.

	$\delta^{18}O_{hum}$ ‰		$\delta^{18}O_s, t = 0$ ‰	$\delta^{18}O_{s,t=end}$ ‰	
	$t = 0$	$t =$ end		$z = 0$ mm	$z = 30$ mm
Experiment (1)	−68.2	−67.5	−10.97	−17.75	−15.72
Experiment (2)	−66.3	−66.1	−11.94	−19.60	−16.60
Experiment (3)	−62.8	−62.2	−10.44	−25.53	−15.00

virtual sieving with different mesh sizes (Haussener et al., 2012).

Three experiments with saturated advective airflow through the snow sample were performed to record the following parameters and analyse their effects: (1) isothermal conditions to analyse the influence of curvature effects (Kaempfer al et., 2007), (2) positive temperature gradient applied to the snow sample where cold air entering the sample is heated while flowing through the sample in order to analyse the influence of sublimation, and (3) negative temperature gradient applied to the snow sample where warm air entering the sample is cooled while flowing across the sample to analyse the influence of net deposition. During the temperature gradient experiments, temperature differences of 1.4 and 1.8 °C were imposed, resulting in gradients of +47 and −60 K m^{-1}, respectively. The runs were performed at atmospheric pressure and with a volume flow rate of 3.0 L min^{-1} corresponding to an average flow speed in the pores of $u_D \approx$ 30 mm s^{-1}. We performed the experiments with airflow velocities in the snow sample at $u_D \approx 30$ mm s^{-1}, which is a factor of 3 higher than calculated by Neumann (2003) for a natural snow pack. However, when looking at the Reynolds number and describing the flow regime inside the pores, our experiments ($Re \approx 0.7$) were in the feasible flow regime (laminar flow) of a natural snow pack ($Re \approx 0.65$). The outlet temperature in experiment (2) and the inlet and the humidifier temperature in experiment (3) were actively controlled using thermo-electric elements. Variations in temperature of up to ±0.8 °C were due to temperature fluctuations inside the cold laboratory, leading to slightly variable temperature gradients and mean temperature in experiment (2) and (3). Table 1 presents a summary of the experimental conditions and the morphological properties of the snow samples. All snow samples were taken from the same snow block with an average density of ≈ 210 kg m^{-3}. The density given in Table 1 was the density of the snow sample in each experiment measured by μCT. At the end of each experiment, the snow sample was cut into five layers of 6 mm height and the isotopic composition of each layer was analysed to examine the spatial $\delta^{18}O$ gradient in the isotopic composition of the snow sample.

A slight increase with a maximum of 0.7 ‰ of $\delta^{18}O$ in the water vapour produced by the humidifier was observed in experiment (1), with lower increases during experiments (2) and (3) (Table 2). This change of ~ 0.7 ‰ is not significant compared to the difference between the isotopic composition of the water vapour and the snow sample in the sample holder of ~ 53 ‰ and the temporal change of the water-vapour isotopes on the back side of the snow sample.

In approximately the first 30 min, the isotopic composition of the measured outflow air $\delta^{18}O_a$ increased from a low $\delta^{18}O$ to a starting value of around −29 ‰ in each experiment. This was due to a memory effect and another possible effect might be condensed water left in the tubes from a prior experiment which had no further impact on the experiments (Penna et al., 2012).

3   Results

3.1   Isothermal condition

Experiment (1) was performed for 24 h at a mean temperature of $T_{mean} = -15.5$ °C. $\delta^{18}O_a$ decreased exponentially in the outlet flow observed throughout the experimental run as shown in Fig. 2. Initially, the $\delta^{18}O_a$ content in the flow was −27.7 ‰ and exponentially decreased to −47.6 ‰ after 24 h. The small fluctuations in the $\delta^{18}O_a$ signal at $t \approx 7$, 17 and 23 h were due to small temperature changes in the cold laboratory.

We observed a strong interaction between the airflow and the snow as manifest by the isotopic composition of the snow. The $\delta^{18}O_s$ signal in the snow decreased by 4.75–7.78 ‰ and an isotopic gradient in the snow was observed after the experimental run, shown in Fig. 3. Initially, the snow had a homogeneous isotopic composition of $\delta^{18}O_s = -10.97$ ‰ but post-experiment sampling showed a decrease in the snow $\delta^{18}O$ at the inlet side to −17.75 ‰ and at the outlet side to −15.72 ‰. The spatial $\delta^{18}O_s$ gradient of the snow had an approximate slope of 0.68 ‰ mm^{-1} at the end of the experimental run. Table 2 shows the $\delta^{18}O$ value in snow at the beginning ($t = 0$) and end ($t = 24$ h) of the experiment.

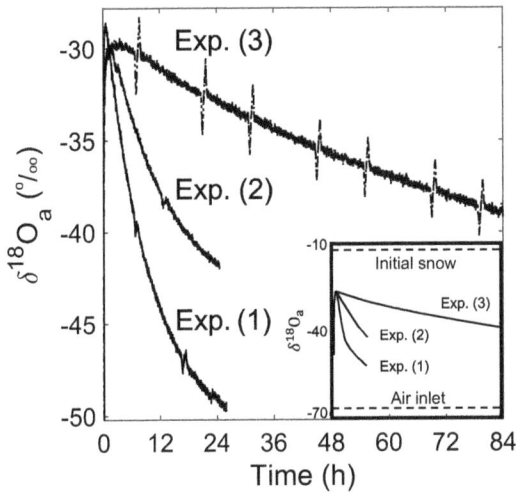

Figure 2. Temporal isotopic composition of $\delta^{18}O$ of the outflow for each of the experimental runs. The spikes in the $\delta^{18}O$ were due to small temperature changes in the cold laboratory (Ebner et al., 2014). Exp. (1) corresponds to the isothermal conditions, Exp. (2) to air warming and Exp. (3) to air cooling in the snow sample. The higher the recrystallization rate of the snow the slower the adaption of $\delta^{18}O$ of the outlet air to the inlet air. The illustration in the lower right corner shows the relation between $\delta^{18}O$ of the initial snow, inlet and outlet of the air.

3.2 Air warming by a positive temperature gradient along the airflow

Experiment (2) was performed over a period of 24 h with an average temperature gradient of approximatively $+47\,\mathrm{K\,m^{-1}}$ (warmer temperatures at the outlet of the snow) and an average mean temperature of $-14.7\,^\circ\mathrm{C}$. As in the isothermal experiment (1), we observed a relaxing exponential decrease of $\delta^{18}O_a$ in the outlet flow throughout the measurement period as shown in Fig. 2, but the decrease was slower compared to the isothermal run. Initially, the $\delta^{18}O_a$ content in the flow coming through the snow disk was $-29.8\permil$ and exponentially decreased to $-41.9\permil$ after 24 h. The small fluctuations in the $\delta^{18}O_a$ signal at $t \approx 2.7$ and 12.7 h were due to small temperature changes in the cold laboratory.

The $\delta^{18}O_s$ signal in the snow decreased by $4.66–7.66\permil$ and a gradient in the isotopic composition of the snow was observed after the experimental run, shown in Fig. 3. Initially, the snow had a homogeneous isotopic composition of $\delta^{18}O_s = -11.94\permil$, but post-experiment sampling showed a decrease at the inlet side to $-19.6\permil$ and at the outlet side to $-16.6\permil$. The spatial $\delta^{18}O_s$ gradient of the snow had an approximate slope of $1.0\permil\,\mathrm{mm^{-1}}$ at the end of the experimental run. Table 2 shows the $\delta^{18}O_s$ values in snow at the beginning ($t = 0$) and end ($t = 24\,\mathrm{h}$) of the experiment.

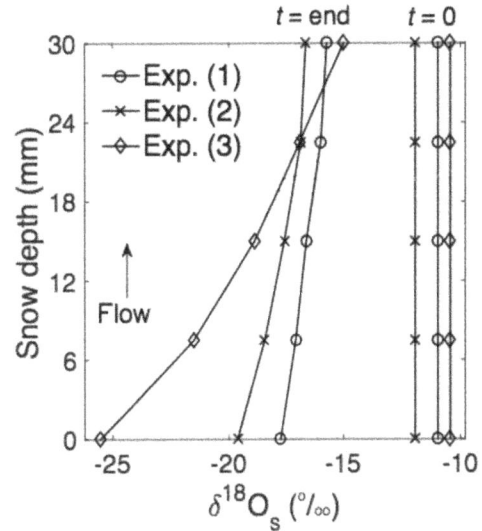

Figure 3. Spatial isotopic composition of $\delta^{18}O$ of the snow sample at the beginning ($t = 0$) and at the end ($t = $ end) for each experiment. The air entered at $z = 0$ mm and exited at $z = 30$ mm. Exp. (1) corresponds to the isothermal conditions Exp. (2) to air warming and Exp. (3) to air cooling in the snow sample.

3.3 Air cooling by a negative temperature gradient along the airflow

Experiment (3) was performed for 84 h instead of 24 h to better estimate the trend in $\delta^{18}O_a$ in the outlet flow. An average temperature gradient of approximately $-60\,\mathrm{K\,m^{-1}}$ (colder temperatures at the outlet of the snow) and an average mean temperature of $-13.2\,^\circ\mathrm{C}$ were observed during the experiment. As in the previous experiments, a relaxing exponential decrease of $\delta^{18}O_a$ in the outlet flow was observed throughout the experimental run as shown in Fig. 2. The decrease was slower compared to experiments (1) and (2). Initially, the $\delta^{18}O_a$ content in the flow was $-29.8\permil$ and exponentially decreased to $-37.7\permil$ after 84 h. The small fluctuations in the $\delta^{18}O_a$ signal at $t \approx 7.3$, 21.3, 31.3, 45.3, 55.3, 69.3 and 79.3 h were due to small temperature changes in the cold laboratory.

The $\delta^{18}O_s$ signal in the snow decreased by $4.46–15.09\permil$ and a gradient in the isotopic composition of the snow was observed after the experimental run, shown in Fig. 3. Initially, the snow had an isotopic composition of $\delta^{18}O_s = -10.44\permil$ but post-experiment sampling showed a decrease at the inlet side to $-25.53\permil$ and at the outlet side to $-15.00\permil$. The spatial $\delta^{18}O_s$ gradient of the snow had an approximate slope of $3.5\permil\,\mathrm{mm^{-1}}$ at the end of the experimental run. Table 2 shows the $\delta^{18}O_s$ value in snow at the beginning ($t = 0$) and end ($t = 84\,\mathrm{h}$) of the experiment.

4 Discussion

All experiments showed a strong exchange in $\delta^{18}O$ between the snow and water-vapour-saturated air, resulting in a significant change in the values of the stable isotopes in the snow. The advective conditions in the experiments were comparable with surface snow layers in Antarctica and Greenland, but at higher temperatures, especially compared to the interior of Antarctica.

The results also showed strong interactions in $\delta^{18}O$ between snow and air depending on the different temperature gradient conditions. The experiments indicate that temperature variation and airflow above and through the snow structures (Sturm and Johnson, 1991; Colbeck, 1989; Albert and Hardy, 1995) seem to be dominant processes affecting water stable isotopes of surface snow. The results also support the statement that an interplay occurs between theoretically expected layer-by-layer sublimation and deposition at the ice-matrix surface and the isotopic content evolution of snow cover due to mass exchange between the snow cover and the atmosphere (Sokratov and Golubev, 2009). The specific surface area of snow exposed to mass exchange (Horita et al., 2008) and by the depth of the snow layer exposed to the mass exchange with the atmosphere (He and Smith, 1999) plays an important role. Our results support the interpretation that changes in surface snow isotopic composition are expected to be significant if large day-to-day surface changes in water vapour occur in between precipitation events, wind pumping is efficient and snow metamorphism is enhanced by temperature gradients in the upper first centimetres of the snow (Steen-Larsen et al., 2014a).

We expect that our findings will lead to an improvement of the interpretation of the water stable-isotope records from ice cores. Classically, ice-core stable-isotope records are interpreted as palaeotemperature, reflecting precipitation-weighted signals. When comparing observations and atmospheric model results for precipitation with ice-core records, such snow–vapour exchanges are normally ignored (e.g. Persson et al., 2011; Fujita and Abe, 2006). However, snow–vapour exchange enhanced by recrystallization rate seems to be an important factor for the high variation in the snow surface $\delta^{18}O$ signal as supported by our experiments. It was hypothesized that the changes in the snow-surface $\delta^{18}O$ reported by Steen-Larsen et al. (2014a) are caused by changes in large-scale wind and moisture advection of the atmospheric water-vapour signal and snow metamorphism. The strong interaction between atmosphere and near-surface snow can modify the ice-core water stable-isotope records.

The rate-limiting step for isotopic exchange in the snow is isotopic equilibration between the pore-space vapour and surrounding ice grains. The relaxing exponential decrease of $\delta^{18}O$ in the outflow of our experiments predicted that full isotopic equilibrium between snow and atmospheric vapour will not be reached at any depth (Waddington et al., 2002; Neumann and Waddington, 2004) but changes move towards equilibrium with the atmospheric state (Steen-Larsen et al., 2013, 2014a).

As snow accumulates, the upper 2 m are advected through the ventilated zone (Neumann and Waddington, 2004; Town et al., 2008). In areas with high accumulation rate (e.g. South Greenland), snow is advected for a short time through the ventilated zone. The snow exposed for a relatively short time to vapour snow exchange would result in higher spatial variability compared to longer exposure. However, the effects of snow ventilation on isotopic composition may become more important as the accumulation rate of the snow decreases ($< 50\,\mathrm{mm\,a^{-1}}$), such that snow remains in the near-surface ventilated zone for many years (Waddington et al., 2002; Hoshina et al., 2014, 2016). As the snow remains for a longer time in the near-surface ventilated zone, a larger $\delta^{18}O$ exchange will occur between snow and atmospheric vapour. Consequently, the isotopic content of layers at sites with high and low accumulation rates can evolve differently, even if the initial snow composition had been equal, and the sites had been subjected to the same histories of air-mass vapour.

Despite a relatively small change in the difference between the isotopic composition of the incoming vapour and the snow, large differences in the isotopic composition of the water vapour at the outlet flow exist for the three different experimental set-ups. Based on the difference in the outlet water-vapour isotopic composition, we hypothesized that different processes are at play for the different experiments. It is obvious that there is a fast isotopic exchange with the surface of the ice crystals and a much slower timescale on which the interior of the ice crystals is altered. Due to the low diffusivity of $H_2{}^{16}O$ and $H_2{}^{18}O$ in ice ($D_{H_2{}^{18}O} \approx D_{H_2{}^{16}O} = \sim 10^{-15}\,\mathrm{m^2\,s^{-1}}$ (Ramseier, 1967; Johnsen et al., 2000), we assumed that the interior of the ice crystals is not altered on the timescale of the experiment. This explained why the net isotopic change of the bulk sample is relatively small compared to the changes in the outlet water-vapour isotopes. The effective "ice-diffusion depth" of the isotopic exchange during the experiments is given as $L_D = \sqrt{D \cdot t}$, where D is the diffusion coefficient of $H_2{}^{16}O$ and $H_2{}^{18}O$ in ice and t is the experimental time. The calculated ice-diffusion depth L_D, is $\sim 9.3\,\mu m$ for experiments (1) and (2), and $\sim 17.4\,\mu m$ for experiment (3), respectively, indicating an expected a minimal change of the interior of the ice crystal. However, snow has a large specific surface area and therefore a high exchange area. This has an effect on the $\delta^{18}O$ snow concentration. The fraction of the total volume V_{tot} of ice that is close enough to the ice surface to be affected by diffusion in time t is then $\rho_{\mathrm{ice}} \cdot SSA \cdot L_D$, where SSA is the specific surface area (area per unit mass), and L_D is the diffusion depth, defined above, for time t. For $t \approx 24\,h$, a large fraction (24 to 43 %) of the total volume V_{tot} of the ice matrix can be accessed through diffusion. It is quite hard to see the total $\delta^{18}O$ snow difference between experiments (1) and (2) after the experiment compared to the $\delta^{18}O$ of the vapour in the air at the outlet.

There is a small but notable difference in the total $\delta^{18}O$ of the snow between experiments (1) and (2). Due to the higher recrystallization rate of experiment (2) the spatial $\delta^{18}O_s$ gradient of the snow ($1.0\,\text{‰}\,\text{mm}^{-1}$) is higher than for experiment (1) ($0.68\,\text{‰}\,\text{mm}^{-1}$). Increasing the experimental time, the $\delta^{18}O$ change in the snow increases (experiment 3). In general, the calculated ice-diffusion depth is realistic under isothermal conditions where diffusion processes are the main factors (Kaempfer and Schneebeli, 2007; Ebner et al., 2015). By applying a temperature gradient, the impact of diffusion is suppressed due to the high recrystallization rate by sublimation and deposition. Due to the low half-life of a few days of the ice matrix, the growth rates are typically of the order of $100\,\mu\text{m}\,\text{day}^{-1}$ (Pinzer et al., 2012). Therefore, this redistribution of ice caused by temperature gradient counteracts the diffusion into the solid ice.

By comparing similarities and differences between the outcomes of the three experimental set-ups we will now discuss the physical processes influencing the interaction and exchange processes within the snowpack between the snow and the advected vapour. We first notice that the final snow isotopic profiles of experiments (1) (isothermal) and (2) (positive temperature gradient along the direction of the flow) are comparable to each other. Despite this similarity, the evolution in the outlet water vapour of experiment (1) showed a significantly stronger depletion compared to experiment (2). For experiment (3) (negative temperature gradient along the direction of the flow) we observed the smallest change in outlet water-vapour isotopes but the largest snowpack isotope gradient after the experiment. However, this change was caused by 84 h flow instead of 24 h.

Curvature effects, temperature gradients and therefore the recrystallization rate influence the mass transfer of $H_2^{16}O/H_2^{18}O$ molecules. The higher the recrystallization rate of the snow the slower the adaption of the outlet air concentration to the inlet air concentration (see in experiments 2 and 3). Under isothermal conditions (experiment 1) the only effect influencing the recrystallization rate is the curvature effect (Kaempfer and Schneebeli, 2007). However, based on the experimental observations (Kaempfer and Schneebeli, 2007) this effect decreases with decreasing temperature and increasing experimental time. Applying an additional temperature gradient to a snow sample causes complex interplays between local sublimation and deposition on surfaces and the interaction of water molecules in the air with the ice matrix due to changing saturation conditions of the airflow. Therefore, the recrystallization rate increases and causes the change in the $\delta^{18}O$ of the air. For experiment (2) there is a complex interplay between sublimation and deposition of water molecules into the interstitial flow (Ebner et al., 2015c), while for experiment (3) there is deposition of molecules carried by the interstitial flow onto the snow crystals (Ebner et al., 2015b). Furthermore, in the beginning of each experiment there is a tendency to sublimate from edges of the individual snow crystals due to the higher curvature.

As the edges were sublimated and deposition occurred in the concavities, the individual snow crystals became more rounded, slowing down the transfer of water molecules into the interstitial airflow. We noticed, for all three experiments, that within the uncertainty of the isotopic composition of the snow, the initial isotopic composition of the vapour was the same and in isotopic equilibrium with the snow. The difference between experiments (1) and (2) lies in the fact that due to the temperature gradient in experiment (2) there is an increased transfer of water vapour with the isotopic composition of the snow into the airflow. Hence the depleted air from the humidifier advected through the snow disk is mixed with a relatively larger vapour flux from the snow crystals. Additionally, we also expected less deposition into the concavities in experiment (2) compared to experiment (1). However, it is interesting to note that the final isotopic profile of the snow disk is similar in experiment (1) and experiment (2). We interpreted this as being a result of two processes acting in opposite directions: although relatively isotope-depleted vapour from the humidifier was deposited onto the ice matrix there was also a higher amount of sublimation of relatively isotope-enriched vapour from the snow disk in experiment (2). Experiment (3) separates itself from the other two experiments in that as the water vapour from the humidifier is advected through the snow disk there is a continuous deposition of very depleted air due to the negative temperature gradient. As for experiments (1) and (2) there was also a constant sublimation of the convexities into the vapour stream in experiment (3). We notice that, despite the fact that experiment (3) ran for 84 h, the snow at the outlet side of the snow disk did not become more isotopically depleted compared to experiments (1) and (2). However, the snow on the inlet side became significantly more isotopically depleted. This observation, together with the fact that the vapour of the outlet of the snow disk is less depleted compared to experiments (1) and (2), leads us to hypothesize that there is a relatively larger deposition of isotopically depleted vapour from the humidifier as the vapour is advected through the snow disk. This means that a relatively larger component of the isotopic composition of the vapour is originating by sublimation from the convexities of the snow disk and a smaller one from the isotopically depleted vapour from the humidifier.

Our results and conclusions indicate that there is a need for additional validation. Specifically, it would be crucial to know the mass balance of the snow disk more precisely, which could be done by reconstructing the entire snow disk following the change in density and morphological properties over the entire height. Ideally, the entire sample would be tomographically measured with a resolution of $4 \times 4 \times 4\,\text{mm}^3$, each cube corresponding to the representative volume. Insights would also be achieved with experiments using snow of the same isotopic composition, but different SSA, as a more precise calculation of the different observed exchange rates would be allowed. Additionally, different and colder background temperatures should be tested to better understand the inland Antarctic environment and the effect of the quasi-liquid layer, which is necessary for the development of a numerical model. Isotopically different combinations of vapour and snow should be performed. In the present paper, vapour with low $\delta^{18}O$ isotopic composition was transported through snow with relative high $\delta^{18}O$ isotopic composition. It would be interesting to reverse the combination and perform experiments with different combinations to provide more insights on mass and isotope exchanges between vapour and snow. Experiments with longer running time help to understand the change in the ice matrix better under low accumulation conditions.

5   Summary and conclusion

Laboratory experimental runs were performed where a transient $\delta^{18}O$ interaction between snow and air was observed. The airflow altered the isotopic composition of the snowpack and supports an improved climatic interpretation of ice core stable water-isotope records. The water-vapour-saturated airflow with an isotopic difference of up to 55‰ changed the original $\delta^{18}O$ isotope signal in the snow by up to 7.64 and 15.06‰ within 24 and 84 h. The disequilibrium between snow and air isotopes led to the observed exchange of isotopes, the rate depending on the temperature gradient conditions. To conclude, increasing the recrystallization rate in the ice matrix causes the temporal change of the $\delta^{18}O$ concentration at the outflow to decrease (experiment 2 and 3). Decreasing the recrystallization rate causes the temporal curve of the outlet concentration to become steeper, reaching the $\delta^{18}O$ inlet concentration of the air faster (experiment 1).

Additionally, the complex interplay of simultaneous diffusion, sublimation and deposition due to the geometrical complexity of snow has a strong effect on the $\delta^{18}O$ signal in the snow and cannot be neglected. A temporal signal can be superimposed on the precipitation signal, (a) if the snow remains near the surface for a long time, i.e. in a low-accumulation area, and (b) if it is exposed to a history of air masses carrying vapour with a significantly different isotopic signature than the precipitated snow.

These are novel measurements and will therefore be important as the basis for further research and experiments. Our results represent direct experimental observations of the interaction between the water isotopic composition of the snow, the water vapour in the air and recrystallization due to temperature gradients. Our results demonstrate that recrystallization and bulk mass exchange must be incorporated into future models of snow and firn evolution. Further studies are required on the influence of temperature and airflow as well as on snow microstructure on the mass transfer phenomena for validating the implementation of stable water isotopes in snow models.

Competing interests.   The authors declare that they have no conflict of interest.

Acknowledgements.   The Swiss National Science Foundation granted financial support under project Nr. 200020-146540. Hans Christian Steen-Larsen was supported by the AXA Research Fund. The authors thank Koji Fujita, Edwin Waddington and an anonymous reviewer for the suggestions and critical review. Matthias Jaggi, Sascha Grimm, Alessandro Schlumpf and Sarah Berben gave technical support. The data for this paper are available by contacting the corresponding author.

Edited by: Benjamin Smith

References

Albert, M. R. and Hardy, J. P.: Ventilation experiments in a seasonal snow cover, in: Biogeochemistry of Seasonally Snow-Covered Catchments, IAHS Publ. 228, edited by: Tonnessen, K. A., Williams, M. W., and Tranter, M., IAHS Press, Wallingford, UK, 41–49, 1995.

Calonne, N., Geindreau, C., Flin, F., Morin, S., Lesaffre, B., Rolland du Roscoat, S., and Charrier, P.: 3-D image-based numerical computations of snow permeability: links to specific surface area, density, and microstructural anisotropy, The Cryosphere, 6, 939–951, https://doi.org/10.5194/tc-6-939-2012, 2012.

Casado, M., Landais, A., Masson-Delmotte, V., Genthon, C., Kerstel, E., Kassi, S., Arnaud, L., Picard, G., Prie, F., Cattani, O., Steen-Larsen, H.-C., Vignon, E., and Cermak, P.: Continuous measurements of isotopic composition of water vapour on the East Antarctic Plateau, Atmos. Chem. Phys., 16, 8521–8538, https://doi.org/10.5194/acp-16-8521-2016, 2016.

Ciais, P. and Jouzel, J.: Deuterium and oxygen 18 in precipitation: Isotopic model, including mixed-cloud processes, J. Geophys. Res., 99, 16793–16803, https://doi.org/10.1029/94JD00412, 1994.

Colbeck, S. C.: Air movement in snow due to windpumping, J. Glaciol., 35, 209–213, 1989.

Craig, H. and Gordon, L. I.: Deuterium and oxygen 18 variations in the ocean and marine atmosphere, in: Proc. Stable Isotopes in Oceanographic Studies and Paleotemperatures, edited by: Toniorgi, E., Spoleto, Italy, 9–130, 1964.

Cuffey, K. M. and Steig, E. J.: Isotopic diffusion in polar firn: implications for interpretation of seasonal climate parameters in ice-core records, with emphasis on central Greenland, J. Glaciol., 44, 273–284, 1998.

Cuffey, K. M., Alley, R. B., Grootes, P. M., Bolzan, J. M., and Anandakrishnan, S.: Calibration of the Delta-O-18 isotopic paleothermometer for central Greenland, using borehole temperatures, J. Glaciol., 40, 341–349, 1994.

Dansgaard, W.: Stable isotopes in precipitation, Tellus, 16, 436–468, 1964.

Ebner, P. P., Grimm, S., Schneebeli, M., and Steinfeld, A.: An instrumented sample holder for time-lapse micro-tomography measurements of snow under advective conditions, Geosci. Instrum. Meth., 3, 179–185, https://doi.org/10.5194/gi-3-179-2014, 2014.

Ebner, P. P., Schneebeli, M., and Steinfeld, A.: Tomography-based monitoring of isothermal snow metamorphism under advective conditions, The Cryosphere, 9, 1363–1371, https://doi.org/10.5194/tc-9-1363-2015, 2015a.

Ebner, P. P., Andreoli, C., Schneebeli, M., and Steinfeld, A.: Tomography-based characterization of ice–air interface dynamics of temperature gradient snow metamorphism under advective conditions, J. Geophys. Res.-Earth, 120, 2437–2451, https://doi.org/10.1002/2015JF003648, 2015b.

Ebner, P. P., Schneebeli, M., and Steinfeld, A.: Metamorphism during temperature gradient with undersaturated advective airflow in a snow sample, The Cryosphere, 10, 791–797, https://doi.org/10.5194/tc-10-791-2016, 2016.

EPICA Members: Eight glacial cycles from an Antarctic ice core, Nature, 429, 623–628, https://doi.org/10.1038/nature02599, 2004.

Gjessing, Y. T.: The filtering effect of snow, in: Isotopes and Impurities in Snow and Ice Symposium, edited by: Oeschger, H., Ambach, W., Junge, C. E., Lorius, C., and Serebryanny, L., IASHAISH Publication, Dorking, 118, 199–203, 1977.

Grootes, P. M., Steig, E., and Stuiver, M.: Taylor Ice Dome study 1993–1994: an ice core to bedrock, Antarct. J. US, 29, 79–81, 1994.

Fisher, D. A. and Koerner, R.: The effect of wind on d(18O) and accumulation given an inferred record of seasonal d amplitude from the Agassiz ice cap, Ellesmere island, Canada, Ann. Glaciol., 10, 34–37, 1988.

Fisher, D. A. and Koerner, R.: Signal and noise in four ice-core records from the Agassiz ice cap, Ellesmere Island, Canada: details of the last millennium for stable isotopes, melt and solid conductivity, Holocene, 4, 113–120, https://doi.org/10.1177/095968369400400201, 1994.

Fisher, D. A., Koerner, R. M., Paterson, W. S. B., Dansgaard, W., Gundestrup, N., and Reeh, N.: Effect of wind scouring on climatic records from ice-core oxygen-isotope profiles, Nature, 301, 205–209, https://doi.org/10.1038/301205a0, 1983.

Friedman, I., Benson, C., and Gleason, J.: Isotopic changes during snow metamorphism, in: Stable Isotope Geochemistry: A Tribute to Samuel Epstein, edited by: O'Neill, J. R. and Kaplan, I. R., Geochemical Society, Washington, D. C., 211–221, 1991.

Haussener, S., Gergely, M., Schneebeli, M., and Steinfeld, A.: Determination of the macroscopic optical properties of snow based on exact morphology and direct pore-level heat transfer modeling, J. Geophys. Res., 117, 1–20, https://doi.org/10.1029/2012JF002332, 2012.

He, H. and Smith, R. B.: An advective-diffusive isotopic evaporation-condensation model, J. Geophys. Res., 104, 18619–18630, https://doi.org/10.1029/1999JD900335, 1999.

Helsen, M. M., van de Wal, R. S. W., van den Broeke, M. R., van As, D., Meijer, H. A. J., and Reijmer, C. H.: Oxygen isotope variability in snow from western Dronning Maud Land, Antarctica and its relation to temperature, Tellus, 57, 423–435, 2005.

Helsen, M. M., van de Wal, R. S. W., van den Broeke, M. R., Masson-Delmotte, V., Meijer, H. A. J., Scheele, M. P., and Werner, M.: Modeling the isotopic composition of Antarctic snow using backward trajectories: Simulation of snow pit records, J. Geophys. Res., 111, D15109, https://doi.org/10.1029/2005JD006524, 2006.

Helsen, M. M., van de Wal, R. S. W., and van den Broeke, M. R.: The isotopic composition of present-day Antarctic snow in a Lagrangian simulation, J. Climate, 20, 739–756, 2007.

Hendricks, M. B., DePaolo, D. J., and Cohen, R. C.: Space and time variation of δ18O and δD in precipitation: can paleotemperature be estimated from ice cores?, Global Biogeochem. Cy., 14, 851–861, https://doi.org/10.1029/1999GB001198, 2000.

Horita, J., Rozanski, K., and Cohen, S.: Isotope effects in the evaporation of water: a status report of the Craig-Gordon model, Isot. Environ. Health Sci., 44, 23–49, https://doi.org/10.1080/10256010801887174, 2008.

Hoshina, Y., Fujita, K., Nakazawa, F., Iizuka, Y., Miyake, T., Hirabayashi, M., Kuramoto, T., Fujita, S., and Motoyama, H.: Effect of accumulation rate on water stable isotopes of near-surface snow in inland Antarctica. J. Geophys. Res.-Atmos., 119, 274–283, https://doi.org/10.1002/2013JD020771, 2014.

Hoshina, Y., Fujita, K., Iizuka, Y., and Motoyama, H.: Inconsistent relations among major ions and water stable isotopes in Antarctica snow under different accumulation environments, Polar Sci., 10, 1–10, https://doi.org/10.1016/j.polar.2015.12.003, 2016.

Johnsen, S. J.: Stable isotope homogenization of polar firn and ice, in: Isotopes and Impurities in Snow and Ice, Proceeding of the Grenoble Symposium, August/September 1975, IAHS AISH Publication, 118, Grenoble, France, 210–219, 1997.

Johnsen, S. J., Clausen, H. B., Cuffey, K. M., Hoffman, G., Schwander, J., and Creyts, T.: Diffusion of stable isotopes in polar firn and ice: the isotope effect in firn diffusion, in: Physics of Ice Core Records, edited by: Hondoh, T., Hokkaido University Press, Sapporo, Japan, 121–140, 2000.

Johnsen, S. J., Dahl-Jensen, D., Gundestrup, N., Steffensen, J. P., Clausen, H. B., Miller, H., Masson-Delmotte, V., Sveinbjörnsdottir A. E., and White, J.: Oxygen isotope and palaeotemperature records from six Greenland ice-core stations: Camp Century, DYE-3, GRIP, GISP2, Renland and NorthGRIP, J. Quaternary Sci., 16, 299–307, https://doi.org/10.1002/jqs.622, 2001.

Jouzel, J. and Merlivat, L.: Deuterium and oxygen 18 in precipitation: modeling of the isotopic effects during snow formation, J. Geophys. Res., 89, 11749–11757, https://doi.org/10.1029/JD089iD07p11749, 1984.

Jouzel, J., Merlivat, L., Petit, J. R., and Lorius, C.: Climatic information over the last century deduced from a detailed isotopic record in the South Pole snow, J. Geophys. Res., 88, 2693–2703, https://doi.org/10.1029/JC088iC04p02693, 1983.

Jouzel, J., Alley, R. B., Cuffey, K. M., Dansgaard, W., Grootes, P., Hoffmann, G., Johnsen, S. J., Koster, R. D., Peel, D., Shuman, C. A., Stievenard, M., Stuiver, M., and White, J.: Validity of the temperature reconstruction from water isotopes in ice cores, J. Geophys. Res., 102, 26471–26487, https://doi.org/10.1029/97JC01283, 1997.

Jouzel, J., Vimeux, F., Caillon, N., Delaygue, G., Hoffman, G., Masson-Delmotte, V., and Parrenin, F.: Magnitude of isotope/temperature scaling for interpretation of central Antarctic ice cores, J. Geophys. Res., 108, 1–6, https://doi.org/10.1029/2002JD002677, 2003.

Kaempfer, T. U. and Schneebeli, M. Observation of isothermal metamorphism of new snow and interpretation as a sintering process, J. Geophys. Res., 112, 1–10, https://doi.org/10.1029/2007JD009047, 2007.

Krinner, G. and Werner, M.: Impact of precipitation seasonality changes on isotopic signals in polar ice cores: A multi-model analysis, Earth Planet. Sc. Lett., 216, 525–538, https://doi.org/10.1016/S0012-821X(03)00550-8, 2003.

Lorius, C., Merlivat, L., Jouzel, J., and Pourchet, M.: A 30,000-yr isotope climatic record from Antarctica ice, Nature, 280, 644–648, https://doi.org/10.1038/280644a0, 1979.

Löwe H., Spiegel, J. K., and Schneebeli, M.: Interfacial and structural relaxations of snow under isothermal conditions, J. Glaciol., 57, 499–510, 2011.

Masson-Delmotte, V., Steen-Larsen, H. C., Ortega, P., Swingedouw, D., Popp, T., Vinther, B. M., Oerter, H., Sveinbjornsdottir, A. E., Gudlaugsdottir, H., Box, J. E., Falourd, S., Fettweis, X., Gallée, H., Garnier, E., Gkinis, V., Jouzel, J., Landais, A., Minster, B., Paradis, N., Orsi, A., Risi, C., Werner, M., and White, J. W. C.: Recent changes in north-west Greenland climate documented by NEEM shallow ice core data and simulations, and implications for past-temperature reconstructions, The Cryosphere, 9, 1481–1504, https://doi.org/10.5194/tc-9-1481-2015, 2015.

Merlivat, L. and Jouzel, J.: Global climatic interpretation of the deuterium-oxygen 18 relationship for precipitation, J. Geophys. Res., 84, 5029–5033, https://doi.org/10.1029/JC084iC08p05029, 1979.

Neumann, T. A.: Effects of firn ventilation on geochemistry of polar snow, PhD thesis, University of Washington, Washington, USA, 2003.

Neumann, T. A. and Waddington, E. D.: Effects of firn ventilation on isotopic exchange, J. Glaciol., 50, 183–194, 2004.

Neumann, T. A., Albert, M. R., Lomonaco, R., Engel, C., Courville, Z., and Perron, F.: Experimental determination of snow sublimation rate and stable-isotopic exchange, Ann. Glaciol., 49, 1–6, 2008.

Otsu, N.: A threshold selection method from gray-level histograms, IEEE T. Syst. Man Cyb., 9, 62–66, 1979.

Penna, D., Stenni, B., Šanda, M., Wrede, S., Bogaard, T. A., Michelini, M., Fischer, B. M. C., Gobbi, A., Mantese, N., Zuecco, G., Borga, M., Bonazza, M., Sobotková, M., Čejková, B., and Wassenaar, L. I.: Technical Note: Evaluation of between-sample memory effects in the analysis of δ^2H and δ^{18}O of water samples measured by laser spectroscopes, Hydrol. Earth Syst. Sci., 16, 3925–3933, https://doi.org/10.5194/hess-16-3925-2012, 2012.

Persson, A., Langen, P. L., Ditlevsen, P., and Vinther, B. M.: The influence of precipitation weighting on interannual variability of stable water isotopes in Greenland, J. Geophys. Res.-Atmos., 116, 1–13, https://doi.org/10.1029/2010JD015517, 2011.

Petit, J. R., Jouzel, J., Raynaud, D., Barkov, N. I., Barnola, J.-M., Basile, I., Bender, M., Chappellaz, J., Davis, M., Delaygue, G., Delmotte, M., Kotlyakov, V. M., Legrand, M., Lorius, C., Pépin, L., Ritz, C., Saltzman, E. S., and Stievenard, M.: Climate and atmospheric history of the past 420,000 years from the Vostok ice core, Antarctica, Nature, 399, 429–436, https://doi.org/10.1038/20859, 1999.

Pinzer, B. R. and Schneebeli, M.: Snow metamorphism under alternating temperature gradients: morphology and recrystallization in surface snow, Geophys. Res. Lett., 36, L23503, https://doi.org/10.1029/2009GL039618, 2009.

Pinzer, B. R., Schneebeli, M., and Kaempfer, T. U.: Vapor flux and recrystallization during dry snow metamor-

phism under a steady temperature gradient as observed by time-lapse micro-tomography, The Cryosphere, 6, 1141–1155, https://doi.org/10.5194/tc-6-1141-2012, 2012.

Ramseier, R. O: Self-diffusion of tritium in natural and synthetic ice monocrystals, J. Appl. Phys., 38, 2553–2556, 1967.

Ritter, F., Steen-Larsen, H. C., Werner, M., Masson-Delmotte, V., Orsi, A., Behrens, M., Birnbaum, G., Freitag, J., Risi, C., and Kipfstuhl, S.: Isotopic exchange on the diurnal scale between near-surface snow and lower atmospheric water vapor at Kohnen station, East Antarctica, The Cryosphere, 10, 1647–1663, https://doi.org/10.5194/tc-10-1647-2016, 2016.

Schleef, S., Jaggi, M., Löwe H., and Schneebeli, M.: Instruments and methods: an improved machine to produce nature-identical snow in the laboratory, J. Glaciol., 60, 94–102, 2014.

Sjolte, J., Hoffmann, G., Johnsen, S. J., Vinther, B. M., Masson-Delmotte, V., and Sturm, C.: Modeling the water isotopes in Greenland precipitation 1959–2001 with the meso-scale model remo-iso, J. Geophys. Res., 116, 1–22, https://doi.org/10.1029/2010JD015287, 2011.

Sokratov, S. A. and Golubev, V. N.: Snow isotopic content change by sublimation. J. Glaciol., 55, 823–828, 2009.

Steen-Larsen, H. C., Masson-Delmotte, V., Sjolte, J., Johnsen, S. J., Vinther, B. M., Breon, F. M., Clausen, H. B., Dahl-Jensen, D., Falourd, S., Fettweis, X., Gallee, H., Jouzel, J., Kageyama, M., Lerche, H., Minster, B., Picard, G., Punge, H. J., Risi, C., Salas, D., Schwander, J., Steffen, K., Sveinbjornsdottir, A. E., Svensson, A., and White, J.: Understanding the climatic signal in the water stable isotope records from the neem shallow firn/ice cores in northwest Greenland, J. Geophys. Res.-Atmos., 116, 1–20, https://doi.org/10.1029/2010JD014311, 2011.

Steen-Larsen, H. C., Johnsen, S. J., Masson-Delmotte, V., Stenni, B., Risi, C., Sodemann, H., Balslev-Clausen, D., Blunier, T., Dahl-Jensen, D., Ellehøj, M. D., Falourd, S., Grindsted, A., Gkinis, V., Jouzel, J., Popp, T., Sheldon, S., Simonsen, S. B., Sjolte, J., Steffensen, J. P., Sperlich, P., Sveinbjörnsdóttir, A. E., Vinther, B. M., and White, J. W. C.: Continuous monitoring of summer surface water vapor isotopic composition above the Greenland Ice Sheet, Atmos. Chem. Phys., 13, 4815–4828, https://doi.org/10.5194/acp-13-4815-2013, 2013.

Steen-Larsen, H. C., Masson-Delmotte, V., Hirabayashi, M., Winkler, R., Satow, K., Prié, F., Bayou, N., Brun, E., Cuffey, K. M., Dahl-Jensen, D., Dumont, M., Guillevic, M., Kipfstuhl, S., Landais, A., Popp, T., Risi, C., Steffen, K., Stenni, B., and Sveinbjörnsdottír, A. E.: What controls the isotopic composition of Greenland surface snow?, Clim. Past, 10, 377–392, https://doi.org/10.5194/cp-10-377-2014, 2014a.

Steen-Larsen, H. C., Sveinbjörnsdottir, A. E., Peters, A. J., Masson-Delmotte, V., Guishard, M. P., Hsiao, G., Jouzel, J., Noone, D., Warren, J. K., and White, J. W. C.: Climatic controls on water vapor deuterium excess in the marine boundary layer of the North Atlantic based on 500 days of in situ, continuous measurements, Atmos. Chem. Phys., 14, 7741–7756, https://doi.org/10.5194/acp-14-7741-2014, 2014b.

Sturm, M. and Johnson, J. B.: Natural convection in the subarctic snow cover, J. Geophys. Res., 96, 11657–11671, https://doi.org/10.1029/91JB00895, 1991.

Town, M. S., Warren, S. G., Walden, V. P., and Waddington, E. D.: Effect of atmospheric water vapor on modification of stable isotopes in near-surface snow on ice sheets, J. Geophys. Res.-Atmos., 113, 1–16, https://doi.org/10.1029/2008JD009852, 2008.

van der Wel, G., Fischer, H., Oerter, H., Meyer, H., and Meijer, H. A. J.: Estimation and calibration of the water isotope differential diffusion length in ice core records, The Cryosphere, 9, 1601–1616, https://doi.org/10.5194/tc-9-1601-2015, 2015.

Waddington, E. D., Cunningham, J., and Harder, S. L.: The effects of snow ventilation on chemical concentrations, in: Chemical Exchange Between the Atmosphere and Polar Snow, edited by: Wolff, E. W. and Bales, R. C., Springer, Berlin, NATO ASI Series, 43, 403–452, 1996.

Waddington, E. D., Steig, E. J., and Neumann, T. A.: Using characteristic times to assess whether stable isotopes in polar snow can be reversibly deposited, Ann. Glaciol., 35, 118–124, 2002.

Werner, M., Langebroek, P. M., Carlsen, T., Herold, M., and Lohmann, G.: Stable water isotopes in the ECHAM5 general circulation model: Toward high-resolution isotope modeling on a global scale, J. Geophys. Res.-Atmos., 116, D15109, https://doi.org/10.1029/2011JD015681, 2011.

White, J. W., Barlow, L. K., Fisher, D., Grootes, P., Jouzel, J., Johnsen, S. J., Stuiver, M., and Clausen, H.: The climate signal in the stable isotopes of snow from Summit, Greenland: Results of comparisons with modern climate observations, J. Geophys. Res., 102, 26425–26439, https://doi.org/10.1029/97JC00162, 1997.

Zermatten, E., Schneebeli, M., Arakawa, H., and Steinfeld, A.: Tomography-based determination of porosity, specific area and permeability of snow and comparison with measurements, Cold Reg. Sci. Technol., 97, 33–40, https://doi.org/10.1016/j.coldregions.2013.09.013, 2014.

3

Evaluation of Greenland near surface air temperature datasets

J. E. Jack Reeves Eyre and Xubin Zeng

Department of Hydrology and Atmospheric Sciences, University of Arizona, Tucson, 85721, USA

Correspondence to: J. E. Jack Reeves Eyre (jeyre@email.arizona.edu)

Abstract. Near-surface air temperature (SAT) over Greenland has important effects on mass balance of the ice sheet, but it is unclear which SAT datasets are reliable in the region. Here extensive in situ SAT measurements (~ 1400 station-years) are used to assess monthly mean SAT from seven global reanalysis datasets, five gridded SAT analyses, one satellite retrieval and three dynamically downscaled reanalyses. Strengths and weaknesses of these products are identified, and their biases are found to vary by season and glaciological regime. MERRA2 reanalysis overall performs best with mean absolute error less than $2\,^\circ\mathrm{C}$ in all months. Ice sheet-average annual mean SAT from different datasets are highly correlated in recent decades, but their 1901–2000 trends differ even in sign. Compared with the MERRA2 climatology combined with gridded SAT analysis anomalies, thirty-one earth system model historical runs from the CMIP5 archive reach $\sim 5\,^\circ\mathrm{C}$ for the 1901–2000 average bias and have opposite trends for a number of sub-periods.

1 Introduction

Near-surface air temperature (SAT) over the Greenland ice sheet (GrIS) is important both for its place in wider climate change and for its effects on mass balance of the ice sheet. Due to its remoteness and extreme climate however, continuous widespread climate monitoring over the GrIS has been carried out for only about the last two decades, and even then with rather sparse coverage in some geographic areas and glaciological regimes. Studies of past climate and surface mass balance (SMB) of the GrIS have used a variety of techniques to achieve complete spatial coverage of SAT, including statistical interpolation, atmospheric reanalysis, dynamic downscaling through regional climate modeling, and satellite remote sensing. Projections of future change in Green-

land climate and ice sheet evolution have used global earth system models, either directly (e.g., Ridley et al., 2005; Vizcaíno et al., 2013) or through dynamical downscaling (e.g., Fettweis et al., 2013; Rae et al., 2012). Many such studies have involved some form of assessment using weather station data (e.g., Box, 2013; Noël et al., 2015; Rae et al., 2012) and inter-comparison of several SAT data sources (e.g., Box, 2013). Here we build on such work to assess and compare a greater number of widely available products, using a more comprehensive set of in situ observations than has customarily been used in previous work. In doing so we hope to guide future dataset and model development over this region and address a number of outstanding questions.

Our main focus here is on global datasets – reanalyses, gridded SAT analyses and earth system models from the CMIP5 archive – though several regional datasets are also included. Regional climate models (RCMs) have been used widely to downscale reanalysis (e.g., Box, 2013; Box et al., 2009; Burgess et al., 2010; Ettema et al., 2010a; Fettweis et al., 2017; Noël et al., 2015) and global climate model output (e.g, Fettweis et al., 2013; Rae et al., 2012). While Noël et al. (2016) demonstrated the benefit of high (< 10 km) resolution downscaling for SMB, the benefit for SAT is less clear: because SAT is strongly elevation-dependent, use of a high resolution model may not lead to a significant improvement compared to a lower resolution model with elevation corrections, as shown by Lucas-Picher et al. (2012) for grid sizes 0.25 and 0.05°. By comparing results from a range of resolutions, including RCMs at relatively high resolutions, we aim to investigate the value added by dynamic downscaling.

Inter-comparison of SMB components has been carried out among different RCMs and between RCMs and global reanalyses (Cullather et al., 2016; Rae et al., 2012; Vernon et al., 2013). The results from these studies point to a wide inter-model spread, which are related to differences in model parameterizations (e.g., snow and ice physics), model ice mask and forcing at the domain lateral boundaries. One goal of this work is to investigate how closely RCM forcing affects SAT representation, by comparing differently forced runs of the same RCM (building on the work of Fettweis et al., 2017), and comparing these runs with results taken directly from the forcing dataset.

Satellite remote sensing data has been key in spatially complete reconstruction of GrIS SAT, whether through direct use (e.g., Hall et al., 2013) or through assimilation into reanalyses. One consequence of this, though, is that only a small proportion of studies extend GrIS SAT back before the satellite era. SMB studies that incorporate centennial scale SAT reconstructions include: Hanna et al. (2011), who combined Twentieth Century Reanalysis (Compo et al., 2011) and ERA–40 reanalysis (Uppala et al., 2005); and Box (2013) who adjusted regional climate model output using in situ observations to reconstruct SAT from 1840 to 2010. The Box (2013) SAT reconstruction was compared to that of Hanna et al. (2011) and found to be cooler over most of the common period, but especially so before about 1930. More recently, Fettweis et al. (2017) investigated the effect on RCM-derived SMB of using different forcing reanalyses and showed that SAT estimates are sensitive to model forcing, with large differences in the first half of the 20th century. By looking at multiple datasets that include the first half of the 20th century (and earlier), we hope to shed light on the climate of the GrIS in this very poorly observed period. In particular, such datasets allow comparison with previous assessments of Greenland SAT climate based on (mainly coastal) station data (e.g., Box, 2002; Chylek et al., 2006; Hanna et al., 2012; Mernild et al., 2014). Long, spatially complete time series also offer the best means of assessing CMIP5 models, without differences introduced by incomplete spatial coverage and short period (~ 30 years) trends and decadal variability.

This paper is structured as follows: in Sect. 2 data sources are described and examples of their past use given; results are broken down into Sect. 3.1, dataset assessment using in situ observations, Sect. 3.2, comparison of long term SAT changes among datasets and Sect. 3.3, further discussion; conclusions are presented in Sect. 4.

2 Data

2.1 Weather station observations

To assess the different SAT products, we use SAT observations made at manned and automatic weather stations (AWSs) from several sources, totalling 17 000 station-months or 1400 station-years. These are briefly described here, and further details are shown in Fig. 1. Coastal station records of monthly mean temperature for 11 stations (stretching as far back as 1784) are compiled by the Danish Mete-

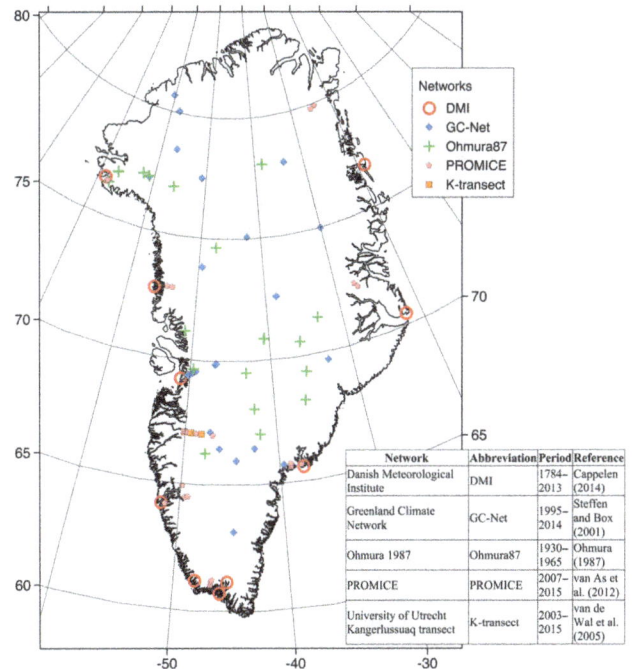

Figure 1. Map of study area and weather stations used in this work. Symbol types represent the different monitoring networks summarized in the inserted table.

orological Institute (DMI; Cappelen, 2014). Thanks to their long records, SAT from these stations has been studied extensively: Box (2002) found a pattern of warming from ~ 1900 to ~ 1940, cooling from ~ 1940 to ~ 1990, and warming from ~ 1990 onwards. In addition, inter-annual variability was found to be closely related to the North Atlantic Oscillation (NAO). Hanna et al. (2012) found similar patterns of warming and cooling using updated SAT data from DMI stations, and concluded that recent temperatures were in excess of SAT from the early 20th century warm period.

In contrast to coastal regions, no long term (e.g., 30 years or more) climate monitoring has occurred on the GrIS. Monthly mean temperatures from mid-20th century expeditions and field camps, concentrated in the 1930s and 1950s, are taken from the Appendix of Ohmura (1987). Since the mid-1990s, the number of SAT observations from the ice sheet has greatly increased. We use records from AWSs operated as part of the Greenland Climate Network (GC–Net), predominantly in the accumulation region of the ice sheet (Steffen and Box, 2001), from the K–transect in western Greenland (operated by the Institute for Marine and Atmospheric Research at the University of Utrecht; van de Wal et al., 2005; van den Broeke et al., 2011) and from AWSs mostly in the ablation region operated by the Geological Survey of Denmark and Greenland (GEUS) under the Program for Monitoring the Greenland Ice Sheet (PROMICE) and Greenland Analogue Project (GAP) programs (Van As et al., 2011). Locations and types of all stations are shown in Fig. 1 and further details are available in Table S1 in the Supplement.

The providers of several of these observational datasets employ quality control tests and/or quality inspection as part of their routine data management. In addition, we remove unrealistic values where our inspection of time series reveals them (e.g., with spikes and step changes). Where data were provided as hourly values, we calculate daily averages (the mean of hourly values) for all days with 20 or more hourly values and monthly averages (the mean of daily values) for all months with 24 or more daily values.

2.2 Gridded SAT products

Most of the datasets assessed here fall into two categories: global reanalysis and interpolated global SAT analyses. The spatial and temporal resolution and length of record (Table 1) vary greatly across these products. It should be noted that even though reanalyses are constrained by (in some cases) remote sensing and some local observations to represent observed synoptic–planetary scale weather, the lack of assimilated SAT observations over Greenland means that the SAT data assessed here are largely the result of modelled atmospheric and surface processes.

Several of the latest generation of global reanalyses are used in this study (Table 1). Most of these are reliant on radio-sonde and satellite data, and thus cover only the period when these are available (1979 onwards; 1958 in one case). In addition, we analyze the Twentieth Century reanalysis version 2c (20CRv2c; Compo et al., 2011) and ERA–20C (Poli et al., 2016), which do not assimilate satellite or radio-sonde data, but instead use a subset of observation types that are available over the 20th century (and earlier) and therefore cover much longer periods. GrIS SAT from reanalyses has been used in SMB modeling: Hanna et al. (2005) used ERA-40, while Hanna et al. (2011) combined ERA–40 with 20CR. However, SAT data from a number of other reanalyses remain untested for such applications. It should be noted that, with the exception of ERA-Interim, SAT from land stations is not assimilated into reanalyses and so the SAT observations described in Sect. 2.1 are indeed an independent verification. In ERA-Interim, SAT is assimilated from land stations by the surface analysis scheme, to update surface fields (such as soil moisture) which have an effect on SAT. To the best of our knowledge, for the period analysed here the only Greenland SAT observations that are assimilated by ERA-Interim are from DMI stations, and so the ice sheet stations still provide independent data.

Reanalysis represents a combination of observations and model. In contrast, several research groups have created gridded SAT datasets based almost entirely on statistical analyses of weather station SAT (we refer to these as *gridded SAT analyses*). Such datasets have not been widely used over Greenland (though see, e.g., Fettweis et al., 2008), and their long time series and temporal homogeneity is a po-

tential strength. For example, some reanalyses are known to suffer from spurious trends as observing networks and processing systems change (e.g., Screen and Simmonds, 2010): comparison between reanalyses and gridded SAT analyses, particularly in the early 20th century, can highlight such problems with reanalyses. Some gridded SAT analyses, due to their analysis methods and requirements for data completeness, have large data gaps over Greenland, e.g., HadCRUT4 (Morice et al., 2012) and NOAAGlobalTemp (Smith et al., 2008; Vose et al., 2012). However, here we use four such datasets that have complete (or very nearly so) coverage over Greenland. Three of these (NASA GISTEMP, University of East Anglia Climatic Research Unit gridded time series data version 3.23 (CRU TS 3.23) and Berkeley Earth; references in Table 1) are widely used global SAT monitoring products, while one (NansenSAT) covers only the Arctic. Note that GISTEMP (Hansen et al., 2010) is provided as anomalies only (relative to 1951–1980 climatology). As the ice sheet weather stations have typically not been operational long enough to calculate a stable climatology, we do not assess GISTEMP using in situ observations; however, we do combine GISTEMP anomalies with MERRA2 climatology to enable assessment of *stationarity* of biases and comparison of long term variability against other datasets.

Recognizing that reanalysis SAT over Greenland is dominated by the model formulation and has relatively coarse horizontal resolutions, a number of researchers have sought to improve results over the GrIS by using reanalysis to force higher resolution regional climate models (RCMs) coupled to comparatively sophisticated snow–ice models. Such models are typically run with grid spacing of 10–20 km. This high resolution (compared to global climate models and most reanalyses) is thought to better resolve the large climate gradients that occur around the margins of the ice sheet. Here we include output from version 3.5.2 of the Modèle Atmosphérique Régional (MAR; Fettweis et al., 2013, 2017) run with 20 km grid spacing, then interpolated to the 5 km polar stereographic grid of Bamber et al. (2001). Three different runs of MAR are used here: one forced by ERA–40 (1958–1978) and ERA–Interim (1979–2015) reanalyses; a second forced by 20CRv2c reanalysis; and a third forced by ERA–20C reanalysis. ERA–40 and ERA–Interim reanalyses have been widely used as forcing data (Box et al., 2009; Ettema et al., 2010a, b; Fettweis et al., 2013); 20CRv2c and ERA–20C have seen more limited use (e.g., Fettweis et al., 2017). It should be noted that the field we use from this model is nominally the 3 m air temperature, whereas most reanalyses output 2 m air temperature (when specified), and the measurement height at weather stations varies as the snow/ice surface changes. We also include an updated version of the SAT reconstruction of Box (2013) which uses statistical relationships between long-running DMI stations and RACMO2 RCM output (e.g., Noël et al., 2015) to estimate Greenland SAT on a 5 km grid from 1840 to 2014. This dataset can

Table 1. Temperature products assessed in this work. Latitude longitude spacing refers to the grids downloaded for this work (not necessarily the native model grid). Maximum output frequency refers to the maximum available – monthly averages are used in the analysis.

Type	Dataset	Center	Latitude longitude spacing [a]	Maximum output frequency	Period	Reference
Reanalysis	MERRA	NASA/GMAO	$0.5° \times 0.667°$	Hourly	1979–2015	Rienecker et al. (2011)
	MERRA2	NASA/GMAO	$0.5° \times 0.625°$	Hourly	1980–2015	Molod et al. (2015)
	CFSR and CFSv2 [b]	NCEP	$0.5° \times 0.5°$	Hourly	1979–2015	Saha et al. (2010, 2014)
	20th Century Reanalysis V2c	NOAA/CIRES	$\sim 1.9° \times 1.875°$	3-hourly	1851–2014	Compo et al. (2011)
	ERA–Interim	ECMWF	$0.75° \times 0.75°$	3-hourly	1979–2015	Dee et al. (2011)
	ERA–20C	ECMWF	$1° \times 1°$	3-hourly	1900–2010	Poli et al. (2016)
	JRA–55	JMA	$\sim 0.56° \times \sim 0.56°$	3-hourly	1958–2014	Kobayashi et al. (2015)
Gridded temperature analysis	GISTEMP	NASA/GISS	$2° \times 2°$	Monthly	1880–2015	Hansen et al. (2010)
	CRU TS 3.23	CRU	$0.5° \times 0.5°$	Monthly	1901–2014	Harris et al. (2014)
	Berkeley Earth Surface temperature	Berkeley Earth	$1° \times 1°$	Monthly	1750–2016	Rohde et al. (2013)
	NansenSAT	Nansen Centers	$2.5° \times 2.5°$	Monthly	1900–2008	Kuzmina et al. (2008)
	Box2013	GEUS	$5\,km \times 5\,km$ [c]	Monthly	1840–2014	Box (2013)
Satellite	AIRS	NASA	$1° \times 1°$	Monthly	2002–2015	Chahine et al. (2006)
Regional down-scaling	MAR–ERA	University of Liège	$5\,km \times 5\,km$ [c]	Monthly	1958–2015	Fettweis et al. (2013, 2017)
	MAR–20CRv2c				1900–2014	
	MAR–ERA–20C				1900–2010	

[a] As downloaded for this study. [b] CFSR, covering 1979–2010, and CFSv2, covering 2011–2015, are appended and referred to together as CFSR in the text. [c] Box2013 and MAR are on the polar stereographic grid of Bamber et al. (2001).

therefore be thought of as a hybrid of an RCM and gridded SAT analysis. We use Box2013 to denote this dataset.

Satellite remote sensing data, in addition to being assimilated by reanalyses, have been used directly to study the GrIS. Several studies have focused on the relationship between SAT and ice sheet surface temperature (IST), and have used data from both microwave (e.g., Shuman et al., 1995, 2001) and infrared sensors (e.g., Comiso et al., 2003; Hall et al., 2008, 2013; Koenig and Hall, 2010). Sounding instruments offer a method to retrieve air temperature more directly, but have received little attention over GrIS. Here we assess SAT from the Atmospheric Infrared Sounder (AIRS; Chahine et al., 2006) on board NASA's AQUA satellite platform. AIRS has been operational since September 2002, providing temperature and humidity retrievals at many vertical levels through the atmosphere. We use the level 3 monthly near surface air temperature from ascending and descending overpasses, taking a weighted average to give a single monthly value at each grid point (further details are given in Table 1). This product is a clear-sky only retrieval: a key part of assessing this product is to understand what effect this has through, for example, seasonally varying cloud amounts and increased wind-driven mixing during winter storms, as discussed in Koenig and Hall (2010).

Earth System Models (ESMs) from the CMIP5 multi-model ensemble archive (Taylor et al., 2011) are included in comparisons of long term areal average SAT. However, comparison of CMIP5 ESMs against in situ observations is not performed because the ESMs are free-running coupled (atmosphere–ocean–land–ice) models, so we do not expect them to have the correct phasing of synoptic weather or interannual or even decadal climate. Apparent biases at station locations would therefore combine bias in the long term average and differences in variability over the relatively short station records. The ice sheet areal averages, compared to the longer reanalyses and gridded SAT analyses, should adequately reveal the first order biases in the ESMs' long term average SAT and its trends. Thirty-one different model configurations from 11 modeling centers are used. We use the first ensemble member (r1i1p1) of historical runs from all model configurations that had the necessary data (SAT and glacial ice fraction). Further details of individual models are given in Table S2. In contrast to other datasets above, CMIP5 ESM SAT data are used on their model native grids, rather than interpolated to a common grid (to be discussed below).

3 Results

Our analysis is based on the monthly mean near-surface air temperature. Except for CMIP5 ESMs and the MAR RCM variants, datasets were spatially interpolated from their native grid to a 5 km equal area grid (the Equal-Area Scalable Earth (EASE) grid of the National Snow and Ice Data Center (NSIDC)) using bilinear interpolation. This resolution is used to attempt to resolve the large SAT gradients that occur over the steep topography at the margin of the ice sheet. Interpolating like this presents some potential problems due to

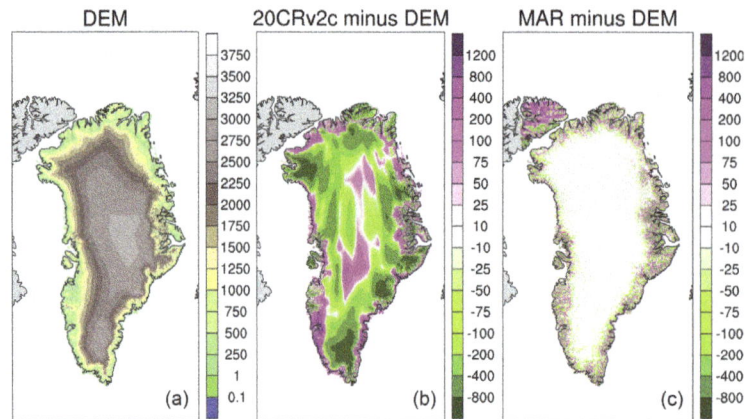

Figure 2. (a) Digital elevation model (DEM) of Bamber et al. (2013) interpolated to EASE 5 km grid; **(b)** bias of 20CRv2c surface elevation field interpolated to EASE grid, relative to Bamber et al. (2013); **(c)** bias of MAR surface elevation field relative to Bamber et al. (2013). Units are meters.

model topography: the surface elevation fields used in many of the datasets here are smoother than the actual topography of Greenland, and this leads to elevation biases as seen in Fig. 2. The relatively low resolution 20CRv2c (Fig. 2b) has mostly positive elevation bias around the edge of the ice sheet and negative bias in the interior; however there are also regions of positive bias close to the center of Greenland. The higher resolution MAR (Fig. 2c) does not have the same magnitude of biases in the interior, but still misses much of the small scale detail, as seen by the speckled pattern of biases of alternating sign. All datasets have a negative mean elevation bias on the ice sheet (Table 2), with MAR the smallest and 20CRv2c the largest. Note that elevation errors are not a monotonic function of resolution: despite a smaller grid spacing than MERRA2 and ERA–Interim, Climate Forecast System Reanalysis (CFSR) still has a larger bias and mean absolute error.

The elevation biases cause the SAT fields to be smoother than in reality, and interpolation of the smooth SAT fields is unlikely to accurately reflect the true SAT gradients, which are strongly influenced by elevation. To account for this, a correction is applied to the reanalysis and AIRS datasets after interpolation to the EASE grid: for each product, the elevation field is also bilinearly interpolated to the EASE grid, and then compared to the digital elevation model (DEM) of Bamber et al. (2013; provided at 1 km grid spacing, and here bilinearly interpolated to the EASE grid). The elevation bias (product minus DEM) is multiplied by the relevant month's lapse rate from Fausto et al. (2009) and their product added to the interpolated SAT field. The importance of this step can be seen by comparing the results below with comparable figures for un-corrected datasets (Figs. S2 and S3 in Supplement). For some datasets in some seasons, the correction leads to a deterioration, but in most cases there is a clear improvement: in many cases, bias and MAE (averaged over all months) are reduced by 50 % or more.

Table 2. Error statistics of model elevation fields (interpolated to EASE grid, except for MAR) relative to the digital elevation model (DEM) of Bamber et al. (2013). Bias and deciles are calculated as (model minus DEM). Averages are taken over all ice sheet grid points, classified using the mask of Bamber et al. (2013).

Dataset	Bias	RMSE	MAE	Lower decile	Upper decile
			(m)		
MERRA	−126.3	290.1	199.7	−466.3	141.7
MERRA2	−48.3	172.3	88.4	−194.6	16.9
ERA–Interim	−67.5	215.0	119.5	−281.3	33.3
ERA–20C	−103.1	274.6	173.0	−380.9	51.2
CFSR	−114.9	262.8	192.0	−422.0	143.1
20CRv2c	−244.3	447.1	337.6	−733.8	151.8
JRA–55	−131.9	272.4	199.2	−439.9	123.5
AIRS	−132.3	274.2	200.3	−440.1	120.2
MAR	−13.4	94.1	37.2	−56.0	18.0

3.1 Monthly mean SAT biases

Comparisons between gridded datasets and in situ observations are made by choosing the nearest EASE grid point (for CRU and Berkeley Earth, which are land-only datasets, the nearest grid point may contain missing data, in which case the nearest non-missing grid point is chosen). Note that an alternative, using bilinear interpolation directly from the native grids to the station locations, gives very similar results. The primary statistics used in the assessment of datasets are mean bias and mean absolute error (MAE). When aggregating results over multiple stations, the average of station-months is taken, rather than averaging over time then over stations. Stations are grouped into coastal (DMI), ice sheet below 1500 m and ice sheet above 1500 m. The elevation of 1500 m is chosen to approximately represent the equilibrium line altitude, as found for the K–transect by van de Wal et al. (2005). The

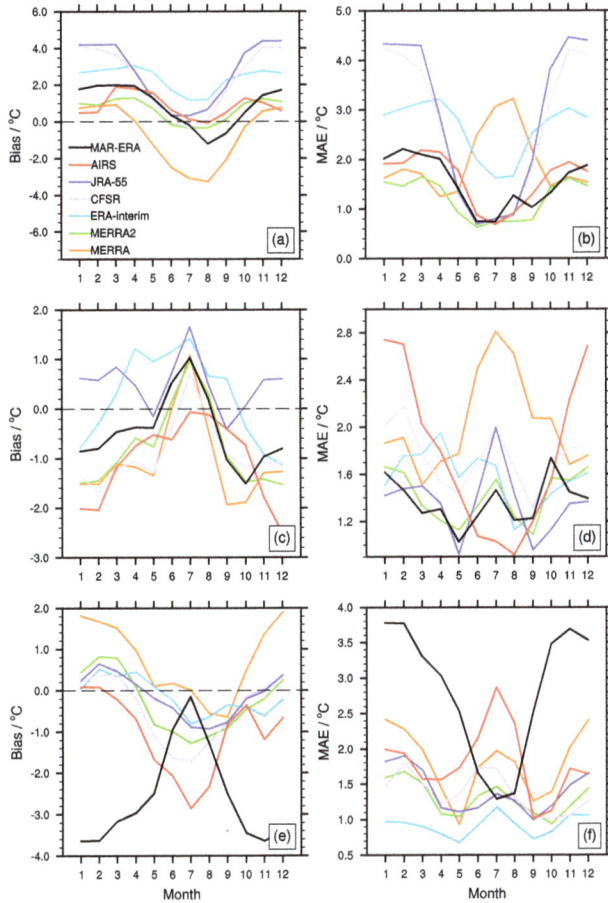

Figure 3. Mean over station-months of bias (**a, c, e**) and absolute error (**b, d, f**) relative to monthly mean SAT at: ice sheet stations above 1500 m (**a, b**); ice sheet stations below 1500 m (**c, d**); and coastal (DMI) stations (**e, f**). Ice sheet stations are from GC–Net, PROMICE and K-transect. All available station months from 1979 onwards are used. All datasets included in this figure are elevation corrected: several shorter reanalyses, AIRS, and MAR–ERA. Note that the vertical scales vary with panels.

Figure 4. As in Fig. 3, but for elevation-corrected long reanalyses, MAR–ERA–20C and MAR–20CRv2C, Box2013 data and three gridded SAT analyses (not elevation corrected).

pattern of biases seen below is largely the same for different separation elevations between 1000 and 2000 m. Aggregating over elevation bands like this can pick out some important aspects of spatial variation in dataset errors, but is likely to miss regional and local patterns of dataset error (see Fig. S1). Note that, when taking the spatial average across the ice sheet, the area above 1500 m dominates: using the DEM and mask of Bamber et al. (2013), Greenland has a total area of 2.16 million km^2, which is 16.5 % ice-free land, 18.6 % ice sheet below 1500 m, and 64.9 % ice sheet above 1500 m.

The seasonal cycle of bias and MAE averaged over all station months from 1979 onwards in Figs. 3 and 4 suggests that many datasets, though not all, show similar seasonal cycles: above 1500 m and at coastal stations, more positive biases in winter and more negative in summer; at ice sheet stations below 1500 m, the opposite cycle. Despite qualitative simi-

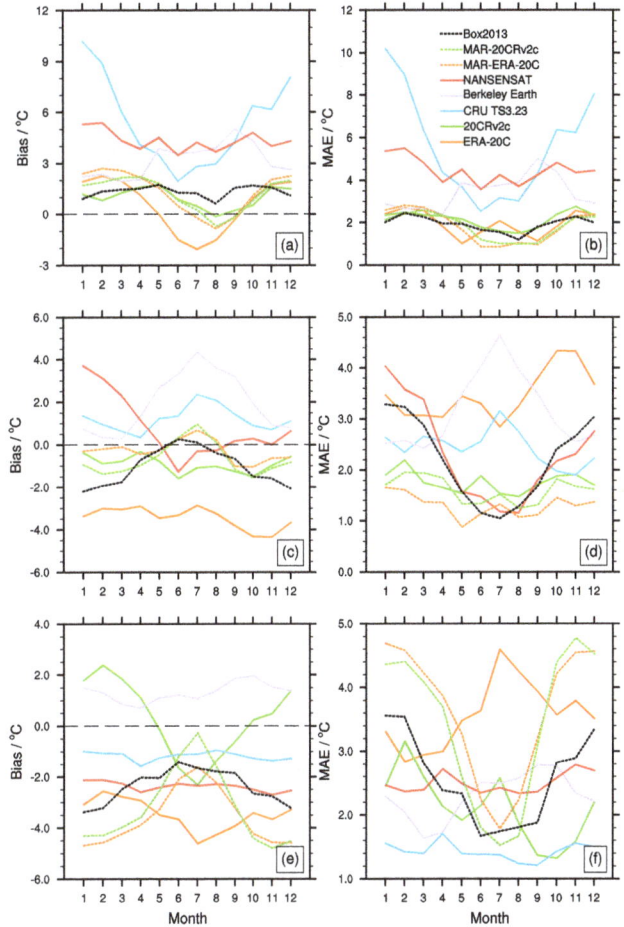

larities, a clear picture of dataset performance emerges. On the ice sheet, MERRA2, MAR (all three versions), Box2013 and 20CRv2c are best. The AIRS satellite product is also very good, except in winter months at stations below 1500 m. NansenSAT is one of the best performers below 1500 m during the summer; however it has large biases and MAE elsewhere and there are concerns with its long term homogeneity (see below). At coastal stations, ERA–Interim performs best (likely related to its assimilation of some of these observations), and JRA–55 and MERRA2 are nearly as good. MAR (all three versions) performs better in summer than in winter. This is thought to be due to the specification of sea ice thickness in the MAR v3.5.2 model: in many regions around the coast of Greenland, sea ice thickness is over-estimated in the model boundary conditions, resulting in a cold bias in adjacent areas (X. Fettweis, personal communication, 2017). Note that without elevation corrections, MAR coastal station errors are larger in summer but smaller in winter (Figs. S2 and S3). CRU and Berkeley Earth results are comparable to the best reanalyses at coastal stations, likely because it is SAT

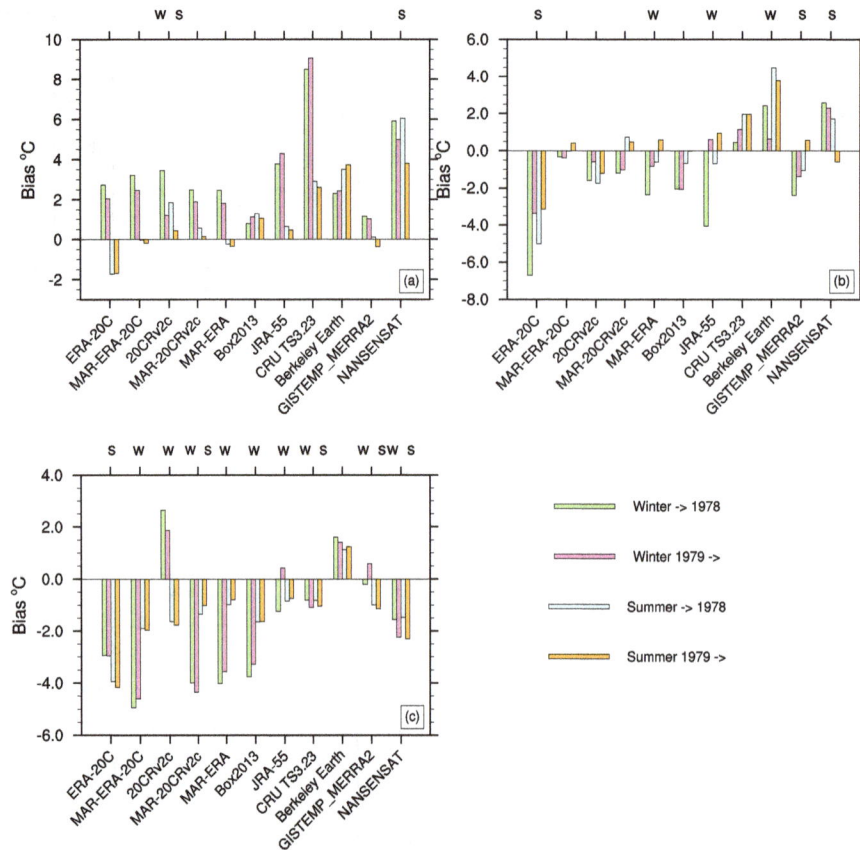

Figure 5. Monthly mean SAT bias for winter (DJF) and summer months (JJA) before and after 1979, for all datasets that extend back before 1979 (elevation-corrected where applicable) at: ice sheet stations above 1500 m (**a**); ice sheet stations below 1500 m (**b**); and coastal (DMI) stations (**c**). Note that these are monthly SAT biases averaged over all months in a season, not biases of seasonal mean SAT. Changes significant at the 99 % level (using Student's t test with unequal variances) are denoted by w (for winter) and s (for summer) on the top axis.

observations from the coastal stations that form the majority of the input data for these datasets. Based on a (rather subjective) assessment of the 12-month average bias (absolute value, to avoid cancellation between months) and MAE (Table 3), the most consistent good performer is MERRA2: the 12-month average biases (absolute values) are approximately equal to or less than 1.0 °C, and 12-month average MAE are less than 1.5 °C, in all regions. MAR (all three versions) and Box2013 have comparable performance across the ice sheet, and in some cases better performance in lower elevations during summer (the most important region and season for SMB modeling), but overall are marred by large winter time biases at coastal stations.

The analysis above aggregates all station months from 1979 onwards. To investigate time variations in biases, Fig. 5 compares mean bias before and after 1979 for those datasets which begin before 1979. Note that the datasets beginning in 1979 show only small changes in bias by decade (not shown). GISTEMP is included here with the MERRA2 elevation-corrected climatology: the absolute values of the biases are highly dependent on the climatology, but here can be ignored

as, for the purpose of assessing the stationarity in GISTEMP bias (and thereby the credibility of its long term variability and trends), we are interested in the *changes* in bias.

Clear differences are apparent for some seasons and datasets. Statistical significance of these differences (using Student's t test for a difference in means with unequal variances, and defining significance at the 1 % level) suggest that a number of the datasets have time-varying biases and so may show spurious long term trends. This is most apparent for the coastal DMI stations, where larger sample sizes give the statistical test greater power. 20CRv2c has negative changes high on the ice sheet and at coastal stations but positive changes in the ablation region. NansenSAT shows negative changes over time in all regions in both winter and summer. Other than NansenSAT, the gridded SAT analyses do not seem more prone to time-varying bias than reanalyses.

3.2 Time series

Areal average (weighted by glacial ice fraction) annual mean temperatures for all datasets show close correlation in recent decades: considering only the period 1979 onwards, the cor-

Table 3. Mean bias and mean absolute error (MAE) for all datasets ranked from smallest (top) to largest (bottom) MAE. These numbers represent an average of results from Figs. 3 and 4, with unweighted average over months and an area-weighted average over glaciological regimes (64.9 % ice sheet above 1500 m, 18.6 % ice sheet below 1500 m and 16.5 % coastal DMI).

Dataset	Bias	MAE
	(°C)	
MERRA2	0.81	1.27
AIRS	0.95	1.67
MAR–ERA	1.38	1.72
MERRA	1.26	1.94
MAR–ERA–20C	1.69	2.04
MAR–20CRv2c	1.59	2.05
20CRv2c	1.04	2.07
Box2013	1.47	2.08
ERA–interim	1.76	2.16
CFSR	1.87	2.28
JRA–55	1.95	2.34
ERA–20C	2.15	2.52
Berkeley Earth	2.67	3.21
NansenSAT	3.42	3.74
CRU TS3.23	3.96	4.34

relation (r) values are in the range 0.71–0.99. In earlier periods, correlations are generally smaller: for the period 1900–1940, we have $r = 0.28$ to 0.98, and for 1940–1980, $r = 0.29$ to 0.97. However, CRU, Berkeley Earth and GISTEMP have pairwise correlation coefficients of 0.90 or greater for all these periods since they are based on a similar set of surface stations, and their correlations with Box2013 are greater than 0.84 in all periods. ERA–20C is highly correlated with MAR–ERA–20C ($r > 0.97$) in all three periods, as the latter is forced by the former. Similarly, 20CRv2c is highly correlated ($r > 0.90$) with MAR–20CRv2c in all these periods. Interestingly, Box2013 is generally more highly correlated with CRU, Berkeley Earth and GISTEMP than with the MAR datasets.

Among the datasets covering the entire 20th century, most have similar inter-decadal variations, with a general pattern of early 20th century warming, up to 1930, followed by cooling to around 1990, then strong warming in recent years (Fig. 6). Nonetheless, differences do exist (Table 4). For instance, NansenSAT shows relatively large early 20th century jumps thought to be caused by changing data sources over this period, indicating this dataset is not suitable for long term monitoring over Greenland. In 20CRv2c, the ∼ 1930 peak is warmer than the most recent years, in contrast to other datasets. Related to the last point, the amount of cooling from 1930 to 1990 varies between datasets, with 20CRv2c showing strongest cooling, and GISTEMP showing least. Anomalies (relative to the 1981–2010 mean; Fig. 6b) reveal some more subtle differences. For example, MAR–ERA and CRU

both show less positive anomalies than other datasets since about 2005. Variability in 20CRv2c matches other datasets closely from 1980 onwards, but before this differs significantly (except in the 1940s). Comparison with anomalies at long-running DMI stations (Fig. S4) suggests it is 20CRv2c that is in error here, as might be expected from the consensus of other datasets. MAR–20CRv2c shows agreement for more of the period, but seems to inherit poor representation of variability before about 1920 and from 1950 to 1980.

Of the datasets that extend back before 1900, Box2013, Berkeley Earth and GISTEMP agree quite closely but show notable differences with 20CRv2c. Box2013, Berkeley Earth and GISTEMP cannot be considered truly independent data sources (as they all rely on similar input data for this period, as suggested by their close correspondence with observations in Fig. S4), and so their consensus is not especially meaningful. However, the fact that their biases are more constant in time (Fig. 5) than those of 20CRv2c suggest that they are more reliable for this period. In common with disparities mentioned above for the first half of the 20th century, users of these SAT datasets should be aware that significant uncertainties exist before 1900, with notable differences in trends and variability (both inter-annual and inter-decadal). We recommend the use of gridded SAT analyses alongside reanalyses and downscaled reanalyses, to assess sensitivity to these differences.

The range of SAT among CMIP5 ESMs is wider than that among the other datasets (Fig. 6), but much of this range comes from a group of four relatively warm models and two relatively cold models: eliminating these gives a range comparable to the gridded analysis and reanalysis datasets. This highlights the fact that choice of verification dataset can have a significant effect on assessments of ESM mean climate. Based on results above, we use GISTEMP with MERRA2 climatology to assess the long term mean temperatures of the CMIP5 ESMs. Using the 1901–2000 mean of ice sheet annual average temperatures, 10 ESMs lie within 1 °C (namely GFDL's CM3 and ESM2G; GISS's E2–H–CC, E2–R and E2–R–CC; IPSL's CM5A–MR, CM5A–LR_historical and CM5A–LR_esmHistorical; CESM1–CAM5; CMCC–CMS; see Table S1 for further details).

The median of the CMIP5 ESM trends (Table 4) is positive for all periods considered – in marked contrast to the other datasets. However, further investigation shows the picture is not so clear: the number of individual ESMs that have positive trends in each period suggest that, with the possible exception of 1990–2005, the models do not give a clear consensus on signs of trends: this may be because inter-decadal climate variability dominates, and the phasing of this variability differs between models. For the 1990–2005 period, 27 out of 31 ESMs have a positive trend and the median is an order of magnitude larger than for the earlier periods (although still smaller than the 1990–2005 trends from the other datasets). Thus the ESMs seem to agree on accelerated warming since 1990. Significance of the trends is tested us-

Table 4. Trends ($^{\circ}$C decade^{-1}) of ice sheet areal average annual mean SAT for given periods. For the CMIP5 ESMs, the trend of the ensemble mean and the median of the individual trends are shown, along with the number of positive, negative and significant trends. Bold type indicates significance at the 0.05 level.

Dataset	Linear trends ($^{\circ}$C decade^{-1})					
	1901–1930	1931–1960	1931–1990	1990–2005	1990–2014	1901–2000
MERRA				**1.495**	**0.823**	
MERRA2				**1.739**	**0.696**	
ERA–Interim				**1.979**	**0.907**	
ERA–20C	0.513	0.023	**−0.183**	**1.637**		0.020
CFSR				**1.526**	**0.589**	
20CRv2c	**0.393**	−0.468	**−0.295**	**2.042**	**1.045**	**−0.156**
JRA–55				**1.799**	**0.949**	
MAR–ERA				**1.358**	**0.559**	
MAR–20CRv2c	0.161	−0.120	**−0.156**	**1.649**	**0.806**	−0.061
MAR–ERA–20C	**0.506**	−0.019	**−0.212**	**1.619**		−0.025
Box (2013)	0.774	−0.402	**−0.312**	**2.196**	**1.343**	0.035
CRU TS3.23	0.411	−0.118	**−0.169**	**1.249**	0.505	0.023
Berkeley Earth	0.628	−0.334	**−0.227**	**1.703**	**0.965**	0.054
GISTEMP	0.361	−0.197	**−0.139**	**1.447**	**0.865**	0.100
NansenSAT	−0.065	−1.003	**−0.497**	**1.066**		**−0.239**
CMIP5: ensemble mean	0.119	0.016	**0.088**	0.485		**0.098**
CMIP5: median	0.081	0.007	0.086	0.407		0.094
CMIP5: number positive (significant)	23 (3)	16 (0)	23 (7)	27 (4)		30 (20)
CMIP5: number negative (significant)	7 (0)	14 (1)	7 (0)	4 (0)		1 (1)

ing the method described in Santer et al. (2000), which is based on a two-tailed Student's t test modified to account for autocorrelation in the time series. Few of the trends are significant, in any of the periods considered. The ensemble mean has a long term average slightly higher than that from MERRA2+GISTEMP, and trends broadly similar to the median of the individual trends (i.e., with accelerated warming since about 1970); however, it does not feature decadal variability that individual CMIP5 ESMs, reanalyses and gridded SAT analyses show, and thus has limitations in representing historical GrIS SAT.

Due to its importance in SMB calculations, we briefly consider summer mean (June–August) ice sheet average SAT (Fig. 6c). Many features are shared with the annual time series, e.g., periods of warming in the years leading up to 1930 and beginning in the 1990s. In addition, we see that the variability in MAR–20CRv2c and MAR–ERA–20C closely follow that in 20CRv2c and ERA–20C respectively. In contrast with the annual mean time series though, the CMIP5 ensemble mean more closely follows the evolution in the observation–based datasets.

3.3 Further discussion

The majority of in situ SAT observations from the ice sheet have been made since 1995. We have used the relatively small number of observations from the mid-20th century to assess the stationarity of biases, and find that several datasets

show significant temporal variations in their bias. At ice sheet stations above 1500 m (the region which dominates in areal averages), 20CRv2c shows large (and significant) changes – becoming more negative with time. 20CRv2c biases also become more negative with time at coastal stations (from which there are many more observations), casting further doubt on the suitability of this dataset for long term trend analysis. ERA–20C has more stable biases in the accumulation region and at coastal stations (though not in the ablation region), as do several of the gridded SAT analyses, suggesting that SAT reconstruction based on anomalies is valid over monthly to centennial time scales. This is not a trivial result, as it is not obvious a priori that conditions driving anomalies at coastal stations will result in a similar, smoothly varying response over different surface types.

Trends among the datasets assessed here (excluding CMIP5 ESMs) generally agree with patterns found in previous studies (e.g., Box, 2002). In addition, interannual variability since 1979 matches closely between datasets. However, differences between longer term trends, along with temporal changes in bias (discussed above), suggest that some datasets have limitations in their representation of early to mid-20th century GrIS SAT. In particular, 20CRv2c shows stronger cooling between 1930 and 1990 than most other datasets, and has a 1930s warm period warmer than the 21st century warm period. Such discrepancies between 20CRv2c and anomaly based SAT datasets have been noted at the

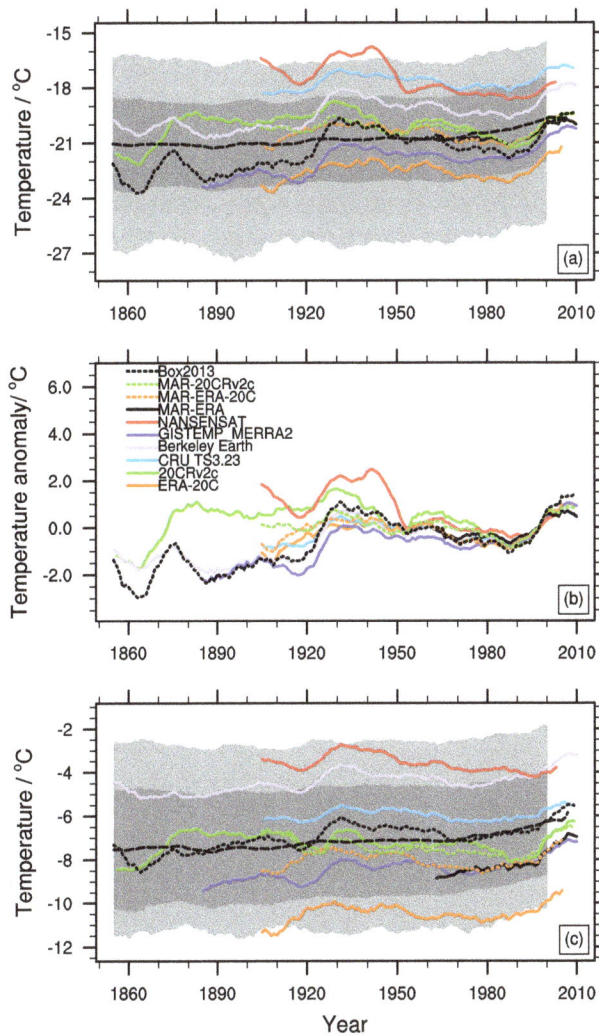

Figure 6. (a) Time series of ice sheet areal average smoothed annual mean SAT for long reanalyses, gridded temperature analyses and all three MAR variants (elevation corrected where applicable; colored lines) and CMIP5 climate models (not elevation corrected; ensemble mean in dashed black line; ±1 standard deviation in dark grey shading; maxima and minima in light grey shading). **(b)** Anomalies of ice sheet areal average smoothed annual mean SAT from long reanalyses, gridded temperature analyses and both MAR variants, relative to 1981–2010 mean. **(c)** As in **(a)** but for June–August mean SAT. In all panels, time series are smoothed using a centered, uniform-weighted 11-year window, to highlight decadal variability and aid legibility.

global scale by Compo et al. (2013), although the differences here are much greater than those for global SAT. Similarity of anomalies among gridded SAT analyses and ERA–20C, along with the greater temporal constancy of their biases, leads us to put greater faith in their representation of long term trends and inter-decadal variability.

While, as noted above, interannual variability in the last 30 years matches closely between datasets, there is variation

in the magnitude of ice sheet average trends (Table 4) and spatial variation in trends (Fig. S5) over this period. Box2013 has the largest recent trends, with largest trends in the west. MERRA2 has its largest trends in the south-west, whilst all three MAR versions have their largest trends in the northeast.

One of our central questions in this study is whether global SAT datasets are as good as RCM-downscaled datasets, which are, at least for SMB modeling, the current state of the art. For MAR–ERA and MAR–ERA–20C, results are generally better than for SAT taken directly from the forcing dataset (even with elevation corrections applied). However, at coastal stations, MAR–ERA performs worse than ERA–Interim. For MAR–20CRv2c, the difference is minimal at ice sheet stations and downscaling is detrimental at coastal stations in winter (though without elevation corrections, MAR–20CRv2c has smaller biases and MAE than 20CRv2c; see Fig. S3). Comparing MAR against all global datasets, we find MERRA2 has biases and MAE comparable to or less than MAR (all three forcings) in all seasons and regions. This is likely due to the comprehensive (relative to other reanalyses) snow/ice model in MERRA2 (Cullather et al., 2014) and reinforces the importance of atmosphere–ice sheet coupling in modeling SAT. In summer, and particularly in the ablation region, MAR and Box2013 are among the best datasets, confirming their suitability for SMB modelling. However, for SAT more generally, the benefits of RCM downscaling seem to be limited.

Another question related to the RCM downscaling is: how closely does the forcing dataset constrain climate variability in the downscaled RCM? Correlations (between 20Crv2c and MAR–20CRv2c, and between ERA–20C and MAR–ERA–20C) of ice sheet annual mean SAT before 1979 suggest that the constraint is close: for example, MAR–20CRv2c has correlation coefficients with 20CRv2c greater than 0.9 for both 1900–1940 and 1940–1980, while its correlation with other datasets is lower (0.54–0.62 for 1900–1940; 0.29–0.82 for 1940–1980). The variability of summer SAT is even more closely constrained (Fig. 6c). Downscaling is able to remedy some large biases shown by reanalysis (e.g., for ERA–20C in summer, Fig 6c), and consideration of anomalies (Fig. 6b) suggests that the downscaling improves representation of climate variability by bringing MAR–20CRv2c more into line with other datasets. Nonetheless, differences remain, particularly before 1920 and between 1950 and 1980, and we consider that MAR–20CRv2c still suffers from some shortcomings in 20CRv2c's representation of variability before 1980.

Although the comparison is for a shorter period than for other datasets, we have found that AIRS gives very good results over the ice sheet in summer – with biases and MAE values among the smallest of any dataset in the ablation region for June, July and August. However, its performance is poor in winter over the ablation region and in summer at coastal stations. The wintertime biases in the accumulation region do not agree (although those in the ablation region do) with the findings of Koenig and Hall (2010) at Summit, that satellite-derived clear-sky only temperatures were lower than all-cloud in situ measurements. They attributed this finding to the fact that clear-sky only retrievals miss winter storms – during which strong winds mix warm air from above an inversion down to the surface – which should lead to negative wintertime biases. The fact that AIRS has positive bias in the accumulation region during winter suggests compensating errors from other sources, for example from retrieval of temperature profiles or from times of day of satellite overpass. Attributing the overall bias to different causes is beyond the scope of this study. In summary, the summertime results suggest AIRS may be a useful dataset for studies of recent SMB, but further investigation is needed into the consequences of clear-sky retrievals, particularly the wintertime discrepancy with previous work and the possibility of compensating errors.

Note that there is a discrepancy between various products in calculating monthly mean SAT. As discussed in Wang and Zeng (2013) the daily mean calculated using 24 hourly values per day is different from that calculated using just maximum and minimum SAT. Comparisons for AWSs on the GrIS suggest the difference for monthly mean temperatures is $\sim 0.2\,^\circ$C, but can exceed $0.5\,^\circ$C in some individual months. Other averaging methods (e.g., mean of 3-hourly values; weighted mean of 0800, 1400 and 2100 local time; Box, 2002) are unlikely to introduce larger errors than the maximum plus minimum method. Overall these relatively small uncertainties are unlikely to affect our conclusions.

Our evaluation of 5 km grid box values using point measurements may also be affected by the sampling errors due to the SAT variation within a grid box (e.g., in grid boxes containing a large range of elevations and different surface types). Quantifying such an error could in principle be done using several stations within the same grid box; we do not have any 5 km grid boxes containing more than one station, however. Instead we look to the variation of elevation, assuming that this is the dominant source of SAT variation at small spatial scales and implicitly neglecting effects of varying surface type and other factors. Elevation variation at any particular location is quantified by taking the standard deviation of elevation values at the nearest and 24 surrounding grid boxes from the 1 km version of the Bamber et al. (2013) DEM. This is then multiplied by a (slightly conservative) lapse rate of $9.0\,^\circ$C km^{-1}, to give a likely range of SAT variation over this elevation range. This formulation gives smaller sampling error over relatively flat terrain: $\sim 0.1\,^\circ$C above 1500 m on the ice sheet. In more variable terrain, around the margins of the ice sheet and in coastal land regions, sampling errors are larger – usually in the range 0.3–$1.0\,^\circ$C. Overall these uncertainties are relatively small in magnitude compared to the large biases and MAEs between various datasets and in situ observations, and hence our conclusions are largely unaffected.

In our assessment of biases and their changes through time we have assumed that all observations are un-biased. Observation biases are likely to exist (e.g., the positive bias of unaspirated thermometer shields in low wind, high solar radiation conditions; Genthon et al., 2011) and are likely to vary in space and time due to differences in station siting, instrumentation and observing practices (e.g., number per day and timing of manual thermometer readings). By breaking down the bias assessment into two altitude bands (below and above 1500 m), our analysis aims to reduce the impact of station siting changes (e.g., a large increase in the proportion of ablation zone observations as the PROMICE network has been set up). Our analysis also, to some extent, isolates different instrument types, as the PROMICE network and K–transect stations are mostly below 1500 m, while GC–Net stations are mostly above 1500 m. Side-by-side comparisons of different instrument types, across different climatological regimes on the ice sheet, is needed for a future study to better understand the spatial and temporal patterns of bias shown here. This could include the replication of historic observing practices and instruments, to better understand, and make the most of, the limited number of mid-20th century ice sheet SAT observations.

4   Conclusions

We have assessed a number of global SAT datasets using in situ observations over Greenland, and found large differences in their performance. Reanalyses generally perform better than gridded SAT analyses – particularly at high elevations on the ice sheet. Simple elevation-based corrections applied to reanalyses lead, in most cases, to improved performance: changes in mean monthly MAE (weighted as in Table 3) vary from a 3 % increase to a 42 % decrease. Considering all regions and seasons, the smallest biases are seen in (elevation-corrected) MERRA2 reanalysis. Biases vary by season and by region of the ice sheet: in the ablation region (demarcated here by the 1500 m elevation contour) during summer, most reanalyses have a $\sim 1\,^\circ$C positive bias (though 20CRv2c and ERA–20C have negative biases) while CRU and Berkeley Earth gridded SAT analyses have larger positive biases. These biases have implications for SMB reconstruction, as this region and season contribute a large proportion of meltwater creation.

Among global datasets that cover the entire 20th century, 20CRv2c generally has the smallest biases and MAEs when comparing against observations made since 1979. However, combining GISTEMP anomalies with the MERRA2 climatology gives slightly better results and, given concerns about spurious long term trends in 20CRv2c (in particular, a warm bias before 1950), we recommend this type of approach (i.e., combining GISTEMP with MERRA2) to represent monthly SAT over the early and mid-20th century. Similarity of anomalies between gridded SAT analyses (except

NansenSAT) suggests that observed biases result from their climatology fields, but their anomalies are suitable alternatives to GISTEMP.

Alongside multi-decadal global SAT datasets, we have analyzed SAT from recent (2002 to present) AIRS satellite retrievals and from RCM-downscaled reanalysis. AIRS has among the smallest biases and MAE in summer months over the ice sheet, but larger errors in winter and when comparing to coastal stations. RCMs are found to reduce biases in comparison to their respective forcing datasets and provide among the best representations of SAT on the ice sheet. However, MERRA2 reanalysis performs comparably on the ice sheet, and better in comparison to coastal stations. The long term variability of RCM SAT closely follows that from the forcing dataset; the shortcomings that we highlight for 20CRv2c thus also persist, to some degree, in the version of MAR forced by 20CRv2c. MAR–ERA–20C has long term variability closer to gridded SAT analyses and long-running DMI stations, but differences remain. The Box2013 dataset, by using spatial information from a similar RCM, has similar patterns of bias to the MAR datasets. However, Box2013 inherits its long term variation from the same SAT observations as used in global SAT analyses, rather than from (as in MAR) reanalysis forcing; thus its anomalies closely follow those from CRU, GISTEMP and especially Berkeley Earth.

We have assessed CMIP5 ESMs by comparing their ice sheet average SAT with that from other datasets. A key finding is that such an assessment depends crucially on the choice of verification dataset. Using GISTEMP combined with MERRA2 climatology (due to its overall good performance in comparison with in situ observations), we find that a large number of the CMIP5 ESMs have similar ice sheet long term annual average SATs (10 within 1 °C, 19 within 2 °C). The 1901–2000 trends from most individual models and the ensemble mean are positive. For a number of sub-periods examined, some individual ESMs have negative trends, though the ensemble mean does not, highlighting the fact that the ensemble mean does not exhibit realistic decadal variability. The 1990–2005 trends are positive and larger than for earlier periods (though mostly not statistically significant) for the majority of CMIP5 ESMs analyzed here, suggesting that forced changes dominate over internal variability in this period.

Our analysis highlights several avenues for future work. Comparison of different instrument types and measurement practices would allow a quantitative assessment of the effects of instrument bias on the results shown here. Such work is also crucial to investigations of GrIS diurnal temperature variation, for example in model assessment and SMB studies using positive degree day methods (Fausto et al., 2011; Rogozhina and Rau, 2014). Results for AIRS retrievals suggest it may provide useful SAT information over the GrIS in summer, but further work is needed on the effects of only sampling clear-sky SAT. Investigation is required to establish the cause of disparities in trends and variability between 20CRv2c and ERA–20C – which are ostensibly formulated in similar ways. Possible causes include different representation of atmospheric circulation and different sea ice and sea surface temperature datasets. While RCM downscaling is currently an important tool in assessing past and future GrIS mass balance changes, our results provide new evidence that results from RCMs are highly dependent on the forcing. For SAT, RCM downscaling can reduce biases and give realistic spatial patterns compared to the forcing dataset, but does not seem to greatly alter the long term evolution of the areal average. It remains to be seen whether the same is true for SMB. The greatest SAT differences between the versions of MAR used here occur before 1980, but there are differences since 2000 too, highlighting that uncertainties in GrIS SMB exist even in the better-observed recent past.

Code and data availability. Most of the data used in this work are freely and publicly available. Full dataset references are given in the Supplement. Derived data fields (e.g., elevation-corrected SAT) and code used to analyze data and plot figures are available from the corresponding author on request.

DMI AWS data were downloaded from http://www.dmi.dk/laer-om/generelt/dmi-publikationer/2013/, GC–Net AWS data from http://cires1.colorado.edu/steffen/gcnet/, and PROMICE data from http://www.promice.dk/. MERRA, MERRA2 and AIRS data were downloaded from the NASA Goddard Earth Sciences Data and Information Services Center (GES DISC). ERA–Interim and ERA–20C were downloaded from the EMCWF website (http://apps.ecmwf.int/datasets/). JRA–55, CFSR and CFSv2 data were obtained from the Research Data Archive at the National Center for Atmospheric Research (NCAR) Computational and Information Systems Laboratory. 20CRv2c data were downloaded from http://www.esrl.noaa.gov/psd/. CRU data were downloaded from the British Atmospheric Data Centre (BADC), Berkeley Earth from http://berkeleyearth.org/data/, GISTEMP from http://data.giss.nasa.gov/gistemp/, and NANSEN SAT from http://www.niersc.spb.ru. MAR data were downloaded from ftp://ftp.climato.be/fettweis/MARv3.5.2/Greenland/. CMIP5 data were obtained from the U.S. Department of Energy's Program for Climate Model Diagnosis and Intercomparison (http://cmip-pcmdi.llnl.gov/cmip5/data_portal.html).

Competing interests. The authors declare that they have no conflict of interest.

Acknowledgements. This research was supported by NASA (NNX14AM02G), DOE (DE-SC0016533), and the Agnese Nelms Haury Program in Environment and Social Justice. We thank Jason Box, Xavier Fettweis and an anonymous reviewer for their constructive comments and suggestions. Chris Castro and Guo-Yue Niu are thanked for useful discussions during the preparation of this manuscript. We also thank the various groups and centers for making their datasets and model results available. We thank C. J. P. P. (Paul) Smeets and the Institute for Marine and Atmospheric Research at the Utrecht University for providing the K-transect data.

Edited by: Marco Tedesco

References

Bamber, J. L., Layberry, R. L., and Gogineni, S. P.: A new ice thickness and bed data set for the Greenland ice sheet: 1. Measurement, data reduction, and errors, J. Geophys. Res., 106, 33773–33780, https://doi.org/10.1029/2001JD900054, 2001.

Bamber, J. L., Griggs, J. A., Hurkmans, R. T. W. L., Dowdeswell, J. A., Gogineni, S. P., Howat, I., Mouginot, J., Paden, J., Palmer, S., Rignot, E., and Steinhage, D.: A new bed elevation dataset for Greenland, The Cryosphere, 7, 499–510, https://doi.org/10.5194/tc-7-499-2013, 2013.

Box, J. E.: Survey of Greenland instrumental temperature records: 1873–2001, Int. J. Climatol., 22, 1829–1847, https://doi.org/10.1002/joc.852, 2002.

Box, J. E.: Greenland Ice Sheet Mass Balance Reconstruction. Part II: Surface Mass Balance (1840–2010), J. Climate, 26, 6974–6989, https://doi.org/10.1175/JCLI-D-12-00518.1, 2013.

Box, J. E., Yang, L., Bromwich, D. H., and Bai, L.-S.: Greenland Ice Sheet Surface Air Temperature Variability: 1840–2007, J. Climate, 22, 4029–4049, https://doi.org/10.1175/2009JCLI2816.1, 2009.

Burgess, E. W., Forster, R. R., Box, J. E., Mosley-Thompson, E., Bromwich, D. H., Bales, R. C., and Smith, L. C.: A spatially calibrated model of annual accumulation rate on the Greenland Ice Sheet (1958–2007), J. Geophys. Res., 115, F02004, https://doi.org/10.1029/2009JF001293, 2010.

Cappelen, J. (Ed.): Greenland – DMI Historical Climate Data Collection 1784-2013, Technical Report 14-04, Danish Meteorological Institute, available from: http://www.dmi.dk/fileadmin/user_upload/Rapporter/TR/2014/tr14-04.pdf, last access: 3 October 2016, 2014.

Chahine, M. T., Pagano, T. S., Aumann, H. H., Atlas, R., Barnet, C., Blaisdell, J., Chen, L., Divakarla, M., Fetzer, E. J., Goldberg, M., Gautier, C., Granger, S., Hannon, S., Irion, F. W., Kakar, R., Kalnay, E., Lambrigtsen, B. H., Lee, S.-Y., Le Marshall, J., McMillan, W. W., McMillin, L., Olsen, E. T., Revercomb, H., Rosenkranz, P., Smith, W. L., Staelin, D., Strow, L. L., Susskind, J., Tobin, D., Wolf, W., and Zhou, L.: AIRS: Improving Weather Forecasting and Providing New Data on Greenhouse Gases, B. Am. Meteorol. Soc., 87, 911–926, https://doi.org/10.1175/BAMS-87-7-911, 2006.

Chylek, P., Dubey, M. K., and Lesins, G.: Greenland warming of 1920–1930 and 1995–2005, Geophys. Res. Lett., 33, L11707, https://doi.org/10.1029/2006GL026510, 2006.

Comiso, J. C., Yang, J., Honjo, S., and Krishfield, R. A.: Detection of change in the Arctic using satellite and in situ data, J. Geophys. Res., 108, 3384, https://doi.org/10.1029/2002JC001347, 2003.

Compo, G. P., Whitaker, J. S., Sardeshmukh, P. D., Matsui, N., Allan, R. J., Yin, X., Gleason, B. E., Vose, R. S., Rutledge, G., Bessemoulin, P., Brönnimann, S., Brunet, M., Crouthamel, R. I., Grant, A. N., Groisman, P. Y., Jones, P. D., Kruk, M. C., Kruger, A. C., Marshall, G. J., Maugeri, M., Mok, H. Y., Nordli, Ø., Ross, T. F., Trigo, R. M., Wang, X. L., Woodruff, S. D., and Worley, S. J.: The Twentieth Century Reanalysis Project, Q. J. Roy. Meteor. Soc., 137, 1–28, https://doi.org/10.1002/qj.776, 2011.

Compo, G. P., Sardeshmukh, P. D., Whitaker, J. S., Brohan, P., Jones, P. D., and McColl, C.: Independent confirmation of global land warming without the use of station temperatures, Geophys. Res. Lett., 40, 3170–3174, https://doi.org/10.1002/grl.50425, 2013.

Cullather, R. I., Nowicki, S. M. J., Zhao, B., and Suarez, M. J.: Evaluation of the Surface Representation of the Greenland Ice Sheet in a General Circulation Model, J. Climate, 27, 4835–4856, https://doi.org/10.1175/JCLI-D-13-00635.1, 2014.

Cullather, R. I., Nowicki, S. M. J., Zhao, B., and Koenig, L. S.: A Characterization of Greenland Ice Sheet Surface Melt and Runoff in Contemporary Reanalyses and a Regional Climate Model, Front. Earth Sci., 4, 10 pp., https://doi.org/10.3389/feart.2016.00010, 2016.

Dee, D. P., Uppala, S. M., Simmons, A. J., Berrisford, P., Poli, P., Kobayashi, S., Andrae, U., Balmaseda, M. A., Balsamo, G., Bauer, P., Bechtold, P., Beljaars, A. C. M., van de Berg, L., Bidlot, J., Bormann, N., Delsol, C., Dragani, R., Fuentes, M., Geer, A. J., Haimberger, L., Healy, S. B., Hersbach, H., Hólm, E. V., Isaksen, L., Kållberg, P., Köhler, M., Matricardi, M., McNally, A. P., Monge-Sanz, B. M., Morcrette, J.-J., Park, B.-K., Peubey, C., de Rosnay, P., Tavolato, C., Thépaut, J.-N., and Vitart, F.: The ERA-Interim reanalysis: configuration and performance of the data assimilation system, Q. J. Roy. Meteor. Soc., 137, 553–597, https://doi.org/10.1002/qj.828, 2011.

Ettema, J., van den Broeke, M. R., van Meijgaard, E., van de Berg, W. J., Box, J. E., and Steffen, K.: Climate of the Greenland ice sheet using a high-resolution climate model – Part 1: Evaluation, The Cryosphere, 4, 511–527, https://doi.org/10.5194/tc-4-511-2010, 2010a.

Ettema, J., van den Broeke, M. R., van Meijgaard, E., and van de Berg, W. J.: Climate of the Greenland ice sheet using a high-resolution climate model – Part 2: Near-surface climate and energy balance, The Cryosphere, 4, 529–544, https://doi.org/10.5194/tc-4-529-2010, 2010b.

Fausto, R. S., Ahlstrøm, A. P., Van As, D., Bøggild, C. E., and Johnsen, S. J.: A new present-day temperature parameterization for Greenland, J. Glaciol., 55, 95–105, 2009.

Fausto, R. S., Ahlstrøm, A. P., Van As, D., and Steffen, K.: Present-day temperature standard deviation parameterization for Greenland, J. Glaciol., 57, 1181–1183, https://doi.org/10.3189/002214311798843377, 2011.

Fettweis, X., Hanna, E., Gallée, H., Huybrechts, P., and Erpicum, M.: Estimation of the Greenland ice sheet surface mass balance for the 20th and 21st centuries, The Cryosphere, 2, 117–129, https://doi.org/10.5194/tc-2-117-2008, 2008.

Fettweis, X., Box, J. E., Agosta, C., Amory, C., Kittel, C., Lang, C., van As, D., Machguth, H., and Gallée, H.: Reconstructions of the 1900–2015 Greenland ice sheet surface mass balance using the regional climate MAR model, The Cryosphere, 11, 1015–1033, https://doi.org/10.5194/tc-11-1015-2017, 2017.

Genthon, C., Six, D., Favier, V., Lazzara, M., and Keller, L.: Atmospheric Temperature Measurement Biases on the Antarctic Plateau, J. Atmos. Ocean. Tech., 28, 1598–1605, https://doi.org/10.1175/JTECH-D-11-00095.1, 2011.

Hall, D. K., Box, J. E., Casey, K. A., Hook, S. J., Shuman, C. A., and Steffen, K.: Comparison of satellite-derived and in-situ observations of ice and snow surface temperatures over Greenland, Remote Sens. Environ., 112, 3739–3749, https://doi.org/10.1016/j.rse.2008.05.007, 2008.

Hall, D. K., Comiso, J. C., DiGirolamo, N. E., Shuman, C. A., Box, J. E., and Koenig, L. S.: Variability in the surface temperature and melt extent of the Greenland ice sheet from MODIS, Geophys. Res. Lett., 40, 2114–2120, https://doi.org/10.1002/grl.50240, 2013.

Hanna, E., Huybrechts, P., Janssens, I., Cappelen, J., Steffen, K., and Stephens, A.: Runoff and mass balance of the Greenland ice sheet: 1958–2003, J. Geophys. Res., 110, D13108, https://doi.org/10.1029/2004JD005641, 2005.

Hanna, E., Huybrechts, P., Cappelen, J., Steffen, K., Bales, R. C., Burgess, E., McConnell, J. R., Peder Steffensen, J., Van den Broeke, M., Wake, L., Bigg, G., Griffiths, M., and Savas, D.: Greenland Ice Sheet surface mass balance 1870 to 2010 based on Twentieth Century Reanalysis, and links with global climate forcing, J. Geophys. Res., 116, D24121, https://doi.org/10.1029/2011JD016387, 2011.

Hanna, E., Mernild, S. H., Cappelen, J., and Steffen, K.: Recent warming in Greenland in a long-term instrumental (1881–2012) climatic context: I. Evaluation of surface air temperature records, Environ. Res. Lett., 7, 45404, https://doi.org/10.1088/1748-9326/7/4/045404, 2012.

Hansen, J., Ruedy, R., Sato, M., and Lo, K.: Global surface temperature change, Rev. Geophys., 48, RG4004, https://doi.org/10.1029/2010RG000345, 2010.

Harris, I., Jones, P. D., Osborn, T. J., and Lister, D. H.: Updated high-resolution grids of monthly climatic observations – the CRU TS3.10 Dataset, Int. J. Climatol., 34, 623–642, https://doi.org/10.1002/joc.3711, 2014.

Kobayashi, S., Ota, Y., Harada, Y., Ebita, A., Moriya, M., Onoda, H., Onogi, K., Kamahori, H., Kobayashi, C., Endo, H., Miyaoka, K., and Takahashi, K.: The JRA-55 Reanalysis: General Specifications and Basic Characteristics, J. Meteorol. Soc. Jpn., Ser. II, 93, 5–48, https://doi.org/10.2151/jmsj.2015-001, 2015.

Koenig, L. S. and Hall, D. K.: Comparison of satellite, thermochron and air temperatures at Summit, Greenland, during the winter of 2008/09, J. Glaciol., 56, 735–741, https://doi.org/10.3189/002214310793146269, 2010.

Kuzmina, S. I., Johannessen, O. M., Bengtsson, L., Aniskina, O. G., and Bobylev, L. P.: High northern latitude surface air temperature: comparison of existing data and creation of a new gridded data set 1900–2000, Tellus A, 60, 289–304, https://doi.org/10.3402/tellusa.v60i2.15260, 2008.

Lucas-Picher, P., Wulff-Nielsen, M., Christensen, J. H., Aðalgeirsdóttir, G., Mottram, R., and Simonsen, S. B.: Very high resolution regional climate model simulations over Greenland: Identifying added value, J. Geophys. Res., 117, D02108, https://doi.org/10.1029/2011JD016267, 2012.

Mernild, S. H., Hanna, E., Yde, J. C., Cappelen, J., and Malmros, J. K.: Coastal Greenland air temperature extremes and trends

1890–2010: annual and monthly analysis, Int. J. Climatol., 34, 1472–1487, https://doi.org/10.1002/joc.3777, 2014.

Molod, A., Takacs, L., Suarez, M., and Bacmeister, J.: Development of the GEOS-5 atmospheric general circulation model: evolution from MERRA to MERRA2, Geosci. Model Dev., 8, 1339–1356, https://doi.org/10.5194/gmd-8-1339-2015, 2015.

Morice, C. P., Kennedy, J. J., Rayner, N. A., and Jones, P. D.: Quantifying uncertainties in global and regional temperature change using an ensemble of observational estimates: The HadCRUT4 data set, J. Geophys. Res., 117, D08101, https://doi.org/10.1029/2011JD017187, 2012.

Noël, B., van de Berg, W. J., van Meijgaard, E., Kuipers Munneke, P., van de Wal, R. S. W., and van den Broeke, M. R.: Evaluation of the updated regional climate model RACMO2.3: summer snowfall impact on the Greenland Ice Sheet, The Cryosphere, 9, 1831–1844, https://doi.org/10.5194/tc-9-1831-2015, 2015.

Noël, B., van de Berg, W. J., Machguth, H., Lhermitte, S., Howat, I., Fettweis, X., and van den Broeke, M. R.: A daily, 1 km resolution data set of downscaled Greenland ice sheet surface mass balance (1958–2015), The Cryosphere, 10, 2361–2377, https://doi.org/10.5194/tc-10-2361-2016, 2016.

Ohmura, A.: New temperature distribution maps for Greenland, Zeitschrift für Gletscherkunde und Glaziolgeologie, 23, 1–45, 1987.

Poli, P., Hersbach, H., Dee, D. P., Berrisford, P., Simmons, A. J., Vitart, F., Laloyaux, P., Tan, D. G. H., Peubey, C., Thépaut, J.-N., Trémolet, Y., Hólm, E. V., Bonavita, M., Isaksen, L., and Fisher, M.: ERA-20C: An Atmospheric Reanalysis of the Twentieth Century, J. Climate, 29, 4083–4097, https://doi.org/10.1175/JCLI-D-15-0556.1, 2016.

Rae, J. G. L., Aðalgeirsdóttir, G., Edwards, T. L., Fettweis, X., Gregory, J. M., Hewitt, H. T., Lowe, J. A., Lucas-Picher, P., Mottram, R. H., Payne, A. J., Ridley, J. K., Shannon, S. R., van de Berg, W. J., van de Wal, R. S. W., and van den Broeke, M. R.: Greenland ice sheet surface mass balance: evaluating simulations and making projections with regional climate models, The Cryosphere, 6, 1275–1294, https://doi.org/10.5194/tc-6-1275-2012, 2012.

Ridley, J. K., Huybrechts, P., Gregory, J. M., and Lowe, J. A.: Elimination of the Greenland ice sheet in a high CO$_2$ climate, J. Climate, 18, 3409–3427, 2005.

Rienecker, M. M., Suarez, M. J., Gelaro, R., Todling, R., Bacmeister, J., Liu, E., Bosilovich, M. G., Schubert, S. D., Takacs, L., Kim, G.-K., Bloom, S., Chen, J., Collins, D., Conaty, A., da Silva, A., Gu, W., Joiner, J., Koster, R. D., Lucchesi, R., Molod, A., Owens, T., Pawson, S., Pegion, P., Redder, C. R., Reichle, R., Robertson, F. R., Ruddick, A. G., Sienkiewicz, M., and Woollen, J.: MERRA: NASA's Modern-Era Retrospective Analysis for Research and Applications, J. Climate, 24, 3624–3648, https://doi.org/10.1175/JCLI-D-11-00015.1, 2011.

Rogozhina, I. and Rau, D.: Vital role of daily temperature variability in surface mass balance parameterizations of the Greenland Ice Sheet, The Cryosphere, 8, 575–585, https://doi.org/10.5194/tc-8-575-2014, 2014.

Rohde, R., Muller, R., Jacobsen, R., Perlmutter, S., Rosenfeld, A., Wurtele, J., Curry, J., Wickham, C., and Mosher, S.: Berkeley Earth Temperature Averaging Process, Geoinformatics & Geostatistics: An Overview, 1, 13 pp., https://doi.org/10.4172/gigs.1000103, 2013.

Saha, S., Moorthi, S., Pan, H.-L., Wu, X., Wang, J., Nadiga, S., Tripp, P., Kistler, R., Woollen, J., Behringer, D., Liu, H., Stokes, D., Grumbine, R., Gayno, G., Wang, J., Hou, Y.-T., Chuang, H.-Y., Juang, H.-M. H., Sela, J., Iredell, M., Treadon, R., Kleist, D., Van Delst, P., Keyser, D., Derber, J., Ek, M., Meng, J., Wei, H., Yang, R., Lord, S., Van Den Dool, H., Kumar, A., Wang, W., Long, C., Chelliah, M., Xue, Y., Huang, B., Schemm, J.-K., Ebisuzaki, W., Lin, R., Xie, P., Chen, M., Zhou, S., Higgins, W., Zou, C.-Z., Liu, Q., Chen, Y., Han, Y., Cucurull, L., Reynolds, R. W., Rutledge, G., and Goldberg, M.: The NCEP Climate Forecast System Reanalysis, B. Am. Meteorol. Soc., 91, 1015–1057, https://doi.org/10.1175/2010BAMS3001.1, 2010.

Saha, S., Moorthi, S., Wu, X., Wang, J., Nadiga, S., Tripp, P., Behringer, D., Hou, Y.-T., Chuang, H., Iredell, M., Ek, M., Meng, J., Yang, R., Mendez, M. P., van den Dool, H., Zhang, Q., Wang, W., Chen, M., and Becker, E.: The NCEP Climate Forecast System Version 2, J. Climate, 27, 2185–2208, https://doi.org/10.1175/JCLI-D-12-00823.1, 2014.

Santer, B. D., Wigley, T. M. L., Boyle, J. S., Gaffen, D. J., Hnilo, J. J., Nychka, D., Parker, D. E., and Taylor, K. E.: Statistical significance of trends and trend differences in layer-average atmospheric temperature time series, J. Geophys. Res., 105, 7337–7356, https://doi.org/10.1029/1999JD901105, 2000.

Screen, J. A. and Simmonds, I.: Erroneous Arctic Temperature Trends in the ERA-40 Reanalysis: A Closer Look, J. Climate, 24, 2620–2627, https://doi.org/10.1175/2010JCLI4054.1, 2010.

Shuman, C. A., Alley, R. B., Anandakrishnan, S., and Stearns, C. R.: An empirical technique for estimating near-surface air temperature trends in central Greenland from SSM/I brightness temperatures, Remote Sens. Environ., 51, 245–252, https://doi.org/10.1016/0034-4257(94)00086-3, 1995.

Shuman, C. A., Steffen, K., Box, J. E., and Stearns, C. R.: A Dozen Years of Temperature Observations at the Summit: Central Greenland Automatic Weather Stations 1987–99, J. Appl. Meteorol., 40, 741–752, https://doi.org/10.1175/1520-0450(2001)040<0741:ADYOTO>2.0.CO;2, 2001.

Smith, T. M., Reynolds, R. W., Peterson, T. C., and Lawrimore, J.: Improvements to NOAA's Historical Merged Land–Ocean Surface Temperature Analysis (1880–2006), J. Climate, 21, 2283–2296, https://doi.org/10.1175/2007JCLI2100.1, 2008.

Steffen, K. and Box, J.: Surface climatology of the Greenland Ice Sheet: Greenland Climate Network 1995–1999, J. Geophys. Res., 106, 33951–33964, https://doi.org/10.1029/2001JD900161, 2001.

Taylor, K. E., Stouffer, R. J., and Meehl, G. A.: An Overview of CMIP5 and the Experiment Design, B. Am. Meteorol. Soc., 93, 485–498, https://doi.org/10.1175/BAMS-D-11-00094.1, 2011.

Uppala, S. M., Kållberg, P. W., Simmons, A. J., Andrae, U., Bechtold, V. D. C., Fiorino, M., Gibson, J. K., Haseler, J., Hernandez, A., Kelly, G. A., Li, X., Onogi, K., Saarinen, S., Sokka, N., Allan, R. P., Andersson, E., Arpe, K., Balmaseda, M. A., Beljaars, A. C. M., Berg, L. V. D., Bidlot, J., Bormann, N., Caires, S., Chevallier, F., Dethof, A., Dragosavac, M., Fisher, M., Fuentes, M., Hagemann, S., Hólm, E., Hoskins, B. J., Isaksen, L., Janssen, P. A. E. M., Jenne, R., Mcnally, A. P., Mahfouf, J.-F., Morcrette, J.-J., Rayner, N. A., Saunders, R. W., Simon, P., Sterl, A., Trenberth, K. E., Untch, A., Vasiljevic, D., Viterbo, P., and Woollen, J.: The ERA-40 re-analysis, Q. J. Roy. Meteor. Soc., 131, 2961–3012, https://doi.org/10.1256/qj.04.176, 2005.

Van As, D., Fausto, R. S., Ahlstrøm, A. P., Andersen, S. B., Andersen, M. L., Citterio, M., Edelvang, K., Gravesen, P., Machguth, H., Nick, F. M., Nielsen, S., and Weidick, A.: Programme for Monitoring of the Greenland Ice Sheet (PROMICE): first temperature and ablation records, Geol. Surv. Den. Greenl., 23, 73–76, 2011.

van den Broeke, M. R., Smeets, C. J. P. P., and van de Wal, R. S. W.: The seasonal cycle and interannual variability of surface energy balance and melt in the ablation zone of the west Greenland ice sheet, The Cryosphere, 5, 377–390, https://doi.org/10.5194/tc-5-377-2011, 2011.

van de Wal, R. S. W., Greuell, W., van den Broeke, M. R., Reijmer, C. H., and Oerlemans, J.: Surface mass-balance observations and automatic weather station data along a transect near Kangerlussuaq, West Greenland, Ann. Glaciol., 42, 311–316, https://doi.org/10.3189/172756405781812529, 2005.

Vernon, C. L., Bamber, J. L., Box, J. E., van den Broeke, M. R., Fettweis, X., Hanna, E., and Huybrechts, P.: Surface mass balance model intercomparison for the Greenland ice sheet, The Cryosphere, 7, 599–614, https://doi.org/10.5194/tc-7-599-2013, 2013.

Vizcaíno, M., Lipscomb, W. H., Sacks, W. J., van Angelen, J. H., Wouters, B., and van den Broeke, M. R.: Greenland Surface Mass Balance as Simulated by the Community Earth System Model. Part I: Model Evaluation and 1850–2005 Results, J. Climate, 26, 7793–7812, https://doi.org/10.1175/JCLI-D-12-00615.1, 2013.

Vose, R. S., Arndt, D., Banzon, V. F., Easterling, D. R., Gleason, B., Huang, B., Kearns, E., Lawrimore, J. H., Menne, M. J., Peterson, T. C., Reynolds, R. W., Smith, T. M., Williams, C. N., and Wuertz, D. B.: NOAA's Merged Land–Ocean Surface Temperature Analysis, B. Am. Meteorol. Soc., 93, 1677–1685, https://doi.org/10.1175/BAMS-D-11-00241.1, 2012.

Wang, A. and Zeng, X.: Development of Global Hourly 0.5° Land Surface Air Temperature Datasets, J. Climate, 26, 7676–7691, https://doi.org/10.1175/JCLI-D-12-00682.1, 2013.

4

The importance of accurate glacier albedo for estimates of surface mass balance on Vatnajökull: evaluating the surface energy budget in a regional climate model with automatic weather station observations

Louise Steffensen Schmidt[1], **Guðfinna Aðalgeirsdóttir**[1], **Sverrir Guðmundsson**[1,2], **Peter L. Langen**[3], **Finnur Pálsson**[1], **Ruth Mottram**[3], **Simon Gascoin**[4], **and Helgi Björnsson**[1]

[1]University of Iceland, Institute of Earth Sciences, Reykjavik, Iceland
[2]Keilir Institute of Technology, Reykjanesbær, Iceland
[3]Danish Meteorological Institute, Copenhagen, Denmark
[4]Centre d'Etudes Spatiales de la Biosphère, Université de Toulouse, CNES/CNRS/IRD/UPS, Toulouse, France

Correspondence to: Louise Steffensen Schmidt (lss7@hi.is)

Abstract. A simulation of the surface climate of Vatnajökull ice cap, Iceland, carried out with the regional climate model HIRHAM5 for the period 1980–2014, is used to estimate the evolution of the glacier surface mass balance (SMB). This simulation uses a new snow albedo parameterization that allows albedo to exponentially decay with time and is surface temperature dependent. The albedo scheme utilizes a new background map of the ice albedo created from observed MODIS data. The simulation is evaluated against observed daily values of weather parameters from five automatic weather stations (AWSs) from the period 2001–2014, as well as in situ SMB measurements from the period 1995–2014. The model agrees well with observations at the AWS sites, albeit with a general underestimation of the net radiation. This is due to an underestimation of the incoming radiation and a general overestimation of the albedo. The average modelled albedo is overestimated in the ablation zone, which we attribute to an overestimation of the thickness of the snow layer and not taking the surface darkening from dirt and volcanic ash deposition during dust storms and volcanic eruptions into account. A comparison with the specific summer, winter, and net mass balance for the whole of Vatnajökull (1995–2014) shows a good overall fit during the summer, with a small mass balance underestimation of 0.04 m w.e. on average, whereas the winter mass balance is overestimated by on average 0.5 m w.e. due to too large precipitation at the highest areas of the ice cap. A simple correction of the accumulation at the highest points of the glacier reduces this to 0.15 m w.e. Here, we use HIRHAM5 to simulate the evolution of the SMB of Vatnajökull for the period 1981–2014 and show that the model provides a reasonable representation of the SMB for this period. However, a major source of uncertainty in the representation of the SMB is the representation of the albedo, and processes currently not accounted for in RCMs, such as dust storms, are an important source of uncertainty in estimates of snow melt rate.

1 Introduction

Worldwide, glaciers and ice caps are losing mass at increasing rates as a response to climate change (e.g. Vaughan et al., 2013). Major changes in the dimensions of glaciers are expected to affect the sea level and climate throughout the world, and it is therefore important to describe and understand the glacier climate. Glacier retreat and mass loss at significantly increasing rates are also observed for Icelandic glaciers (Björnsson et al., 2013), which could potentially contribute to the rise in sea level by 1 cm (Björnsson and Pálsson, 2008; Björnsson et al., 2013). The runoff from Vatnajökull ice cap is economically important to hydropower production in Iceland and the present and future mass balance is thus of keen interest. Numerical high-resolution regional climate models (RCMs), such as MAR (Gallée and Schayes, 1994), RACMO2 (Meijgaard et al., 2008), or HIRHAM5 (Christensen et al., 2006), are valuable tools for estimating the meteorological parameters and mass balance variability at the surface of glaciers. However, to carry out reliable future projections, or reconstruct the past climate, it is important to evaluate how well models simulate the present climate

Evaluation of RCMs is important, not only because it reveals possible biases in the model but also because it could yield recommendations for model improvements. Much work has gone into evaluating RCMs over Greenland (e.g. Box and Rinke, 2003; Noël et al., 2015; Rae et al., 2012; Langen et al., 2017; Fettweis et al., 2017) and Antarctica (e.g. Lenaerts and Van Den Broeke, 2012; Agosta et al., 2015), but less effort has gone into evaluating them over Iceland (e.g. Ágústsson et al., 2013; Nawri, 2014).

However, a long-term meteorological monitoring programme has been conducted on Icelandic glaciers since the 1991–1992 glaciological year (e.g. Björnsson et al., 1998). Therefore, Icelandic glaciers are excellent candidates for evaluating modelled meteorological and SMB components. Compared to Greenland, observations are recorded in a relatively small area, offering a good opportunity to evaluate the spatial and temporal variability of the HIRHAM5 model on a regional scale. As albedo in Iceland is significantly different from that of Greenland or Antarctica, e.g. due to frequent dust storms and occasional volcanic eruptions, model evaluations over Iceland provides important insight into the effect of albedo changes on the glacier energy balance.

Due to the large spatial and temporal variation in albedo of Icelandic glaciers (spanning from less than 0.1 for dirty ice in the ablation zone to 0.9–0.95 for new snow), and the large sensitivity of melt to variations in albedo, it is crucial to have correct estimates of the albedo when modelling the surface mass balance. However, accurate modelling of the albedo can be challenging. For example, volcanic eruptions and dust storms can significantly lower the glacier albedo, and thus increase the amount of melt (e.g. Conway et al., 1996; Gascoin et al., 2017; Wittmann et al., 2017), but are difficult to include in albedo models. Accurate simulations of the ice albedo is also problematic, as for some glaciers it varies with elevation (e.g. Knap et al., 1999) but not for others (e.g. Greuell et al., 1997). In addition, the ice albedo may decrease with time (e.g. Reijmer et al., 1999), increase with time (e.g. Oerlemans and Knap, 1998), or remain constant (e.g. Greuell et al., 1997) depending on the glacier.

Here we present a 1981–2014 SMB data set of Vatnajökull ice cap modelled by HIRHAM5 at 5.5 km resolution. HIRHAM5 is a state-of-the-art, high-resolution RCM that has been well validated over Greenland (e.g. Box and Rinke, 2003; Lucas-Picher et al., 2012; Rae et al., 2012; Langen et al., 2017). In this study, HIRHAM5 incorporates an up-

dated albedo scheme, using a background MODIS ice albedo field, in the aim of capturing the effect of dust and tephra on ice albedo in the ablation zone. This method of determining the ice albedo has previously been used by, for example, van Angelen et al. (2012). Model simulation results are compared to observations from automatic weather stations (AWSs) and in situ mass balance observations, in an effort to improve the performance of the model. The possible physical reasons for any model biases are discussed, and recommendations for corrections are made where possible.

2 Model description

2.1 HIRHAM5

In this study we employed the RCM HIRHAM5 (Christensen et al., 2006), which was developed at the Danish Meteorological Institute. It is a hydrostatic RCM which combines the dynamical core of the HIRLAM7 numerical forecasting model (Eerola, 2006) and physics schemes from the ECHAM5 general circulation model (Roeckner et al., 2003). Model simulations have been successfully validated over Greenland using AWS and ice core data (e.g. Box and Rinke, 2003; Stendel et al., 2008; Lucas-Picher et al., 2012; Langen et al., 2015; Rae et al., 2012; Langen et al., 2017).

While the original HIRHAM5, as described in Christensen et al. (2006), used unchanged ECHAM physics, an updated model version, which includes a dynamic surface scheme that explicitly calculates the surface mass budget on the surface of glaciers and ice sheets, is used in this study. This new scheme takes melting of snow and bare ice into account and resolves the retention and refreezing of liquid water in the snow pack (Langen et al., 2015, 2017). In addition, the five-layer surface scheme in ECHAM has been expanded to 25 layers.

2.1.1 New albedo parametrization

The updated model also features a more sophisticated snow albedo scheme (Nielsen-Englyst, 2015) than that used in the original HIRHAM5; whereas the previous scheme was purely temperature dependent, the new scheme depends both on the age of the snow and the surface temperature. The scheme is similar to that used in Oerlemans and Knap (1998), which assumes that the albedo decays exponentially as it ages, but in this study an additional temperature component is applied. If there is snow on the surface, the change in the snow albedo from one time step to the next depends on whether the surface is in a dry (< 271 K) or wet regime (≥ 271 K). In the dry regime, the surface temperature is too low for any melting to occur, while in the wet regime the temperature in the surface layer is high enough for the surface to be melting. The snow albedo changes over a time step, δt, as

$$\alpha_{\text{snow}}^{t} = (\alpha_{\text{snow}}^{t-1} - \alpha_{\text{mx}}) \cdot e^{-\delta t / \tau_x} + \alpha_{\text{mx}}, \qquad (1)$$

where α_{mx} is the minimum snow albedo value that can be reached from ageing of the snow and τ_x is a timescale which determines how fast the albedo reaches its minimum value. These two variables take on different values depending on whether the snow is in the dry (d) or wet (w) regime.

Observations from the AC and ELA stations were used to determine α_{mx} and τ_x. The optimal variables were found by minimizing the weighted mean RMSE between the modelled and measured albedo by varying the values of α_{mx} and τ_x. The best-fit values were found to be $\alpha_{md} = 0.65$, $\alpha_{mw}=0.41$, $\tau_{md} = 5$ days, and $\tau_{mw} = 10$ days.

Albedo is only refreshed to the maximum value if snowfall constitutes more than 95 % of the total precipitation. A partial refreshment is possible as the albedo is only reset to the maximum allowed value if the amount of snowfall on that day (S_0) is higher than 0.03 m w.e. This threshold was chosen to provide the best fit with the AWS observations. The rate of refreshment b is given by

$$b = \min\left[1, \frac{S_f}{S_0}\right], \tag{2}$$

where S_f is the amount of snowfall during the model time step in m w.e. and S_0 is the critical amount of snowfall in m w.e. per model time step needed to completely refresh the albedo. Using this rate, the albedo is then refreshed using

$$\alpha_{snow}^{t+1} = \alpha_{snow}^{t} + b \cdot (\alpha_{max} - \alpha_{snow}^{t}), \tag{3}$$

where α_{max} is the maximum albedo for freshly fallen snow, set equal to 0.85 as this provides the best average fit with the observations.

In the case of shallow snow cover, the surface albedo will be affected by the albedo of the underlying ice. A smooth transition between the snow and bare ice albedo is therefore implemented, and the final albedo is thus expressed as

$$\alpha^{t+1} = \alpha_{snow}^{t+1} + (\alpha_{ice} - \alpha_{snow}^{t+1}) \cdot \exp\left(\frac{-d^{t+1}}{d_s}\right), \tag{4}$$

where d is the snow depth, and d_s is a characteristic scale for snow depth. Following Oerlemans and Knap (1998), the characteristic scale is set to 3.2 cm snow depth. If no snow is present, the albedo is set to the bare ice albedo. The bare ice albedo is determined from a background ice albedo map which was created using MODIS observations from the period 2001–2012. How this map was created is described in Sect. 3.

The extent to which this bare ice MODIS albedo map improves the simulations will be estimated by comparing the results with those from a model simulation using a constant ice albedo in Sect. 4.8.

2.1.2 Experimental design

In this study, HIRHAM5 is run at a resolution of 0.05° (equivalent to ~ 5.5 km) on a rotated pole grid for the period

1980–2014. The model uses 31 irregularly spaced vertical atmospheric levels from the surface to 10 hPa with a model time step of 90 s in the dynamical scheme. The model is configured for a domain containing all of Greenland and Iceland. The model is forced at the lateral and lower boundaries by the ECMWF ERA-Interim reanalysis data set (Dee et al., 2011), which uses observations from satellites, weather balloons, and ground stations to create a comprehensive reanalysis of the atmosphere. The model is forced by temperature, wind, relative humidity, and surface pressure at the lateral boundary, and sea surface temperature and sea ice fraction at the lower boundary at 6 h intervals.

The new snow/ice surface scheme discussed above is run offline in this study, meaning that the subsurface scheme is run separately from the atmospheric code. This is done by forcing the subsurface scheme every 6 h by radiative and turbulent surface fluxes, as well as snow, rain, evaporation, and sublimation data from a HIRHAM5 experiment (Mottram et al., 2016) with a previous version of the albedo and refreezing schemes (e.g. Langen et al., 2017). While a full, high-resolution HIRHAM5 run is computationally very expensive, the offline model offers a fast and flexible option to test new model implementations and allows for a quick and thorough spin-up of the subsurface. The offline model was initialized with values from a previous offline model run with a different albedo scheme and then a model spin-up was performed by integrating the model for 150 years repeating the forcing from 1980. The largest adjustments occurred during the first 75 years of the spin-up, after which the variation was much smaller than the interannual variability. At the end of the run, the solar radiation, surface mass balance, runoff, snow depth, and refreezing had all converged, as had the temperature, liquid and snow content in all 25 subsurface layers. The final state of the spin-up was then used as the initial condition for the 1980–2014 model simulation. The reported values of albedo, upward longwave and shortwave radiation, and surface mass balance in the following are all from the offline run.

A disadvantage of this method is that it neglects feedbacks between the atmospheric circulation and the surface conditions like the albedo and temperature. However, since the surface temperature of Vatnajökull is typically near the melting point during the summer, both in reality and in the model, changes in the albedo should not have a large effect on upward longwave radiation and the turbulent fluxes. Thus, while the updated surface scheme is important for the mass balance components, the error due to the neglected feedbacks is likely small in the model calculations.

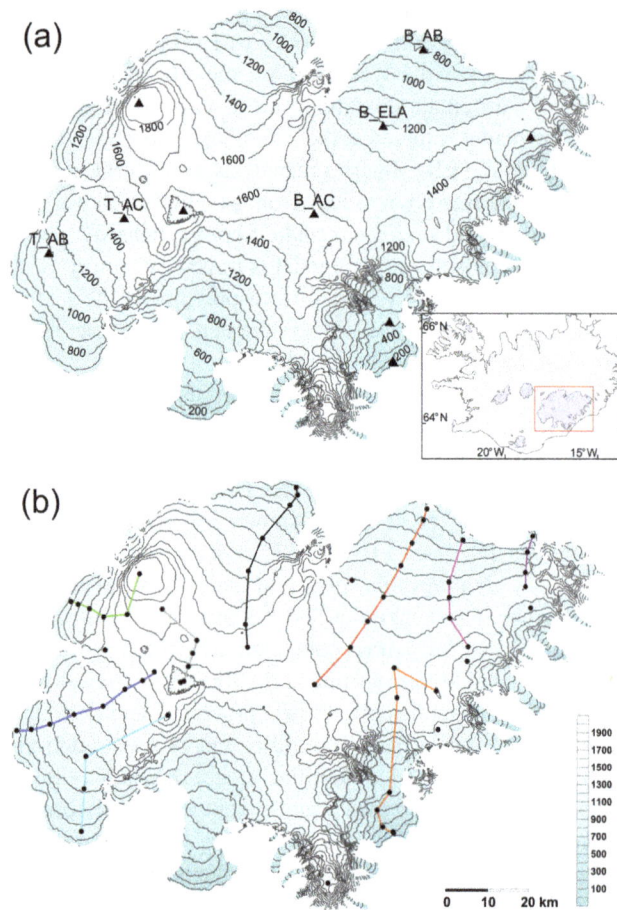

Figure 1. (a) The average location of the AWS sites. Only the labelled sites were used in this study. **(b)** The average location of the mass balance sites from 1995 to 2014. The coloured lines connect mass balance sites along a transect. Not all mass balance sites were measured every year.

Table 1. Average measured elevation and average bias of the interpolated HIRHAM5 elevation at each station for 2001–2014.

Station	Average elevation (m)	Average model elevation bias (m)
B_{AB}	839	22
T_{AB}	1089	47
B_{ELA}	1205	31
B_{AC}	1526	17
T_{AC}	1457	13

2.1.3 Model uncertainty

Due to nonlinearities in the HIRHAM5's model dynamics and physics, it has an implicit uncertainty due to internal model variability originating from nonlinear processes (e.g. Giorgi and Bi, 2000; de Elía et al., 2002). This variability is caused by numerical sensitivity, uncertainty in the boundary and initial conditions, and errors due to model parametrizations (e.g. Box and Rinke, 2003), including, for example, the albedo parameterization, the vertical gradients in the boundary layer, or cloud radiative effects. In addition, using a constant value of z_0 for both snow and bare ice could lead to large errors in the turbulent fluxes (e.g. Brock et al., 2000).

3 Observational data

The primary observational data set used in this study was collected by AWSs at selected locations on Vatnajökull. Since

1994, 1–13 stations have been operated on the ice cap during the summer months (e.g. Oerlemans et al., 1999; Guðmundsson et al., 2006). The temperature, relative humidity, wind speed, and wind direction at 2 m above the surface have been measured during the entire period (1992–present), while the radiation components have been measured since 1996. For this study, data from five AWSs were considered – three on Brúarjökull (B) and two on Tungnaárjökull (T) (see Fig. 1). Both Brúarjökull and Tungnaárjökull are outlet glaciers of Vatnajökull ice cap. Two stations are situated in the ablation zone (henceforth referred to as the AB stations), one station is situated near the equilibrium line altitude (ELA station), and two stations are in the accumulation zone (AC stations). The average elevation of each station is shown in Table 1. All five stations have been operated on the glacier every year during the period 2001–2014. Observations of 2 m temperature, humidity, wind speed, and radiative fluxes were used to validate HIRHAM5 over Vatnajökull.

The uncertainties of the AWS observations vary depending on the sensor. The temperature and humidity sensors have an accuracy of 0.2 K and 2 % for temperature and humidity, respectively, while the accuracy of the wind speed is $0.2 \, \mathrm{m\,s^{-1}}$ (Guðmundsson et al., 2009). The radiative fluxes were measured using either Kipp and Zonen CM14, CNR1 or CNR4 sensors that have a maximum manufacturer-reported uncertainty of ± 10 % for daily totals (e.g. Kipp and Zonen, 2002). However, the uncertainty has independently been evaluated to be lower (3–5 %) when used in an ice sheet environment (van den Broeke et al., 2004; Guðmundsson et al., 2009). The turbulent fluxes, combining sensible and latent heat fluxes, and surface pressure were not measured at the stations, but were estimated using the methods described in Sect. 3.1.

In addition to AWS data, in situ mass balance measurements were used to evaluate the simulated surface mass balance (SMB) at several sites on Vatnajökull. Conventional in situ mass balance measurements have been carried out every glaciological year since 1991–1992, with 60 stations measured each year on average. The measurement sites are shown in Fig. 1. The uncertainty of the mass balance measurements has been estimated to be ± 0.3 m w.e.

The SMB measurements are conducted at the beginning and end of the accumulation season in order to measure both the winter and summer balance. The winter balance is measured in the beginning of the melt season by drilling down to the previous summer layer and weighting the snow column. The summer surface is used as the reference level even if some snow accumulation had occurred by the time the summer balance measurements were conducted. The snow thickness on top of the summer surface at the time of the autumn survey has been measured since 1995. This is needed when comparing with the simulation of snow accumulation.

Observations of the broadband albedo in the shortwave spectrum (0.3–5.0 μm) from the MODerate Resolution Imaging Spectroradiometer (MODIS) were used to create a background map of the ice albedo at all glacier grid points in HIRHAM5, which was used in the implemented HIRHAM5 albedo scheme. MODIS product MCD43A3 v006 was used for the background map (Schaaf, 2015). The MODIS estimates of the albedo on Vatnajökull are in good agreement with AWS data (Gascoin et al., 2017). The MODIS data were extracted in geographical coordinates (long–lat) at a resolution of 0.005°, i.e. close to the original MODIS resolution of 500 m. This was done using the MODIS reprojection tool with the bilinear interpolation method. These MODIS data in latitude-longitude coordinates were then resampled to match the rotated HIRHAM5 long–lat grid coordinates by bilinear interpolation using MATLAB's interpn function (MATLAB, 2015).

In order to determine the bare ice albedo at each grid point, daily MODIS data over Iceland from the period 2001–2012 were used. Years with volcanic eruptions were discarded, as the volcanic ash lowered the albedo values far below the average. The minimum autumn albedo value was then determined in each grid point using values from July–September and that value used to create a bare ice albedo map of the glaciers. The final albedo map had ice albedo values in the range 0.03–0.3 for Vatnajökull. The spectral properties of ice in the ablation zone are controlled by tephra layers in the ice, which are exposed as the glacier melts (Larsen et al., 1996). Additional tephra or dust deposition will therefore only have a small effect on the spectral properties of the ice, as the ice surface is already covered in dark bands. In addition, field observations suggest that the new particles are generally washed off from year to year. Applying one background map for the entire period should therefore provide the same results as applying a map created for each year. In addition, it allows us to run the model for years where no MODIS observations are available or where the amount of observations over the ice cap are sparse due to, for example, clouds.

3.1 AWS point models

The turbulent energy fluxes were calculated from AWS measurements using a one-level eddy flux model (Björnsson, 1972; Guðmundsson et al., 2009) which uses Monin–Obukhov similarity theory (Monin and Obukhov, 1954) and implements different roughness lengths for the vertical

profiles of wind, temperature, and water vapour (Andreas, 1987). The model is described in detail in Guðmundsson et al. (2009). Uncertainties of this model for example pertain to the aerodynamic roughness length for momentum z_0. The majority of z_0 values recorded over melting glacier surfaces vary over 2 orders of magnitude (between 1 and 10 mm), but over fresh snow or smooth ice surfaces the roughness length is generally around 0.1 mm (Brock et al., 2006). An order of magnitude increase in z_0 can more than double the estimated turbulent fluxes (Brock et al., 2000), so the chosen roughness length parametrization can greatly affect the performance of the model. Generally, a constant value of z_0 is prescribed for snow and/or ice surfaces (Brock et al., 2006), which is an oversimplification as the roughness may vary significantly over the ablation season (e.g. Grainger and Lister, 1966).

However, since measurements of the evolution of z_0 over the entire measurement period are not available, a constant roughness length of 1 mm was chosen in the calculation of the non-radiative fluxes. Sensitivity tests were conducted to estimate how large an error this choice of roughness length could lead to at the used AWS sites. A roughness length of 0.1 mm would decrease the calculated turbulent fluxes by 16–22 %, while using a roughness length of 10 mm would increase the calculated fluxes by 10–19 %, depending on the station. Since the contribution of the turbulent fluxes to the total energy balance is generally low, this translates into an increase or a decrease in the total energy balance at the stations by a maximum of 7 %.

The surface air pressure at the station is also needed to calculate the turbulent fluxes, but it is not measured at the AWS sites. Instead it is estimated at the relevant elevation h using synoptic observations from meteorological stations operated by the Icelandic Met Office and the following relationship:

$$P(h) = P(h_0)\left(1 - \frac{0.0065(h - h_0)}{T(h_0)}\right)^{5.25}, \qquad (5)$$

where $P(h_0)$ and $T(h_0)$ are the air pressure and air temperature, respectively, observed at an elevation h_0 (e.g Wallace et al., 2006). This method has previously been applied successfully at various locations on Vatnajökull and Langjökull (e.g. Guðmundsson et al., 2006, 2009).

3.2 Validation method

AWS data from the period 2001–2014 for three Brúarjökull stations and two Tungnaárjökull stations are considered, as well as SMB point measurements from 1995 to 2014. All stations were operated during the summer months, but since 2006 the lowest Brúarjökull station has been operated year round. Comparisons are made between daily averages from the HIRHAM5 model and the in situ observations collected at the AWSs. HIRHAM5 daily means are calculated from 6-hourly outputs, while the AWS daily means are calculated from observations at 10 min intervals.

Comparisons between station values and model values are made by bilinearly interpolating the model output to the measurement position using the four closest model grid points and using only glacier-surface type grid cells.

In order to remove the effect of seasonally varying magnitudes of the energy balance components, the percentage errors listed in Tables 2–4 are calculated as the root mean square error (RMSE) divided by the observations.

HIRHAM5 uses an elevation model over Iceland which has been interpolated onto the 5.5 km model grid. Since errors in the elevation of the glacier surface can introduce significant biases in temperature and pressure which are not caused by physical model errors (Box and Rinke, 2003), any elevation bias in the model has to be taken into account before validating the results. The elevation bias was calculated as the difference between the model elevation and GPS observations at each site (Table 1).

The temperature was corrected for the elevation bias in order to compare the model results to the AWS measurements at AWS locations. This was done using a constant lapse rate of $6.5\,\mathrm{K\,km^{-1}}$, which resulted in temperature corrections on the order of 0.1–0.3 K. Pressure is corrected using Eq. (5) decreasing the bias down to 0.1 to 0.5 hPa. Thus, although the HIRHAM5 elevation is consistently overestimated, the resulting differences are not large enough to introduce significant biases in temperature and surface pressure.

4   Results and discussion

4.1   Meteorological variables

As the sensible and latent heat fluxes are computed using the surface pressure p_{sl}, air temperature at 2 m, T_{2m}, relative humidity r_{2m}, and wind speed u, these model variables were evaluated at all five stations at the measurement height. How well these variables are simulated should indicate the model's ability to simulate the turbulent fluxes.

The comparison of modelled and observed mean daily values during the summer months from the period 2001–2014 is shown in Fig. 2 and Table 2. The surface pressure, p_{sl}, which was not observed at the stations but estimated using Eq. (5), is generally forecast with only a small error. At each station there is a high positive correlation ($r > 0.9$) between modelled and estimated pressure (Eq. 5), for the entire time series and for each individual year.

The model also captures the 2 m temperatures, T_{2m}, satisfactorily. The largest deviation from the observations is found at the B_{AB} station, which underestimates the temperature by 0.8 K on average. The temperature is also underestimated at the four other stations, but by at most 0.6 K. The model simulates the variation in temperature well; for example, it captures the temperature dampening over a melting glacier surface. This is expressed in the high correlation values for all five stations ($r \sim 0.9$).

The measured relative humidity, r_{2m}, at all five stations is generally high, with only 1–3 % of the data points at each station falling below 70 %, and the minimum daily value between 42 and 58 %. The model simulates a lower mean humidity than the measured at all five stations, with 8–20 % of the points at each stations having values lower than 70 % and minimum daily values between 18 and 30 %. Since the exchange coefficient for moisture is a function of the atmospheric temperature profile, the underestimation of the relative humidity could be due to a too low temperature gradient between the atmosphere and the surface. This is consistent with the underestimation found in the 2 m temperature. The correlation of between 0.68 and 0.7 indicates that the model simulates the humidity fluctuations satisfactorily.

The lowest wind speed level in HIRHAM5 is at 10 m and the AWS wind speeds are measured at between 2 and 4 m, depending on the year, the HIRHAM5 wind speed is extrapolated to the measurement height using a logarithmic profile with a roughness length of 1 mm. At all five locations, HIRHAM5 simulates winds that are too weak on average. This could be due to the uncertainty arising from the interpolation of the model winds from second-lowest level (30 m) to the lowest level (10 m) under stable conditions, as the wind speed can change significantly over the 20 m interval.

4.2   Longwave radiation

As shown above, HIRHAM5 underestimates the temperature at all five stations, with the largest underestimation at the B_{AB} station. As a result, a similar underestimation of incoming longwave radiation is obtained at all five stations, with the largest difference occurring at the B_{AB} station 3. The average percentage difference is approximately 8 % for all five locations (see Table 3), and falls well within the 10 % uncertainty of the AWS observations. However, Fig. 3a also shows that 25–30 % of the simulated days have errors larger than 10 %.

The incoming LW radiation is mainly emitted from clouds and atmospheric greenhouse gases, and therefore a source of the underestimation could be that the model underrates cloud formation and/or simulates clouds that are too optically thin in the LW region of the spectrum. An underestimation of the temperature in the atmosphere could also be causing the underestimation.

Figure 3b shows the comparison of the modelled and measured outgoing LW radiation. There is a small overestimation at the T_{AC} station, and a small underestimation of the other four stations, but in general the model reproduces the daily values well ($r \sim 0.76$). The average percentage deviation between the modelled and measured values is only around 3 %, combined with between 0.5 and 2 % of the HIRHAM5 data points having deviations larger than 10 %.

Due to an underestimation of the incoming LW radiation, and only small negative or positive biases in the outgoing

Table 2. Comparison of the surface pressure p_{sl}, air temperature at 2 m, T_{2m}, relative humidity r_{2m}, and wind speed u, from HIRHAM5 simulations and AWS measurements during the summer months (April–October) for the period 2001–2014. The HIRHAM5 bias (HIRHAM5-AWS), the root-mean-square error (RMSE), the percentage error, and the correlation (r) are shown.

Parameter	Station	AWS value	HIRHAM5 bias	RMSE	% error	r
p_{sl} (hPa)	B_{AB}	911.9	−0.2	2.8	0.3	0.96
	T_{AB}	884.2	−0.4	3.0	0.3	0.95
	B_{ELA}	872.1	−0.6	2.9	0.3	0.95
	B_{AC}	837.0	0.1	2.2	0.3	0.97
	T_{AC}	845.1	−0.9	2.7	0.3	0.96
T_{2m} (K)	B_{AB}	274.1	−0.8	1.5	0.6	0.94
	T_{AB}	274.0	−0.6	1.3	0.5	0.89
	B_{ELA}	272.9	−0.1	1.1	0.4	0.91
	B_{AC}	271.6	−0.1	1.4	0.5	0.90
	T_{AC}	272.1	0.0	1.2	0.5	0.91
r_{2m}	B_{AB}	87.9	−6.2	12.2	13.9	0.68
	T_{AB}	89.6	−6.1	11.5	12.9	0.76
	B_{ELA}	91.8	−3.8	9.8	10.7	0.73
	B_{AC}	93.9	−3.5	9.6	10.2	0.68
	T_{AC}	90.0	−2.6	9.7	10.7	0.72
u (m s^{-1})	B_{AB}	5.1	−1.2	2.0	39.0	0.80
	T_{AB}	5.3	−0.3	1.8	33.0	0.87
	B_{ELA}	4.4	−0.1	1.8	41.1	0.82
	B_{AC}	5.9	−0.7	1.8	30.8	0.86
	T_{AC}	5.2	−0.1	2.0	38.9	0.82

LW, the net LW (incoming–outgoing) radiation has a mean negative bias at all AWS locations ($-7.9\,\mathrm{W\,m^{-2}}$).

4.3 Shortwave radiation and albedo

Figure 4 and Table 3 show the comparisons of the modelled and measured components of the shortwave (SW) radiation as well as the surface albedo. On average, the incoming SW radiation is underestimated at all five stations. This underestimation is also present in the means at all five stations for most years, except in 2002, 2004, 2005, and 2014 at the B_{AB} station. This suggests that there are errors in either the modelling of the clouds, e.g. due to an overestimation of the cloud fraction, the amount of cloud formation, or the optical thickness of the clouds in the shortwave region, and/or because of errors in the clear-sky fluxes.

The albedo comparison is shown in Fig. 4b. The modelled albedo at the two AB stations has the largest deviation from the observations; this is partly due to the modelled snow cover, which either does not completely disappear or disappears later in the year than the AWS data show. At the B_{AB} station, the ice layer is generally exposed in the model (except in 2001 and 2011–2013), although the snow cover always persists longer than in reality. One exception occurs in 2001, where the modelled albedo never drops down to the ice value, whereas observations show albedo values as low as 0.03. This one year therefore highly contributes to the aver-

age overestimation of the albedo. This very low albedo value could be due to a layer of dust or tephra beneath the station, so it may not represent the ice albedo. However, very low ice albedo values down to 0.05 are not uncommon in the ablation zone of Vatnajökull (e.g. Gascoin et al., 2017). Comparisons with the mass balance measurements (discussed in Sect. 4.6.1) show that the winter balance is overestimated during approximately half of the measured years, which contributes to delay the albedo drop in the model.

At the T_{AB} station, a too-thick modelled snow cover in winter is also the cause of some of the discrepancy. Comparisons with mass balance measurements (Sect. 4.6.1) show that the winter balance is always overestimated at this station. An overestimation of the snow thickness at the beginning of summer, combined with an underestimation in the radiation and turbulent fluxes, leads to persistent snow cover at the end of summer. As a result, the ice surface is never exposed in the model during any of the modelled years, and the albedo never drops much below 0.4 (the minimum snow albedo), even though the AWS data shows that the ice surface was exposed during all but two years, i.e. 2008 and 2010. During these two years, the simulated albedo fits well with observations.

Another issue which affects both stations is that the MODIS albedo at these points is not as low as the measured albedo. The MODIS ice albedo at these stations is 0.10 (B_{AB}) and 0.16 (T_{AB}), whereas the observations show the albedo

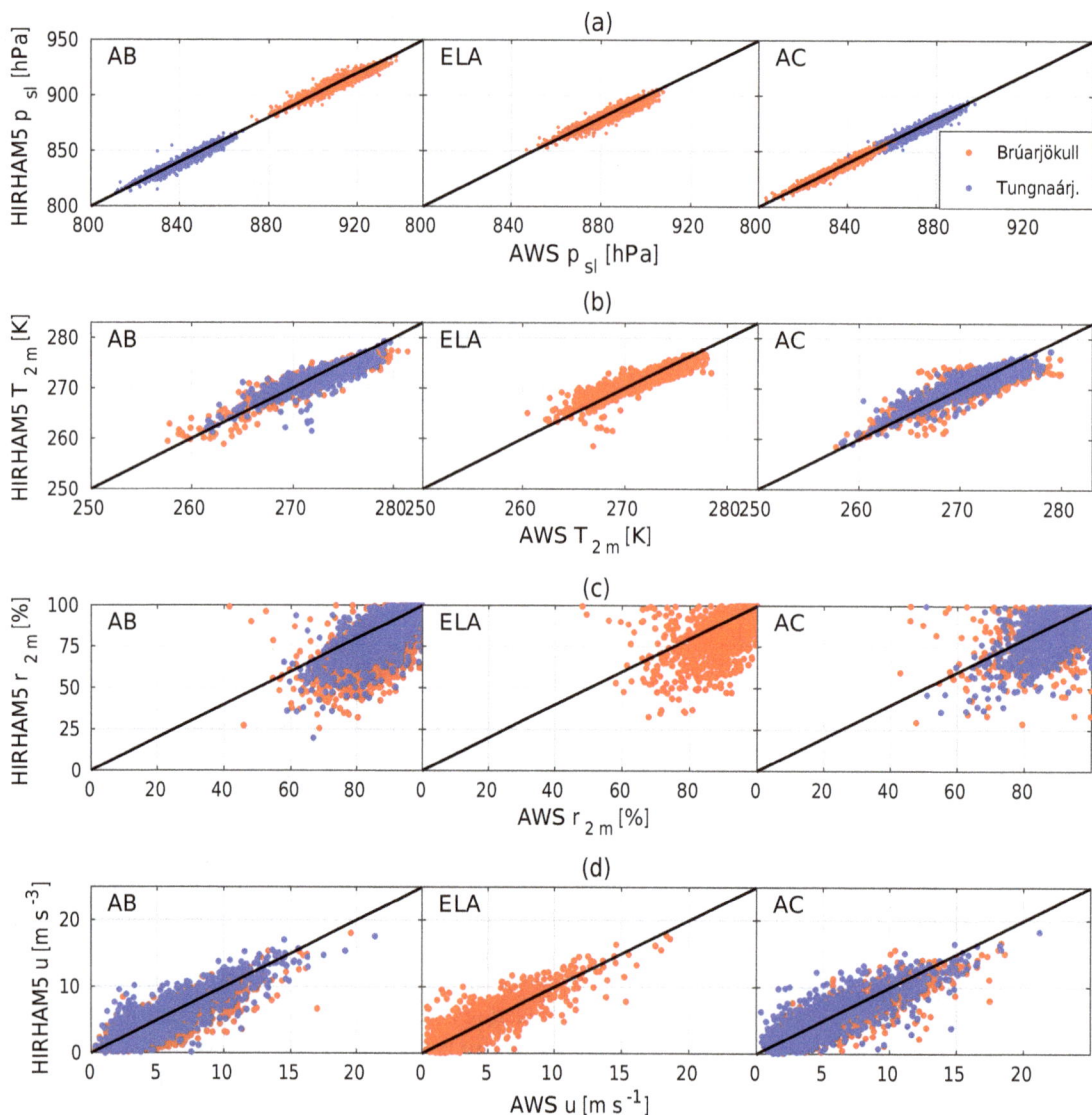

Figure 2. Scatter plots of the measured (**a**) surface pressure, (**b**), air temperature at 2 m, (**c**) relative humidity at 2 m, and (**d**) wind speed at 2 m, by stations on Bruarjökull (red) and Tungnaárjökull (blue) versus the same components simulated by HIRHAM5 at the same locations.

can drop as low as 0.01 at both stations. The albedo drops below the MODIS value every year at the B_{AB}, and during 2001–2005 and 2011 at the T_{AB} stations. This is presumably due to the heterogeneity of the albedo in the ablation zone, which means that a low in situ albedo value at a point cannot be captured at the current HIRHAM5 resolution.

At the ELA station, the mean albedo value is underestimated (Table 3). Close to the equilibrium line, the albedo is highly variable both temporally and spatially; for example, there is a large difference in albedo depending on whether the previous year's summer surface was exposed or not. In general, the model overestimates the albedo during years where the summer surface was exposed, and underestimates the albedo during years where it was not. In addition, the winter mass balance at this station is always underestimated

(Sect. 4.6.1), meaning that the thickness of snow layer in spring is underestimated and the effect of the underlying ice layer will therefore be overestimated, leading to the underestimation in albedo.

The smallest difference between modelled and observed albedo is found at the two AC stations. The B_{AC} station generally provides the best fit with the observations, while the model tends to underestimate the albedo at the T_{AC} station. An exception to this is found in 2010 and 2011, where the albedo was overestimated by the model at both stations due to ash deposition from the Eyjafjallajökull and Grímsvötn eruptions (e.g. Gudmundsson et al., 2012).

A general reason for the model overestimating the albedo is that it does not take the albedo changes due to dust storms or volcanic dust deposition into account. For instance, the

Table 3. Comparison of incoming and outgoing long- and shortwave radiation, albedo (α), turbulent fluxes (H_{s+1}), and total energy (E) from HIRHAM5 simulations and AWS measurements during summer months (April–October) from the period 2001–2014. The HIRHAM5 bias (HIRHAM5-AWS), the root-mean-square error (RMSE), the percentage error, and the correlation (r) are shown.

Parameter	Station	AWS value	HIRHAM5 bias	RMSE	% error	r
LW↓ (W m^{-2})	B_{AB}	290.6	−16.9	26.3	9.1	0.79
	T_{AB}	287.3	−7.0	20.9	7.3	0.80
	B_{ELA}	283.9	−9.0	21.7	7.7	0.79
	B_{AC}	280.9	−8.5	24.4	8.7	0.79
	T_{AC}	274.1	−3.8	20.4	7.4	0.83
LW↑ (W m^{-2})	B_{AB}	309.2	−1.9	7.3	2.4	0.87
	T_{AB}	311.9	−2.5	7.4	2.4	0.78
	B_{ELA}	309.9	−3.3	10.5	3.4	0.70
	B_{AC}	299.9	−1.5	12.9	4.3	0.76
	T_{AC}	301.4	2.6	11.6	3.9	0.68
SW↓ (W m^{-2})	B_{AB}	189.1	−4.0	55.5	29.3	0.81
	T_{AB}	220.8	−35.2	72.2	32.7	0.79
	B_{ELA}	229.3	−36.2	64.6	28.1	0.83
	B_{AC}	236.8	−43.7	69.9	29.5	0.82
	T_{AC}	247.2	−41.9	72.5	29.2	0.79
SW↑ (W m^{-2})	B_{AB}	86.6	18.1	61.0	70.4	0.64
	T_{AB}	112.5	−6.9	54.7	48.7	0.73
	B_{ELA}	146.1	−29.9	59.2	40.5	0.75
	B_{AC}	173.2.9	−31.3	56.4	32.6	0.79
	T_{AC}	173.5	−33.4	65.6	37.8	0.68
α (%)	B_{AB}	34.6	12.7	23.6	68.2	0.75
	T_{AB}	44.5	9.96	21.0	47.2	0.68
	B_{ELA}	60.7	−2.9	18.4	30.2	0.57
	B_{AC}	72.2	0.8	10.5	14.5	0.62
	T_{AC}	70.1	−2.2	16.1	22.9	0.47
H_{s+1} (W m^{-2})	B_{AB}	34.7	−5.0	28.6	116	0.71
	T_{AB}	36.2	−3.8	25.2	69.6	0.79
	B_{ELA}	24.5	−2.0	26.2	107	0.71
	B_{AC}	20.7	−12.3	28.2	136	0.31
	T_{AC}	20.8	−6.3	23.0	110	0.49
E (W m^{-2})	B_{AB}	131.6	−44.4	82.8	62.9	0.67
	T_{AB}	120.1	−36.7	98.0	72.3	0.58
	B_{ELA}	84.4	−13.4	49.6	58.8	0.68
	B_{AC}	64.8	−28.6	50.3	77.5	0.53
	T_{AC}	67.7	−21.2	78.6	89.7	0.43

very low albedo values obtained at the T_{AC} station (Fig. 4b) are due to tephra deposition on the glacier during the 2010 eruption of Eyjafjallajökull (e.g. Gudmundsson et al., 2012; Gascoin et al., 2017). Even though dust events do not cause as large changes in albedo as a volcanic eruption, they can still significantly lower the albedo (e.g. Painter et al., 2007; Wittmann et al., 2017) . As previously mentioned, the albedo in HIRHAM5 often reaches its yearly minimum value later in the summer than the observed. Such discrepancy could be explained by dust events, advancing or delaying the drop in surface albedo. Wittmann et al. (2017) investigated 10 dust

events which occurred at the B_{ELA} station in 2012, and found a lowering in the albedo during all events and showed that the dust storms have a significant effect on the resulting energy balance.

The error in the outgoing shortwave radiation is caused by errors in the albedo and the incoming SW. At the B_{AB} station, the incoming radiation is slightly underestimated but the albedo is overestimated; hence, the outgoing SW is overestimated. The values at the four other stations are all underestimated, due to larger underestimations of the incoming SW radiation and lower albedo errors.

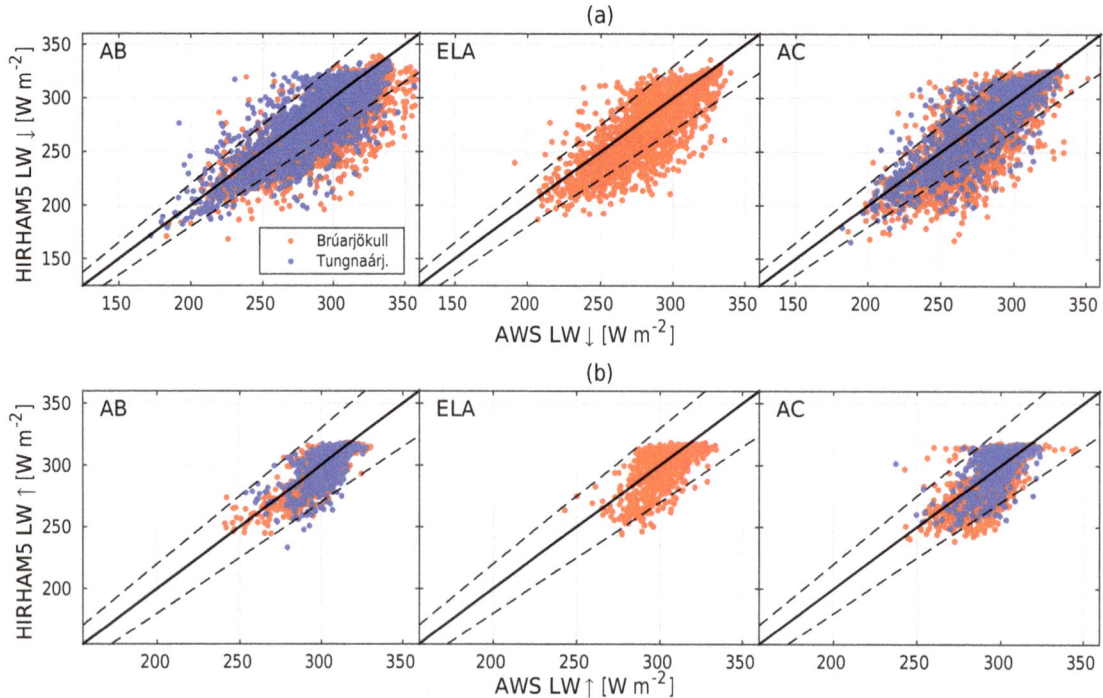

Figure 3. Scatter plots of the measured longwave radiation components, LW↓ and LW↑, by stations on Brúarjökull (red) and Tungnaárjökull (blue) versus the LW radiation components simulated by HIRHAM5 at the same locations. The dashed line corresponds to ±10 %, i.e. the manufacturer-reported uncertainty of the AWS measurements.

As both the incoming and outgoing SW radiation are underestimated at most stations, the net SW shows a negative bias of ~ -6 to $-12\,\mathrm{W\,m^{-2}}$ at the AC and ELA stations, and of -22 and $-28\,\mathrm{W\,m^{-2}}$ at the two AB stations. The resulting average model error at all five stations is $-15.5\,\mathrm{W\,m^{-2}}$.

4.4 Turbulent fluxes

As HIRHAM5 underestimates meteorological variables at all stations, similar underestimation is obtained for the turbulent fluxes (Table 3 and Fig. 5). The two AC stations have the largest differences and also the lowest correlation (0.45 and 0.49) between the AWS estimate and the HIRHAM simulation. The other three stations also have significantly lower values in the HIRHAM5 model than in the AWS model, but with higher correlation coefficients (0.69–0.73).

It is important to bear in mind that this comparison is a model–model comparison, so while the eddy flux model may give a good estimate of the turbulent fluxes, model errors still affect the results, e.g. due to the use of a constant roughness length.

4.5 Total energy balance

After the simulated components of the energy balance were evaluated against AWS observations, the total energy balance was estimated (see Table 3). The energy balance (E) is found

using

$$E = \mathrm{LW_{net}} + \mathrm{SW_{net}} + H_{s+l}, \qquad (6)$$

where $\mathrm{LW_{net}}$ is the net LW radiation, $\mathrm{SW_{net}}$ is the net SW radiation, and H_{s+l} are the turbulent fluxes. Overall, the melt energy is underestimated, owing to all elements of the energy balance generally being underestimated. This is in large part due to the underestimation of the modelled incoming radiation. We attribute this to an error in the modelling of the clouds, but since both the incoming SW and LW radiation are underestimated, inaccurate cloud representation cannot be the the only source of the error. Errors in the interaction of clouds and radiation, e.g. error in the optical thickness of the clouds, or in the clear-sky fluxes, could partly explain these discrepancies. The underestimation of the incoming LW radiation could also be due to errors in the vertical atmospheric temperature gradient.

Since the simulated outgoing LW radiation generally only has a small negative bias, the deviation in net LW radiation is governed by the incoming radiation. Errors in the simulated albedo mean that both the in- and outgoing SW radiation greatly contribute to the deviation in net SW radiation. These errors can be partly attributed to ash and dust deposition during volcanic eruptions and dust storms, which are not taken into account in HIRHAM5. In addition, errors in the simulated albedo also stem from snow cover that disappears too slowly compared to AWS records in the ablation

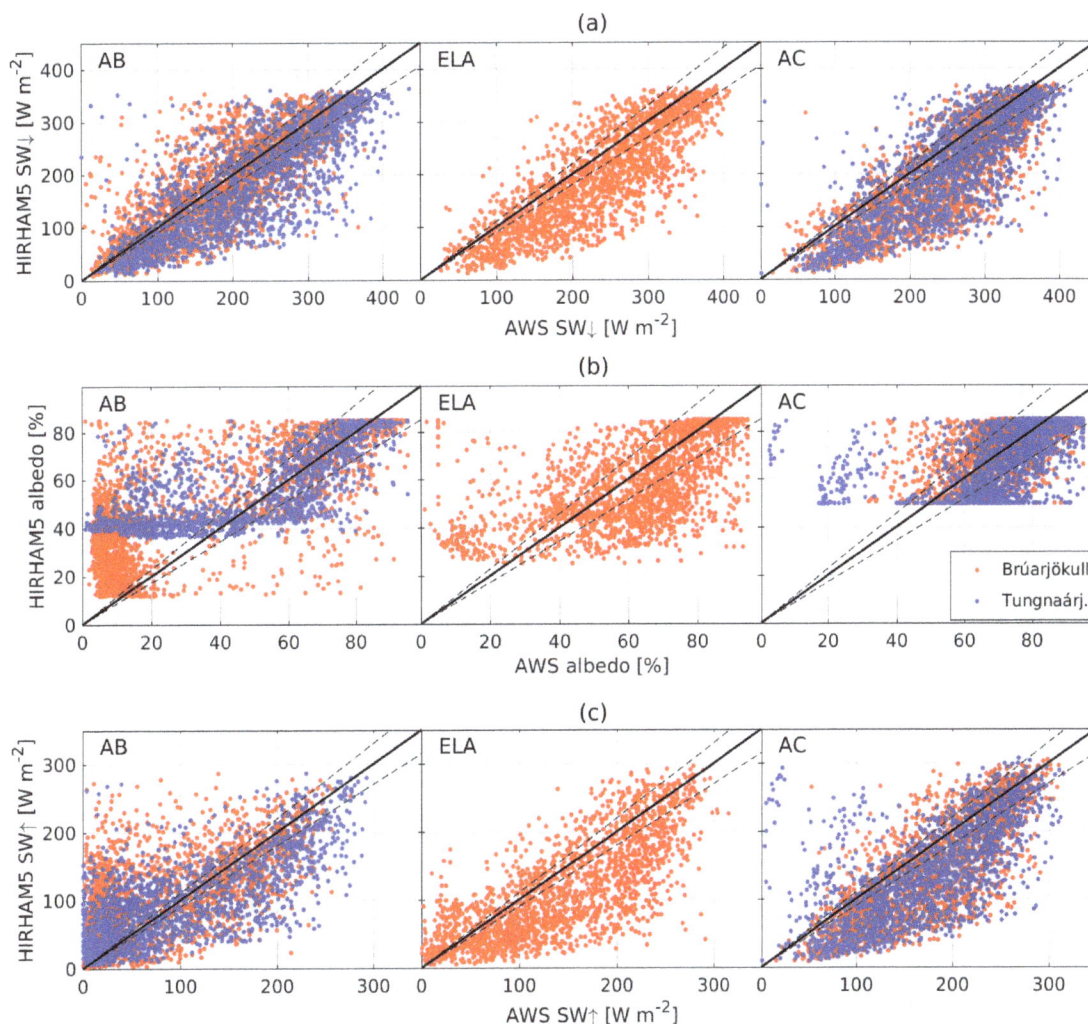

Figure 4. Scatter plots of the measured shortwave radiation components, **(a)** SW↓, **(b)** albedo, and **(c)** SW↑, by stations on Bruarjökull (red) and Tungnaárjökull (blue) versus the shortwave radiation components simulated by HIRHAM5 at the same locations. The dashed line corresponds to the uncertainty of the measured AWS components.

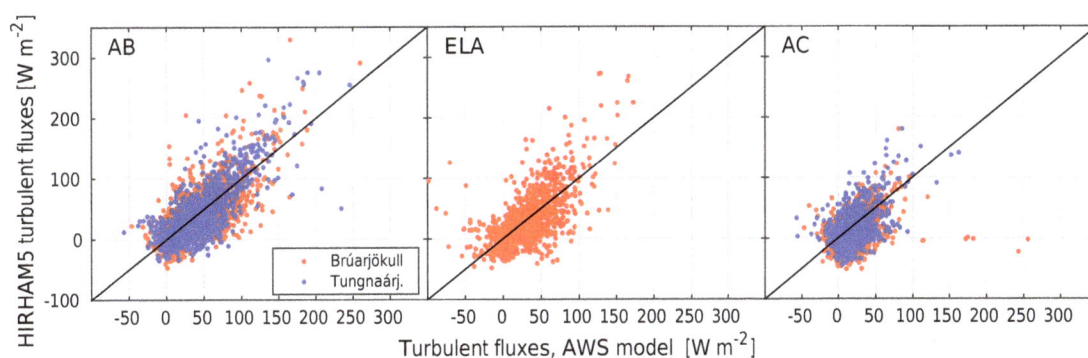

Figure 5. The total turbulent fluxes calculated from AWS stations using the one-level flux model versus the HIRHAM5 simulated values.

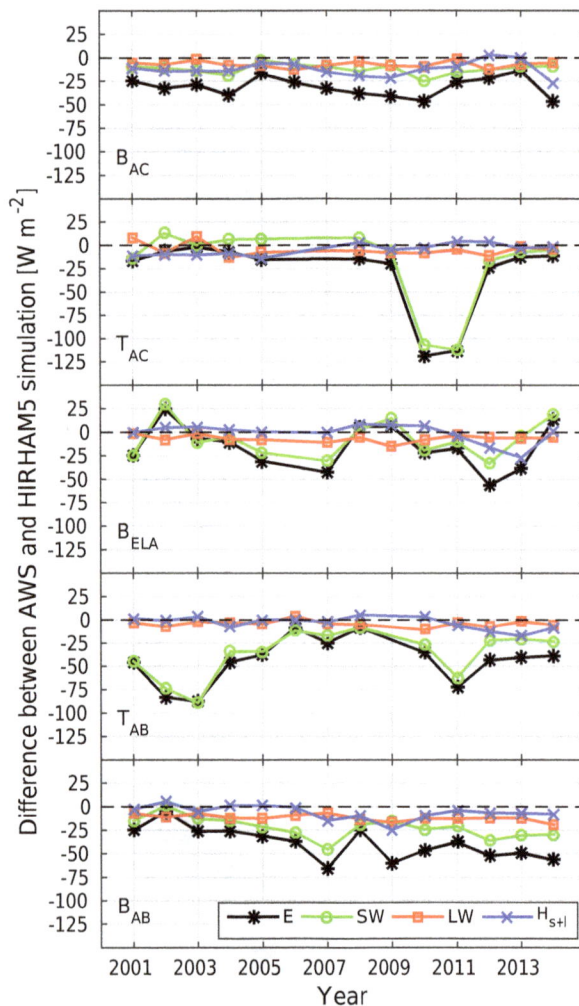

Figure 6. The average summer (April–October) bias of each energy balance component for the measurement period at each AWS site. The large deviation in the SW radiation at the Tunaárjökull sites in 2010–2011 is due to deposition of ash on the glacier during the 2010 Eyjafjallajökull and 2011 Grímsvötn eruptions.

zone. As a result, modelled albedo drops too slowly compared to the measured albedo. The underestimation of the net SW and LW radiation and the turbulent fluxes leads to underestimated melt energy. This contributes to the overestimation of the modelled snow thickness.

In order to estimate how much the different components contribute to the energy difference on a year-to-year basis, the mean difference between modelled and observed energy components during each summer (April–October) is shown for each station (Fig. 6).

At the B_{AC} station, the contribution of the long- and shortwave radiation and turbulent fluxes to the energy difference is consistent for the entire period, with the error of each component being almost equal, varying between -25 and $0\,W\,m^{-2}$. At the T_{AC} station, the error due to the three components is also of the same order of magnitude, except in 2010 and

2011 where the error in the net SW radiation is much larger than that in the other components. This is due to a large drop in the albedo as a result of the Eyjafjallajökull (2010) and Grímsvötn (2011) eruptions. The mean difference between observations and the simulations of the SW radiation for non-eruption years is $-3\,W\,m^{-2}$, whereas the radiation difference in 2010 is $-106\,W\,m^{-2}$. Assuming the larger deviation from the mean in 2010 is only due to the volcanic eruption, the increase in available energy due to the eruption is $103\,W\,m^{-2}$. If it is further assumed that the surface was always at melting point, the increase in melt due to the 2010 Eyjafjallajökull eruption over the 128-day measuring period would be $\sim 3.1\,m$ w.e. at this station.

At the ELA site, the contribution from the modelled turbulent fluxes to the energy balance deviation generally varies between $\pm 10\,W\,m^{-2}$, except in 2013 where the bias is around $-25\,W\,m^{-2}$. Modelled longwave radiation is consistently underestimated by $-10\,W\,m^{-2}$. The deviation in the shortwave radiation is more variable, as expected from the results of the albedo comparison. Depending on whether bare ice was exposed or not, the albedo is generally either over- or underestimated. For example, at B_{ELA}, the ice surface was reached in, for example, 2007 and 2012, resulting in an overestimation of the albedo. In, for example, 2002 and 2009, however, the albedo was high the entire summer as no ice was exposed, resulting in an underestimation of the predicted albedo.

At the T_{AB} station, both the net LW radiation and the turbulent fluxes agree well with observations for the entire period. The net SW radiation, however, is always underestimated, especially in the period 2001–2003 and 2011. These years, the measured albedo at the station goes below 0.1, while the HIRHAM5 albedo stays around 0.4. As previously discussed, this albedo bias, and hence underestimated net SW radiation, occurs because of an overestimation of the snow cover at the station due to an overestimation of the winter accumulation and possibly also the proximity of the equilibrium line. An underestimation of the incoming SW radiation, which we attribute to an error in cloud cover amount of clear-sky fluxes, also contributes to this error.

At the B_{AB} station, the longwave radiation bias is relatively constant with values close to $0\,W\,m^{-2}$ for much of the measurement period. The absolute deviation due to the turbulent fluxes is less than $10\,W\,m^{-2}$ for most of the period, although with slightly larger deviations from the period 2007–2010. The SW radiation is always underestimated at this station, mostly due to the previously discussed overestimation of the albedo.

4.6   Surface mass balance

4.6.1   At AWS sites

Scatter plots of measured and HIRHAM5 simulated SMB are shown in Fig. 7 and the average deviations are shown in Table 4.

The winter mass balance comparison allows to evaluate of the winter precipitation in HIRHAM5. The simulated mass balance at the B_{ELA} and B_{AC} are always underestimated, while the T_{AC} stations is underestimated during all years but one (2012). The simulated value at the T_{AB} station is overestimated over the whole period. The modelled mass balance at the B_{AB} station has an almost equal amount of years which are over- and underestimated. Apparently the model either carries too much precipitation when the clouds reach the glacier, resulting in too much precipitation at the ice sheet margin, or more melting occurs at the ablation area stations during the winter months than the model estimates.

The summer SMB results are in good agreement with the results of the energy balance calculations. The summer SMB is generally overestimated, although it is underestimated occasionally at all stations except T_{AB}. The ELA station has the largest amount of underestimated points, which is consistent with the findings from the energy balance calculations. Besides the errors introduced due to the underestimation of the energy balance, possible over- or underestimations of the modelled summer accumulation contribute to these errors as well.

Due to the difference in the summer and winter balance, the net balance at the B_{AC}, T_{AC}, and B_{ELA} stations is generally underestimated in HIRHAM5, while the balance at the two AB stations is generally overestimated. This is due to a general overestimation of the winter balance in the ablation area, due to either an underestimation of the winter melt or an overestimation of precipitation, as discussed above.

4.6.2   At all measurement sites

SMB is also measured at 25–120 non-AWS sites, depending on the year (Fig. 1). In order to estimate how well the model represents the SMB at non-AWS sites, the data from all the sites between 1995 and 2014 were compared with the HIRHAM5 simulation (Fig. 8; Table 4).

The winter balance at all measured points is slightly overestimated by HIRHAM5 on average. However, this is mostly due to a large difference between measured and simulated SMB at the ice-covered, high-elevation, central volcano Öræfajökull (the white dots in Fig. 8). Only one site has been measured on this glacier for a few years only (Guðmundsson, 2000), in a spot that always receives a large amount of precipitation. However, since HIRHAM5 consistently overestimates the accumulation by 100–200 %, this one point has a large effect on the mean error. This is a well-known issue with hydrostatic models like HIRHAM5, as they char-

acteristically overestimate the precipitation on the upslope and peaks in complex terrain. The reason for this is that the precipitation is calculated as a diagnostic variable – i.e. it is not governed by an equation that is a derivative of time, meaning that when the required conditions for precipitation are met in the local atmosphere, the precipitation appears instantaneously on the surface. Thus, the scheme does not allow horizontal advection of snow and rain by atmospheric winds, which is a key process in complex terrain, as it can force the precipitation downslope (e.g. Forbes et al., 2011). Without this effect, precipitation is generally overestimated at high peaks like Öræfajökull. Removing this location from the comparison, the total difference drops to one-third the difference with respect to the AWS sites only (-0.09 m w.e.). The reason the difference is smaller than for the AWS sites only is that more sites close to the edge of the ice cap are included. The winter balance at the measurement points at the outer parts of the icecap generally is overestimated in the model, and therefore these points partly offset the underestimation in the middle of the ice cap.

On average, the summer ablation is underestimated, which is consistent with the findings from the AWS stations that there is an average underestimation of the energy available for melt. The mean error and RMSE is only slightly larger than at the AWS sites.

The mean net balance is overestimated by approximately the same amount as the summer balance, partly due to the low mean deviation in the winter SMB. Due to the large deviation at Öræfajökull in the winter SMB, the Öræfajökull points clearly have the largest bias. If these points are excluded, a RMSE closer to that for the AWS locations is found (1.1 m).

4.7   Reconstructing the SMB of Vatnajökull

Spatial maps of the (uncorrected) average winter, summer, and net SMB from the 1980–1981 glaciological year until 2013–2014 are shown in Fig. 9. The approximate location of the average ELA is marked in the figure. The model captures the position of the ELA fairly well, but at, for example, Brúarjökull, where the average ELA is at 1200 m, the position of the average ELA is at a too high elevation. The average deviation between observation and model over the observation period at each measurement location is also shown in Fig. 9, in order to give an indication of the average error of the model at different parts of the ice cap. The winter balance (Fig. 9e) is generally overestimated at low elevations and underestimated at high elevations, except for at Öræfajökull, where there is a large overestimation of the winter balance, as discussed in the previous section. As can be seen in Fig. 9e, there is generally a low SMB bias at high elevations and a high SMB bias at low elevations during the summer. This is consistent with the comparisons with AWS stations, as we found that the bias in the energy available for melt was smaller at high elevation than at low elevation (see Table 3).

Table 4. Comparison of HIRHAM5 and mass balance measurements, both at AWS sites and for all measuring sites on Vatnajökull.

	Season	AWS value	HIRHAM5 bias	RMSE	% error
AWS locations	Winter	1.37	−0.26	0.71	51.6
	Summer	−2.34	0.48	0.81	−34.6
	Total	−0.98	0.23	1.15	−118
All locations	Winter	1.46	0.04	1.21	82.9
	Summer	−2.28	0.52	0.94	−41.1
	Total	−0.83	0.56	1.56	−186

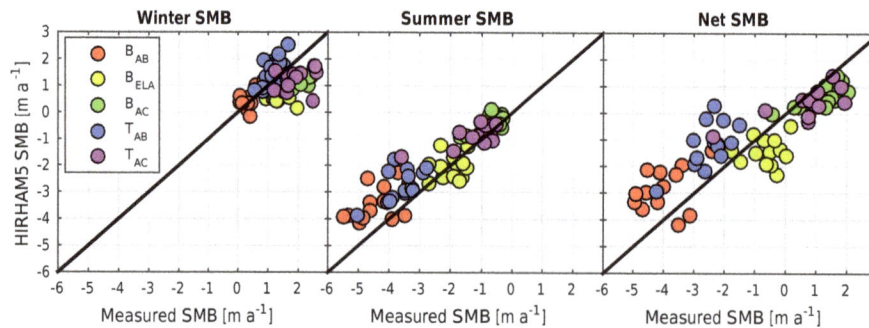

Figure 7. Comparison of the winter, summer, and net mass balance from the period 1995–2014 between the mass balance measurements at the five AWS sites and the HIRHAM5 simulation.

This was partly due to a larger albedo bias for stations in the ablation zone than for stations in the accumulation zone.

In addition to the spatial maps, the winter, summer, and net mass balances of Vatnajökull were calculated for the entire simulation period, and the results were compared with an estimate of the specific balance from 1995 to 2014, created by interpolation of the mass balance measurements (e.g. Pálsson et al., 2015); see Fig. 10. The model prediction of the mean specific summer mass balance generally fits well with the interpolated observations, with an overall difference of only 0.06 m w.e. The largest deviations are obtained in 1995, where ablation is overestimated in the simulation, and in 1997, 2005, and 2010–2012, where ablation is underestimated, most likely due to ash depositions on the glacier following the 1996 Gjálp eruption, the 2004 and 2011 Grímsvötn eruptions or the 2010 Eyjafjallajökull eruption, which are not taken into account in the model.

Excluding the years where the albedo was affected by volcanic eruptions, the average difference becomes smaller but the model also predicts slightly too much ablation, as the difference becomes −0.02 m w.e.

There is a shift in the summer and annual mass balance calculated by the model and the in situ MB measurements around 1996, with a generally more negative mass balance after 1996 than before. This is consistent with the increase of the annual mean temperature of Iceland in the mid-1990s, which resulted in a mean annual temperature ~ 1 K higher in the decade after than the decade prior to 1995. This is likely linked with atmospheric and ocean circulation changes

around Iceland, as there was a rapid increase in ocean temperatures off the southern coast in 1996 (Björnsson et al., 2013).

The specific winter mass balance is overestimated in HIRHAM5 for the entire measurement period with an average of 0.54 m w.e. Due to this difference, and only the small negative mean difference in summer mass balance, the annual mass balance of Vatnajökull is overestimated every year with an average difference of 0.50 m w.e.

However, this is mostly due to the large overestimation of the winter accumulation on Öræfajökull; comparison with the mass balance measurements showed that the model overestimated the winter accumulation by 100–200 % compared with the observations. In an attempt to estimate how much this error affects the results, a simple correction was added to the Öræfajökull points by reducing the simulated winter SMB by 50 %. The correction was added to four model grid points around Öræfajökull, due to the high (> 10 m yr^{-1}) annual specific mass balance in these points (see Fig. 9a). The resulting modelled winter and annual specific balance are shown in Fig. 11. The winter balance is still overestimated, but the difference between modelled and interpolated values has been reduced to only 0.1 m w.e. In addition, the average difference between the HIRHAM5 and interpolated annual SMB drops to only 0.08 m w.e.

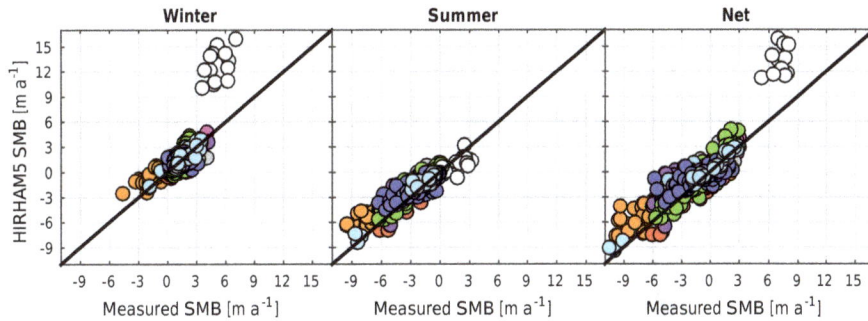

Figure 8. Comparison of SMB measurements from Vatnajökull ice cap from 1995 to 2014 and HIRHAM5 simulated values. Different colours represent different outlet glaciers; see Fig. 1b. The white dots are from a point on Öræfajökull.

4.8 Comparison with constant ice albedo simulation

In order to quantify the changes in the model performance resulting from the new albedo scheme used in this study, which utilizes an albedo map based on MODIS data (Gascoin et al., 2017), the results are compared to those of a previous run using a constant ice albedo of 0.3. The average difference in albedo and mass balance over the period 2001–2014 in each grid point are shown in Fig. 12, as well as the position of the AWS stations.

There is little to no difference between the two runs in the accumulation zone, due to the year-round snow cover. In the ablation zone, however, using the MODIS ice albedo map has a large effect on the simulated albedo. The largest differences are found on the southern outlet glacier Skeiðarárjökull, which is unfortunately a glacier where no mass balance or AWS measurements have been conducted. The B_{AB} and B_{ELA} stations are located in areas that are affected by the ice albedo, either because ice is exposed (B_{AB}) or because the underlying surface contributes to the albedo (B_{ELA}). The T_{AB} station is located in the ablation area, but the ice surface is never exposed in the model due to an overestimation of the winter accumulation. The albedo estimate at this station was therefore not improved by using the MODIS albedo.

When the model is run with the constant ice albedo of 0.3, the amount of ablation will be lower and thus the specific summer balance will be higher. Compared to the simulation using the MODIS map (Fig. 11), the constant ice albedo simulation results in an increase in the specific summer SMB by an average of 0.37 m w.e., or 18 %, per year for the period 1995–2014. The increase in the summer SMB ranges from 14 cm (in 2014) to 85 cm (in 2001) and the percentage increase varies between 8 % (in 2011) and 39 % (in 1995). As the winter balance is not dependant on the ice albedo, there are no changes in the specific winter SMB between the two simulations.

5 Conclusions

The comparison of a HIRHAM5 simulation with data from five AWSs on Vatnajökull ice cap allows us to evaluate the model performance. By comparing observations from April to October with model output, it was found that the model simulates the surface energy balance components and surface mass balance well, albeit with general underestimations of the energy balance components. Even though the energy balance was generally underestimated, the model simulated the near-surface temperature well. The reason for this is that the comparisons only use observations from the summer months, where the glacier surface is generally at the melting point, and thus the energy is used for melting and not for raising the temperature of the surface.

The modelled incoming radiation is underestimated on average in both the shortwave and longwave spectrum, which we suggest is due to biases in the modelling of the cloud cover combined with errors in the optical thickness in the short- or longwave spectrum, or errors in the clear-sky fluxes.

Whereas the modelled outgoing LW radiation component is within the uncertainty of the LW observations at the five stations, which is consistent with the ability of the model to capture surface temperatures, there was a larger difference between the modelled and measured outgoing SW radiation. This is partly due to the underestimation of the incoming SW radiation and partly due to inaccuracies in the simulated albedo. The albedo was simulated using an iterative, temperature-based albedo scheme (Nielsen-Englyst, 2015) with a bare ice albedo determined from MODIS data (Gascoin et al., 2017). The simulated albedo was generally overestimated during the summer and did not reach the lowest yearly value as early in the year as the measured albedo, particularly in the ablation zone. This was attributed to an overestimation of the snow cover in the ablation zone, an overestimation in the MODIS ice albedo compared with AWS observations, and the fact that the model does not account for the effect of volcanic dust deposition during eruptions and dust events on the albedo. A possible means of capturing dust storms or eruptions into the model is to implement a stochas-

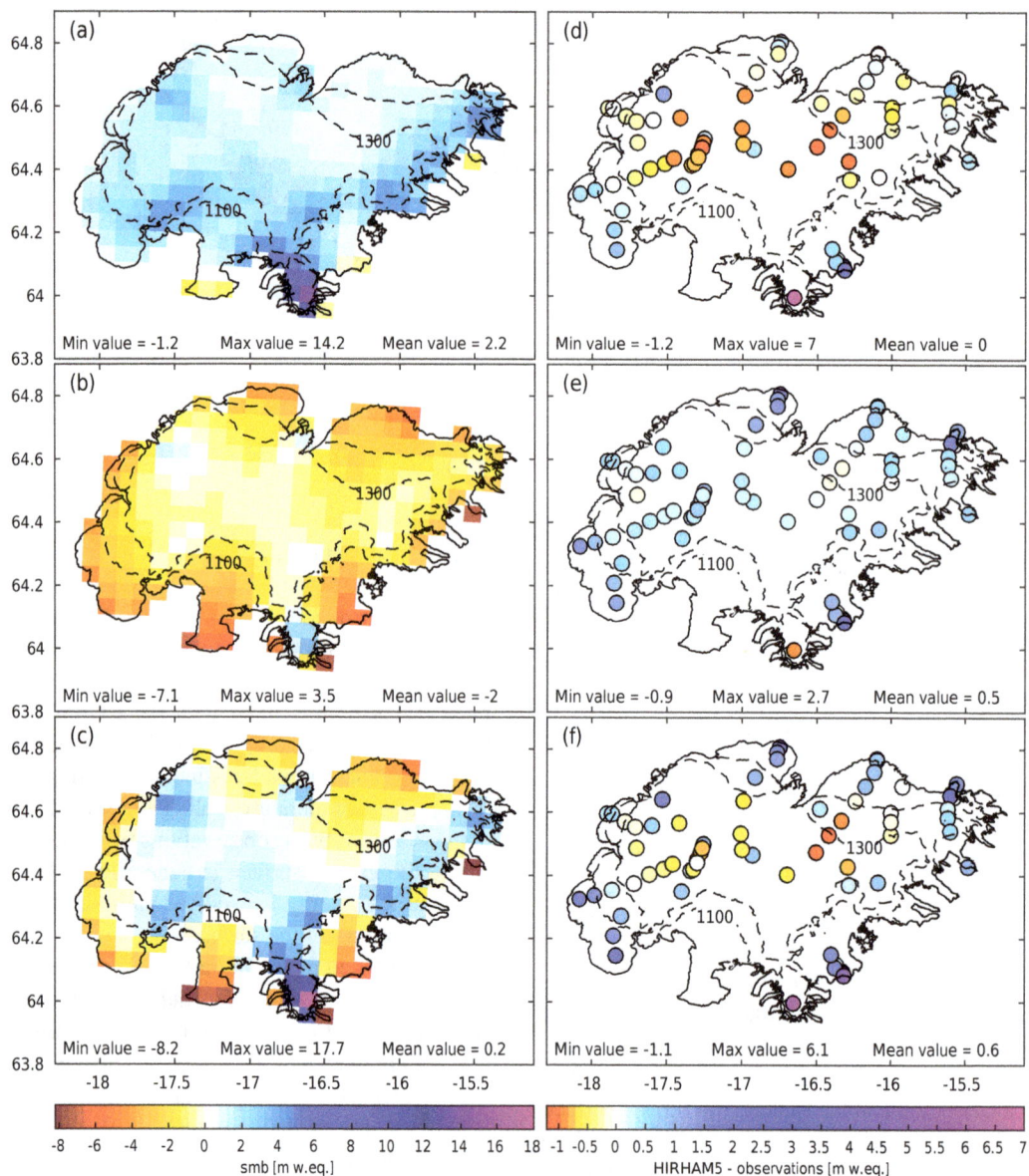

Figure 9. The average **(a)** winter, **(b)** summer, and **(c)** net SMB simulated by HIRHAM5 from the 1980–1981 glaciological year to 2013–2014. The contour lines marks the approximate location of the ELA, which generally lies between approximately 1100 and 1300 m elevation. Panels **(d)**–**(f)** show the average deviation between model and observations over the observation period (1992–2014) for each measurement location for the **(d)** winter, **(e)** summer, and **(f)** whole glaciological year.

tic ashes or dust generator, which distributes dust onto the glacier. Including simulations of dust depositions and concentrations from a dust mobilization model could also be an option, as Wittmann et al. (2017), for example, used the model FLEXDUST to simulate dust events on Vatnajökull in 2012, and found that the modelled dust events correspond well with albedo drops at two AWSs on Brúarjökull.

Due to the general underestimation of the energy balance components, the ablation during the summer months is underestimated on average. Comparison with mass balance measurements from the AWS sites and from sites scattered

across Vatnajökull shows an overall overestimation of the summer balance by about 0.5 m w.e. The overestimation is largest in the ablation zone. The winter balance is on average underestimated at the survey sites, albeit with the highest measuring site (on Öræfajökull) having a large overestimation of the winter balance.

The mean specific summer, winter, and net mass balances are reconstructed for all of Vatnajökull from the period 1981–2014, and estimates of the specific SMB based on in situ SMB measurements are compared to the reconstructed specific SMB for the period 1995–2014. The summer balance is

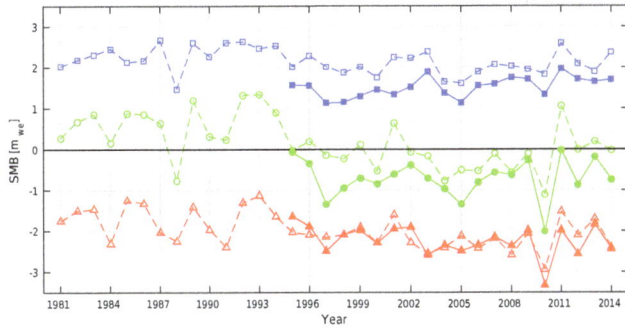

Figure 10. Average summer (red lines), winter (blue lines), and net (green lines) specific surface mass balance for the whole of Vatnajökull. The solid lines are the mass balance of Vatnajökull based on mass balance measurements and manual interpolation, while the dashed lines are the mass balance as simulated by HIRHAM5.

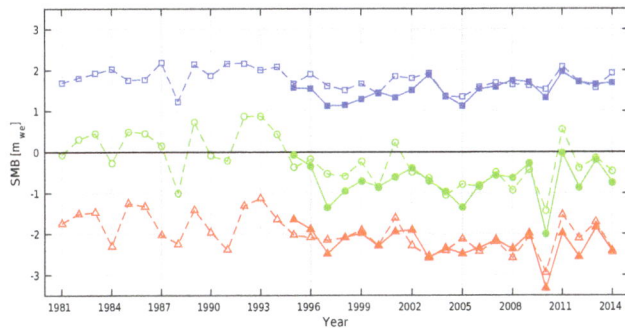

Figure 11. Same as Fig. 10, but corrected at the Öræfajökull area by reducing the HIRHAM5 simulated winter balance with 50 %.

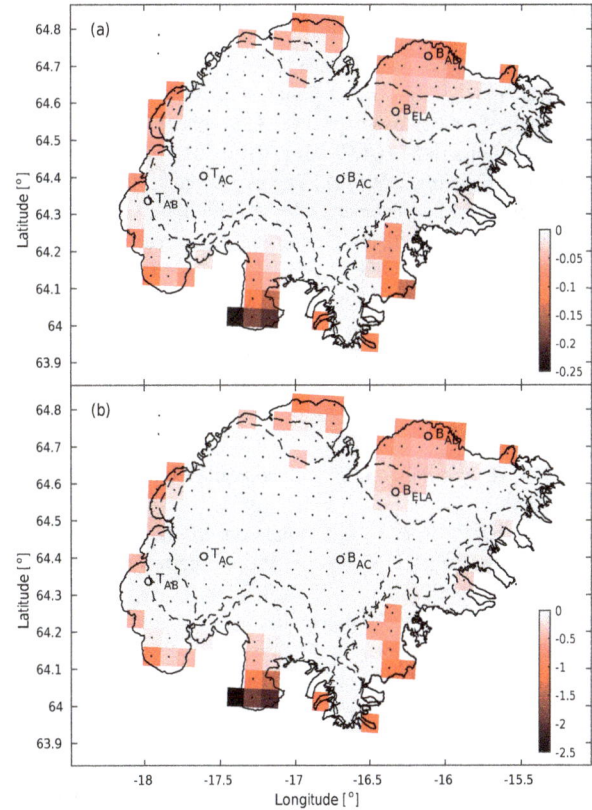

Figure 12. Difference in (a) mean albedo and (b) mean SMB in m w.e. for the period 2001–2014 between two runs with HIRHAM5, one using a MODIS bare ice albedo map and the other with a constant ice albedo. The locations of the AWS stations used in this study are shown with black circles.

overestimated by 0.06 m w.e. on average – i.e. there is generally too little ablation in the summer, with too much ablation in 1995, and too little ablation in years with, or following, volcanic eruptions. The winter balance is overestimated by 0.5 m w.e., mostly due to a large overestimation at the high elevation glacier Öræfajökull. This overestimation of accumulation at high elevation is characteristic for hydrostatic RCMs (Forbes et al., 2011). If the overestimation at these points is corrected, we estimate that the simulated winter balance would fit well with the observations, as the overestimation of the balance would drop to around 0.1 m w.e.

That the model catches the changes in the specific mass balance well over the mass balance measurement period, and also captures the shift in mass balance in the mid-1990s, gives us confidence that the model estimates the specific mass balance of Vatnajökull well over the entire simulated period from 1980 to 2014. HIRHAM5 is therefore a useful tool to expand the time series of the specific SMB beyond the measurement years. However, as ERA-Interim reanalysis data only go back to 1979, the model would need to be forced at the lateral, for example, by output of a general circulation model. However, using other reanalysis data probably leads to different errors; this needs further investigation. The model

could also be a useful tool to estimate the future evolution of the SMB of the ice cap, but this would also require a different forcing at the lateral boundary like general circulation model output. This would most likely introduce larger biases than the ones found using ERA-Interim, and the magnitude of these biases would need to be estimated and corrected before using the model for future projections.

Competing interests. The authors declare that they have no conflict of interest.

Acknowledgements. This work is supported by project SAMAR, funded by the Icelandic Research Fund (RANNIS, Grant no. 140920-051), as well as the National Power Company of Iceland (Landsvirkjun). Measurements from automatic weather stations and in situ mass balance surveys are from joint projects of the National Power Company and the Glaciology group at the Institute of Earth Sciences, University of Iceland.

Edited by: Xavier Fettweis

References

Agosta, C., Fettweis, X., and Datta, R.: Evaluation of the CMIP5 models in the aim of regional modelling of the Antarctic surface mass balance, The Cryosphere, 9, 2311–2321, https://doi.org/10.5194/tc-9-2311-2015, 2015.

Ágústsson, H., Hannesdóttir, H., Thorsteinsson, T., Pálsson, F., and Oddsson, B.: Mass balance of Mýrdalsjökull ice cap accumulation area and comparison of observed winter balance with simulated precipitation, JÖKULL, 63, 91–104, 2013.

Andreas, E. L.: A theory for the scalar roughness and the scalar transfer coefficients over snow and sea ice, Bound.-Lay. Meteorol., 38, 159–184, https://doi.org/10.1007/BF00121562, 1987.

Björnsson, H.: Bægisárjökull, North Iceland. result of glaciological investigations 1967-1968. Part II. The energy balance, Jökull, 22, 44–61, 1972.

Björnsson, H. and Pálsson, F.: Icelandic glaciers, Jökull, 58, 365–386, 2008.

Björnsson, H., Pálsson, F., Guðmundsson, M. T., and Haraldsson, H. H.: Mass balance of western and northern Vatnajökull, Iceland, 1991–1995, Jökull, 45, 35–58, 1998.

Björnsson, H., Pálsson, F., Guðmundsson, S., Magnússon, E., Aðalgeirsdóttir, G., Jóhannesson, T., Berthier, E., Sigurdsson, O., and Thorsteinsson, T.: Contribution of Icelandic ice caps to sea level rise: Trends and variability since the Little Ice Age, Geophys. Res. Lett., 40, 1546–1550, https://doi.org/10.1002/grl.50278, 2013.

Box, J. and Rinke, A.: Evaluation of Greenland Ice Sheet Surface Climate in the HIRHAM Regional Climate Model Using Automatic Weather Station Data., J. Climate, 16, 1302–1319, 2003.

Brock, B. W., Willis, I. C., Sharp, M. J., and Arnold, N. S.: Modelling seasonal and spatial variations in the surface energy balance of Haut Glacier d'Arolla, Switzerland, Ann. Glaciol., 31, 53–62, 2000.

Brock, B. W., Willis, I. C., and Sharp, M.: Measurement and parametrisation of surface roughness variations at Haut Glacier d'Arolla, J. Glaciol., 52, 281–297, 2006.

Christensen, O. B., Drews, M., Christensen, J. H., Dethloff, K., Ketelsen, K., Hebestadt, I., and Rinke, A.: The HIRHAM Regional Climate model Version 5., Tech. rep., Danish Meteorological institute, 2006.

Conway, H., Gades, A., and Raymond, C. F.: Albedo of dirty snow during conditions of melt, Water Resour. Res., 32, 1713–1718, https://doi.org/10.1029/96WR00712, 1996.

Dee, D. P., Uppala, S. M., Simmons, A. J., Berrisford, P., Poli, P., Kobayashi, S., Andrae, U., Balmaseda, M. A., Balsamo, G., Bauer, P., Bechtold, P., Beljaars, A. C. M., Bidlot, J., Bormann, N., Delsol, C., Dragani, R., Fuentes, M., Geer, A. J., Isaksen, L., Haimberger, L., Healy, S. B., Hersbach, H., Matricardi, M., Mcnally, A. P., Peubey, C., Rosnay, P. D., Tavolato, C., and Vitart, F.: The ERA-Interim reanalysis: configuration and performance of the data assimilation system, Q. J. Roy. Meteor. Soc., 137, 553–597, 2011.

de Elía, R., Laprise, R., and Denis, B.: Forecasting Skill Limits of Nested, Limited-Area Models: A Perfect-Model Approach, Mon. Weather Rev., 130, 2006–2023, https://doi.org/10.1175/1520-0493(2002)130<2006:FSLONL>2.0.CO;2, 2002.

Eerola, K.: About the performance of HIRLAM version 7.0, HIRLAM Newsletter, 51, 93–102, 2006.

Fettweis, X., Box, J. E., Agosta, C., Amory, C., Kittel, C., Lang, C.,

van As, D., Machguth, H., and Gallée, H.: Reconstructions of the 1900–2015 Greenland ice sheet surface mass balance using the regional climate MAR model, The Cryosphere, 11, 1015–1033, https://doi.org/10.5194/tc-11-1015-2017, 2017.

Forbes, R., Tompkins, A. M., and Untch, A.: A new prognostic bulk microphysics scheme for the IFS, ECMWF Technical Memoranda, available at: https://www.ecmwf.int/sites/default/files/elibrary/2011/9441-new-prognostic-bulk-microphysics-scheme-ifs.pdf, 2011.

Gallée, H. and Schayes, G.: Development of a Three-Dimensional Meso-γ Primitive Equation Model: Katabatic Winds Simulation in the Area of Terra Nova Bay, Antarctica, Mon. Weather Rev., 122, 671, https://doi.org/10.1175/1520-0493(1994)122<0671:DOATDM>2.0.CO;2, 1994.

Gascoin, S., Guðmundsson, S., Aðalgeirsdóttir, G., Pálsson, F., Schmidt, L. S., and Berthier, E.: Evaluation of MODIS albedo product over ice caps in Iceland and impact of volcanic eruptions on albedo, Remote Sens., 9, 399, https://doi.org/10.3390/rs9050399, 2017.

Giorgi, F. and Bi, X.: A study of internal variability of a regional climate model, J. Geophys. Res., 105, 29503–29522, https://doi.org/10.1029/2000JD900269, 2000.

Grainger, M. E. and Lister, H.: Wind speed, stability and eddy viscosity over melting ice surfaces, J. Glaciol., 6, 101–127, 1966.

Greuell, W., Knap, W. H., and Smeets, P. C.: Elevational changes in meteorological variables along a midlatitude glacier during summer, J. Geophys. Res., 102, 25941, https://doi.org/10.1029/97JD02083, 1997.

Guðmundsson, M. T.: Mass balance and precipitation on the summit plateau of Öræfajökull, SE-Iceland, Jökull, 48, 49–54, 2000.

Guðmundsson, S., Björnsson, H., Pálsson, F., and Haraldsson, H. H.: Energy balance of Brúarjökull and circumstances leading to the August 2004 floods in the river Jökla, N-Vatnajökull, Jökull, 55, 121–138, 2006.

Guðmundsson, S., Björnsson, H., Pálsson, F., and Haraldsson, H. H.: Comparison of energy balance and degree-day models of summer ablation on the Langjökull ice cap, SW-Iceland, Jökull, 59, 1–18, 2009.

Gudmundsson, M. T., Thordarson, T., Höskuldsson, Á., Larsen, G., Björnsson, H., Prata, F. J., Oddsson, B., Magnússon, E., Högnadóttir, T., Petersen, G. N., Hayward, C. L., Stevenson, J. a., and Jónsdóttir, I.: Ash generation and distribution from the April-May 2010 eruption of Eyjafjallajökull, Iceland, Scientific Reports, 2, 1–12, https://doi.org/10.1038/srep00572, 2012.

Kipp and Zonen: CNR1 Net Radiometer Instruction Manual, available at: http://www.kippzonen.com/Download/87/CNR-1-Net-Radiometer-Brochure (last access: 3 July 2017), 2002.

Knap, W. H., Brock, B. W., Oerlemans, J., and Willis, I. C.: Comparison of Landsat TM-derived and ground-based albedos of Haut Glacier d'Arolla, Switzerland, Int. J. Remote Sens., 20, 3293–3310, https://doi.org/10.1080/014311699211345, 1999.

Langen, P. L., Mottram, R. H., Christensen, J. H., Boberg, F., Rodehacke, C. B., Stendel, M., van As, D., Ahlstrøm, A. P., Mortensen, J., Rysgaard, S., Petersen, D., Svendsen, K. H., Aðalgeirsdóttir, G., and Cappelen, J.: Quantifying energy and mass fluxes controlling godthåbsfjord freshwater input in a 5-km simulation (1991–2012), J. Climate, 28, 3694–3713, 2015.

Langen, P. L., Fausto, R. S., Vandecrux, B., Mottram, R. H., and

Box, J. E.: Liquid Water Flow and Retention on the Greenland Ice Sheet in the Regional Climate Model HIRHAM5: Local and Large-Scale Impacts, Front. Earth Sci., 4, 10, https://doi.org/10.3389/feart.2016.00110, 2017.

Larsen, G., Guðmundsson, M. T., and Björnsson, H.: Tephrastratigraphy of Ablation Areas of Vatnajökull Ice Cap, Iceland, Glaciers, Ice Sheets and Volcanoes: A Tribute to Mark F. Meier, p. 75, 1996.

Lenaerts, J. T. M. and Van Den Broeke, M. R.: Modeling drifting snow in Antarctica with a regional climate model: 2. Results, J. Geophys. Res. Atmos., 117, D5, https://doi.org/10.1029/2010JD015419, 2012.

Lucas-Picher, P., Wulff-Nielsen, M., Christensen, J. H., Aðalgeirsdóttir, G., Mottram, R. H., and Simonsen, S. B.: Very high resolution regional climate model simulations over Greenland: Identifying added value, J. Geophys. Res., 117, 2108, https://doi.org/10.1029/2011JD016267, 2012.

MATLAB: version 8.5.0 (R2015a), The MathWorks Inc., 2015.

Meijgaard, E. V., Ulft, L. H. V., Bosveld, F. C., Lenderink, G., and Siebesma, A. P.: The KNMI regional atmospheric climate model RACMO version 2.1, Technical report; TR – 302, p. 43, 2008.

Monin, A. S. and Obukhov, A. M.: Basic laws of turbulent mixing in the surface layer of the atmosphere, Contrib. Geophys. Inst. Acad. Sci. USSR, 24, 163–187, 1954.

Mottram, R., Boberg, F., and Langen, P.: HIRHAM5 GL2 simulation dataset, available at: http://prudence.dmi.dk/data/temp/RUM/HIRHAM/GL2, 2016.

Nawri, N.: Evaluation of HARMONIE reanalyses of surface air temperature and wind speed over Iceland, Tech. rep., Veðurstofa Íslands, available at: http://www.vedur.is/media/vedurstofan/utgafa/skyrslur/2014/VI_2014_005.pdf (last access: 3 July 2017), 2014.

Nielsen-Englyst, P.: Impact of albedo parameterizations on surface mass balance and melt extent on the Greenland Ice Sheet, Master's thesis, 2015.

Noël, B., van de Berg, W. J., van Meijgaard, E., Kuipers Munneke, P., van de Wal, R. S. W., and van den Broeke, M. R.: Evaluation of the updated regional climate model RACMO2.3: summer snowfall impact on the Greenland Ice Sheet, The Cryosphere, 9, 1831–1844, https://doi.org/10.5194/tc-9-1831-2015, 2015.

Oerlemans, J. and Knap, W. H.: A 1-year record of global radiation and albedo in the ablation zone of Marteratschgletscher, Switzerland, J. Glaciol., 44, 231–238, 1998.

Oerlemans, J., Björnsson, H., Kuhn, M., Obleitner, F., Palsson, F., Smeets, C., Vugts, H. F., and Wolde, J. D.: Glacio-Meteorological Investigations On Vatnajökull, Iceland, Summer 1996: An Overview, Bound.-Lay. Meteorol., 92, 3–24, https://doi.org/10.1023/A:1001856114941, 1999.

Painter, T. H., Barrett, A. P., Landry, C. C., Neff, J. C., Cassidy, M. P., Lawrence, C. R., McBride, K. E., and Farmer, G. L.: Impact of disturbed desert soils on duration of mountain snow cover, Geophys. Res. Lett., 34, 12, https://doi.org/10.1029/2007GL030284, 2007.

Pálsson, F., Gunnarsson, A., Jónsson, Þ., Steinþórsson, S., and Pálsson, H. S.: Vatnajökull: Mass balance, meltwater drainage and surface velocity of the glacial year 2014_15, Tech. rep., Institute of Earth Sciences, University of Iceland and National Power Company, RH-06-2015, 2015.

Rae, J. G. L., Aðalgeirsdóttir, G., Edwards, T. L., Fettweis, X., Gregory, J. M., Hewitt, H. T., Lowe, J. A., Lucas-Picher, P., Mottram, R. H., Payne, A. J., Ridley, J. K., Shannon, S. R., van de Berg, W. J., van de Wal, R. S. W., and van den Broeke, M. R.: Greenland ice sheet surface mass balance: evaluating simulations and making projections with regional climate models, The Cryosphere, 6, 1275–1294, https://doi.org/10.5194/tc-6-1275-2012, 2012.

Reijmer, C. H., Knap, W. H., and Oerlemans, J.: The Surface Albedo Of The Vatnajökull Ice Cap, Iceland: A Comparison Between Satellite-Derived And Ground-Based Measurements, Bound.-Lay. Meteorol., 92, 123–143, https://doi.org/10.1023/A:1001816014650, 1999.

Roeckner, E., Bäuml, G., Bonaventura, L., Brokopf, R., Esch, M., Giorgetta, M., Hagemann, S., Kirchner, I., Kornblueh, L., Manzini, E., Rhodin, A., Schlese, U., Schulzweida, U., and Tompkins, A.: The atmospheric general circulation model ECHAM 5 PART I: Model description, Tech. Rep. 349, Report/MPI für Meteorologie, 2003.

Schaaf, Z. W.: MCD43A3 MODIS/Terra+Aqua BRDF/Albedo Daily L3 Global – 500m V006. NASA EOSDIS Land Processes DAAC., https://doi.org/10.5067/modis/mcd43a3.006, 2015.

Stendel, M., Christensen, J. H., and Petersen, D.: High-Arctic Ecosystem Dynamics in a Changing Climate, vol. 40 of Advances in Ecological Research, Elsevier, https://doi.org/10.1016/S0065-2504(07)00002-5, 2008.

van Angelen, J. H., Lenaerts, J. T. M., Lhermitte, S., Fettweis, X., Kuipers Munneke, P., van den Broeke, M. R., van Meijgaard, E., and Smeets, C. J. P. P.: Sensitivity of Greenland Ice Sheet surface mass balance to surface albedo parameterization: a study with a regional climate model, The Cryosphere, 6, 1175–1186, https://doi.org/10.5194/tc-6-1175-2012, 2012.

van den Broeke, M., van As, D., Reijmer, C., and van de Wal, R.: Assessing and improving the quality of unattended radiation observations in Antarctica, J. Atmos. Ocean. Tech., 21, 1417–1431, 2004.

Vaughan, D. G., Comiso, J. C., Allison, I., Carrasco, J., Kaser, G., Kwok, R., Mote, P., Murray, T., Paul, F., Ren, J., Rignot, E., Solomina, O., Steffen, K., and Zhang, T.: Observations: Cryosphere, in: Climate Change 2013: The Physical Science Basis. Contribution of Working Group I to the Fifth Assessment Report of the Intergovernmental Panel on Climate Change, edited by: Stocker, T. F., Qin, D., Plattner, G.-K., Tignor, M., Allen, S. K., Boschung, J., Nauels, A., Xia, Y., Bex, V., and Midgley, P. M., Cambridge University Press, Cambridge, United Kingdom and New York, NY, USA, Cambridge, 2013.

Wallace, J. M., Hobbs, P. V., Wallace, J. M., and Hobbs, P. V.: 3. Atmospheric Thermodynamics, in: Atmospheric Science, Elsevier, 2nd Edn., 63–111, 2006.

Wittmann, M., Groot Zwaaftink, C. D., Steffensen Schmidt, L., Guðmundsson, S., Pálsson, F., Arnalds, O., Björnsson, H., Thorsteinsson, T., and Stohl, A.: Impact of dust deposition on the albedo of Vatnajökull ice cap, Iceland, The Cryosphere, 11, 741–754, https://doi.org/10.5194/tc-11-741-2017, 2017.

5

Measurements of precipitation in Dumont d'Urville, Adélie Land, East Antarctica

Jacopo Grazioli[1,2]**, Christophe Genthon**[3]**, Brice Boudevillain**[3]**, Claudio Duran-Alarcon**[3]**, Massimo Del Guasta**[4]**, Jean-Baptiste Madeleine**[5,6]**, and Alexis Berne**[1]

[1]Environmental Remote Sensing Laboratory (LTE), École Polytechnique Fédérale de Lausanne (EPFL), Lausanne, Switzerland
[2]MeteoSwiss, Locarno-Monti, Switzerland
[3]Univ. Grenoble Alpes, CNRS, IGE, 38000 Grenoble, France
[4]Istituto nazionale di Ottica, INO-CNR, Italy
[5]Sorbonne Universités, UPMC Univ Paris 06, UMR 8539, Laboratoire de Météorologie Dynamique (IPSL), Paris, France
[6]CNRS, UMR 8539, Laboratoire de Météorologie Dynamique (LMD), IPSL Climate Modeling Center, Paris, France

Correspondence to: Jacopo Grazioli (jacopo.grazioli@epfl.ch)

Abstract. The first results of a campaign of intensive observation of precipitation in Dumont d'Urville, Antarctica, are presented. Several instruments collected data from November 2015 to February 2016 or longer, including a polarimetric radar (MXPol), a Micro Rain Radar (MRR), a weighing gauge (Pluvio^2), and a Multi-Angle Snowflake Camera (MASC). These instruments collected the first ground-based measurements of precipitation in the region of Adélie Land (Terre Adélie), including precipitation microphysics. Microphysical observations during the austral summer 2015/2016 showed that, close to the ground level, aggregates are the dominant hydrometeor type, together with small ice particles (mostly originating from blowing snow), and that riming is a recurring process. Eleven percent of the measured particles were fully developed graupel, and aggregates had a mean riming degree of about 30 %. Spurious precipitation in the Pluvio^2 measurements in windy conditions, leading to phantom accumulations, is observed and partly removed through synergistic use of MRR data. The yearly accumulated precipitation of snow (300 m above ground), obtained by means of a local conversion relation of MRR data, trained on the Pluvio^2 measurement of the summer period, is estimated to be 815 mm of water equivalent, with a confidence interval ranging between 739.5 and 989 mm. Data obtained in previous research from satellite-borne radars, and the ERA-Interim reanalysis of the European Centre for Medium-Range Weather Forecasts (ECMWF) provide lower yearly totals: 655 mm for ERA-Interim and 679 mm for the climatological data over DDU. ERA-Interim overestimates the occurrence of low-intensity precipitation events especially in summer, but it compensates for them by underestimating the snowfall amounts carried by the most intense events. Overall, this paper provides insightful examples of the added values of precipitation monitoring in Antarctica with a synergistic use of in situ and remote sensing measurements.

1 Introduction

The ice sheets of Antarctica contain about 90 % of the world's ice and thus their evolution has potential impacts at a global scale. They condition the evolution of the sea level height (Rignot et al., 2011; DeConto and Pollard, 2016) and the radiative budget of the lower atmosphere. In this context, the quantification and prediction of the surface mass balance (SMB) of the Antarctic ice cap is a pressing scientific topic of investigation which is carried out in order to understand whether the continent is losing or gaining ice and at what rate (Vaughan et al., 1999; Lenaerts et al., 2016).

Precipitation is an important component of the SMB as it represents, together with vapor deposition, the only net input of water and ice at the continental scale (Krinner et al., 2007).

Precipitation is unfortunately also very difficult to monitor at high latitudes. The major problems hampering classical measurement techniques in Antarctica are, in the interior, the sparsity of human installations over a very large area; the extremely low temperatures and low precipitation amounts; and on the coasts the very strong katabatic winds blowing from the interior. Additionally, the complex logistics of Antarctic installations causes further difficulties and limitations for measurements to be conducted.

Until recently, information about precipitation was obtained indirectly by analyzing moisture transports, glaciological surface-based observations (Bromwich, 1990) and reanalysis based on numerical weather prediction models (Bromwich et al., 2011). Additionally, long-running but qualitative human observation records of clouds and precipitation have been collected at some scientific stations by staff dedicated to meteorological measurements (e.g., König-Langlo et al., 1998). Recent research proposed a climatology of precipitation over a large part of the continent (Palerme et al., 2014, 2016) by exploiting the potential of the profiling radar on board the CloudSat satellite, which is able to sample large horizontal areas but limited by the inability to measure precipitation at altitudes below a so-called "blind-range" above ground (1200 m above the surface for CloudSat).

In order to validate and to improve the performance of the models, and to constrain satellite-based measurements, it is necessary to establish and maintain some in situ observation sites in the medium to long term, instrumented with precipitation measurement devices that are as autonomous and accurate as possible. There is therefore the need for accurate measurements of precipitation, including at a very local scale (Frezzotti et al., 2004; Schlosser et al., 2010; Welker et al., 2014). A recent effort in this direction was the establishment of an observatory in the escarpment zone of Dronning Maud Land, East Antarctica (Gorodetskaya et al., 2015). The synergy of in situ and remote sensing measurements allowed the very first statistics of cloud and precipitation (Gorodetskaya et al., 2014, 2015) which showed that a few intense precipitation events govern the SMB in the area; measurement combinations have also been used to evaluate the quality of satellite-based precipitation products (Maahn et al., 2014) provided by CloudSat. It has been shown that the blind range of CloudSat in the area of the measurements can lead to an underestimation of precipitation amount of the order of 10 % and an underestimation of the occurrence frequency of the order of 5 %. The installation in DML can be considered the first well-documented observatory in Antarctica to include precipitation measurements from remote sensing and in situ instruments. An earlier effort involved co-located measurements of precipitation using radar and precipitation gauges, and was conducted at the Showa[1] Japanese station (Konishi et al., 1998), but very limited information about the outcome of those measurements

is yet available in the literature. A more recent effort is currently taking place in the McMurdo base, in the framework of the AWARE project Witze (2016), starting from November 2015.

In this work we present the results of an intensive observation campaign during the austral summer 2015–2016 and a first year of precipitation measurements conducted in the French base Dumont d'Urville, Adélie Land, from November 2015 to November 2016 (and still ongoing). The data were collected in the framework of the *APRES3* project (Antarctic Precipitation, Remote Sensing from Surface and Space, see http://apres3.osug.fr). We provide statistics of precipitation quantity and occurrence, and we compare them with model reanalyses and with the visual observations collected by the French meteorological office (Météo France) throughout the year. The main scientific objectives of this work are to contribute to a better quantification of precipitation in Antarctica (also by evaluating the products of numerical weather models) and to underline the innovative and promising aspects of the data collected until now, which may serve as an example for long-term monitoring of precipitation in other Antarctic regions. The paper is structured as follows: Sect. 2 describes the precipitation measurements, Sect. 3 lists the most relevant results, which are discussed and put into perspective in Sect. 4. Section 5 provides the summary and conclusions of the paper.

2 Methods

Here we present data collected at a coastal location of Antarctica: the station Dumont d'Urville (DDU). The base is situated in Adélie Land, $-66.6628°$ S, $140.0014°$ E, (41 m a.s.l.), on a coastal location highlighted in Fig. 1a. This region is located at the transition between the Antarctic continent and the Southern Ocean, where the terrain, which slopes downward from the inner continent to the coast, meets the ocean.

2.1 Climate and operational measurements

The climate at DDU is relatively mild in terms of temperatures, with minima rarely below $-30°$C, and maxima above $0°$C in January and December, as illustrated in Fig. 1b. On the contrary, the wind regime is more extreme: in the low layer of the atmosphere the dominant winds are katabatic, coming from the inner continent, and the dominant wind origins are always between 90° (east) and 180° (south), as illustrated in the wind rose of Fig. 1b. Because of the intensity and persistence of the winds, which are able to reach hurricane force, Adélie Land has often been described as the windiest place on planet Earth (e.g., Wendler et al., 1997). Standard measurements of atmospheric variables (temperature, wind speed, wind direction, relative and specific humidity, atmospheric pressure) are collected regularly all year

---

[1] Sometimes spelled "Syowa"

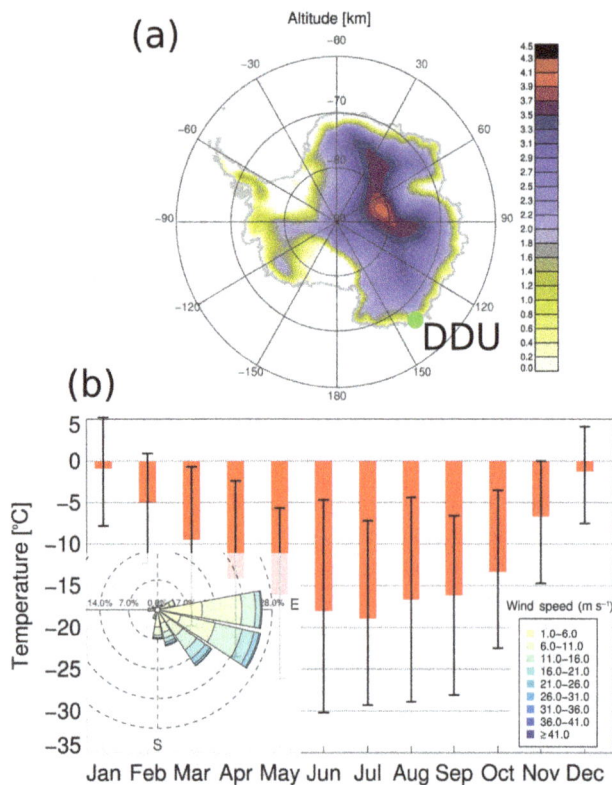

Figure 1. (a) Map of Antarctica with digital elevation model (domain south of 60° S). A green filled circle locates the station Dumont d'Urville (DDU). **(b)** Temperature statistics in Dumont d'Urville, based on data collected at 1 min time resolution in the period 2011–2015. The red bars locate the mean value and the black error bars are used to highlight the 1 and 99 % quantiles. Overlayed: wind rose (origin and intensity) statistics.

long by the French meteorological service (Météo France), and a balloon radiosounding is launched daily at 00:00 UTC. Balloon soundings have been regularly conducted since 1956 at DDU. Visual observations of cloud, precipitation, and present weather are recorded as well. It is worth noting that at this location the visual observations are the only daily in situ archive of past precipitation occurrence in DDU before the measurements described in this paper.

2.2 APRES3 instruments

Several instruments were deployed at DDU, starting from November 2015. The instruments were deployed as illustrated in Fig. 2, and they are listed in Table 1. A Micro Rain Radar (MRR) was installed within an existing radome and has collected uninterrupted measurements since 22 November 2015. This radar system is used to vertically profile precipitation with a resolution of 100 m at height levels ranging from 341 to 3141 m a.s.l.[2]. The processed data were col-

[2] 300 m is the third range gate of the MRR, where the first valid measurements are available, and 41 m is the altitude of DDU.

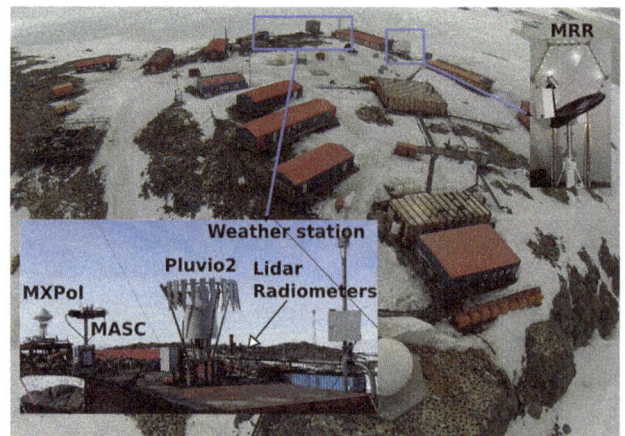

Figure 2. Main instruments deployed at DDU over the time period ranging from November 2015 to November 2016 (the MRR is, however, still collecting observations at the time of publication).

lected with a temporal resolution of 1 min. The potential of the MRR to monitor polar regions has already been highlighted by the works of Maahn et al. (2014) and Gorodetskaya et al. (2015). The simplicity of its deployment and operation makes it an attractive tool for long-term measurements in places with complex logistics and with limited possibility of support, in the case of instrumental failures. The raw K-band reflectivity measurements collected by the MRR were first processed with the method proposed by Maahn and Kollias (2012), then converted to X-band reflectivities and in a third step to snowfall intensities. Additional information about the processing of the MRR data is provided in Sect. 2.2.1.

A second radar, named MXPol (Mobile X-band dual-Polarization) collected measurements in the months of December 2015 and January 2016. This system, described in Schneebeli et al. (2013) and in Scipion et al. (2013), is a scanning dual-polarization Doppler radar. During its operation period at DDU, it was mainly collecting data at 75 m radial resolution and a maximum radial distance of 30 km, mostly conducting different types of scans within a repeating scanning sequence of 5 min: (i) plan position indicator (PPI) scans, i.e., quasi-horizontal slices of the atmosphere, (ii) range height indicator (RHI) scans, i.e., vertical slices of the atmosphere, and (iii) static vertical profiles, such as the ones performed by the MRR.

A depolarization lidar (e.g., Del Guasta et al., 1993), deployed at a distance of about 200 m from MXPol, collected data in December 2015 and January 2016, as a test-bed for future long-term installation of a similar device. Lidar measurements allow for the discrimination of the phase of the tropospheric clouds and detection of the occurrence of supercooled liquid water, and they complement the observations of ground-based radars, which are often not sensitive to these particles. An example is given in Fig. 3, where the time

Table 1. Nonexhaustive list of the instruments deployed at DDU in the framework of APRES3. Only the APRES3 instruments with a certain relevance for precipitation monitoring are listed here.

Name	Deployment period	Instrument type	Measurement	Reference
MRR	21 Nov 2015–ongoing	FMCW[a] radar profiler, 24 GHz	Clouds/precipitation	Maahn and Kollias (2012)
MXPol	7 Dec 2015–31 Jan 2016	Dual-pol Doppler radar, 9.41 GHz	Clouds/precipitation	Schneebeli et al. (2013)
Lidar	15 Dec 2015–29 Jan 2016	Depolarization lidar	Clouds/precipitation	Del Guasta et al. (1993)
MASC	11 Nov 2015–31 Jan 2016	Snowflake imager	Precipitation/blowing Snow	Garrett et al. (2012)
Pluvio[2]	17 Nov 2015–31 Jan 2016	Weighing gauge	Precipitation	Colli et al. (2014)
Biral VPF-730[b]	3 Dec 2015–25 Dec 2015	Present weather sensor	Visibility/present weather	–
Vaisala Weather Transmitter WXT 520[c]	11 Nov 2015–31 Jan 2016	Weather station	T, RH, Wind	–

[a]: Frequency modulation continuous wave.

[b]: For the rest of the time this instrument was (and is) deployed on the Antarctic continent, about 5 km away from DDU.

[c]: The co-located weather station of Météo France is providing data all year long uninterruptedly.

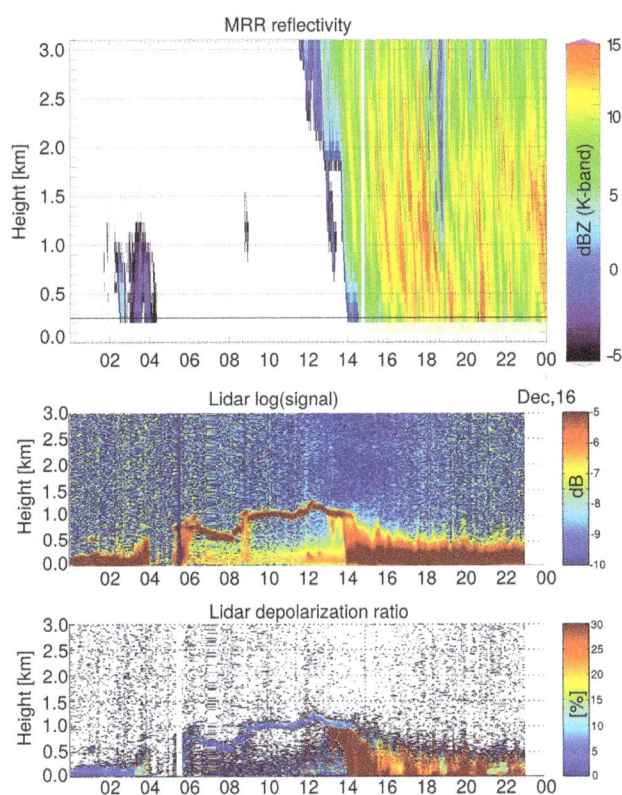

Figure 3. Example of a time series (time–height image) of MRR data and lidar data for the 15 December 2015.

series of MRR reflectivity, lidar signal and depolarization ratio are shown for the 15 December 2015. Supercooled liquid water appears in the lidar data as a layer of enhanced signal and low depolarization ratio (e.g., Del Guasta et al., 1993; Hogan et al., 2003), often when no MRR signal is visible. On the contrary, when precipitation occurs, (around 04:00 UTC, and from 14:00 to 24:00 UTC) the lidar signal gets fully attenuated in the lowest 500 m while the MRR is still able to sample the vertical precipitation column.

A weighing precipitation gauge (Pluvio[2], manufactured by OTT) was deployed from November 2015 to January 2016. This instrument provides the liquid water equivalent of snowfall falling within its measurement area at a time resolution of 1 min. To avoid excessive contamination of precipitation signals by blowing snow, the Pluvio[2] was installed at a height of about 3 m above ground and its inlet was protected by a standard wind fence designed by the same manufacturer as the instrument. It must be noted that this wind shield is not sufficient to avoid the adverse effect of strong wind (frequently occurring at DDU).

Located close to the weighing gauge, a multi-angle snowflake camera (MASC) was deployed, also during the period from November 2015 to January 2016. This instrument collects high-resolution stereoscopic photographs of snowflakes in free fall, while they cross its sampling area (Garrett et al., 2012), thus providing information about snowfall microphysics and particle fall velocity. The MASC used three identical 2448×2048 pixels cameras (with common focal point) with apertures and exposure times adjusted to trade off the contrast on snowflakes photographs and motion blur effects, and a resolution of about 33 μm per pixel. The cameras are triggered when a falling particle crosses two series of near-infrared sensors. A detailed description of the system and its calibration can be found in Garrett et al. (2012), and Praz et al. (2017). To complete the set of in situ measurements, a weather station (Vaisala Weather Transmitter WXT 520) was installed close to the Pluvio[2] and the MASC to sample the environmental conditions in the close proximity of their measurements, as illustrated in Fig. 2.

2.2.1 Pre-processing of MRR data

The MRR was co-located with MXPol for the period of the summer campaign 2015/2016 (See Table 1). The purpose of the former instrument at DDU is long-term monitoring, which involves exposure to the extremely windy winter conditions. It was decided, in order to avoid failures during the winter when no member of the scientific team is on site, to install the MRR inside an existing radome previously used in the base for satellite communications, as shown in Fig. 2. Although this installation ensures protection and easy access to the instrument, it adds an unknown amount of attenuation to the measurements. For this reason the co-located MXPol measurements collected during the summer period are used to map the radome-affected reflectivity data provided at K-band (MRR) into X-band reflectivities.

The scatter plot in Fig. 4 shows the comparison of reflectivity values measured by the MRR and by MXPol for data collected during the period of co-location of the instruments. Because overall the relation between the two sets of measurements is close to linear ($\rho^2 \approx 0.88$), and almost equivalent to a simple offset subtraction, we can hypothesize that the eventual non-Rayleigh effects, due to centimeter-size snowflakes, were similar at the two frequencies and the following conversion has been applied to MRR data:

$$Z_X = 0.99 Z_K + 6.14 \pm \epsilon, \tag{1}$$

where Z_X (Z_K) [dBZ] is used to indicate reflectivity at X-(K-)band, and ϵ is the measure of uncertainty of the linear relation with respect to the scatter plot of Fig. 4 (whose standard deviation of the residuals is 1.9 dB). It is worth mentioning that Z_K is originally obtained with the method of Maahn and Kollias (2012), who proposed an improved and innovative processing chain for MRR data collected in snow. Once mapped to X-band, reflectivity can be converted to snowfall rate S rate by means of Z–S power laws available in the literature. For example, the six relations proposed by Matrosov et al. (2009) and listed in Table 2 can be used. These relations were obtained by combining two different snowflake size distribution data sets, and three different mass-to-size relations. The error component of Eq. (1) and the large variability of the Z–S relations lead to very uncertain retrievals of snowfall rate. For this reason, we optimized a local power law, by fitting its two parameters in the Z–S space given by the MRR measurements at the lowest available height and the Pluvio^2 measurements collected close to the ground, during the summer period 2015/2016. The parameters (intercept and exponent) of the power law are obtained by means of nonlinear least square estimation. The local relation, also listed in Table 2, takes the form of $Z = 76 S^{0.91}$. In order to mitigate the difference in sampling volume of the two instruments, it has been derived for hourly data. The 95 % confidence intervals for the two parameters are 69–83 (prefactor) and 0.78–1.09 (exponent).

2.2.2 Pre-processing of Pluvio^2 data

It has been observed that occasionally the values of equivalent water of the Pluvio^2 show a "phantom" accumulation (similar to that reported by World Meteorological Organization, 2014). In such cases, no precipitation was observed by the researchers that were present on site and no precipitation signal was visible in the MRR data but the content of

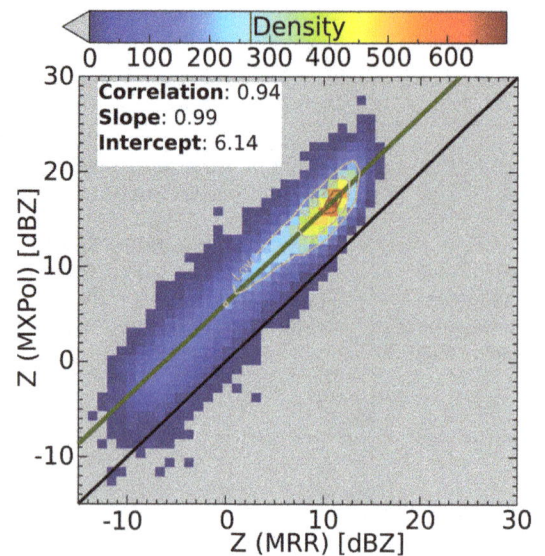

Figure 4. Scatter plot of reflectivity values at 9.41 GHz (X-band, measured by MXPol) and at 24.3 GHz (K-band, measured by the MRR) during the summer campaign 2015/2016. The data correspond to time steps at which both radar were profiling (PPI and RHI scans of MXPol do not contribute).

Table 2. Parameters of the six X-band conversion relations between radar reflectivity Z and snowfall intensity S (mm h^{-1}) of Matrosov et al., 2009, and the local relation, obtained using the instruments at DDU. In these relations, the radar reflectivity (Z) must be used in linear units [mm^6 m^{-3}]. The six X-band relations originate from two different data sets (B90, ground-based, and W08, from in situ aircraft measurements), and three different mass to diameter relations, as detailed in Matrosov et al. (2009).

Relation*	Equation
B90A (1)	$Z = 67 S^{1.28}$
B90B (2)	$Z = 114 S^{1.39}$
B90C (3)	$Z = 136 S^{1.30}$
W08A (4)	$Z = 28 S^{1.44}$
W08B (5)	$Z = 36 S^{1.56}$
W08C (6)	$Z = 48 S^{1.45}$
Local DDU	$Z = 76 S^{0.91}$

* Parentheses indicate the way the relations were numbered in Matrosov et al. (2009).

the Pluvio^2 bucket increased. In order to discard these cases, we combined the information coming from remote sensing (MRR) and in situ data (Pluvio^2). More precisely, time steps at which no signal was recorded by the MRR at its lowest available gate (300 m a.g.l.) are considered precipitation free and any increase in the cumulative precipitation records of the Pluvio^2 is thus related to external contaminations. The assumption is that precipitation is extremely unlikely to completely develop in the lowest 300 m of the atmosphere. An

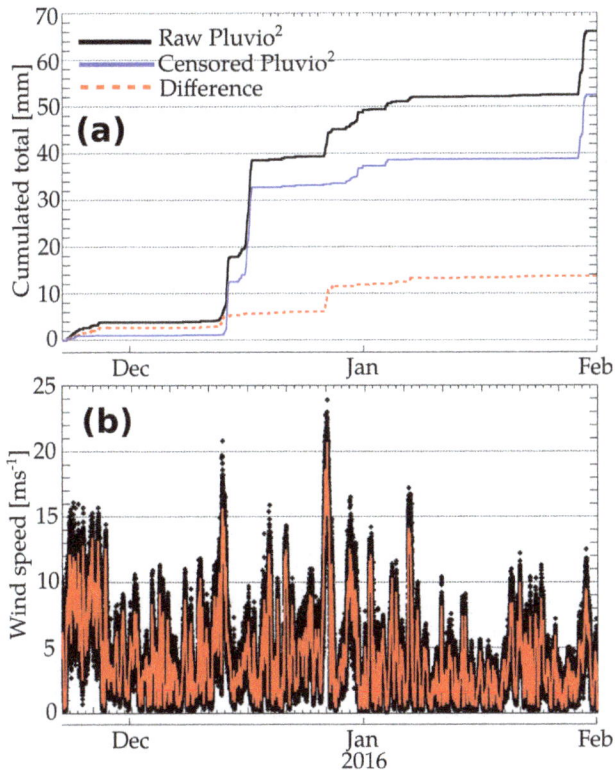

Figure 5. (a) Time series of Pluvio^2 untreated data, censored data (taking the MRR measurements as occurrence indicators), and their difference. **(b)** Wind speed measured in the near proximity (≤ 2 m) of the Pluvio^2 inlet, at the temporal resolution of 1 min.

example of the behavior of this simple censoring filter can be found in Fig. 5a. From the end of October 2015 to the end of January 2016, about 14 mm of liquid water equivalent snowfall have been removed, corresponding to about 21 % of the uncensored data.

Figure 5b shows the evolution of wind speed in the near proximity of the Pluvio^2 inlet and illustrates that the most intense phantom accumulations occur when the strongest wind peaks are observed. Because the cameras of the co-located MASC were not triggered by hydrometeors during the censored time steps, we ruled out the possibility that phantom accumulation is in this case due to clear-sky blowing snow, and we hypothesize that it is caused by wind-induced vibrations of the instrument. It must be noted that this simple pretreatment cannot compensate for the contribution of snowfall mixed with blowing snow when the positive contribution of blowing snow and precipitation, and the negative contribution due to wind-induced loss of catching efficiency occur together.

2.3 Additional data

Due to the lack of both short- and long-term precipitation measurements, net precipitation estimates in Antarc-

tica have been obtained from numerical weather prediction models (e.g., Cullather et al., 1998; Schlosser et al., 2010). Among the available model-based products, the ERA-Interim global reanalysis provided by the European Centre for Medium-Range Weather Forecasts (ECMWF) is taken as a reference as it is considered to provide the best representation of precipitation variability (Bromwich et al., 2011; Palerme et al., 2014) and the best agreement with satellite-borne measurements (Behrangi et al., 2016; Palerme et al., 2016). ERA-Interim reanalysis is used here for this reason, and because of its global coverage and easy access. The analyses at 00:00 and 12:00 UTC, and forecast time steps of 6, and 12 h are used in the present work for the grid point which is the closest to DDU. The spatial resolution of ERA-Interim is $0.75° \times 0.75°$. To quantify precipitation, the model variable tp (total precipitation) is used here.

3 Results

3.1 Microphysical observations during summer 2015/2016

The period between November 2015 and January 2016 was heavily instrumented with devices that are able to provide microphysical information about precipitation; thus microphysical aspects are better documented during the summer months. While a complete investigation of the dominant microphysical processes and the small-scale dynamics of precipitation in this region is beyond the scope of this paper, it is worth investigating an important microphysical parameter: the hydrometeor type.

Hydrometeor types have been recorded near the ground level (about 2.5 m above ground) by the MASC instrument through the classification of individual particle pictures with the recently developed method of Praz et al. (2017), which is able to classify individual hydrometeors into six classes (and melting snow) and assign to them a continuous riming degree index ranging from 0 to 1, with 1 corresponding to fully developed graupel. The riming degree is textural information obtained by supervised classification originating from a manually labeled training set including almost 3400 images, as detailed in Praz et al. (2017). The choice of the available classes is based on the widely used scheme of Magono and Lee (1966). Because the instruments are deployed at a height lower than 3 m a.g.l., both precipitation and blowing snow particles are recorded and classified.

A second classification method is obtained from the polarimetric data of MXPol, which can be converted into hydrometeor measurements with an hydrometeor classification algorithm (Grazioli et al., 2015). This algorithm was developed by partitioning a large number of radar observations into spatially coherent clusters by means of data mining techniques and then assigning to each cluster a dominant hydrometeor type by means of scattering simulations, interpretation of po-

Hydrometeor classification
2015-12-29 00:05

Crystals Rimed Aggregates

PPI: elevation 5°

RHI: azimuth 293°

RHI: azimuth 203°

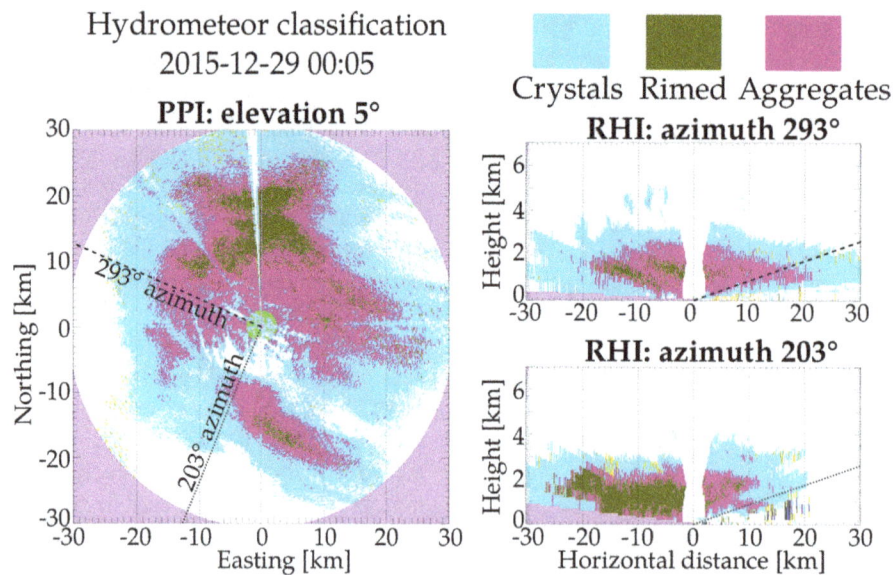

Figure 6. Example of a PPI and two RHI scans collected by MXPol on the 29 December around 00:14 UTC. The variable displayed in the image is the hydrometeor classification, obtained with the method of Grazioli et al. (2015). Noise in the classification at the lowest elevation angles is due to ground clutter. Range gates closer than 2 km with respect to the radar location have been censored to allow reliable polarimetric variables to be computed. Elevation angles larger than 45° have been censored as well in order to limit the geometric reduction of the intensity of polarimetric signature with increasing elevation angles (Ryzhkov et al., 2005).

larimetric signatures, and comparison with in situ data. Despite the drawbacks of it being an indirect method and not being able to carry out retrievals at near-ground heights (because of ground clutter contamination in the radar data), it has the advantage of providing hydrometeor types over large domains and at different height levels, as shown in Fig. 6. Figure 6 illustrates PPI and RHI scans of the hydrometeor classification for a case where all its ice-phase hydrometeor classes are observed. This classification method discriminates pure snowfall into three categories: crystals, aggregates, and rimed particles. Figure 7 illustrates the statistical distribution of those three classes for the period of operation of MXPol, as a function of height. Below 2000 m, the proportion of the three hydrometeor types is relatively constant with about 10 % of rimed snowflakes, 40 % of aggregates, and 50 % of crystals. With increasing height and closeness to the cloud top, aggregates and rimed snowfall rapidly disappear while crystals constitute the dominant hydrometeor class.

The classification obtained with the MASC and the method by Praz et al. (2017) is summarized in Fig. 8. At ground level, the majority of the particles (54 %) are classified as small, indicating that hydrometeors are too small for their geometry and texture to be properly classified by the MASC. This proportion is three times higher than similar measurements collected in a wind-sheltered location in the Swiss Alps, while the proportion of the other hydrometeors is similar among the two different locations (not shown here). The occurrence of strong katabatic winds being a major dif-

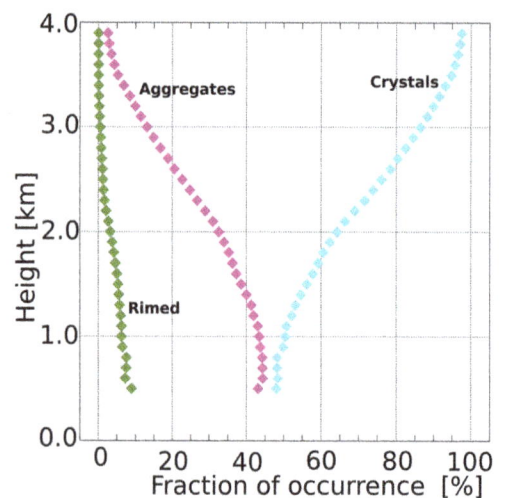

Figure 7. Fraction of occurrence of different hydrometeor types as a function of height above ground, obtained over the period of operation of MXPol, with the hydrometeor classification algorithm of Grazioli et al. (2015).

ference between the sites, it can be assumed that the large majority of these small particles observed at DDU are associated with blowing snow. During blowing snow events with strong winds (identified from visual and MRR observations on site), the number of images collected by the MASC is very large. The majority of those are classified as small particles, and this results in a large percentage of this hydrometeor type in the final statistics. Also from this classification, based on

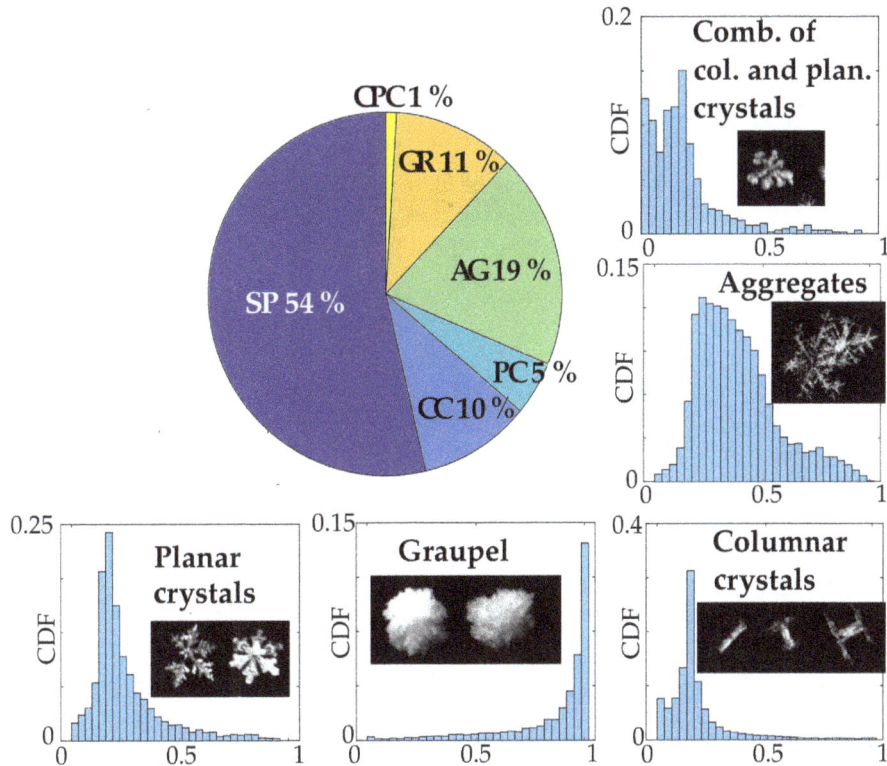

Figure 8. Pie chart of the hydrometeor types classified by the MASC instrument in the period from the 21 November 2015 to the end of January 2016. The histograms shows the distribution of a riming index, ranging from 0 (unrimed) to 1 (fully rimed), for the particles of each hydrometeor class. The riming index is undefined for small particles, i.e., particles that are too small to be identified as a particular hydrometeor class. The classes of the chart are small particles (SP), columnar crystals (CC), aggregates (AG), planar crystals (PC), graupel (GR), combination of columnar and planar crystals (CPC), as described in Praz et al. (2017).

the MASC, we observe that riming occurs. In fact, 11 % of the particles are fully rimed (graupel), while all the other hydrometeor types have a riming degree ranging mostly from 0.1 to 0.5, and are sometimes larger than 0.5 for the aggregates.

While the outcomes of the classification from MXPol and the MASC are not directly comparable because of the differences in measurement height, sampling volume, and available classes, it must be underlined that radar measurements are very sensitive to the size of the hydrometeors. Thus, a few large aggregates within a radar sampling volume will dominate and overcome the signal coming from smaller hydrometeors. This can partially explain the different proportion of aggregates observed by MXPol (about 40 % at a 400 m height), and by the MASC (19 %). A second contribution to this difference may be the low-level mechanical breakup of the aggregates (e.g., Vardiman, 1978). A third, and very likely, contribution is the contamination of blowing snow in the MASC measurements, namely in the small particle hydrometeor class. If, assuming that most of the small particles originate from blowing snow, they are removed from the statistics, then aggregates account for 41 % of the hydrom-

eteors, a value much closer to the 40 % obtained with the classification of MXPol.

3.2 One year of MRR precipitation data

The MRR instrument collected precipitation data uninterruptedly, covering the evolution of precipitation over the entire year. It therefore offers an interesting ground-based (but remotely sensed) set of data to compare with model-based data and with available human observations. Figure 9 shows the estimates coming from the MRR and other available sources of information over a year of measurements. As expected, the agreement of the local MRR relation with the Pluvio^2 is good over the summer period (December–January), during which the relation was obtained. Also in this period the estimate of ERA-Interim provides a total cumulated precipitation within the envelope of values of the optimized Z–S relation, even though the curves show some differences in precipitation occurrence. The optimized Z–S relation provides estimates that are close to the B90A relation of Matrosov et al. (2009).

The months with the highest accumulated precipitation were the late fall and winter months of May and June, and

Table 3. Monthly accumulated precipitation of snow (mm of liquid water equivalent) from the MRR, using the locally optimized Z–S relation and the confidence interval of its parameters, and ERA-Interim data. The mean, minimum, and maximum snowfall of each month for ERA-Interim data from 1995 to 2015 are also shown.

Month	$MRR_{2015/16}^{Min-Max}$ [mm]	$ERA_{2015/16}$ [mm]	$ERA_{1995-2015}^{Mean}$ [mm]	$ERA_{1995-2015}^{Min-Max}$ [mm]
Jan	13.6–19.0	27.6	52.9	17.7–106.4
Feb	35.3–45.7	33.4	44.9	18.8–81.5
Mar	76.0–93.0	80.9	55.3	9.9–203.3
Apr	49.9–77.3	35.4	51.5	11.8–114.3
May	126.0–160.0	113.3	42.7	5.7–108.7
Jun	115.0–158.8	80.2	36.5	4.9–81.7
Jul	28.6–34.3	36.5	48.6	2.6–96.6
Aug	37.6–46.5	27.6	61.8	16.9–113.6
Sep	147.6–208.9	113.3	44.2	4.4–75.2
Oct	3.4–4.7	8.3	30.9	0.1–117.4
Nov	75.2–100.5	72.9	22.5	1.6–59.6
Dec	31.3–40.3	25.4	51.4	17.5–131.4
Total	739.5–989.0	654.8	543.1	392.8–702.5

the month of September. Seasonally[3], Summer was the driest season, contributing only 11 % of the yearly total, compared to values close to 30 % for spring, 34 % for fall, and 25 % for winter (Table 3). The ERA-Interim totals of each month of the comparison period are within what could be observed in the period 1995–2015, with the exception of September, which was the snowiest since 1995.

3.3 Precipitation occurrence

Long-term precipitation data records in Antarctica are often only visual observations of precipitation occurrence. For this reason, comparing precipitation occurrence measurements is a way to better understand the quality of this source of information. For the year 2015–2016, we can compare in terms of occurrence the information coming from ERA-Interim, Pluvio[2], MRR, and the visual observations archived by Météo France. We deal at first with occurrence at the daily scale, and we define it for the MRR and ERA-Interim as precipitation exceeding a given threshold over a given duration. A threshold of 0.07 mm over 6 h was proposed by Palerme et al. (2014), and we thus take a value of 0.28 mm d^{-1} as a first guess. However, the choice of a unique threshold is delicate, and we also apply a minimum (maximum) threshold of 0.001 mm d^{-1} (1 mm d^{-1}) to cover any value that appears reasonable to assume. Figure 10 shows the number of days with precipitation recorded during each month of the measurement period.

As a past reference, the historical record of precipitation occurrences from visual daily observations for the preceding

---

[3]We refer here to the seasons of the midlatitudes of the Southern Hemisphere. Summer: December, January, and February. Fall: March, April, May. Winter: June, July, and August. Spring: September, October, and November.

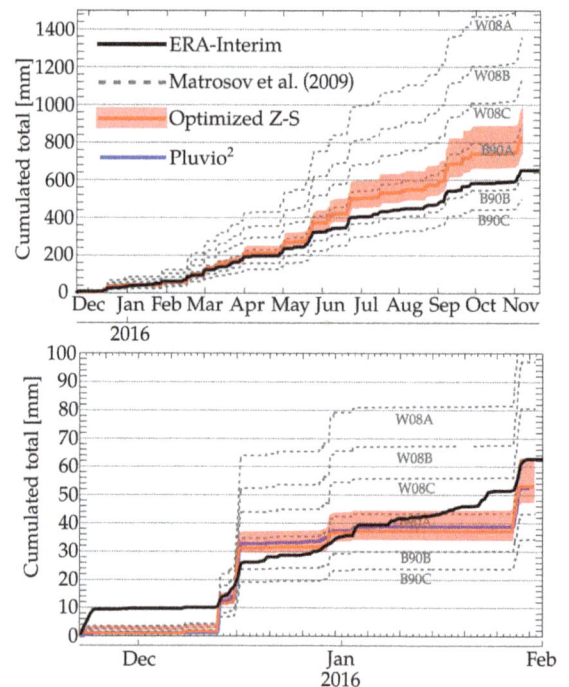

Figure 9. Time series of accumulated snowfall liquid water equivalent. The relations are obtained from Pluvio[2] (in blue, for availability periods), censored from phantom precipitation, MRR (in grey the curves corresponding to the relations of Table 2, and ERA-Interim data (in black). Top panel: data corresponding to the year of measurements from November 2015 to November 2016. Bottom panel: data corresponding to the summer campaign 2015/2016 from November 2015 to February 2016.

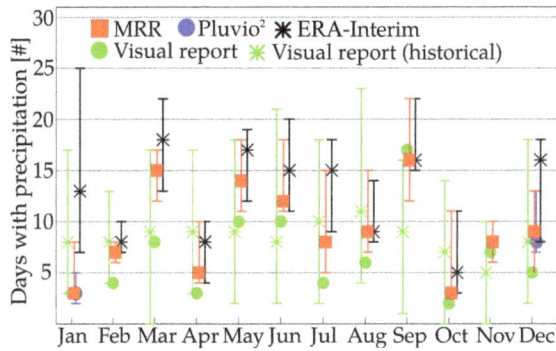

Figure 10. Precipitation occurrence at the daily scale. The error bars (where applicable) come from the use of a threshold of $0.001\,\mathrm{mm\,d^{-1}}$ (upper limit), and $1\,\mathrm{mm\,d^{-1}}$ (lower limit), while the central points are calculated with a threshold of $0.28\,\mathrm{mm\,d^{-1}}$ following the threshold of Palerme et al. (2014). The bars of the historical visual reports indicate instead the minimum and maximum occurrences in the period 1981–2015.

years (1981 to 2015) are also shown in green with a variability range. The year under investigation had an extremely dry January, and an extremely snowy September (in term of occurrence), while the other months are within the range of past occurrences. Overall, ERA-Interim mostly overestimates precipitation occurrence with respect to the MRR, especially in summer, while the visual observations underestimate it. For January and December, when the Pluvio^2 was in operation, it is in agreement with the MRR. Given the measurement correction principle based on false detection described in Sect. 2.2.2, this implies that no misdetection is evident.

A good example of the overestimation of occurrence by ERA-Interim is shown in Fig. 9, bottom. The period between the 10 and 25 January is seen as dry by the MRR and Pluvio^2, while several low-intensity precipitation events appear in the ERA-Interim time series. The overestimation of occurrence compensates for the underestimation of the most intense snowfall events, such that at the end of January the total accumulated precipitation of ERA-Interim gets close to that of the Pluvio^2. As a result, the contribution of lower snowfall rates to the total accumulated snowfall is much larger for ERA-Interim, with respect to the measurements collected by the MRR and Pluvio^2, and this difference is particularly pronounced in the summer period (as shown in Fig. 11).

Figure 12 shows, at a 6 h timescale (here we consider quantitative precipitation, thus we focus on a higher temporal resolution), the evolution of precipitation occurrence as a function of a given average precipitation intensity threshold, for the full year of observations (top panel) and for the summer campaign (bottom panel). Also here the overestimation of precipitation events by ERA-Interim is evident, especially in summer. The curves of Pluvio^2 and MRR are relatively close. At the lowest thresholds the minimum inten-

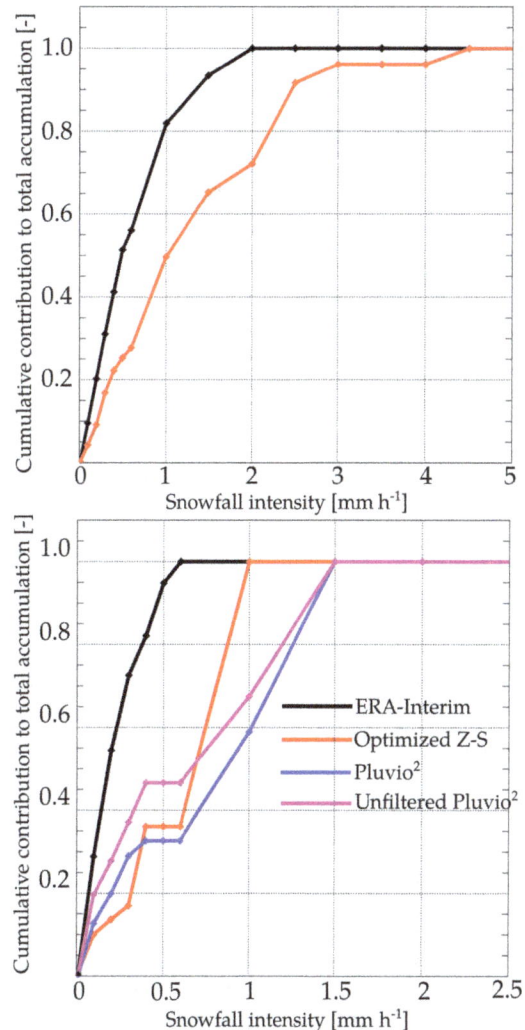

Figure 11. Cumulative contribution of increasing snowfall rates to the accumulated snowfall for the full year (top panel) and summer period (bottom panel). The timescale of snowfall intensity is also 6 h for MRR and Pluvio^2 data, to be consistent with the temporal resolution of ERA-Interim.

sity recordable by the Pluvio^2 becomes a limitation due to the quantization effect. The black curve of ERA-Interim is above the red curve of the MRR for most of the precipitation thresholds. The two curves cross each other where the threshold is approximately $0.5\,\mathrm{mm\,h^{-1}}$. At 6 h timescale, the yearly snowfall amount is entirely associated to snowfall intensities lower than $2\,\mathrm{mm\,h^{-1}}$ for ERA-Interim, while intensities up to $4.4\,\mathrm{mm\,h^{-1}}$ have been measured with the MRR. (Fig. 11)

Visual observations provided by Météo France are not limited to precipitation, and several present weather codes are archived. At DDU, those are SYNOP codes belonging to the group 7wwW1W2. In the codes recurring at DDU, three types of phenomena are mostly documented: clouds (codes 1–3), blowing snow (codes 36–39), and snow (codes 22 and 71). We consider here the codes related to snow and blowing

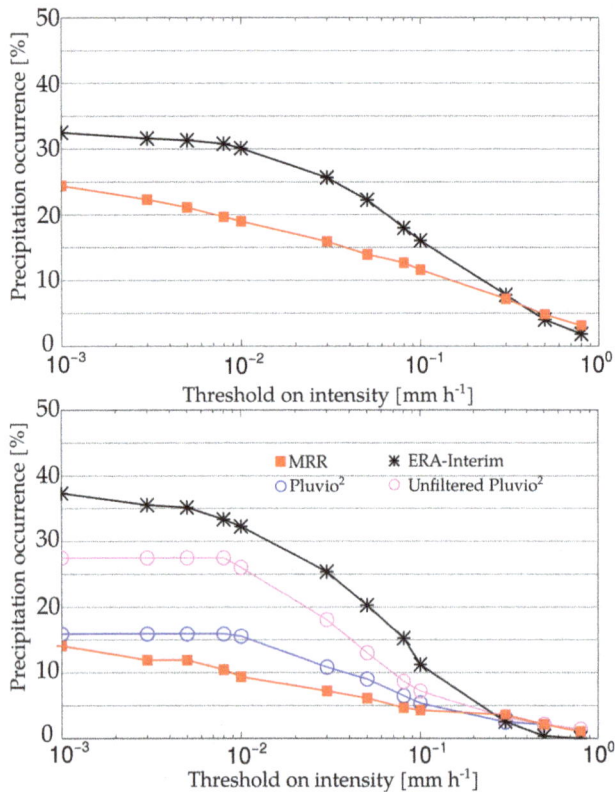

Figure 12. Precipitation occurrence as a function of the threshold on precipitation intensity for the full year (top) and summer period (bottom) respectively. The precipitation occurrence timescale is 6 h.

Table 4. Contingency table between the occurrence of near-ground (300 m) MRR signal (columns) and visual observations of snow and blowing snow conducted by Météo France (rows). A threshold on the MRR data of 10^{-2} mm h^{-1} is used to discriminate between dry sky and precipitation. The elements of the contingency table are normalized to the total number of observations and sum up to 100 %.

	Precipitation (MRR)	Dry (MRR)
Blowing snow	14	44.4
Snow	34.9	6.7

snow and we disregard the observations of clouds. The observations are conducted during each day, on average every 5 h (with higher frequency during day hours), and we compare them here with the MRR measurements. This is shown in Table 4, where the MRR observations at 300 m height are compared with the visual observations of snow and blowing snow.

Given the intrinsic difference between those observations, it is not possible to take one as an overall reference in the confusion matrix. However, it can be assumed that visual observations are better at reporting blowing snow, because they are conducted at the ground level, while MRR measurements at 300 m above ground are better at reporting snow-

fall. The most interesting outcomes of this comparison are the following ones. First, there is a good correspondence between occurrence of snow according to the MRR and visual observations of snow. Second, blowing snow occurrence is not well captured by the MRR. When visual observations report blowing snow, the MRR mostly does not report any occurrence. 44.4 % of the visual reports analyzed here correspond to cases where blowing snow has been observed at the ground level, but no valid signal has been recorded by the MRR. This comparison is to a certain degree dependent on the threshold used to discriminate between dry weather and precipitation in the MRR data. However, similar results have been obtained for various threshold levels (not shown here). This result, as discussed in the next section, is not unexpected because blowing snow rarely exceeds heights of 100 m above terrain (Palm et al., 2011).

4 Discussion

4.1 Microphysical observations

The microphysical observations, collected during the austral summer 2015–2016, and illustrated in Figs. 7 and 8 suggest that even at this location on the Antarctic coasts, riming is an important microphysical process. From radar retrievals close to the ground level, about 10 % of precipitation is rimed. According to the MASC classification, 11 % of the hydrometeors are fully rimed (graupel) and most of the other hydrometeor types have a degree of riming greater than 0, in particular aggregates that tend to be larger and easier targets for riming (e.g., Houze and Medina, 2005). The presence of riming indirectly shows that mixed-phase clouds often occur and that supercooled liquid water is available in the regions of precipitation formation. This has been documented in the past at this location by Del Guasta et al. (1993), and it could be observed also in the test data collected with the depolarization lidar (see Table 1, and Fig. 3) during the summer period.

At the ground level the large majority of hydrometeors recorded by the MASC were small particles of nondiscernible habit and nondefinable riming degree (given the MASC resolution), with an occurrence three times higher than similar measurements conducted in Alpine locations. This is probably the signature of the significant contribution of blowing snow to the near-ground snow flux, which is particularly effective in recirculating small and light particles (e.g., Mann et al., 2000; Gordon et al., 2009), but it could partly be the result of the fragmentation of aggregates in the lower levels of the atmosphere, where strong katabatic winds blow.

4.2 Blowing snow and wind effect

The contribution of blowing snow was visible in the observations collected with the MASC, generating very large number of small hydrometeors, and strong winds affected Pluvio^2 measurements, generating phantom accumulations probably due to the vibrations of the weighing gauge. As illustrated in Fig. 5, periods with wind speeds exceeding roughly 15 ms^{-1} at the proximity of the inlet of the instrument generate phantom accumulations of precipitation that are removed if the MRR does not receive any signal at the same time at its lowest gate. The lowest gate of the MRR (300 m in this case) is considered to be high enough to be above the height of any wind-blown snow layer (Gordon et al., 2009; Scarchilli et al., 2010; Palm et al., 2011), which rarely exceeds 200 m of vertical development.

This combination of ground-based and in situ instruments should be proposed again, and maintained for at least a full year to cover all the seasons. The main limitation of the filter of Pluvio^2 data is that they cannot detect cases when blowing snow occurs together with precipitation. In the case of our measurements, the total accumulated precipitation of the Pluvio^2 in the summer period drops from 66 to 52 mm, after the censoring is performed. The removed portion is then 21 % of the total raw accumulated precipitation.

Regarding blowing snow, the comparison between the occurrence of signal at the lowest MRR gate and the visual weather reports conducted by Météo France, summarized in Table 4, confirms that the current MRR configuration does not capture the occurrence of blowing snow events. Lower range gate spacing should be employed (lower than 100 m used in the measurements shown here) if blowing snow is of interest.

4.3 Precipitation quantification

The quantification of precipitation in the coastal regions of Antarctica remains a difficult task, affected by significant uncertainty. This study provides some estimates that help contextualize the information available until now. Figure 9 and the summary in Table 3 shows that MRR estimates of total accumulated precipitation at the yearly scale can diverge significantly (from 484 to 1581 mm) if a range of standard Z–S relations is used, while the use of a local Z–S relation, calibrated in the summer season, allows for a significant reduction of this range of values (from 740 to 989 mm). In this case, however, an important assumption is made; i.e., that this relation can be considered representative of the other seasons.

ERA-Interim provides a yearly estimate of 655 mm for the measurement period from November 2015 to November 2016, about 10 % lower than the lowest estimate obtained with the local Z–S relation. It must be underlined once more that the estimates provided by the MRR at DDU correspond to a minimum height of 300 ($\pm$50 m above ground corresponding to a range gate spacing of 100 m). This, together with the large grid of ERA-Interim, may contribute to the differences observed with the MRR. Interestingly, ERA-Interim is initially in agreement with the MRR precipitation values

until March 2016, but later on it underestimates them. This may also be due to a seasonal change in snowfall type, which is no longer representative of the summer snowfall events used to build the Z–S relation.

As an external reference, the mean climatological estimate proposed by Palerme et al. (2014) is of 679 mm over DDU (climatology obtained for the period 2006 to 2011), a value not very far from the 2015/2016 measurements.

The year of measurements (2015/2016) was characterized by a significant interseasonal and intermonth variability. However, according to ERA-Interim records, the monthly totals are within what has been observed since 1995, with the exception of the snowy month of September 2016.

4.4 Precipitation occurrence

Occurrence is an interesting parameter, because it is the only precipitation-related measurement that has been collected on the DDU base for a long time. In terms of precipitation occurrence we take the MRR as a reference, because visual reports are discontinuous and affected by the limitations of visibility that can occur near the ground. Figure 12 shows that the year under investigation had some peculiarities: the month of September had the highest occurrence of precipitation since 1981, while January, February, and April all took records of the lowest monthly occurrence for the same period.

ERA-Interim generally overestimates the occurrence of precipitation, which could be caused by a sampling effect due to the much larger grid size of ERA-Interim with respect to the local MRR measurements, despite temporal integration to 6 h to reduce this effect. This overestimation is evident mostly in summer, in particular in December and January and it is well depicted by the time series of Fig. 9, bottom. With respect to the MRR, ERA-Interim tends to overestimate the occurrence of low-intensity events and underestimate the occurrence of high-intensity events (illustrated in Fig. 12). An optimal threshold that matches the two occurrences over the year of measurement (at 6 h scale) is between 0.1 and 0.5 mm h^{-1}. The events of highest intensity, which can contribute to the major part of the yearly snowfall accumulation (e.g., Gorodetskaya et al., 2014, 2015) in East Antarctica, did not occur in summer. This can explain the underestimation of ERA-Interim amounts starting in March 2016 (Fig. 9).

Visual observations tend to underestimate the occurrence of precipitation, as shown in Fig. 12, but they rarely produce false alarms of precipitation, as visible in Table 4. In other words, when visual observations report precipitation, they are generally correct, but they can miss some occurrences, probably due to visibility limitations, human errors, confusion with wind-blown snow, and due to the discontinuous nature of human observations. In Fig. 12, we can observe no clear seasonality in the underestimation of occurrence. For example a large underestimation is observed both in March and in July, while in June, April, and September the occurrence is very close between MRR and visual reports. This is to a certain extent surprising, as a larger missed detection rate could be expected during the dark winter months, when the reduced visibility may affect human observations.

5   Summary and conclusions

In this paper we present unprecedented observations of precipitation collected at a coastal location of East Antarctica since October 2015. Several remote sensing and in situ instruments collected measurements during summer 2015/2016, and one (the MRR) has been operating continuously since then. These instruments have provided an insightful example of their usefulness to monitor precipitation on the Antarctic continent. It has been shown that radar data can be used to remove phantom accumulations from in situ weighing gauge observations. These accumulations, occurring in high-wind conditions and tracked down to be due mostly to vibrations, accounted for 21 % of the total accumulated precipitation of the summer period. Microphysical observations at the ground level, collected by the MASC in summer, showed that the large majority of hydrometeors (54 %) were small ice particles of nondefined habit, probably resulting from blowing snow, followed by aggregates (19 %), and other hydrometeor types. Both from radar-based hydrometeor classification and from MASC measurements, it appeared that riming is a significantly active process. About 10 % of the radar measurements at low-level were classified as containing rimed hydrometeors, 11 % of the hydrometeors were classified as fully developed graupel (23 % if small particles are not considered), and most of the other hydrometeors classified with the MASC showed riming degrees even larger than 0.5. The presence of supercooled liquid water, a necessary ingredient for riming, has been reported at DDU by previous studies and was evident in the lidar measurements collected in 2015.

One year of MRR data allowed for the estimation of the total yearly precipitation, from October 2015 to October 2016, giving values ranging between 740 and 989 mm, at least 10 % larger than that provided by ERA-Interim reanalysis (655 mm). The MRR estimates were based on a local reflectivity-to-snowfall rate relation, obtained for summer snowfall data only. An important assumption, which will need to be verified or improved, is that we considered this relation representative for the entire year of MRR measurements. Precipitation occurrence was generally overestimated by ERA-Interim with respect to the MRR, especially in the summer period, and was underestimated by the visual reports collected by Météo France. The overestimation of occurrence by ERA-Interim could be due to its microphysical parameterization or to a spatial resolution that is very different from the one of the point measurements used as a reference. On the contrary, the underestimation of occurrence by visual reports is probably due to their discontinuous nature and the difficulties in discriminating, at the ground, pure precipitation and blowing snow. Even though they underestimate occurrence, visual observations had a very low false alarm rate on occurrence. It is worth underlining that the overestimation of occurrence by ERA-Interim partially compensates for an observed underestimation of snowfall amounts for the most significant snowfall events. This compensation, over long time periods, may lead to an overestimate of the performance of the model for individual precipitation events.

It was shown that the MRR, whose lowest measurements are about 300 m above ground (third gate with a 100 m resolution), is not able to detect blowing snow. This means that a configuration with a higher range resolution, at the expense of a lower maximum sampled height, must be used if this instrument is required to monitor blowing snow.

The measurements collected at DDU and illustrated in this paper show the potential of ground-based instruments to complement and validate satellite and numerical weather prediction model products related to precipitation. Such measurements can also provide information about the microphysical aspects of precipitation, like the dominant hydrometeor types and their degree of riming in the present case. The synergy between remote sensing and in situ instruments has the potential to improve the quantification of snowfall amounts in conditions where strong winds affect ground-based measurements, even though much remains to be done in cases when precipitation and blowing snow occur at the same time. The installation and long-term operation of a similar combination of instruments should be conducted again at DDU and at other locations in Antarctica. Efforts will be devoted to develop a better long-term constraint for radar-based snowfall estimations by means of in situ measurements of precipitation in synergy with microphysical observations and retrievals, because the relation used in this study was built on summer data only. Future work should also focus on better discriminating between snowfall and blowing snow, on the validation of satellite-based snowfall retrievals since it is of great interest to monitor the entire Antarctic continent, and in further validating ERA-Interim reanalyses and other weather and climate models.

Competing interests. The authors declare that they have no conflict of interest.

Acknowledgements. The authors are thankful to Météo France, and in particular to the team of Dumont d'Urville who provided the access to their in situ measurements and observations. We thank the French Polar Institute (IPEV), in particular Gregory Tran,

Doris Thuillier, and Patrice Godon, who allowed the APRES3 measurement campaign to take place. We thank Paul Dufay, overwinterer at DDU, who provided crucial assistance for the operation of the MRR during the winter season. The first author, Jacopo Grazioli, thanks the Swiss National Science foundation SNF for the grant 200021_163287, financing his participation to the project. The authors also acknowledge the support of the French National Research Agency (ANR) to the APRES3 project and also of CNES/TOSCA, program EECLAT. For the remote technical support provided, we want to thank Andrew Pazmany and Johnatan Leachman (Prosensing Inc., manufacturer of MXPol). Jean-Baptiste Madeleine also thanks UPMC university for financial assistance. We are thankful to Tim Raupach and Christophe Praz (EPFL LTE) for the help in proofreading and in the revision of the manuscript.

Edited by: Michiel van den Broeke

References

Behrangi, A., Christensen, M., Richardson, M., Lebsock, M., Stephens, G., Huffman, G. J., Bolvin, D., Adler, R. F., Gardner, A., Lambrigtsen, B., and Fetzer, E.: Status of high-latitude precipitation estimates from observations and reanalyses, J. Geophys. Res.-Atmos., 121, 4468–4486, https://doi.org/10.1002/2015JD024546, 2016.

Bromwich, D. H.: Estimates of Antarctic Precipitation, Nature, 343, 627–629, https://doi.org/10.1038/343627a0, 1990.

Bromwich, D. H., Nicolas, J. P., and Monaghan, A. J.: An assessment of precipitation changes over Antarctica and the Southern Ocean since 1989 in contemporary global reanalyses, J. Climate, 24, 4189–4209, https://doi.org/10.1175/2011JCLI4074.1, 2011.

Colli, M., Lanza, L. G., La Barbera, P., and Chan, P. W.: Measurement accuracy of weighing and tipping-bucket rainfall intensity gauges under dynamic laboratory testing, Atmos. Res., 144, 186–194, https://doi.org/10.1016/j.atmosres.2013.08.007, 2014.

Cullather, R. I., Bromwich, D. H., and Van Woert, M. L.: Spatial and temporal variability of Antarctic precipitation from atmospheric methods, J. Climate, 11, 334–367, https://doi.org/10.1175/1520-0442(1998)011<0334:SATVOA>2.0.CO;2, 1998.

DeConto, R. M. and Pollard, D.: Contribution of Antarctica to past and future sea-level rise, Nature, 531, 591–597, https://doi.org/10.1038/nature17145, 2016.

Del Guasta, M., Morandi, M., Stefanutti, L., Brechet, J., and Piquad, J.: One-year of cloud lidar data from Dumont-Durville (Antarctica). 1. General overview of geometrical and optical-properties, J. Geophys. Res., 98, 18575–18587, https://doi.org/10.1029/93JD01476, 1993.

Frezzotti, M., Pourchet, M., Flora, O., Gandolfi, S., Gay, M., Urbini, S., Vincent, C., Becagli, S., Gragnani, R., Proposito, M., Severi, M., Traversi, R., Udisti, R., and Fily, M.: New estimations of precipitation and surface sublimation in East Antarctica from snow accumulation measurements, Clim. Dynam., 23, 803–813, https://doi.org/10.1007/s00382-004-0462-5, 2004.

Garrett, T. J., Fallgatter, C., Shkurko, K., and Howlett, D.: Fall speed measurement and high-resolution multi-angle photography of hydrometeors in free fall, Atmos. Meas. Tech., 5, 2625–2633, https://doi.org/10.5194/amt-5-2625-2012, 2012.

Gordon, M., S. and Taylor, P. A.: Measurements of blowing snow, Part I: Particle shape, size distribution, velocity, and number flux at Churchill, Manitoba, Canada, Cold Reg. Sci. Technol., 55, 63–74, https://doi.org/10.1016/j.coldregions.2008.05.001, 2009.

Gorodetskaya, I. V., Tsukernik, M., Claes, K., Ralph, M. F., Neff, W. D., and Van Lipzig, N. P. M.: The role of at-mospheric rivers in anomalous snow accumulation in East Antarctica, Geophys. Res. Lett., 41, 6199–6206, https://doi.org/10.1002/2014GL060881, 2014.

Gorodetskaya, I. V., Kneifel, S., Maahn, M., Van Tricht, K., Thiery, W., Schween, J. H., Mangold, A., Crewell, S., and Van Lipzig, N. P. M.: Cloud and precipitation properties from ground-based remote-sensing instruments in East Antarctica, The Cryosphere, 9, 285–304, https://doi.org/10.5194/tc-9-285-2015, 2015.

Grazioli, J., Tuia, D., and Berne, A.: Hydrometeor classification from polarimetric radar measurements: a clustering approach, Atmos. Meas. Tech., 8, 149–170, https://doi.org/10.5194/amt-8-149-2015, 2015.

Hogan, R. J., Francis, P. N., Flentje, H., Illingworth, A. J., Quante, M., and Pelon, J.: Characteristics of mixed-phase clouds. I: Lidar, radar and aircraft observations from CLARE'98, Q. J. Roy. Meteor. Soc., 129, 2089–2116, https://doi.org/10.1256/qj.01.208, 2003.

Houze, R. A. and Medina, S.: Turbulence as a mechanism for orographic precipitation enhancement, J. Atmos. Sci., 62, 3599–3623, 2005.

König-Langlo, G., King, J. C., and Pettré, P.: Climatology of the three coastal Antarctic stations Dumont d'Urville, Neumayer, and Halley, J. Geophys. Res., 103, 10935–10946, https://doi.org/10.1029/97JD00527, 1998.

Konishi, H., Wada, M., and Endoh, T.: Seasonal variations of cloud and precipitation at Syowa station, Antarctica, Ann. Glaciol., 27, 597–602, 1998.

Krinner, G., Magand, O., Simmonds, I., Genthon, C., and Dufresne, J. L.: Simulated Antarctic precipitation and surface mass balance at the end of the twentieth and twenty-first centuries, Clim. Dynam., 28, 215–230, https://doi.org/10.1007/s00382-006-0177-x, 2007.

Lenaerts, J. T. M., Vizcaino, M., Fyke, J., van Kampenhout, L., and van den Broeke, M. R.: Present-day and future Antarctic ice sheet climate and surface mass balance in the Community Earth System Model, Clim. Dynam., 47, 1367–1381, https://doi.org/10.1007/s00382-015-2907-4, 2016.

Maahn, M. and Kollias, P.: Improved Micro Rain Radar snow measurements using Doppler spectra post-processing, Atmos. Meas. Tech., 5, 2661–2673, https://doi.org/10.5194/amt-5-2661-2012, 2012.

Maahn, M., Burgard, C., Crewell, S., Gorodetskaya, I. V., Kneifel, S., Lhermitte, S., Van Tricht, K., and van Lipzig, N. P. M.: How does the spaceborne radar blind zone affect derived surface snowfall statistics in polar regions?, J. Geophys. Res., 119, 13604–13620, https://doi.org/10.1002/2014JD022079, 2014.

Magono, C. and Lee, C. W.: Meteorological classification of natural snow crystals, J. Fac. Sci., Hokkaido Univ., Series VII, 2, 321–335, 1966.

Mann, G., Anderson, P., and Mobbs, S.: Profile measurements of blowing snow at Halley, Antarctica, J. Geophys. Res., 105, 24491–24508, https://doi.org/10.1029/2000JD900247, 2000.

Matrosov, S. Y., Campbell, C., Kingsmill, D., and Sukovich, E.: Assessing snowfall rates from X-band Radar reflectivity measurements, J. Atmos. Ocean. Tech., 26, 2324–2339, https://doi.org/10.1175/2009JTECHA1238.1, 2009.

Palerme, C., Kay, J. E., Genthon, C., L'Ecuyer, T., Wood, N. B., and Claud, C.: How much snow falls on the Antarctic ice sheet?, The

Cryosphere, 8, 1577–1587, https://doi.org/10.5194/tc-8-1577-2014, 2014.

Palerme, C., Genthon, C., Claud, C., Kay, J. E., Wood, N. B., and L'Ecuyer, T.: Evaluation of current and projected Antarctic precipitation in CMIP5 models, Clim. Dynam., 48, 1–15, https://doi.org/10.1007/s00382-016-3071-1, 2016.

Palm, S. P., Yang, Y., Spinhirne, J. D., and Marshak, A.: Satellite remote sensing of blowing snow properties over Antarctica, J. Geophys. Res., 116, D16123, https://doi.org/10.1029/2011JD015828, 2011.

Praz, C., Roulet, Y.-A., and Berne, A.: Solid hydrometeor classification and riming degree estimation from pictures collected with a Multi-Angle Snowflake Camera, Atmos. Meas. Tech., 10, 1335–1357, https://doi.org/10.5194/amt-10-1335-2017, 2017.

Rignot, E., Velicogna, I., van den Broeke, M. R., Monaghan, A., and Lenaerts, J. T. M.: Acceleration of the contribution of the Greenland and Antarctic ice sheets to sea level rise, Geophys. Res. Lett., 38, L05503, https://doi.org/10.1029/2011GL046583, 2011.

Ryzhkov, A. V., Giangrande, S. E., Melnikov, V. M., and Schuur, T. J.: Calibration issues of dual-polarization radar measurements, J. Atmos. Ocean. Tech., 22, 1138–1155, https://doi.org/10.1175/JTECH1772.1, 2005.

Scarchilli, C., Frezzotti, M., Grigioni, P., De Silvestri, L., Agnoletto, L., and Dolci, S.: Extraordinary blowing snow transport events in East Antarctica, Clim. Dynam., 34, 1195–1206, https://doi.org/10.1007/s00382-009-0601-0, 2010.

Schlosser, E., Powers, J. G., Duda, M. G., Manning, K. W., Reijmer, C. H., and Van Den Broeke, M. R.: An extreme precipitation event in Dronning Maud Land, Antarctica: a case study with the Antarctic Mesoscale Prediction System, Phys. Rev. E, 29, 330–344, https://doi.org/10.1111/j.1751-8369.2010.00164.x, 2010.

Schneebeli, M., Dawes, N., Lehning, M., and Berne, A.: High-resolution vertical profiles of polarimetric X-band weather radar observables during snowfall in the Swiss Alps, J. Appl. Meteorol. Clim., 52, 378–394, https://doi.org/10.1175/JAMC-D-12-015.1, 2013.

Scipion, D., Mott, R., Lehning, M., Schneebeli, M., and Berne, A.: Seasonal small-scale spatial variability in alpine snowfall and snow accumulation, Water Resour. Res., 49, 1446–1457, https://doi.org/10.1002/wrcr.20135, 2013.

Vardiman, L.: The generation of secondary ice particles in clouds by crystal-crystal collision, J. Atmos. Sci., 35, 2168–2180, 1978.

Vaughan, D. G., L., B. J., Giovinetto, M., Russell, J., and Cooper, A. P. R.: Reassessment of the net surface mass balance in Antarctica, J. Climate, 12, 933–946, 1999.

Welker, C., Martius, O., Froidevaux, P., Reijmer, C. H., and Fischer, H.: A climatological analysis of high-precipitation events in Dronning Maud Land, Antarctica, and associated large-scale atmospheric conditions, J. Geophys. Res., 119, 11932–11954, https://doi.org/10.1002/2014JD022259, 2014.

Wendler, G., Stearns, C., Weidner, G., Dargaud, G., and Parish, T.: On the extraordinary katabatic winds of Adelie Land, J. Geophys. Res.-Atmos ., 102, 4463–4474, https://doi.org/10.1029/96JD03438, 1997.

Witze, A.: CLIMATE SCIENCE Antarctic cloud study takes off, Nature, 529, p. 12, 2016.

World Meteorological Organization: Project team and (reduced) international organizing committee for the WMO solid precipitation intercomparison experiment. Fifth Session. Final Report, WMO-No.168, World Meteorological Organization, Sodankyla, Finland, 5th edn., 2014.

Unmanned aerial system nadir reflectance and MODIS nadir BRDF-adjusted surface reflectances intercompared over Greenland

John Faulkner Burkhart[1,2]**, Arve Kylling**[3]**, Crystal B. Schaaf**[4]**, Zhuosen Wang**[5,6]**, Wiley Bogren**[7]**, Rune Storvold**[8]**, Stian Solbø**[8]**, Christina A. Pedersen**[9]**, and Sebastian Gerland**[9]

[1]Department of Geosciences, University of Oslo, Oslo, Norway
[2]University of California, Merced, CA, USA
[3]Norwegian Institute for Air Research, Kjeller, Norway
[4]School for the Environment, University of Massachusetts Boston, Boston, MA, USA
[5]NASA Goddard Space Flight Center, Greenbelt, MD, USA
[6]Earth System Science Interdisciplinary Center, University of Maryland, College Park, MD, USA
[7]U.S. Geological Survey, Flagstaff, AZ, USA
[8]Norut-Northern Research Institute, Tromsø, Norway
[9]Norwegian Polar Institute, Fram Centre, Tromsø, Norway

Correspondence to: John Faulkner Burkhart (john.burkhart@geo.uio.no)

Abstract. Albedo is a fundamental parameter in earth sciences, and many analyses utilize the Moderate Resolution Imaging Spectroradiometer (MODIS) bidirectional reflectance distribution function (BRDF)/albedo (MCD43) algorithms. While derivative albedo products have been evaluated over Greenland, we present a novel, direct comparison with nadir surface reflectance collected from an unmanned aerial system (UAS). The UAS was flown from Summit, Greenland, on 210 km transects coincident with the MODIS sensor overpass on board the Aqua and Terra satellites on 5 and 6 August 2010. Clear-sky acquisitions were available from the overpasses within 2 h of the UAS flights. The UAS was equipped with upward- and downward-looking spectrometers (300–920 nm) with a spectral resolution of 10 nm, allowing for direct integration into the MODIS bands 1, 3, and 4. The data provide a unique opportunity to directly compare UAS nadir reflectance with the MODIS nadir BRDF-adjusted surface reflectance (NBAR) products. The data show UAS measurements are slightly higher than the MODIS NBARs for all bands but agree within their stated uncertainties. Differences in variability are observed as expected due to different footprints of the platforms. The UAS data demonstrate potentially large sub-pixel variability of MODIS reflectance products and the potential to explore this variability using the UAS as a platform. It is also found that, even at the low elevations flown typically by a UAS, reflectance measurements may be influenced by haze if present at and/or below the flight altitude of the UAS. This impact could explain some differences between data from the two platforms and should be considered in any use of airborne platforms.

1 Introduction

Albedo, the ratio of reflected to incident energy at the surface of the earth, is a fundamental parameter in energy balance computations, and therefore any prediction of climate must account for albedo through a parameterization process (Henderson-Sellers and Wilson, 1983). Generally climate models rely on simplified estimations of albedo as single-value climatological means of broadband albedo that are a function of seasonal changes in surface characteristics and the presence of snow (Curry and Schramm, 2001). For modeling snow and ice melt processes on the earth's surface, albedo is a critical parameter, providing the most coarse adjustment with respect to available energy to drive melt. Satellite instruments play an important role in providing a characterization of albedo of the surface of earth that is relevant for climate and earth system modeling.

Stroeve et al. (2005) and more recently Stroeve et al. (2013) have carefully evaluated the Moderate Resolution Imaging Spectroradiometer (MODIS) albedo products over Greenland. Several further studies have evaluated presently available satellite products and compared these products with ground-based observations. This recent body of work has been largely spurred by the 2012 melt events on the Greenland Ice Sheet (GrIS). These events, recorded by MODIS satellite observations, have been linked to albedo feedback stemming from thermodynamic processes (Box et al., 2012). Dumont et al. (2014), Goelles et al. (2015), and Keegan et al. (2014) attribute some of the darkening to deposition of soot from forest fires, pollution, and dust, while others have linked the changes predominately to delivery of warm water vapor and low-level clouds (Bennartz et al., 2013; Miller et al., 2015).

Due to the impact on snow and ice albedo, the Intergovernmental Panel on Climate Change (IPCC) identified black carbon on snow as an important process driving changes in the cryospheric energy balance with significant associated uncertainty (IPCC, 2013). These findings are based on research that has focused on the theoretical response of snow to black carbon deposition (Warren and Wiscombe, 1980; Hansen and Nazarenko, 2004). Subsequent studies have shown that small changes in snow albedo globally may have significant impact on the top-of-atmosphere forcing and could be driving a component of the Arctic warming witnessed today (Flanner et al., 2009). However, presenting a distinct challenge, Warren (2012) suggests that the changes anticipated from this effect are below the present-day measurement capabilities.

Numerous studies have used ground-based measurements to compare and validate satellite sensor data, a necessary process to assess accuracy of the observations, and particularly to understand the variability that may be missed by different sensor footprint scales. A seminal study is that of Salomonson and Marlatt (1971), who conducted an evaluation of surface reflectance conditions for application to retrievals from the Medium-Resolution Infrared Radiometer (MRIR) instrument aboard the Nimbus II and III satellites. The study focused on the measurement of bidirectional reflectance distribution function (BRDF) over a variety of terrestrial surfaces. Appreciable anisotropy on all the surfaces was found, leading to the conclusion of the importance of using a BRDF model for satellite retrievals – a standard application today (Schaaf et al., 2002; Jin, 2003; Román et al., 2009; Ju et al., 2010; Wang et al., 2014).

Wright et al. (2014) intercompare ground-based spectral observations from an ASD FieldSpec spectroradiometer at Summit station with both the MODIS Collection 5 (C5) and Collection 6 (C6) data, and show a marked improvement of the MODIS C6 retrieval. Prior studies relied predominately on existing GrIS fixed station data from the Greenland Climate Network (GC-Net) of automatic weather stations (Box et al., 2012; Stroeve et al., 2005, 2006, 2013). To our knowledge most investigations have compared ground-based measurements of albedo with satellite albedo products. While this is valuable, one must recognize that albedo products are developed through a processing chain of models and therefore do not represent a direct measurement, making comparisons complicated.

Recently, the advent of relatively low-cost unmanned airborne systems (UASs) has created a rush to utilize this novel platform to provide unique data sets otherwise unobtainable without manned flight. Furthermore, UASs provide a unique niche in the ability to characterize cryospheric surfaces in relatively localized regions at higher resolutions than may be possible with traditional aircraft, and they certainly offer the potential to extend the observational range of a traditional ground-based campaign (Bhardwaj et al., 2016). In one of the first applications of a UAS for cryospheric characterization, Hakala et al. (2014) demonstrated the potential for BRDF measurements from a simple quadcopter. Immerzeel et al. (2014) used a UAS to characterize glacial dynamics in the Himalaya, while numerous other have recently applied structure from motion photogrammetry to several applications related to snow and ice surfaces (e.g., Jagt et al., 2015; Ryan et al., 2015; Rippin et al., 2015). Most recently, Ryan et al. (2017) have attempted to use cameras to measure albedo from a UAS on board a fixed-wing platform on the perimeter of Greenland.

Herein we provide a first-of-a-kind, "apples-to-apples" evaluation of the accuracy of the MODIS nadir BRDF-adjusted reflectance (NBAR) retrievals through intercomparison with reflectance observed from a UAS platform over Greenland. The advent of UASs presents an immense opportunity to spatially assess the accuracy of satellite sensors, versus simple validation against ground point observations. However, as discussed in this work, there are a host of complications that must be considered. We explore those further herein.

In this study, spectral reflectance measurements made from a UAS flying in the dry snow region near Summit, Greenland, are used to evaluate sub-pixel-scale variability of the MODIS NBAR retrievals. The campaign was conducted in 2010 and provides data for two transects on separate days coincident with the near nadir subtrack of the MODIS instrument overpasses. Due to the pristine nature of the snow surface in this area, and the limited influence of aerosols and warm temperatures, albedo and reflectance variability in this region is expected to be less than 10%, and potentially as low as 3% – within the 5% stated accuracy of the MODIS data sets. The standard MODIS products retrieve narrowband reflectance and then use a narrowband-to-broadband algorithm to convert the discrete narrowband measurements into a broadband albedo (Schaaf et al., 2002; Stroeve et al., 2005). Our approach allows a direct comparison with MODIS, as the UAS observations have a high spectral resolution and therefore may be integrated directly to narrowband reflectance values using the MODIS response functions for bands 1, 3, and 4 (see Sect. 5).

The "Variability of Albedo Using Unmanned Aerial Vehicles" (VAUUAV) project had a primary objective of evaluating whether present-day satellite observations allow the capacity to evaluate the impact of aerosols on albedo variability across a cryospheric landscape and to provide input for validation of theoretical modeling. However, through our work, we discovered that the application of UAS to obtain data suitable for validation of MODIS data sets is greatly complicated by aspects of the platform that to date have not been addressed, particularly with respect to albedo. Therefore, as we have the capability, we have chosen to conduct the comparison with reflectance – providing a direct evaluation of the platform capabilities. Further we have attempted to address several of the complex issues that result from the UAS in this analysis and otherwise highlight the potential for uncertainty in the observations. In Sect. 2 we present the UAS measurement platform. Section 3 describes radiative transfer calculations, while Sect. 4 describes the data selection, and Sect. 5 describes the MODIS data used for this study. A comparison and discussion of the MODIS and UAS data are presented in the Sect. 6 followed by a summary and conclusion.

2 Surface reflectance measurements from an unmanned aerial system

The Cryowing (see Sect. 2.1) UAS performed several flights during the summer of 2010 in the region of Summit, Greenland, at the Greenland Environmental Observatory (http://www.geosummit.org). The flights were designed to measure the downwelling irradiance and the upwelling nadir radiance as discussed further in Sect. 2.2.

Two of the flights were specifically designed to be closely aligned with MODIS overpasses and were flown as close in time as possible to the satellite overpasses. On 5 and 6 August 2010, the UAS completed a flight pattern with coverage over a region which was nearly coincident with the MODIS sensor overpass on board the Aqua and Terra satellites. On both days, clear-sky acquisitions are available from overpasses within 2 h of the UAS flights. The flight pattern covered 210 km ground distance and was completed autonomously for a duration of over 2 h. From the UAS observations we develop a nadir data set suitable for direct comparison with the MODIS NBAR products with a reduced reliance on a model chain.

2.1 The Cryowing UAS

The Cryowing UAS is an autonomous fixed-wing airborne sensor platform developed in Norway. It has a maximum takeoff weight of 30 kg, payload capacity of 15 kg including fuel, and a wingspan of 3.8 m. The Cryowing is powered by a two-stroke engine, fueled by a petrol–oil mixture. The normal cruising speed is 100–120 km h^{-1}, with a range of up to 500 km or 5 h flight. The Cryowing has a 2500 m dynamic altitude range, with a 5000 m absolute altitude cap. A dedicated GPS independent from the payload is used for navigation and autopilot control. While the Cryowing is capable of autonomous control for the full period of a flight, in practice a skilled technician is present to control launch and landing via radio control. Once stable flight is achieved, communication with the UAS is maintained for the duration of the flight using a radio modem or Iridium satellite modem.

A standard suite of instruments is deployed on the Cryowing. This includes a meteorological package, two inertial measurement units (IMUs), and two GPS systems, as well as the computer and communications systems responsible for flight control and ground station contact. The meteorological package measures air pressure, temperature, and humidity, while the flight computer and systems record aircraft position and altitude. Position is recorded as altitude in meters along with latitude and longitude, while attitude is recorded in quaternion form. In this analysis, platform attitude was converted to the azimuth and zenith angles for subsequent radiative transfer modeling as discussed in Sect. 3. Position and attitude variables were recorded at a frequency of 100 Hz by the IMU.

2.2 UAS-based surface reflectance

In the VAUUAV payload configuration, the Cryowing is equipped with two TriOS RAMSES spectroradiometers to measure reflected and incoming radiation in the visible spectrum (VIS). The upward-facing sensor, measuring incoming radiation, has a cosine-corrected fore-optic made of synthetic fused silica, transparent to 190 nm, to measure full-sky hemisphere irradiance. The nadir-facing sensor has a Gershun tube restricted 7° field-of-view (FOV) fore-optic, measuring reflected radiance emanating from a footprint beneath the plane. At a cruise altitude of 250 m the footprint is on the order of 30 m in diameter.

Details of the TriOS sensors and the configuration used can be found in Nicolaus et al. (2010). For completeness, we describe the essential characteristics of the spectral radiometers here. The TriOS RAMSES ACC-2 VIS hyper-spectral radiometers are based on a miniature spectrometer with a wavelength range from 310 to 1100 nm and spectral resolution and accuracy of 3.3 and 0.3 nm, respectively. TriOS uses the VIS–near-infrared (NIR) specification of the spectroradiometers (wavelength range from 360 to 900 nm) to post-calibrate the instruments to a wavelength range from 320 to

950 nm. Software controlling the instruments enables an automatic adjustment of the integration time for each measurement, ranging between 4 and 8192 ms.

There are two versions of the sensors, one containing an inclination and pressure sensor, and one without. For our purposes, these additional components were not required, as a part of the standard suite of measurements aboard the UAS includes highly accurate inclination from the IMU. As the sensors were initially designed for water quality applications, they are built to be water resistent to a depth of 300 m. This creates additional weight due to the robust design of the casing and sealed body of the sensors. In order to use the sensors in the UAS, modifications from the sensors described in Nicolaus et al. (2010) were required. To reduce the length and weight of the sensors as available from the manufacturer (and described in Nicolaus et al., 2010), the solid-steel casing was removed and replaced with a lightweight aluminum version. This reduced the weight of the sensors significantly from the initial 833 g to less than 400 g.

The in situ observations presented in this study provide the downwelling irradiance ($E(\theta_i)$) and the upwelling radiance $L_r(\theta_i, \phi_i; \theta_r, \phi_r)$ with a field of view of $7°$ in the direction $\theta_r = \pi/2$, $\phi_r = 0$ for incident zenith (azimuth) angle θ_i (ϕ_i).

The nadir reflectance measured by the UAS is

$$\rho = \frac{\pi L_r\left(\theta_i, \phi_i; \pi/2, 0\right)}{E(\theta_i)}, \tag{1}$$

which may be directly compared with the NBAR from MODIS (Schaaf et al., 2002). All wavelength dependence in Eq. (1) has been omitted for clarity.

Our instrument measures spectral reflectance from 320 to 950 nm with 3.3 nm per pixel resolution with a 3 nm oversampling, making an effective 10 nm resolution, allowing us to integrate across the MODIS bands 1, 3, and 4 for a direct intercomparison. The spectral response for MODIS bands 1, 3, and 4 are shown in Fig. 1. The wavelengths covered by these bands are provided from the MODIS specifications website: http://modis.gsfc.nasa.gov/about/specifications.php. For comparison with the MODIS NBAR data (see Sect. 5), the UAS spectra were multiplied with the MODIS spectral band functions (Fig. 1), and the respective NBARs calculated according to Eq. (1).

2.3 Radiance offset correction

To establish the relative sensitivity of the radiance and irradiance sensors, measurements were made on the ground with the two UAS sensors co-located together with a reference irradiance sensor looking skyward. The relative sensitivity of the radiance and irradiance sensors was calculated, and a third-order polynomial was fit to this ratio in the wavelength region relevant for comparison of UAS and MODIS data. All measured radiance spectra were corrected for the wavelength-dependent offset using the polynomial fit.

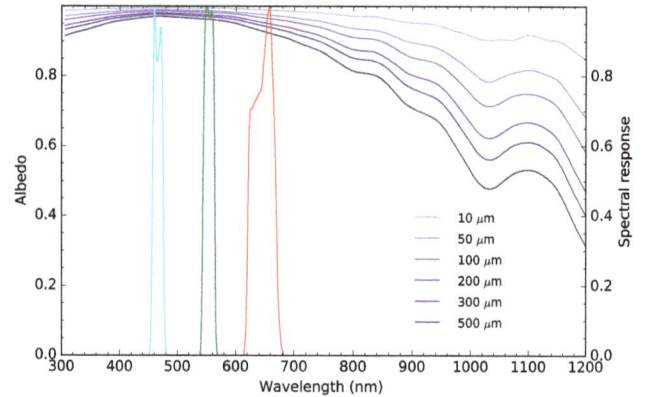

Figure 1. The pure-snow albedo as a function of wavelength for snow grain sizes between 10 and 500 µm. Also shown in the thicker lines is the MODIS spectral response for band 3 (blue), 4 (green), and 1 (red). The spectral response data were obtained from http://mcst.gsfc.nasa.gov/calibration/parameters.

2.4 Cosine error correction

The uplooking sensor measures the irradiance, requiring a detector with a hemispheric cosine response fore-optic. In reality the angular response of cosine detector deviates from a cosine shape. Cosine error corrections have been thoroughly investigated for UV spectrometers (see for example Bais et al., 1998). The cosine error correction depends on the atmospheric state when the measurements were made. The deviations typically become larger as the incidence angle increases, implying that measured irradiance is underestimated compared with an instrument with a perfect angular response. This underestimate may be corrected for providing that the sky conditions during the measurements are known and that the angular response of the instrument is known (Bais et al., 1998).

The TriOS sensors are laboratory certified and have undergone calibration by the manufacturer prior to each field season. For the zenith angles encountered during the flights ($< 70°$) we expect the deviation from a perfect cosine response to be less than 2 %.

2.5 Angular sensitivity

The uplooking detector must be properly leveled to allow accurate measurements of the downwelling irradiance (Bogren et al., 2016). This may be achieved, for example, by stabilizing the measurement platform (Wendisch et al., 2001) or by mounting the instrument with a tilt such that the instrument is leveled during flight. The latter approach was adopted with the UAS; however, this requires that the platform is stable during flight. To estimate the effect of angular changes on a fixed detector, radiative transfer simulations were performed as described below in Sect. 3. The roll angle of the aircraft, and thus the detector, was changed between 0 and 10°, while the yaw angle was changed from from 0 to 360°. The radiation field was simulated for a cloudless sky over a snow-covered surface. The response relative to a leveled detector is shown in Fig. 3 for a solar zenith angle of 55.66° and azimuth of 0°. If the detector has a roll angle of 10.0° and yaw angle 90° with respect to the sun, implying that it is facing away from the sun, the detector will measure only about 80 % of the radiation of a leveled detector.

Similarly, a detector shifted such that it faces the sun will overestimate the radiation compared to a leveled detector. The results presented in Fig. 3 are for a cloudless sky. For an overcast sky, with the aircraft flying below the cloud, the change in angular response is negligible with given azimuth and roll angles (Bogren et al., 2016). The effect of roll and yaw angles on the measurements will change with solar zenith angle, surface albedo, sky conditions, and wavelength. As such they are challenging to correct for when the aircraft is moving around due to changes in the flying directions or changing wind conditions. For the analysis below, UAS data were screened and selected for stable flight conditions. In addition a tilt correction was applied to the direct portion of the irradiance impinging the upward-facing sensor. As presented by Bogren et al. (2016), the response of a sensor tilted θ_t degrees and rotated ϕ degrees relative to the sun is

$$R^t(\theta_t, \phi) = \cos\left(\theta_0 - \theta_t \cos(\phi)\right), \qquad (2)$$

where θ_0 is the solar zenith angle. For a leveled sensor $R^l = \cos(\theta_0)$. The tilt error correction is largest for the direct part of the irradiance and negligible for the diffuse part (Bogren et al., 2016). We thus tilt-correct the measured downwelling irradiance E_m as follows:

$$E = f E_m R^l / R^t + (1 - f) E_m. \qquad (3)$$

Here the first term on the right side is the tilt correct direct contribution, and the second term is the uncorrected diffuse contribution. Furthermore, f is the wavelength-dependent direct / global irradiance ratio. It was estimated by the libRadtran model, described in Sect. 3, to be 0.98, 0.92, and 0.85 for MODIS bands 1, 4, and 3, respectively.

Due to dismounting and remounting for maintenance, or from the thrust of the catapult at launch, the instrument package may become slightly disoriented. Thus θ_t and ϕ may be offset. During the analysis it was found that θ_t should be reduced by 0.7°, and a 10° azimuth offset added.

2.6 Atmospheric corrections

During the flights the altitude of the UAS varied between 270 and 320 m above the surface. The atmosphere between the surface and the aircraft may influence the aircraft nadir measurements. To estimate the impact of the intervening atmosphere, UAS radiance and irradiance spectra were simulated for noon (solar zenith angle of 55.66°) at Summit for elevations between 270 and 320 m a.g.l. in steps of 10 m. The simulated spectra were multiplied with the MODIS band 1, 3, and 4 response functions (Fig. 1), and the corresponding UAS nadir measurements were integrated to the corresponding narrow bandwidths. It was found that the atmosphere between the aircraft and surface caused less than 0.2 % changes in the band 1 nadir reflectance and less than 0.04 % difference in the bands 3 and 4 nadir reflectance. Thus, the nadir reflectance derived from the UAS were not corrected for the intervening atmosphere.

2.7 Error estimate

Estimates of the measurement error are inherently difficult to make and require time resources often not available in the field. However, best estimates of the measurment error from various sources are assumed and used to calculate a total error as summarized below.

Ideally NBAR measurements should be made over flat surfaces. As shown in Siegfried et al. (2011), the region around Summit is sufficiently flat with a slope of less than 2 m km^{-1} in the east–west direction and less than 0.5 m km^{-1} in the north–south orientation. Our flights were further north from Summit than measured by Siegfried et al. (2011), but data from available digital elevations maps from the Greenland Ice Mapping Project (Howat et al., 2014) confirm the area covered by the UAS flights is indeed flat. However, small-scale wind-blown snow features can not be ruled out. This is potentially the largest source of error in the analysis. Sustruggi structures on the snow surface can cause strong scattering and geometric optical effects. This variability is difficult to resolve or explicitly model. However, effects of scattering will be reduced by integrating over the footprint of the measurement. An uncertainty of 0.5 % is assumed due to measurements over non-flat surfaces.

The offset between the up- and downlooking sensors was measured and corrected for as explained in Sect. 2.3. A remaining error of 0.2 % is assumed for the offset correction.

The tilt error has been corrected for as described in Sect. 2.5. The attitude is specified to have an uncertainty of 2 %. The uncertainty in the data due to remaining tilt error and assumption about the direct / global radiation ratio is thus taken to be 2 %.

According to the manufacturer, the cosine error is better than 6–10 %, depending on wavelength, while for the 7° detector the angular response is better than 6 %. The cosine error typically increases with zenith angle in addition to wavelength. The error will thus be largest for large solar zenith angles as found at high latitude. No cosine response measurements were available for a detailed assessment of the cosine error. A 2 % cosine error correction has been applied to the uplooking sensor (Sect. 2.4). A 2 % cosine error uncertainty is assigned to the measurements. The downlooking sensor is exposed to diffuse radiation, and the error in the angular response is of less concern.

MODIS Band: band 3   band 4   band 1

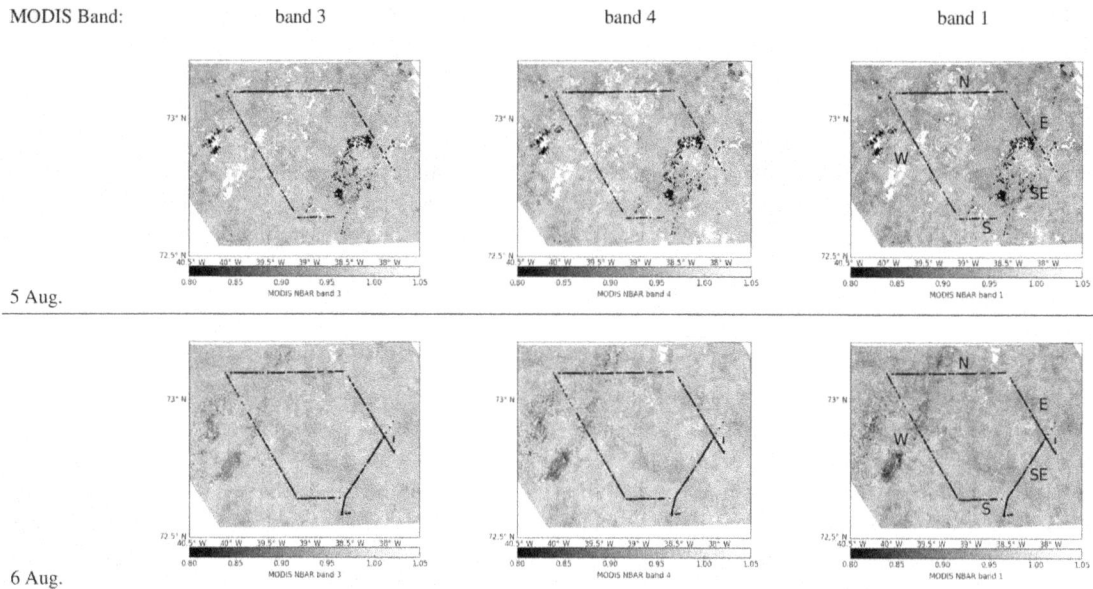

Figure 2. MODIS blue (band 3), green (band 4), and red (band 1) NBARs for 5 August (upper row) and 6 August (lower row) 2010. The black dots represent UAS data which were recorded during stable flight conditions. White areas indicate missing data.

For an integration time of 8 s the manufacturer gives a noise equivalent irradiance (NEI) of $0.4\,\mu\mathrm{W\,m^{-2}\,nm^{-1}}$ at 400 and 500 nm, and $0.6\,\mu\mathrm{W\,m^{-2}\,nm^{-1}}$ at 700 nm for the cosine response detector. For the $7°$ detector the NEI is $0.25\,\mu\mathrm{W\,m^{-2}\,nm^{-1}}$. The integration times during the flights were shorter; thus a conservative estimate of the NEI during the flights is 0.5 %.

We assume that all errors are independent of wavelength. Squaring the errors gives a total error in the UAS reflectance of 2.9 %.

3   Radiative transfer simulations

As a part of the data reduction and analysis process, and to assess the sensitivity to radiative processes in the atmosphere (Sect. 6), we conducted radiative transfer simulations. These were performed to test the UAS sensitivity to changes in pitch, roll, and yaw angles (Sect. 2.5 above), and to simulate cloudless shortwave broadband radiation at Summit (Sect. 4 below). The libRadtran radiative transfer package was utilized for these calculations (Mayer and Kylling, 2005; Emde et al., 2016). The molecular absorption was parameterized with the LOWTRAN band model (Pierluissi and Peng, 1985), as adopted from the SBDART code (Ricchiazzi et al., 1998). The C version of the DISORT radiative transfer solver (Stamnes et al., 1988; Buras et al., 2011) was utilized. The snow albedo model of Wiscombe and Warren (1980) as implemented in the libRadtran software package was used to calculate the spectral surface reflectance as shown in Fig. 9. The sub-Arctic summer atmosphere (Anderson et al., 1986) was used, and the surface altitude set to 3126 m.

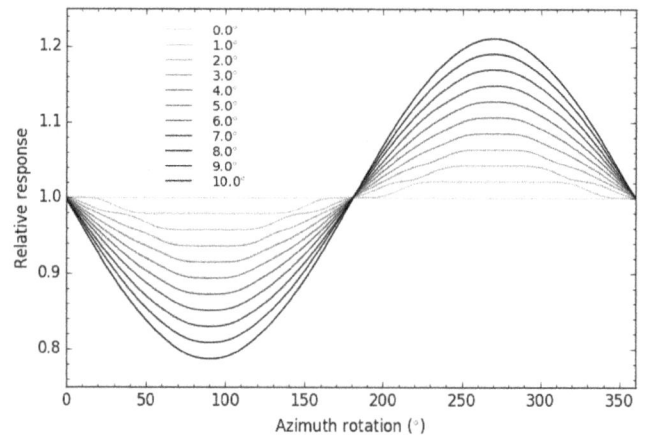

Figure 3. The relative response when tilting (from 0 to $10°$) and rotating around azimuth an irradiance sensor on board a platform 300 m above a snow surface. The solar zenith (azimuth) angle is $55.66°$ ($0°$). The wavelength is 465 nm.

4   Measurements selected for analysis

On 5 and 6 August 2010, the UAS flew a 210 km pattern designed such that one of the flight legs would be centered on the MODIS granule as close in time as possible to the MODIS overpass. The flight pattern flown is shown in Fig. 2. On 5 August the sky at Summit was overcast with some blue patches at takeoff. The cloud deck thickened during the flight to a uniform diffuse cover. This development is readily visible in the global shortwave measurements recorded at Summit (blue line, Fig. 4). The measured shortwave radiation is clearly below the cloudless simulated shortwave

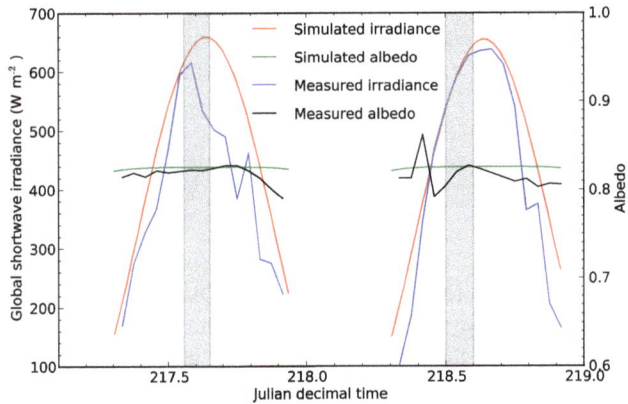

Figure 4. The measured (blue) and the simulated cloudless sky (red) downwelling shortwave radiation for Summit, Greenland, on 5 August (day 217) and 6 August (day 218) 2010. The measured (black) and simulated (green) shortwave broadband albedos are also shown. The grey shaded areas indicate the time when the UAS was flying.

radiation, and its behavior indicates the presence of clouds. The wind was blowing from the southeast at speeds around 6–7 m s^{-1} (Table 2). On 6 August the sky was mostly clear during most of the flight, with some thin layer cirrus forming at the end of the flight. The wind was more gentle on 6 August, with speeds increasing from about 1.5 to 3.5 m s^{-1} during the flight. At the same time the wind direction changed from southeast to a more southerly direction.

The pitch and roll angles of the UAS had nonzero offsets, indicating that the aircraft was flying in a non-leveled manner due to impact of wind, fuel-load, and placement of the center of mass. The offsets varied between the various flights and are given in Table 2. UAS data that are within $\pm0.5°$ of the mean pitch and roll angles were included in the analysis. Furthermore, due to the sensitivity of the measurements to the orientation of the uplooking instrument discussed above, only data for which the yaw angle was stable were included in the analysis. For 5 and 6 August the instrument azimuth (blue dots) and the corresponding tilt correction (red dots) are shown in Figs. 5a and 6a, respectively.

MODIS bands 1, 3, and 4 NBARs (see Sect. 5) were extracted from the MODIS pixels that coincide with the Cryowing UAS data points. The blue, green, and red MODIS NBARs are shown in the second, third, and fourth panels, respectively, of Figs. 5 and 6. Data that are flagged as high quality are in black, while lower-quality flagged MODIS data are in yellow.

5   MODIS nadir NBAR measurements

The MODIS NBAR product from MCD43 Collection 6 is used for intercomparison in this analysis. The MODIS instrument (Justice et al., 1998) measures radiance at the top of the atmosphere. These measurements must first be cloud-

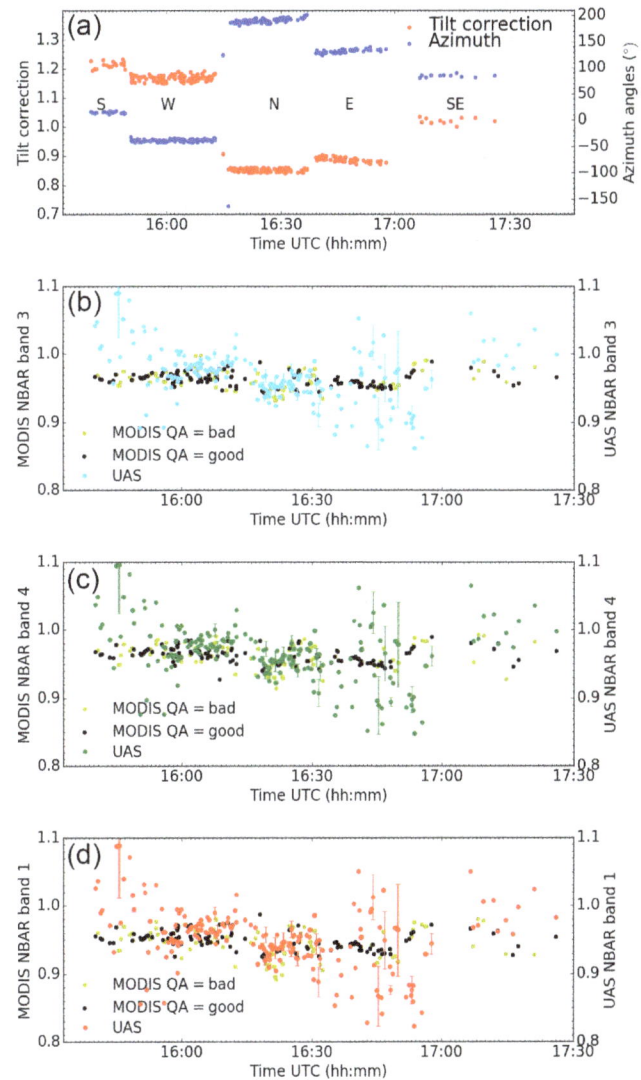

Figure 5. (a) The azimuth angle of the UAS (blue dots) and the tilt correction factor R^l/R^t, Eq. (3) (red dots). The various flight track elements are identified by letters and correspond to the flight tracks in Fig. 2. **(b–d)** The MODIS (black dots: good quality; yellow dots: low quality) and UAS band 3 (blue), 4 (green), and 1 (red) NBARs as a function of time. In the case of several UAS data points within one MODIS pixel, the UAS data have been grouped together and presented as a dot with standard deviation. All data from the flight on 5 August 2010.

cleared and atmospherically corrected, following which multiangle directional reflectances from both the Terra and Aqua MODIS sensors, over a period of 16 days, are accumulated for a location. From these directional reflectances, an appropriate RossThick LiSparse Reciprocal empirical kernel-based BRDF model is estimated. The MODIS BRDF/albedo product is widely used and has been described by Lucht et al. (2000), Schaaf et al. (2002), and Stroeve et al. (2005). The retrieved BRDF is then integrated over all view zenith angles to calculate an intrinsic directional hemispherical reflectance

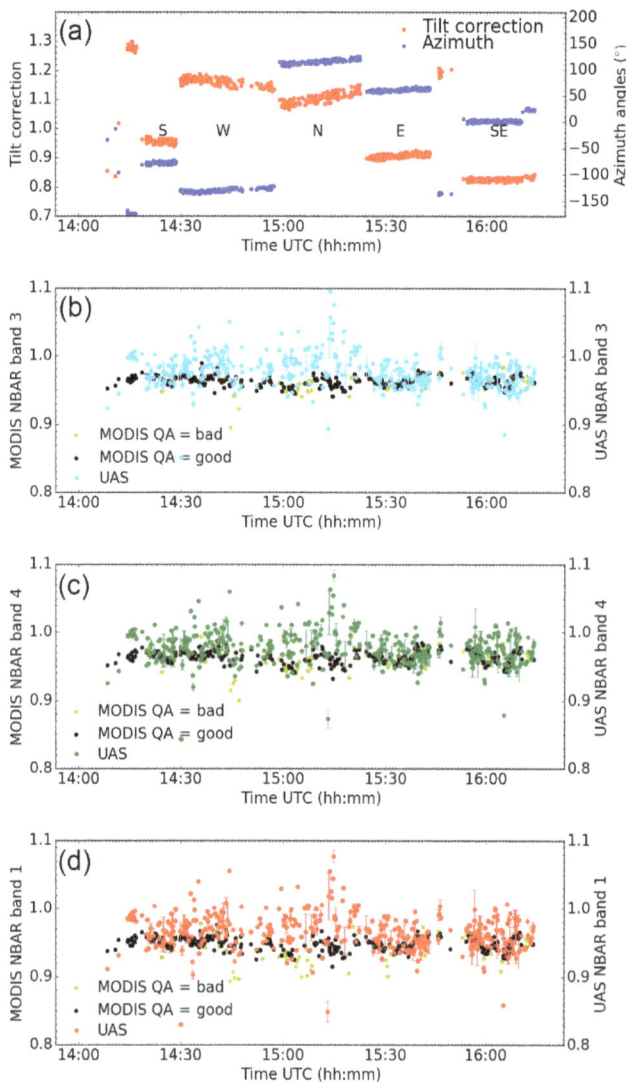

Figure 6. (a) The azimuth angle of the UAS (blue dots) and the tilt correction factor R^l/R^t, Eq. (3) (red dots). The various flight track elements are identified by letters and correspond to the flight tracks in Fig. 2. **(b–d)** The MODIS (black dots: good quality; yellow dots: low quality) and UAS band 3 (blue), 4 (green), and 1 (red) NBARs as a function of time. In the case of several UAS data points within one MODIS pixel, the UAS data have been grouped together and presented as a dot with standard deviation. All data from the flight on 6 August 2010.

(a black-sky albedo) for the seven MODIS land bands. The BRDF model is further integrated over all possible illumination angles to produce a bihemispherical reflectance (or white-sky albedo). The latest reprocessed operational Collection 6 MODIS daily BRDF/albedo/NBAR products improve the temporal aggregation of snow observations (Wang et al., 2012, 2014) by using a daily measurement and triangulated filter to emphasize the nearest-day observations. The snow/non-snow status of the day of interest is utilized for retrievals in Collection 6 instead of the previous Collection 5

strategy of only capturing snow measurements when snow cover represented the majority situation over the 16-day retrieval period. Of the seven available, bands 3, 4, and 1 are used here.

6 Discussion

In the following we evaluate differences between the UAS-measured nadir reflectance and the MODIS NBAR product. Unless otherwise noted, the data refer to high-quality flagged Collection 6 of the MODIS daily NBAR product (MCD43). We also include data from Collection 5 to demonstrate some of the marked improvements as well as evaluate the importance of quality flagging.

The MODIS albedo product has been compared with in situ measurements in Greenland (e.g., Stroeve et al., 2005) and a number of other snow-covered locations (Wang et al., 2014). The root mean square error between MODIS and in situ measurements was within ±0.04 (±0.07) for high-quality (poor-quality) flagged MODIS albedos. A high-quality flag indicates that sufficiently high-quality surface reflectances to adequately sample the full angular hemisphere were acquired and that a high-quality full inversion BRDF model was able to be developed to produce NBAR and the intrinsic surface albedo quantities. For poor-quality flagged retrievals a BRDF model could not be retrieved, and a backup algorithm with a predetermined BRDF for that location had to be utilized. These intrinsic surface quantities are related to the surface structure, and the albedos represent fully direct and fully diffuse values. Critically, the direct and diffuse values need to be combined as a function of optical thickness to simulate the blue-sky albedos routinely captured with albedometers at surface tower locations (Lucht et al., 2000; Schaaf et al., 2002; Román et al., 2010). In order to incorporate the full atmospheric effects, the full multiple scattering of the Román et al. (2010) formation needs to be used over snow surfaces. While these analyses have been clearly valuable to the community, the intercomparison is complicated due to the nature of the measurement–model chain required to derive albedo. We reduce, by at least one degree, the required modeling for the comparison by evaluating NBAR, which both platforms measure directly, rather than albedo, which must be derived.

In general the agreement between the UAS nadir reflectance and MODIS NBARs is within the measurement uncertainties (Figs. 5 and 6). Systematically, the UAS measurements are higher than those of the MODIS NBARs, albeit well within 3 % of each other. The mean NBAR values for each flight are provided in Table 1. Percentage differences based on the mean value of the platforms show that on 5 August the differences between the UAS and MODIS data were within 1 %, despite greater variability on this day. On 6 August the differences were greater, while still less than 2 %. As expected from the refractive index of ice and NBAR calcu-

Table 1. Mean NBAR values from MODIS and the UAS on 5 and 6 August for bands 3, 4, and 1. Percentage differences are calculated based on the mean of the two values.

Band	5 August			6 August		
	UAS	MODIS	%	UAS	MODIS	%
3	0.971	0.967	0.41	0.978	0.965	1.34
4	0.974	0.966	0.83	0.980	0.965	1.54
1	0.956	0.952	0.42	0.967	0.950	1.77

lations (see Fig. 1), there is little wavelength dependence in the NBARs of bands 3 and 4, whereas there is an expected decrease in NBAR for band 1.

In Table 2 we also include a summary of the MODIS Collection 5 data, which are only retrieved once every 8 days and not filtered to weight the day of interest. The agreement between the UAS and MODIS version 6 is better than between the UAS and MODIS version 5 for bands 3 and 1. For band 4 there is better agreement between the UAS and MODIS version 5. However, the differences are within the uncertainties; see below. Further, the standard deviation is generally smaller for the version 6 data.

Variability in the measurements reflects the conditions at time of acquisition. Consistently, the standard deviations for all products are slightly larger for the flight on 5 August due to the more turbid atmospheric conditions. The variability is a product not only of the cloud cover but also of the wind speed, which potentially increased turbulence for the aircraft. We further see strong support for the quality flagging of the MODIS products. The variations of the MODIS NBARs are larger for the pixels identified as low-quality retrievals than for the good-quality retrievals (see standard deviations in parentheses in Table 2).

The UAS measures the instantaneous upwelling radiance within 7° and the full-hemisphere downwelling irradiance, with the ratio of the two providing the reflectance, whereas the radiances measured by MODIS at the top of the atmosphere are atmospherically corrected to surface reflectance by means of radiative transfer modeling. These directional surface reflectances at a location – time weighted to the day of interest – are gathered over a 16-day period from both Terra and Aqua, and used to derive the BRDF for the full range of solar and viewing angles. The BRDF is then used to calculate an NBAR or an albedo for the seven MODIS bands. Here we have used the BRDF to calculate a solar-noon nadir reflectance for comparison with the UAS. Hence, it is important to recognize the MODIS products are not instantaneous nadir reflectance measurements but rather a calculation guided by data acquisition. Given the difference between the derived MODIS data product and instantaneous UAS measurement, the correspondence is impressive. The standard deviation in the UAS data varies between 0.025 and 0.065. The standard deviation in high (low)-quality MODIS

data varies from 0.008 (0.009) to 0.015 (0.024). Given the smaller footprint, instabilities in the platform, and uncertainties related to the snow surface roughness, it is expected for the UAS-derived reflectances to have higher standard deviation. Further, as these data are instantaneous measurements rather than an integrated model product, one should expect greater variance. This greater variance and sub-pixel variability potentially contribute to the fact that, despite more stable flight conditions on 6 August, the difference between the two platforms is slightly greater (cf. Table 1).

The footprint of the MODIS reflectance shown in this study is $500\,m^2$ at nadir but has an effective footprint of $833\,m \times 613\,m$ at the latitudes of this study (Campagnolo et al., 2016). The footprint of the UAS is circular with a diameter of about 30 m. Thus the UAS may be used to investigate MODIS sub-pixel variability. Where several UAS measurements are available within a pixel, there is considerable variability within a MODIS pixels; see data points with error bars in Figs. 5 and 6. However, this variability is within the uncertainty in the UAS data. Given more stable flight conditions, measurement of MODIS sub-pixel variability should be fully feasible with a UAS.

Optical satellite instruments require cloud-free conditions to make NBAR estimates. This clearly limits the number of days available for NBAR measurements. The UAS is not limited to a cloud-free sky but may be used to measure nadir reflectance also under cloudy conditions. However, as discussed below, the UAS must be below the cloud layer and not in it or a haze layer.

In Fig. 7 we evaluate the differences between Collection 5 and 6 MODIS products. Two distinct features stand out. First, there is an improvement in the poorer-quality magnitude estimates in Collection 6 as demonstrated by the systematic shift leftward of the data cluster from Collection 5 to 6. Most data fall below a reflectance of 1.0, whereas in Collection 5 several values were greater than 1. We note a value of greater than 1 is not impossible, and in fact quite apparent for the UAS data, likely resulting from forward scattering driven by the sustruggi and expected at the scale of these measurements (30 m footprint). For the MODIS data, however, covering a square kilometer (km^2) footprint, one would expect reflectances to be more smoothed, and values greater than 1 are expectedly rare. The second feature is a clear decrease in the variability of the data, both in terms of the overall spread and for the flagged values. There are fewer poor-quality retrievals in Collection 6, resulting from the improved temporal retrieval frequency in the algorithm.

Regarding the quality and no data flagging, we find the differences noted between 5 and 6 August clearly important (Fig. 8). The flight conditions were better (fewer clouds) on 6 August, and more MODIS data points have a good retrieval flag. Nevertheless the data points flagged as low quality have a spatial variation in agreement with the good-quality points; compare yellow (low) and black (good) dots in Figs. 5 and 6 and the scatterplots shown in Fig. 7. Particularly for Collec-

MODIS Band: Band 3   Band 4   Band 1

5 Aug.

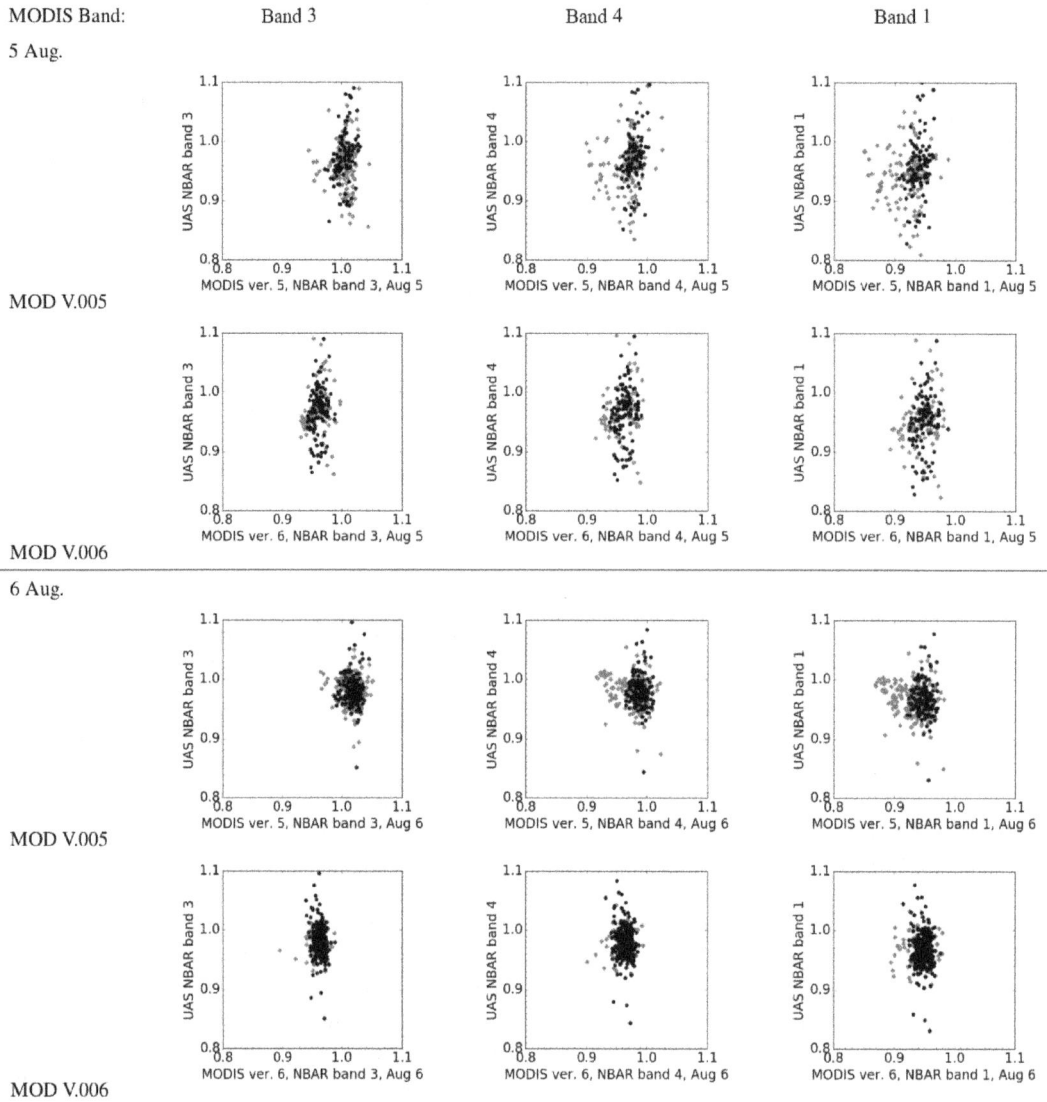

MOD V.005

MOD V.006

6 Aug.

MOD V.005

MOD V.006

Figure 7. The MODIS (black: high quality; grey: low quality) versus UAS NBAR for bands 3, 4, and 1. Left column is data for band 3, middle column is data for band 4, and right column is data for band 1. Rows 1 and 3 are MODIS version 5, and rows 2 and 4 are MODIS version 6. Rows 1–2 (3–4) are data from the flight on 5 (6) August.

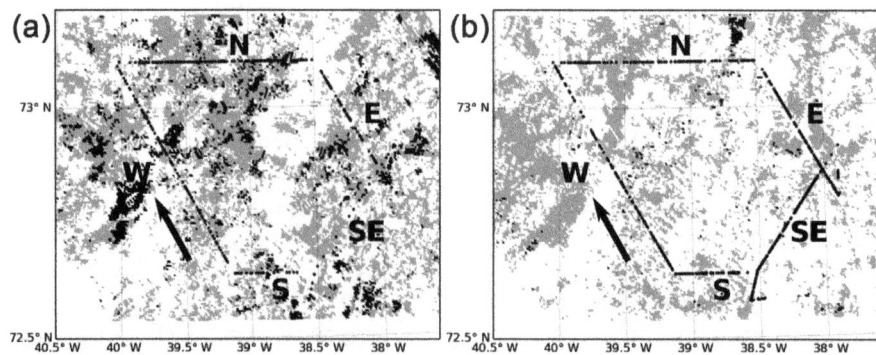

Figure 8. MODIS quality flags for 5 August (**a**) and 6 August (**b**) 2010. Grey color indicates low-quality retrieval; black indicates no retrieval. White areas have high-quality retrieval flags. Flight transects are as labeled in Figs. 5 and 6. The arrow indicates direction of flight. Flights started with the S transect and were flown clockwise.

Table 2. Information for the two UAS MODIS route overpass flights. For the NBAR values the average value is given together with the standard deviation in parentheses.

	5 August		6 August	
Start of flight (hh:mm:ss, UTC)	15:22:55		14:00:28	
End of flight (hh:mm:ss, UTC)	17:37:27		16:23:34	
Mean pitch (SD) angle (°)	6.05 (0.55)		7.32 (0.52)	
Mean roll (SD) angle (°)	−4.43 (0.48)		5.40 (0.54)	
Weather	Overcast		Mostly cloudless	
Solar zenith angle (°)	60–56		55–58	
Solar azimuth angle (°)	−46 to −17		11 to −27	
Wind speed ($m\,s^{-1}$)	6–7		1.5–3.5	
Wind direction (°)	≈ 135		140–170	
# UAS spectra	497		882	
MODIS quality flag	0 (Good)	1 (Low)	0 (Good)	1 (Low)
MODIS ver 5, NBAR band 3 (blue)	1.005 (0.012)	1.005 (0.015)	1.017 (0.011)	1.017 (0.012)
MODIS ver 6, NBAR band 3 (blue)	0.967 (0.010)	0.965 (0.014)	0.965 (0.008)	0.966 (0.009)
UAS band 3	0.971 (0.056)		0.978 (0.025)	
MODIS ver 5, NBAR band 4 (green)	0.978 (0.011)	0.967 (0.024)	0.988 (0.010)	0.982 (0.019)
MODIS ver 6, NBAR band 4 (green)	0.966 (0.013)	0.964 (0.016)	0.965 (0.009)	0.965 (0.012)
UAS band 4	0.974 (0.064)		0.980 (0.026)	
MODIS ver 5, NBAR band 1 (red)	0.939 (0.0123)	0.956 (0.027)	0.948 (0.011)	0.937 (0.022)
MODIS ver 6, NBAR band 1 (red)	0.952 (0.0146)	0.947 (0.020)	0.950 (0.010)	0.949 (0.014)
UAS band 1	0.956 (0.065)		0.967 (0.028)	

tion 5, there is larger variation in the low-quality MODIS data, yet there is no support for this variation in UAS measurements. Thus overall, the MODIS algorithm appears to correctly discriminate good- and low-quality data.

We note significantly greater variability in the 5 August flight data. This is to be anticipated given the sky conditions, but the data provide a valuable reference for comparison with the MODIS data. Some features stand out from the flight, the first being what appears to be a consistent decrease in reflectance from the start of the flight until 17:00 UTC, when the flight initiates the SE leg. Overall the reflectance decreases in this period by almost 10 %. The flight on 6 August followed the same flight pattern, but no such drop in reflectance is present. There are several plausible explanations for the drop, including a drop in surface reflectance, measurement error, or presence of slightly absorbing particles in the atmosphere. As the drop was only seen on 5 August and not on the subsequent day, and, further, the drop is not seen in the good-quality MODIS data, we therefore rule out a change in the surface reflectance. The drop could be due to incorrect tilt correction. However we investigated this thoroughly and find no feasible explanation why this error would be introduced on the 5 August flight but not on the 6 August flight. Additionally, as the drop occurred continuously through multiple legs and is then not seen at the end of the flight, we also rule out this explanation.

From Fig. 4 it is evident that some clouds were present during part of the flight on 5 August. While these could have an impact on reflectance due to shadowing, a non-absorbing cloud will not change the surface reflectance as measured by the UAS. On the other hand, if the UAS encounters an optically thin, slightly absorbing haze layer, the UAS measured reflectance will drop. To quantify this drop, radiative transfer calculations were made of the UAS reflectance with the UAS being at different flight altitudes. Cloudless and various haze conditions were considered. As shown in Fig. 9, the reflectance on a cloudless day (red line) does not depend on the altitude of the UAS. If an optically thin (optical depth: 0.5) and slightly absorbing (single-scattering albedo: 0.95) haze layer of 1 km vertical thickness is included, the UAS measured reflectance will be lower than the surface reflectance (green line), and the difference will increase with increasing altitude.

The haze layer optical property values used are representative for those reported for Arctic haze (cf. Tsay et al., 1989; Hess et al., 1998; Quinn et al., 2007). Increasing the absorption (single-scattering albedo: 0.9) increases the difference even more (dash–dot line). The average flight altitude of UAS on 5 August is indicated by the horizontal dotted line in Fig. 9. The drop seen in the UAS reflectance on 5 August may thus be explained by the UAS entering an optically thin and slightly absorbing haze layer. However, due to a lack of additional measurements (aerosol properties) we can not prove

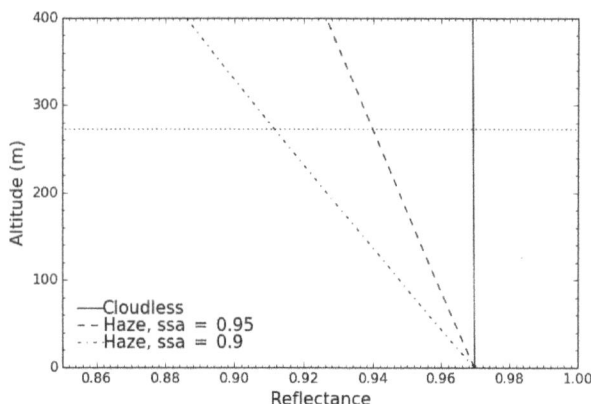

Figure 9. Simulated UAS reflectance for MODIS band 1 as a function of flight altitude for various sky conditions. See text for details. The dotted horizontal line indicates the average flight altitude of the UAS on 5 August. The solar zenith angle is 58°, corresponding to the values during the flight on 5 August (see Table 2).

this. But it is noted that during haze conditions reflectance measured by an airborne platform may be affected by the atmosphere between the platform and the surface. An ideal platform for surface reflectance observations would include aerosol observations such as those presented in Bates et al. (2013).

Covering the same flight path on 6 August, the UAS data and MODIS products agree remarkably well. As mentioned earlier, we note greater variability in the UAS data resulting from several factors. For one, the UAS is measuring a smaller area, and thus there is far less smoothing of the data. Certainly surface roughness and forward scattering play a role, but despite best efforts to select only stable periods of flight, there is likely error introduced to the data from platform stability. Nonetheless, the consistent feature is that the UAS values are systematically larger than the MODIS products (Table 1). Given the nature and temporal smoothing of the MODIS products, we believe this is a real artifact and that the MODIS data may in fact provide values slightly lower than actual reflectances that would be observed instantaneously. We also note that the UAS platform has immense potential to provide greater insight into the sub-pixel variability of the MODIS products – which is likely significant.

7 Conclusions

During July–August 2010, several UAS flights were made over Summit, Greenland. Two of the flights were designed to cover the MODIS swath and were made close in time to the MODIS overpass. The UAS measured the up- and downwelling radiation between 320 and 950 nm with a resolution of 3.3 nm. In this analysis we have made a direct comparison of reflectance as measured by the UAS with the MODIS NBAR product. The main findings are as follows:

– The UAS and MODIS reflectances for bands 3, 4, and 1 agree within their uncertainties. However, due to the larger footprint and temporal smoothing of MODIS, the product provides a slightly lower overall reflectance.

– Sub-pixel variability of MODIS reflectance products is potentially significant. Further work should be conducted to evaluate the magnitude and consequent impacts on modeling of greater variability at the sub-kilometer scale.

– Consistent with theory, the UAS and MODIS reflectance measurements show a decrease between bands 3 and 4 and band 1. This wavelength dependence agrees with that expected from the refractive index of ice.

– Even at the low elevations flown typically by a UAS, reflectance measurements may be influenced by haze if present at and/or below the flight altitude of the UAS.

– The UAS platform is proven as a capable resource to collect reflectance measurements over an extensive region and provides a reliable resource for evaluating spatial variability of reflectance for Summit, Greenland, and the surrounding area.

Of significance in this evaluation, and any intercomparison, is the concept of "truth". Neither of the platforms presented provide a perfect measure of NBAR, but this is a much more direct intercomparison than would be with albedo, which would require several further assumptions. MODIS and the UAS platforms attempt to provide an accurate characterization of NBAR and nadir reflectance, respectively. However, MODIS, while collecting direct radiance measurements, requires a model and data assimilation chain to provide a product – ultimately produced from a BRDF model. The UAS is challenged due to platform instabilities and expected instrumental errors. The nature of reflectance observations is further challenged by the fact that it is not a simple property of a surface (e.g., snow), but rather a product of a system. The system includes not only many temporally varying factors such as solar zenith angle and azimuth but also more dynamic processes including atmospheric scattering, surface roughness, and snow conditions. The timescales of variability of these processes differ and are not trivial to represent consistently. Given the importance of this parameter and the general derivation of albedo from these observations, it is critical that we understand well the expected variability. The UAS platform provides a unique opportunity to collect observations that are more representative spatially for applications where point-based measurements are used, for instance, the validation of satellite-based remotely sensed measurements and "grid"-based climate and earth system models. This work not only demonstrates the feasibility of collecting these observations but also exemplifies the challenges associated with benchmarking different observations. Furthermore, the observations presented herein were collected over a range offering relatively low variability in reflectance. To further increase our current understanding of the reflectance of the cryosphere and its development, more UAS measurement campaigns at other locations and surface conditions are warranted.

Competing interests. The authors declare that they have no conflict of interest.

Acknowledgements. This work was conducted within the Norwegian Research Council's NORKLIMA program under the Variability of Albedo Using Unmanned Aerial Vehicles (VAUUAV; NFR no. 184724) and Hydrologic sensitivity to Cryosphere-Aerosol interaction in Mountain Processes (HyCAMP; NFR no. 222195) projects, and with additional support from NFR no. 196204 (RISCC) and NFR no. 195143 (Arctic-EO). Computational and data storage resources were provided by NOTUR/NORSTORE projects NS9333K and NN9333K. Data processing was conducted with open-source Python tools, namely numpy (Ascher et al., 1999) and xarray (Hoyer and Hamman, 2017). We acknowledge the Greenland Home Rule Government for permission to work in Greenland; the 109th Airlift Wing of the New York Air National Guard (NY ANG) for unsurpassed air transport; CH2M HILL Polar Services and the on-site science technicians for superb logistical support; the GEOSummit Science Coordination Office (SCO, PLR-1042531) for providing contacts and access to data; and overall the support from the Division of Polar Programs at the National Science Foundation. Konrad Steffen provided the shortwave radiation data from Summit, while NOAA-ESRL (Brian Vasel and Tom Mefford) provided wind speed and direction.

Edited by: Edward Hanna

References

Anderson, G. P., Clough, S. A., Kneizys, F. X., Chetwynd, J. H., and Shettle, E. P.: AFGL Atmospheric Constituent Profiles (0.120 km) Environmental research papers, Accession Number: ADA175173, AIR FORCE GEOPHYSICS LAB HANSCOM AFB MA, Defense Technical Information Center, available at: http://www.dtic.mil/docs/citations/ADA175173 (last access: 28 June 2017), 1986.

Ascher, D., Dubois, P. F., Hinsen, K., Hugunin, J., and Oliphant, T.: Numerical Python, Lawrence Livermore National Laboratory, Livermore, CA, ucrl-ma-128569 Edn., 1999.

Bais, A. F., Kazadzis, S., Balis, D., Zerefos, C. S., and Blumthaler, M.: Correcting Global Solar Ultraviolet Spectra Recorded by a Brewer Spectroradiometer for its Angular Response Error, Appl. Optics, 37, 6339, https://doi.org/10.1364/AO.37.006339, 1998.

Bates, T. S., Quinn, P. K., Johnson, J. E., Corless, A., Brechtel, F. J., Stalin, S. E., Meinig, C., and Burkhart, J. F.: Measurements of atmospheric aerosol vertical distributions above Svalbard, Norway, using unmanned aerial systems (UAS), Atmos. Meas. Tech., 6, 2115–2120, https://doi.org/10.5194/amt-6-2115-2013, 2013.

Bennartz, R., Shupe, M. D., Turner, D. D., Walden, V. P., Steffen, K., Cox, C. J., Kulie, M. S., Miller, N. B., and Pettersen, C.: July 2012 Greenland melt extent enhanced by low-level liquid clouds, Nature, 496, 83–86, https://doi.org/10.1038/nature12002, 2013.

Bhardwaj, A., Sam, L., Akanksha, Martín-Torres, F. J., and Kumar, R.: UAVs as remote sensing platform in glaciology: Present applications and future prospects, Remote Sens. Environ., 175, 196–204, https://doi.org/10.1016/j.rse.2015.12.029, 2016.

Bogren, W. S., Burkhart, J. F., and Kylling, A.: Tilt error in cryospheric surface radiation measurements at high latitudes: a model study, The Cryosphere, 10, 613–622, https://doi.org/10.5194/tc-10-613-2016, 2016.

Box, J. E., Fettweis, X., Stroeve, J. C., Tedesco, M., Hall, D. K., and Steffen, K.: Greenland ice sheet albedo feedback: thermodynamics and atmospheric drivers, The Cryosphere, 6, 821–839, https://doi.org/10.5194/tc-6-821-2012, 2012.

Buras, R., Dowling, T., and Emde, C.: New secondary-scattering correction in DISORT with increased efficiency for forward scattering, J. Quant. Spectrosc. Ra., 112, 2028–2034, https://doi.org/10.1016/j.jqsrt.2011.03.019, 2011.

Campagnolo, M. L., Sun, Q., Liu, Y., Schaaf, C., Wang, Z., and Román, M. O.: Estimating the effective spatial resolution of the operational BRDF, albedo, and nadir reflectance products from MODIS and VIIRS, Remote Sens. Environ., 175, 52–64, https://doi.org/10.1016/j.rse.2015.12.033, 2016.

Curry, J. and Schramm, J.: Applications of SHEBA/FIRE data to evaluation of snow/ice albedo parameterizations, J. Geophys. Res., 106, 15345–15355, https://doi.org/10.1029/2000JD900311, 2001.

Dumont, M., Brun, E., Picard, G., Michou, M., Libois, Q., Petit, J.-R., Geyer, M., Morin, S., and Josse, B.: Contribution of light-absorbing impurities in snow to Greenland's darkening since 2009, Nat. Geosci., 7, 509–512, https://doi.org/10.1038/NGEO2180, 2014.

Emde, C., Buras-Schnell, R., Kylling, A., Mayer, B., Gasteiger, J., Hamann, U., Kylling, J., Richter, B., Pause, C., Dowling, T., and Bugliaro, L.: The libRadtran software package for radiative transfer calculations (version 2.0.1), Geosci. Model Dev., 9, 1647–1672, https://doi.org/10.5194/gmd-9-1647-2016, 2016.

Flanner, M. G., Zender, C. S., Hess, P. G., Mahowald, N. M., Painter, T. H., Ramanathan, V., and Rasch, P. J.: Springtime warming and reduced snow cover from carbonaceous particles, Atmos. Chem. Phys., 9, 2481–2497, https://doi.org/10.5194/acp-9-2481-2009, 2009.

Goelles, T., Bøggild, C. E., and Greve, R.: Ice sheet mass loss caused by dust and black carbon accumulation, The Cryosphere, 9, 1845–1856, https://doi.org/10.5194/tc-9-1845-2015, 2015.

Hakala, T., Riihelä, A., Lahtinen, P., and Peltoniemi, J. I.: Hemispherical-directional reflectance factor measurements of snow on the Greenland Ice Sheet during the Radiation, Snow Characteristics and Albedo at Summit (RASCALS) campaign, J. Quant. Spectrosc. Ra., 146, 280–289, https://doi.org/10.1016/j.jqsrt.2014.04.010, 2014.

Hansen, J. and Nazarenko, L.: Soot climate forcing via snow and ice albedos, P. Natl. Acad. Sci. USA, 101, 423–428, https://doi.org/10.1073/pnas.2237157100, 2004.

Henderson-Sellers, A. and Wilson, M.: Surface Albedo Data for Climatic Modeling, Rev. Geophys., 21, 1743–1778, https://doi.org/10.1029/RG021i008p01743, 1983.

Hess, M., Koepke, P., and Schult, I.: Optical properties of aerosols and clouds: The software package OPAC, B. Am. Meteorol. Soc., 79, 831–844, 1998.

Howat, I. M., Negrete, A., and Smith, B. E.: The Greenland Ice Mapping Project (GIMP) land classification and surface elevation data sets, The Cryosphere, 8, 1509–1518, https://doi.org/10.5194/tc-8-1509-2014, 2014.

Hoyer, S. and Hamman, J.: xarray: N-D labeled arrays and datasets in Python, Journal of Open Research Software, 5, 10, https://doi.org/10.5334/jors.148, 2017.

Immerzeel, W., Kraaijenbrink, P., Shea, J., Shrestha, A., Pellicciotti, F., Bierkens, M., and de Jong, S.: High-resolution monitoring of Himalayan glacier dynamics using unmanned aerial vehicles, Remote Sens. Environ., 150, 93–103, https://doi.org/10.1016/j.rse.2014.04.025, 2014.

IPCC: Climate Change 2013: The Physical Science Basis. Contribution of Working Group I to the Fifth Assessment Report of the Intergovernmental Panel on Climate Change, edited by: Stocker, T. F., Qin, D., Plattner, G.-K., Tignor, M., Allen, S. K., Boschung, J., Nauels, A., Xia, Y., Bex, V., and Midgley, P. M., Cambridge University Press, Cambridge, United Kingdom and New York, NY, USA, 1535 pp., 2013.

Jagt, B., Lucieer, A., Wallace, L., Turner, D., and Durand, M.: Snow Depth Retrieval with UAS Using Photogrammetric Techniques, Geosciences, 5, 264–285, https://doi.org/10.3390/geosciences5030264, 2015.

Jin, Y.: Consistency of MODIS surface bidirectional reflectance distribution function and albedo retrievals: 2. Validation, J. Geophys. Res., 108, 1–15, https://doi.org/10.1029/2002JD002804, 2003.

Ju, J., Roy, D. P., Shuai, Y., and Schaaf, C.: Remote Sensing of Environment Development of an approach for generation of temporally complete daily nadir MODIS reflectance time series, Remote Sens. Environ., 114, 1–20, https://doi.org/10.1016/j.rse.2009.05.022, 2010.

Justice, C. O., Vermote, E., Townshend, J. R. G., Defries, R., Roy, D. P., Hall, D. K., Salomonson, V. V., Privette, J. L., Riggs, G., Strahler, A., Lucht, W., Myneni, R. B., Knyazikhin, Y., Running, S. W., Nemani, R. R., Wan, Z., Huete, A. R., van Leeuwen, W., Wolfe, R. E., Giglio, L., Muller, J., Lewis, P., and Barnsley, M. J.: The Moderate Resolution Imaging Spectroradiometer (MODIS): land remote sensing for global change research, IEEE T. Geosci. Remote, 36, 1228–1249, https://doi.org/10.1109/36.701075, 1998.

Keegan, K. M., Albert, M. R., McConnell, J. R., and Baker, I.: Climate change and forest fires synergistically drive widespread melt events of the Greenland Ice Sheet, P. Natl. Acad. Sci. USA, 111, 7964–7967, https://doi.org/10.1073/pnas.1405397111, 2014.

Lucht, W., Schaaf, C. B., and Strahler, A. H.: An algorithm for the retrieval of albedo from space using semiempirical BRDF models, IEEE T. Geosci. Remote, 38, 977–998, https://doi.org/10.1109/36.841980, 2000.

Mayer, B. and Kylling, A.: Technical note: The libRadtran software package for radiative transfer calculations – description and examples of use, Atmos. Chem. Phys., 5, 1855–1877, https://doi.org/10.5194/acp-5-1855-2005, 2005.

Miller, N. B., Shupe, M. D., Cox, C. J., Walden, V. P., Turner, D. D., and Steffen, K.: Cloud Radiative Forcing at Summit, Greenland,

J. Climate, 28, 6267–6280, https://doi.org/10.1175/JCLI-D-15-0076.1, 2015.

Nicolaus, M., Hudson, S. R., Gerland, S., and Munderloh, K.: A modern concept for autonomous and continuous measurements of spectral albedo and transmittance of sea ice, Cold Reg. Sci. Technol., 62, 14–28, https://doi.org/10.1016/j.coldregions.2010.03.001, 2010.

Pierluissi, J. H. and Peng, G.-S.: New Molecular Transmission Band Models For LOWTRAN, Opt. Eng., 24, 243541, https://doi.org/10.1117/12.7973523, 1985.

Quinn, P. K., Shaw, G., Andrews, E., Dutton, E. G., Ruoho-Airola, T., and Gong, S. L.: Arctic haze: current trends and knowledge gaps, Tellus B, 59, 99–114, https://doi.org/10.1111/j.1600-0889.2006.00238.x, 2007.

Ricchiazzi, P., Yang, S., Gautier, C., and Sowle, D.: SBDART: A Research and Teaching Software Tool for Plane-Parallel Radiative Transfer in the Earth's Atmosphere, B. Am. Meteorol. Soc., 79, 2101–2114, 1998.

Rippin, D. M., Pomfret, A., and King, N.: High resolution mapping of supra-glacial drainage pathways reveals link between micro-channel drainage density, surface roughness and surface reflectance, Earth Surf. Proc. Land., 40, 1279–1290, https://doi.org/10.1002/esp.3719, 2015.

Román, M. O., Schaaf, C. B., Woodcock, C. E., Strahler, A. H., Yang, X., Braswell, R. H., Curtis, P. S., Davis, K. J., Dragoni, D., Goulden, M. L., Gu, L., Hollinger, D. Y., Kolb, T. E., Meyers, T. P., Munger, J. W., Privette, J. L., Richardson, A. D., Wilson, T. B., and Wofsy, S. C.: Remote Sensing of Environment The MODIS (Collection V005) BRDF/albedo product: Assessment of spatial representativeness over forested landscapes, Remote Sens. Environ., 113, 2476–2498, https://doi.org/10.1016/j.rse.2009.07.009, 2009.

Román, M. O., Schaaf, C. B., Lewis, P., Gao, F., Anderson, G. P., Privette, J. L., Strahler, A. H., Woodcock, C. E., and Barnsley, M.: Assessing the coupling between surface albedo derived from MODIS and the fraction of diffuse skylight over spatially-characterized landscapes, Remote Sens. Environ., 114, 738–760, https://doi.org/10.1016/j.rse.2009.11.014, 2010.

Ryan, J. C., Hubbard, A. L., Box, J. E., Todd, J., Christoffersen, P., Carr, J. R., Holt, T. O., and Snooke, N.: UAV photogrammetry and structure from motion to assess calving dynamics at Store Glacier, a large outlet draining the Greenland ice sheet, The Cryosphere, 9, 1–11, https://doi.org/10.5194/tc-9-1-2015, 2015.

Ryan, J. C., Hubbard, A., Box, J. E., Brough, S., Cameron, K., Cook, J. M., Cooper, M., Doyle, S. H., Edwards, A., Holt, T., Irvine-Fynn, T., Jones, C., Pitcher, L. H., Rennermalm, A. K., Smith, L. C., Stibal, M., and Snooke, N.: Derivation of High Spatial Resolution Albedo from UAV Digital Imagery: Application over the Greenland Ice Sheet, Frontiers in Earth Science, 5, 40, https://doi.org/10.3389/feart.2017.00040, 2017.

Salomonson, V. V. and Marlatt, W.: Airborne Measurements of Reflected Solar Radiation, Remote Sens. Environ., 2, 1–8, https://doi.org/10.1016/0034-4257(71)90072-1, 1971.

Schaaf, C. B., Gao, F., Strahler, A. H., Lucht, W., Li, X., Tsang, T., Strugnell, N. C., Zhang, X., Jin, Y., Muller, J.-P., Lewis, P., Barnsley, M., Hobson, P., Disney, M., Roberts, G., Dunderdale, M., Doll, C., D'Entremont, R. P., Hu, B., Liang, S., Privette, J. L., and Roy, D.: First operational BRDF, albedo nadir reflectance products from MODIS, Remote Sens. Environ., 83, 135–148, https://doi.org/10.1016/S0034-4257(02)00091-3, 2002.

Siegfried, M., Hawley, R., and Burkhart, J.: High-Resolution Ground-Based GPS Measurements Show Intercampaign Bias in ICESat Elevation Data Near Summit, Greenland, IEEE T. Geosci. Remote, 49, 3393–3400, https://doi.org/10.1109/TGRS.2011.2127483, 2011.

Stamnes, K., Tsay, S. C., Wiscombe, W., and Jayaweera, K.: Numerically stable algorithm for discrete-ordinate-method radiative transfer in multiple scattering and emitting layered media, Appl. Optics, 27, 2502–2509, 1988.

Stroeve, J., Box, J. E., Wang, Z., Schaaf, C., and Barrett, A.: Re-evaluation of MODIS MCD43 Greenland albedo accuracy and trends, Remote Sens. Environ., 138, 199–214, https://doi.org/10.1016/j.rse.2013.07.023, 2013.

Stroeve, J. C., Box, J. E., Gao, F., Liang, S., Nolin, A., and Schaaf, C.: Accuracy assessment of the MODIS 16-day albedo product for snow: comparisons with Greenland in situ measurements, Remote Sens. Environ., 94, 46–60, https://doi.org/10.1016/j.rse.2004.09.001, 2005.

Stroeve, J. C., Box, J. E., and Haran, T.: Evaluation of the MODIS (MOD10A1) daily snow albedo product over the Greenland ice sheet, Remote Sens. Environ., 105, 155–171, https://doi.org/10.1016/j.rse.2006.06.009,2006.

Tsay, S.-C., Stamnes, K., and Jayaweera, K.: Radiative energy budget in the cloudy and hazy arctic, J. Atmos. Sci., 46, 1002–1018, 1989.

Wang, Z., Schaaf, C. B., Chopping, M. J., Strahler, A. H., Wang, J., O.Román, M., Rocha, A. V., Woodcock, C. E., and Shuai, Y.: Evaluation of Moderate-resolution Imaging Spectroradiometer (MODIS) snow albedo product (MCD43A) over tundra, Remote Sens. Environ., 117, 264–280, 2012.

Wang, Z., Schaaf, C. B., Strahler, A. H., Chopping, M. J., Román, M. O., Shuai, Y., Woodcock, C. E., Hollinger, D. Y., and Fitzjarrald, D. R.: Evaluation of MODIS albedo product (MCD43A) over grassland, agriculture and forest surface types during dormant and snow-covered periods, Remote Sens. Environ., 140, 60–77, https://doi.org/10.1016/j.rse.2013.08.025, 2014.

Warren, S. G.: Can black carbon in snow be detected by remote sensing?, J. Geophys. Res.-Atmos., 118, 779–786, https://doi.org/10.1029/2012JD018476, 2012.

Warren, S. G. and Wiscombe, W. J.: A Model for the spectral albedo of Snow. II: Snow Containing Atmospheric Aerosols, J. Atmos. Sci., 37, 2734–2745, 1980.

Wendisch, M., Müller, D., Schell, D., and Heintzenberg, J.: An Airborne Spectral Albedometer with Active Horizontal Stabilization, J. Atmos. Ocean. Tech., 18, 1856–1866, https://doi.org/10.1175/1520-0426(2001)018<1856:AASAWA>2.0.CO;2, 2001.

Wiscombe, W. and Warren, S.: A model for the spectral albedo of snow. I: Pure snow, J. Atmos. Sci., 37, 2712–2733, 1980.

Wright, P., Bergin, M., Dibb, J., Lefer, B., Domine, F., Carman, T., Carmagnola, C., Dumont, M., Courville, Z., Schaaf, C., and Wang, Z.: Comparing MODIS daily snow albedo to spectral albedo field measurements in Central Greenland, Remote Sens. Environ., 140, 118–129, https://doi.org/10.1016/j.rse.2013.08.044, 2014.

7

Exceptional retreat of Novaya Zemlya's marine-terminating outlet glaciers between 2000 and 2013

J. Rachel Carr[1], Heather Bell[2], Rebecca Killick[3], and Tom Holt[4]

[1]School of Geography, Politics and Sociology, Newcastle University, Newcastle-upon-Tyne, NE1 7RU, UK
[2]Department of Geography, Durham University, Durham, DH13TQ, UK
[3]Department of Mathematics & Statistics, Lancaster University, Lancaster, LA1 4YF, UK
[4]Centre for Glaciology, Department of Geography and Earth Sciences, Aberystwyth University, Aberystwyth, SY23 4RQ, UK

Correspondence to: J. Rachel Carr (rachel.carr@newcastle.ac.uk)

Abstract. Novaya Zemlya (NVZ) has experienced rapid ice loss and accelerated marine-terminating glacier retreat during the past 2 decades. However, it is unknown whether this retreat is exceptional longer term and/or whether it has persisted since 2010. Investigating this is vital, as dynamic thinning may contribute substantially to ice loss from NVZ, but is not currently included in sea level rise predictions. Here, we use remotely sensed data to assess controls on NVZ glacier retreat between 1973/76 and 2015. Glaciers that terminate into lakes or the ocean receded 3.5 times faster than those that terminate on land. Between 2000 and 2013, retreat rates were significantly higher on marine-terminating outlet glaciers than during the previous 27 years, and we observe widespread slowdown in retreat, and even advance, between 2013 and 2015. There were some common patterns in the timing of glacier retreat, but the magnitude varied between individual glaciers. Rapid retreat between 2000 and 2013 corresponds to a period of significantly warmer air temperatures and reduced sea ice concentrations, and to changes in the North Atlantic Oscillation (NAO) and Atlantic Multidecadal Oscillation (AMO). We need to assess the impact of this accelerated retreat on dynamic ice losses from NVZ to accurately quantify its future sea level rise contribution.

1 Introduction

Glaciers and ice caps are the main cryospheric source of global sea level rise and contributed approximately $-215 \pm 26 \, \mathrm{Gt \, yr^{-1}}$ between 2003 and 2009 (Gardner et al., 2013). This ice loss is predicted to continue during the 21st century (Meier et al., 2007; Radić et al., 2014), and changes are expected to be particularly marked in the Arctic, where warming of up to 8 °C is forecast (IPCC, 2013). Outside of the Greenland Ice Sheet, the Russian high Arctic (RHA) accounts for approximately 20 % of Arctic glacier ice (Dowdeswell and Williams, 1997; Radić et al., 2014) and is, therefore, a major ice reservoir. It comprises three main archipelagos: Novaya Zemlya (NVZ; glacier area = 21 200 km²), Severnaya Zemlya (16 700 km²), and Franz Josef Land (12 700 km²) (Moholdt et al., 2012). Between 2003 and 2009, these glaciated regions lost ice at a rate of between 9.1 Gt a^{-1} (Moholdt et al., 2012) and 11 Gt a^{-1} (Gardner et al., 2013), with over 80 % of mass loss coming from Novaya Zemlya (NVZ) (Moholdt et al., 2012). This much larger contribution from NVZ has been attributed to it experiencing longer melt seasons and high snowmelt variability between 1995 and 2011 (Zhao et al., 2014). More recent estimates suggest that the mass balance of the RHA was -6.9 ± 7.4 Gt between 2004 and 2012 (Matsuo and Heki, 2013) and that thinning rates increased to $-0.40 \pm 0.09 \, \mathrm{m \, a^{-1}}$ between 2012/13 and 2014, compared to the long-term average of $-0.23 \pm 0.04 \, \mathrm{m \, a^{-1}}$ (1952 and 2014) (Melkonian et al., 2016). The RHA is, therefore, following the Arctic-wide pattern of negative mass balance (Gardner et al., 2013) and glacier retreat that has been observed in Greenland (Enderlin et al., 2014; McMillan et al., 2016), Svalbard (Moholdt et al., 2010a, b; Nuth et al., 2010), and the Canadian Arctic (Enderlin et al., 2014; McMillan et al., 2016). However, the RHA has been studied far less than other Arctic regions, despite its large ice volumes. Furthermore, assessment of 21st-century glacier volume loss highlights the RHA as one of the largest sources of future ice loss and contribution to sea level rise, with an estimated loss of 20–28 mm of sea level rise equivalent by 2100 (Radić et al., 2014).

Arctic ice loss occurs via two main mechanisms: a net increase in surface melting, relative to surface accumulation, and accelerated discharge from marine-terminating outlet glaciers (e.g. Enderlin et al., 2014; van den Broeke et al., 2009). These marine-terminating outlets allow ice caps to respond rapidly to climatic change, both immediately through calving and frontal retreat (e.g. Blaszczyk et al., 2009; Carr et al., 2014; McNabb and Hock, 2014; Moon and Joughin, 2008) and also through long-term drawdown of inland ice, often referred to as "dynamic thinning" (e.g. Price et al., 2011; Pritchard et al., 2009). During the 2000s, widespread marine-terminating glacier retreat was observed across the Arctic (e.g. Blaszczyk et al., 2009; Howat et al., 2008; McNabb and Hock, 2014; Moon and Joughin, 2008; Nuth et al., 2007), and substantial retreat occurred on Novaya Zemlya between 2000 and 2010 (Carr et al., 2014): retreat rates increased markedly from around 2000 on the Barents Sea coast and from 2003 on the Kara Sea (Carr et al., 2014). Between 1992 and 2010, retreat rates on NVZ were an order of magnitude higher on marine-terminating glaciers ($-52.1\,\mathrm{m\,a^{-1}}$) than on those terminating on land ($-4.8\,\mathrm{m\,a^{-1}}$) (Carr et al., 2014), which mirrors patterns observed on other Arctic ice masses (e.g. Dowdeswell et al., 2008; Moon and Joughin, 2008; Pritchard et al., 2009; Sole et al., 2008) and was linked to changes in sea ice concentrations (Carr et al., 2014). However, the pattern of frontal-position changes on NVZ prior to 1992 is uncertain, and previous results indicate different trends, dependant on the study period: some studies suggest glaciers were comparatively stable or retreating slowly between 1964 and 1993 (Zeeberg and Forman, 2001), whilst others indicate large reductions in both the volume (Kotlyakov et al., 2010) and the length of the ice coast (Sharov, 2005) from ∼ 1950 to 2000, and longer-term retreat (Chizov et al., 1968; Koryakin, 2013; Shumsky, 1949). Consequently, it is difficult to contextualize the observed period of rapid retreat from ∼ 2000 until 2010 (Carr et al., 2014) and to determine if it was exceptional or part of an ongoing trend. Furthermore, it is unclear whether glacier retreat has continued to accelerate after 2010, and hence further increased its contribution to sea level rise, or whether it has persisted at a similar rate. This paper aims to address these limitations, by extending the time series of glacier frontal-position data on NVZ to include the period 1973/76 to 2015, which represents the limits of available satellite data.

Initially, surface elevation change data from NVZ suggested that there was no significant difference in thinning rates between marine- and land-terminating outlet glacier catchments between 2003 and 2009 (Moholdt et al., 2012). This contrasted markedly with results from Greenland (e.g.

Price et al., 2011; Sole et al., 2008) but was similar to the Canadian Arctic, where the vast majority of recent ice loss occurred via increased surface melting (∼ 92 % of total ice loss), rather than accelerated glacier discharge (∼ 8 %) (Gardner et al., 2011). This implied that outlet glacier retreat was having a limited and/or delayed impact on inland ice or that available data were not adequately capturing surface elevation change in outlet glacier basins (Carr et al., 2014). More recent results demonstrate that thinning rates on marine-terminating glaciers on the Barents Sea coast are much higher than on their land-terminating neighbours, suggesting that glacier retreat and calving do promote inland, dynamic thinning (Melkonian et al., 2016). However, higher melt rates also contributed to surface lowering, evidenced by the concurrent increase in thinning observed on land-terminating outlets (Melkonian et al., 2016). High rates of dynamic thinning have also been identified on Severnaya Zemlya, following the collapse of the Matusevich Ice Shelf in 2012 (Willis et al., 2015). Here, thinning rates increased to 3–4 times above the long-term average (1984–2014), following the ice-shelf collapse in summer 2012, and outlet glaciers feeding into the ice shelf accelerated by up to 200 % (Willis et al., 2015). The most recent evidence, therefore, suggests that NVZ and other Russian high Arctic ice masses are vulnerable to dynamic thinning, following glacier retreat and/or ice-shelf collapse. Consequently, it is important to understand the longer-term retreat history on NVZ in order to evaluate its impact on future dynamic thinning. Furthermore, we need to assess whether the high glacier retreat rates observed on NVZ during the 2000s have continued and/or increased, as this may lead to much larger losses in the future and may indicate that a step change in glacier behaviour occurred in ∼ 2000.

In this paper, we use remotely sensed data to assess glacier frontal-position change for all major (> 1 km wide) Novaya Zemlya outlet glaciers (Fig. 1). This includes all outlets from the ice cap of the northern island (hereafter referred to as the northern island ice cap for brevity) and its subsidiary ice fields (Fig. 1). We were unable to find the names of these subsidiary ice fields in the literature, so we name them Sub 1 and Sub 2 (Fig. 1). A total of 54 outlet glaciers were investigated, which allowed us to assess the impact of different glaciological, climatic and oceanic settings on retreat (Table S1 in the Supplement). Specifically, we assessed the impact of coast (Barents Sea versus Kara Sea on the northern ice mass), ice mass (northern island ice cap, Sub 1, or Sub 2), terminus type (marine-, lake-, and land-terminating), and latitude (Table 1). The two coasts of Novaya Zemlya are characterized by very different climatic and oceanic conditions: the Barents Sea coast is influenced by water from the North Atlantic (Loeng, 1991; Pfirman et al., 1994; Politova et al., 2012) and subject to Atlantic cyclonic systems (Zeeberg and Forman, 2001), which results in warmer air and ocean temperatures as well as higher precipitation (Przybylak and Wyszyński, 2016; Zeeberg and Forman, 2001). In

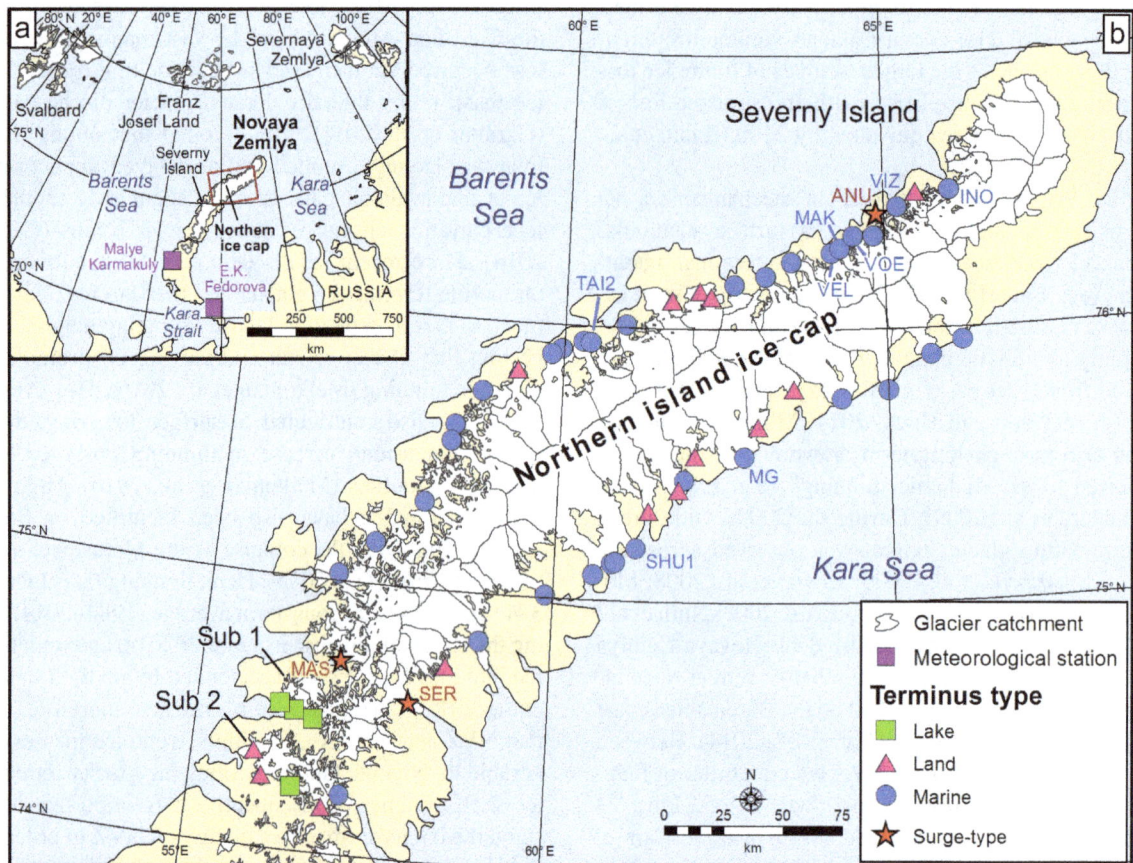

Figure 1. Location map, showing the study area and outlet glaciers. **(a)** Location of Novaya Zemlya, in relation to major land and water masses. Meteorological stations where air temperature data were acquired are indicated by a purple square (Malye Karmakuly, WMO ID: 20744; E. K. Fedorova, WMO ID: 20946). **(b)** Study glacier locations and main glacier catchments (provided by G. Moholdt and available via GLIMS database). Glaciers are symbolized according to terminus type: marine-terminating (blue circle); land-terminating (pink triangle); lake-terminating (green square); and observed surging during the study period (red star). Glaciers observed to surge are Anuchina (ANU), Mashigina (MAS), and Serp i Molot (SER).

contrast, the Kara Sea coast is isolated from North Atlantic weather systems, by the topographic barrier of NVZ (Pavlov and Pfirman, 1995), and is subject to cold, Arctic-derived water, along with much higher sea ice concentrations (Zeeberg and Forman, 2001). We therefore aim to investigate whether these differing climatic and oceanic conditions lead to major differences in glacier retreat between the two coasts. Glaciers identified as surge type (Grant et al., 2009) were excluded from the retreat calculations and analysis. However, frontal-position data are presented separately for three glaciers that were actively surging during the study period. Glacier retreat was assessed from 1973/76 to 2015 in order to provide the greatest temporal coverage possible from satellite imagery. We use these data to address the following questions:

1. At multi-decadal timescales, is there a significant difference in glacier retreat rates according to (i) terminus type (land-, lake- or marine-terminating); (ii) coast

(Barents Sea versus Kara Sea coast); (iii) ice mass (northern ice mass, Sub 1, or Sub 2); and (iv) latitude?

2. Are outlet glacier retreat rates observed between 2000 and 2010 on NVZ exceptional during the past ∼ 40 years?

3. Is glacier retreat accelerating, decelerating, or persisting at the same rate?

4. Can we link observed retreat to changes in external forcing (air temperatures, sea ice, and/or ocean temperatures)?

2   Methods

2.1   Study area

This paper focuses on the ice masses located on Severny Island, which is the northern island of the Novaya Zemlya

Table 1. Number of outlet glaciers contained within each category used to assess spatial variations in retreat rate, specifically coast, ice mass, and terminus type.

Characteristic	Category	Number of glaciers
Coast	Barents Sea	27
	Kara Sea	18
Ice mass	Northern island ice cap	45
	Subsidiary ice mass 1	4
	Subsidiary ice mass 2	5
Terminus type	Marine	34
	Lake	6
	Land	14

archipelago (Fig. 1). The northern island ice cap contains the vast majority of ice (19 841 km^2) and the majority of the main outlet glaciers (Fig. 1). The northern island also has two smaller ice fields, Sub 1 and Sub 2, which are much smaller in area (1010 and 705 km^2 respectively) and have far fewer, smaller outlet glaciers (Sub 1 = 4; Sub 2 = 5) (Fig. 1). All glaciers that have been previously identified as surge type and those smaller than 1 km in width were excluded from our main analysis of glacier retreat rates and response to climate forcing. However, we also observed three glaciers surging during the study period: Anuchina (ANU), Mashigina (MAS), and Serp i Molot (SER) (Fig. 1). MAS and SER have been previously identified as surge type (Grant et al., 2009), but our data provide better constraints on the duration and timing of these surges. ANU was identified as potentially surge type, on the basis of looped moraines (Grant et al., 2009). Our study confirms it as surge type and provides information on the surge timing and duration. These three glaciers are not included in the assessment of NVZ glacier response to climate change, as surging can occur impudently of climate forcing (Meier and Post, 1969), but are discussed separately to improve our knowledge of NVZ surge characteristics. This resulted in a total of 54 outlet glaciers, which were located in a variety of settings and hence allowed us to assess spatial controls on glacier retreat (Table 1). Where available glacier names and World Glacier Inventory IDs are given in Table S1 in the Supplement, along with glacier acronyms used in this paper. The impact of coast could only be assessed for the main ice mass, as the glaciers on the smaller ice masses, Sub 1 and Sub 2, are located on the southern ice margin and so do not fall on either coast (Fig. 1).

2.2 Glacier frontal position

Outlet glacier frontal positions were acquired predominantly from Landsat imagery. These data have a spatial resolution of 30 m and were obtained freely via the United States Geological Survey (USGS) Global Visualization Viewer (GloVis) (http://glovis.usgs.gov/). The frequency of available im-

agery varied considerably during the study period. Data were available annually from 1999 to 2015 and between 1985 and 1989, although georeferencing issues during the latter time period meant that imagery needed to be re-coregistered manually using stable, off-ice locations as tie points. Prior to 1985, the only available Landsat scenes dated from 1973, and these also needed to be manually georeferenced. We verified all images that required georeferencing against Landsat 8 data, which should have the most accurate location data of the imagery time series. We did this by comparing the location of features that should be static between images (e.g. large rock fractures) and also checking for any unrealistic changes in the lateral glacier margins, over and above what could be expected by glacier melting. Any images where we saw changes in the location of static features above the image resolution were not used. As such, orthorectification was not required for these images, as the terrain is relatively gentle on NVZ, and our verification process showed that the images were co-located with the Landsat 8 imagery to within a pixel using just georeferencing. Hexagon KH-9 imagery was used to determine frontal positions in 1976 and 1977, but full coverage of the study area was not available for either year. The data resolution is 20 to 30 ft ($\sim$ 6–9 m). The earliest common date for which we have frontal positions for all glaciers is 1986, and so we calculate total retreat rates for the period 1986–2015 and use these values to assess spatial variability in glacier recession across the study region. All glacier frontal positions are calculated relative to 1986 (i.e. the frontal position in 1986 = 0 m) to allow for direct comparison.

Due to the lack of Landsat imagery during the 1990s, we use synthetic aperture radar (SAR) image mode precision data during this period. The data were provided by the European Space Agency, and we use European Remote-sensing Satellite-1 (ERS-1) and ERS-2 products (https://earth.esa.int/web/guest/data-access/browse-data-products/-/asset_publisher/y8Qb/content/sar-precision-image-product-1477). Following Carr et al. (2013b), the ERS scenes were first co-registered with Envisat imagery and then processed using the following steps: apply precise orbital state vectors; radiometric calibration; multi-look; and terrain correction. This gave an output resolution of 37.5 m, which is comparable to Landsat. For each year and data type, imagery was acquired as close as possible to 31 July to minimize the impact of seasonal variability. However, this is unlikely to substantially effect results, as previous studies suggest that seasonal variability in terminus position is very limited on NVZ ($\sim$ 100 m a^{-1}) (Carr et al., 2014) and is therefore much less than the interannual and inter-decadal variability we observe here. Glacier frontal-position change was calculated using the box method: the terminus was repeatedly digitized from successive images, within a fixed reference box, and the resultant change in area is divided by the reference box width to get frontal-position change (e.g. Moon and Joughin, 2008). Following previous studies (Carr et al., 2014), we determined the frontal-position errors for marine- and lake-terminating outlets glaciers by digitizing 10 sections of rock coastline from six images, evenly spread through the time series (1976, 1986, 2000, 2005, 2010, and 2015) and across NVZ. The resultant error was 17.5 m, which equates to a retreat rate error of $1.75\,\mathrm{m\,a^{-1}}$ at the decadal time intervals discussed here. The terminus is much harder to identify on land-terminating outlet glaciers due to the similarity between the debris-covered ice margins and the surrounding land, which adds an additional source of error. We quantified this by re-digitizing a sub-sample of six land-terminating glaciers in each of the six images, which were spread across NVZ. The additional error for land-terminating glaciers was 66.1 m, giving a total error of 68.4 m, which equates to a retreat rate error of $6.86\,\mathrm{m\,a^{-1}}$ for decadal intervals.

2.3 Climate and ocean data

Air temperature data were obtained from meteorological stations located on, and proximal to, Novaya Zemlya (Fig. 1). Directly measured meteorological data are very sparse on NVZ, and there are large gaps in the time series for many stations. We use data from two stations, Malye Karmakuly (WMO ID: 20744) and E. K. Fedorova (WMO ID: 20946), as these are the closest stations to the study glaciers that have a comprehensive (although still not complete) record during the study period (Supplement Table S2). The data were obtained from the Hydrometeorological Information – World Data Centre Baseline Climatological Data Sets (http://meteo.ru/english/climate/cl_data.php) and were provided at a monthly temporal resolution. For each station, we calculated meteorological seasonal means (December–February, March–May, June–August, September–November) in order to assess the timing of any changes in air temperature, as warming in certain seasons would have a different impact on glacier retreat rates. Seasonal and annual means were only calculated if values were available for all months. Due to data gaps, particularly from 2013 onwards (Supplement Table S2), we also assess changes in air temperature using ERA-Interim reanalysis data (http://www.ecmwf.int/en/research/climate-reanalysis/era-interim). We use temperature data from the surface (2 m elevation) and 850 hPa pressure level, as these are likely to be a good proxy for meltwater availability (X. Fettweis, personal communication, 2017). We use the "monthly means of daily means" product for all months between 1979 and 2015. As with the meteorological stations, we calculate means for the meteorological seasons and annual means.

Sea ice data were acquired from the Nimbus-7 SMMR and DMSP SSM/I-SSMIS Passive Microwave data set (https://nsidc.org/data/docs/daac/nsidc0051_gsfc_seaice.gd.html). The data provide information on the percentage of the ocean covered by sea ice, and this is measured using brightness temperatures from microwave sensors. The data have a spatial resolution of 25×25 km, and we use the monthly-averaged product. This data set was selected due to its long temporal coverage, which extends from 26 October 1978 to 31 December 2015 and thus provides a consistent data set throughout our study period. NVZ glaciers are not located within long fjords and are relatively exposed to the open ocean (Fig. 1). Consequently, sea ice conditions within 25 km of the glacier fronts (i.e. the data resolution) are likely to be reasonably representative of the overall sea ice trends experienced by the glaciers, particularly at the decadal timescales assessed here. However, it should be noted that the data cannot provide detailed information on sea ice conditions specific to each glacier front, but they are used here as they comprise the only data set available for the entire study period. Monthly sea ice concentrations were sampled from the grid squares closest to the study glaciers and were split according to coast (i.e. Barents Sea and Kara Sea). From the monthly data, we calculated seasonal means and the number of ice-free months, which we define as the number of months where the mean monthly sea ice cover is less than 10 %.

Data on the North Atlantic Oscillation (NAO) were obtained from the Climatic Research Unit (https://crudata.uea.ac.uk/cru/data/nao/), and the monthly product was used. This records the normalized pressure difference between Iceland and the Azores (Hurrell, 1995). Arctic Oscillation (AO) data were acquired from the Climate Prediction Center (http://www.cpc.noaa.gov/products/precip/CWlink/daily_ao_index/teleconnections.shtml). The AO is characterized by winds at 55° N, which circulate anti-clockwise around the Arctic (e.g. Higgins et al., 2000; Zhou et al., 2001). The AO index is calculated by projecting the AO loading pattern onto the daily anomaly 1000 mbar height field, at 20–90° N latitude (Zhou et al., 2001). The Atlantic Multidecadal Oscillation (AMO) is a mode of variability associated with averaged, de-trended sea surface temperatures (SSTs) in the North Atlantic and varies over timescales of 60 to 80 years (Drinkwater et al., 2013; Sutton and Hodson, 2005). Monthly data were downloaded from the National Oceanic and Atmospheric Administration (https://www.esrl.noaa.gov/psd/data/timeseries/AMO/).

We use ocean temperature data from the "Climatological Atlas of the Nordic Seas and Northern North Atlantic" (Hurrell, 1995; Korablev et al., 2014) (https://www.nodc.noaa.gov/OC5/nordic-seas/). The atlas compiles data from over 500 000 oceanographic stations, located across the Nordic Seas, between 1900 and 2012. It provides gridded climatologies of water temperature, salinity, and density, at a range of depths (surface to 3500 m), for the region bounded by 83.875 to 71.875° N and 47.125° W to 57.875° E. Here, we use data from the surface and 100 m depth to capture changes in ocean temperatures at different depths: surface warming may influence glacier behaviour through changes in sea ice and/or undercutting at the waterline (Benn et al., 2007), whereas warming in the deeper layers can enhance sub-aqueous melting (Sutherland et al., 2013). A depth of

100 m was chosen, as it is the deepest level that includes the majority of the continental shelf immediately offshore of Novaya Zemlya. Further details of the data set production and error values are given in Korablev et al. (2014). We use the decadal ocean temperature product to identify broad-scale changes, which is provided at the following time intervals: 1971–1980, 1981–1990, 1991–2000, and 2001–2012. We use the decadal product as there are few observations offshore of Novaya Zemlya during the 2000s, whereas the data coverage is much denser in the 1980s and 1990s (a full inventory of the number and location of observations for each month and year is provided here: https://www.nodc.noaa.gov/OC5/nordic-seas/atlas/inventory.html). As a result, maps of temperature changes in the 2000s are produced using comparatively data few points, meaning that they may not be representative of conditions in the region and that directly comparing data at a shorter temporal resolution (e.g. annual data) may be inaccurate. Furthermore, the input data were measured offshore of Novaya Zemlya and not within the glacier fjords. Consequently, there is uncertainty over the extent to which offshore warming is transmitted to the glacier front and/or the degree of modification due to complexities in the circulation and water properties within glacial fjords. We therefore use decadal-scale data to gain an overview of oceanic changes in the region, but we do not attempt to use them for detailed analysis of the impact of ocean warming at the glacier front, nor for statistical testing.

2.4 Statistical analysis

We used a Kruskal–Wallis test to investigate statistical differences in total retreat rate (1986–2015) for the different categories of outlet glacier within our study population, i.e. terminus type (marine-, land-, and lake-terminating), coast (Barents Sea and Kara Sea), and ice mass (northern island ice cap, Sub 1, and Sub 2). The Kruskal–Wallis test is a non-parametric version of the one-way ANOVA (analysis of variance) test and analyses the variance using the ranks of the data values, as opposed to the actual data. Consequently, it does not assume normality in the data, which is required here, as Kolmogorov–Smirnov tests indicate that total retreat rate (1986–2015) is not normally distributed for any of the glacier categories (e.g. terminus type). This is also the case when we test for normality at each of the four time intervals discussed below (1973/76–1986, 1986–2000, 2000–2013, and 2013–2015). The Kruskal–Wallis test gives a p value for the null hypothesis that two or more data samples come from the same population. As such, a large p value suggests it is likely that the samples come from the same population, whereas a small value indicates that this is unlikely. We follow convention and use a significance value of 0.05, meaning that a p value of less than or equal to 0.05 indicates that the data samples are significantly different.

We assessed the influence of glacier latitude on total retreat rate (1986–2015), using simple linear regression. This fits a line to the data points and gives an R^2 value and a p value for this relationship. The R^2 value indicates how well the line describes the data: if all points fell exactly on the line, the R^2 would equal 1, whereas if the points were randomly distributed about the line, the R^2 would equal 0. The p-value tests the null hypothesis that the regression coefficient is equal to zero, i.e. that the predictor variable (e.g. glacier catchment size) has no relationship to the response variable (e.g. total glacier retreat rate). A p value of 0.05 or less therefore indicates that the null hypothesis can be rejected and that the predictor variable is related to the response variable (e.g. glacier latitude is related to glacier retreat rate). The residuals for these regressions were normally distributed. However, we also regressed catchment area against total retreat rate, and the regression residuals were not normally distributed, indicating that it is not appropriate to use regression in this case. Consequently, we used Spearman's rank correlation coefficient, which is non-parametric and therefore does not require the data to be normally distributed. Catchments were obtained from Moholdt et al. (2012).

Wilcoxon tests were used to assess significant differences in mean glacier retreat rates between four time intervals: 1973/76–1986, 1986–2000, 2000–2013, and 2013–2015. These intervals were chosen through manual assessment of apparent breaks in the data. For each interval, data were split according to terminus type (marine, land, and lake), and marine-terminating glaciers were further subdivided by coast (Barents Sea and Kara Sea). For each category, we then used the Wilcoxon test to determine whether mean retreat rates for all of the glaciers during one time period (e.g. 1986–2000) were significantly different from those for another time period (e.g. 2000–2013). The Wilcoxon test was selected as it is non-parametric and our retreat data are not normally distributed, and it is suitable for testing statistical difference between data from two time periods (Miles et al., 2013). As with the Kruskal–Wallis test, a p value of less than or equal to 0.05 is taken as significant and indicates that the two time periods are significantly different. We also used the Wilcoxon test to identify any significant differences in mean air temperatures and sea ice conditions for the same time intervals as glacier retreat to allow for direct comparison. For the first time interval (1973/76–1986), we use air temperature data from 1976 to 1986 from the meteorological stations, but the sea ice and ERA-Interim data are only available from 1979. The statistical analysis was done separately for sea ice on the Barents Sea and Kara Sea coast and using meteorological data from Malye Karmakuly and E. K. Fedorova (Fig. 1). ERA-Interim data were analysed as a whole, as the spatial resolution of the data does not allow us to distinguish between the two coasts. In each case, we compared seasonal means for each year of a certain time period with the seasonal means for the other time period (e.g. 1976–1985 versus 2000–2012). For the sea ice data, we used calendar seasons (January–March, April–June, July–September, October–December), which fits with the Arctic sea ice minima in September and maxima in March. For the air temperature data, meteorological seasons (December–February, March–May, June–August, September–November) are more appropriate. We also tested mean annual air temperatures and the number of sea-ice-free months.

In order to further investigate the temporal pattern of retreat on Novaya Zemlya, we use statistical change-point analysis (Eckley et al., 2011). We applied this to our frontal-position data for marine- and lake-terminating glaciers, and to the sea ice and air temperature data. Land-terminating glaciers are not included, due to the much higher error margins compared to any trends, which could lead to erroneous change-points being identified. Change-point analysis allows us to automatically identify significant changes in the time series data and whether there has been a shift from one mode of behaviour to another (e.g. from slower to more rapid retreat) (Eckley et al., 2011). Formally, a change-point is a point in time where the statistical properties of prior data are different from the statistical properties of subsequent data; the data between two change-points are a segment. There are various ways that one can determine when a change-point should occur, but the most appropriate approach for our data is to consider changes in regression.

In order to automate the process, we use the cpt.reg function in the R EnvCpt package (Killick et al., 2016) with a minimum number of four data points between changes. This function uses the pruned exact linear time (PELT) algorithm (Killick et al., 2012) from the change-point package (Killick and Eckley, 2015) for fast and exact detection of multiple changes. The function returns change-point locations and estimates of the intercept and slope of the regression lines between changes. We give the algorithm no information on when we might be expecting a change or how large it may be, allowing it to automatically determine statistically different parts of the data. In this way, we use the analysis to determine whether, and when, retreat rates change significantly on each of the marine- and lake-terminating glaciers on NVZ, and whether there are any significant breaks in our sea ice and air temperature data. We also apply the change-point analysis to the number of sea-ice-free months, but as the data do not contain a trend, we identify breaks using significant changes in the mean, rather than a change in regression. Thus, we can identify any common behaviour between glaciers, determine the timing of any common changes, and compare this to any significant changes in atmospheric temperatures and sea ice concentrations.

3 Results

3.1 Spatial controls on glacier retreat

The Kruskal–Wallis test was used to identify significant differences in total retreat rate (1986–2015) for glaciers located in different settings. First, terminus type was investigated. Results demonstrated that total retreat rates (1986–2015) were significantly higher on lake- and marine-terminating glaciers than those terminating on land, at a very high confidence interval (< 0.001) (Fig. 2). Retreat rates were 3.5 times higher on glaciers terminating in water (lake $= -49.1 \, \mathrm{m\,a^{-1}}$; marine $= -46.9 \, \mathrm{m\,a^{-1}}$) than those ending on land ($-13.8 \, \mathrm{m\,a^{-1}}$) (Fig. 2). In contrast, there was no significant difference between lake- and marine-terminating glaciers (Fig. 2). Next, we assessed the role of coastal setting (i.e. Barents Sea versus Kara Sea) as climatic and oceanic conditions differ markedly between the two coasts. When comparing glaciers with the same terminus type, there was no significant difference in retreat rates between the two coasts (Fig. 2: p value $= 0.178$ for marine-terminating glaciers, and p value $= 1$ for land-terminating glaciers). Retreat rates on land-terminating glaciers were very similar on both coasts: Barents Sea $= -6.5 \, \mathrm{m\,a^{-1}}$, and Kara Sea $= -9.0 \, \mathrm{m\,a^{-1}}$ (Fig. 2). For marine-terminating outlets, retreat rates being higher on the Barents Sea confirmed that the significant difference in total retreat rates between land- and marine-terminating glaciers persists when individual coasts are considered (Fig. 2). Finally, we tested for differences in retreat rate between the ice masses of Novaya Zemlya, specifically the northern island ice cap, which is by far the largest, and the two smaller subsidiary ice fields, Sub 1 and Sub 2. Here, we found no significant difference in retreat rates between the ice masses (Fig. 2). Retreat rates were highest on Sub 2, followed by the northern island ice cap, and lowest on Sub 1 (Fig. 2). Our results therefore demonstrate that the only significant difference in total retreat rates (1986–2015) relates to glacier terminus type, with land-terminating outlets retreating 3.5 times slower than those ending in lakes or the ocean (Fig. 2).

We used simple linear regression to assess the relationship between total retreat rate (1986–2015) and latitude, as there is a strong north–south gradient in climatic conditions on NVZ, but no significant linear relationship was apparent ($R^2 = 0.001$, $p = 0.819$) (Fig. 3). However, if we divide the glaciers according to terminus type, total retreat rate shows a significant positive relationship for land-terminating glaciers ($R^2 = 0.363$, $p = 0.023$), although the R^2 value is comparatively small (Fig. 3). This indicates that more southerly land-terminating outlets are retreating more rapidly than those in the north. Conversely, total retreat rate for lake-terminating glaciers has a significant inverse relationship with total retreat rate ($R^2 = 0.811$, $p = 0.014$), suggesting that glaciers at high latitudes retreat more rapidly (Fig. 3). No linear relationship is apparent between latitude and total retreat rate for marine-terminating glaciers, and the data show considerable scatter, particularly in the north (Fig. 3). We find no significant relationship between catchment area and total retreat rate ($\rho = -0.149$, $p = 0.339$), which demonstrates that observed retreat patterns are not simply a function of glacier size (i.e. that larger glacier retreat more simply because they are bigger).

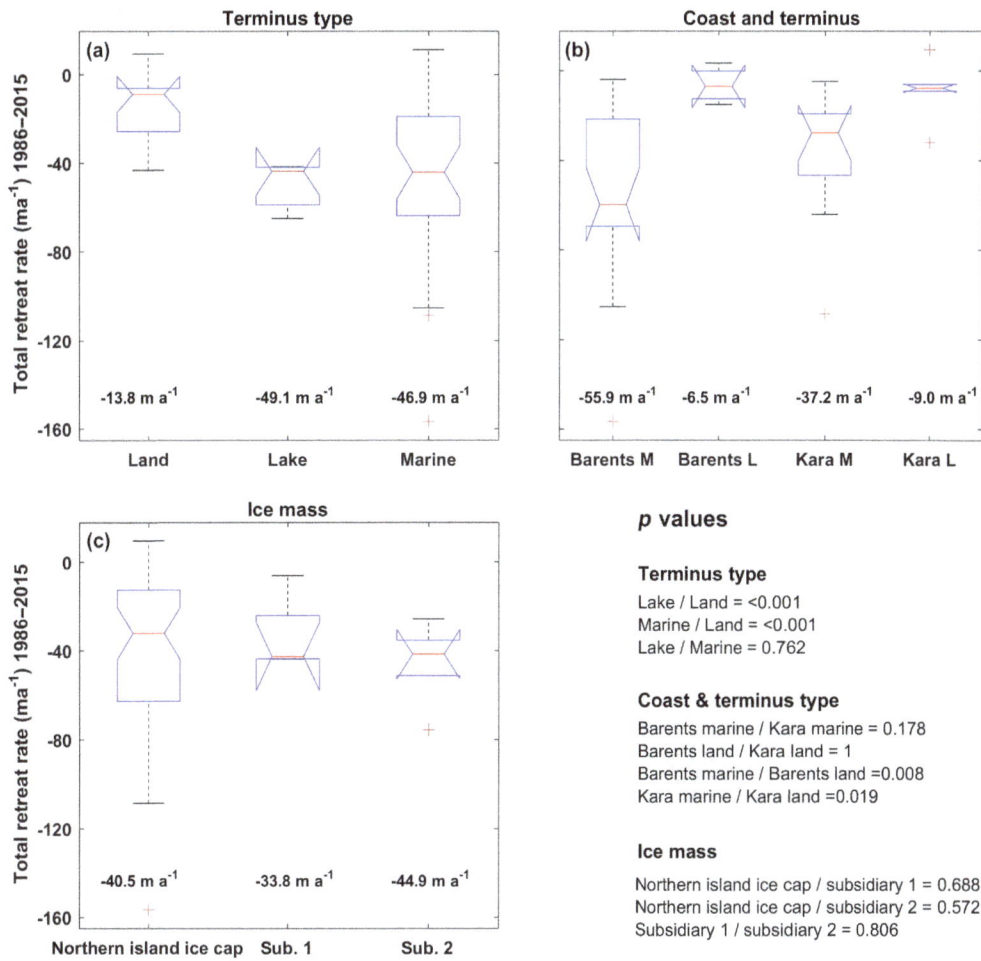

Figure 2. Box plots and Kruskal–Wallis test results for different glacier terminus settings for **(a)** terminus type; **(b)** coast and terminus: L stands for land-terminating, and m for marine-terminating; and **(c)** ice mass, specifically the northern island ice cap and subsidiary ice fields 1 and 2. See Fig. 1 for ice mass locations. In all cases, total retreat rate (1986–2015) is used to test for significant differences between the classes. Mean total retreat rates for each class are given on each plot, below the associated box plot. For each box plot, the red central line represents the median; the blue lines represent the upper and lower quartile; red crosses are outliers (a value more than 1.5 times the interquartile range above/below the interquartile values); and the black lines are the whiskers, which extend from the interquartile ranges to the maximum values that are not classed as outliers. P values for each Kruskal–Wallis test are given on the right of the plot.

3.2 Temporal change

Based on an initial assessment of the temporal pattern of retreat for individual glaciers, we manually identified major break points in the data and divided glacier retreat rates into four time intervals: 1973/76 to 1986, 1986 to 2000, 2000 to 2013, and 2013 to 2015 (Fig. 4). Data were separated according to terminus type and, in the case of marine-terminating glaciers, according to coast. We then used the Wilcoxon test to evaluate the statistical difference between these time periods for each category (Table 2). For land- and lake-terminating glaciers, there were no significant differences in retreat rates between any of the time periods (Fig. 4; Table 2). Indeed, retreat rates on lake-terminating glaciers were remarkably consistent between 1986 and 2015, both

over time and between glaciers (Figs. 4 and 5). For marine-terminating glaciers on the Barents Sea coast, the periods 1973/76–1986 and 1986–2000 were not significantly different from each other, and mean retreat rates were comparatively low (-20.5 and $-22.3\,\mathrm{m\,a^{-1}}$ respectively). In contrast, the periods 2000–2013 and 2013–2015 were both significantly different to all other time intervals (Fig. 4; Table 2). Between 2000 and 2013, retreat rates were much higher than at any other time ($-85.4\,\mathrm{m\,a^{-1}}$). Conversely, the average frontal-position change between 2013 and 2015 was positive, giving a mean advance of $+11.6\,\mathrm{m\,a^{-1}}$ (Fig. 4). On the Kara Sea coast, marine-terminating outlet glacier retreat rates were significantly higher between 2000 and 2013 than any other time period ($-64.8\,\mathrm{m\,a^{-1}}$) (Fig. 4; Table 2). Retreat rates reduced substantially during the period 2013–2015

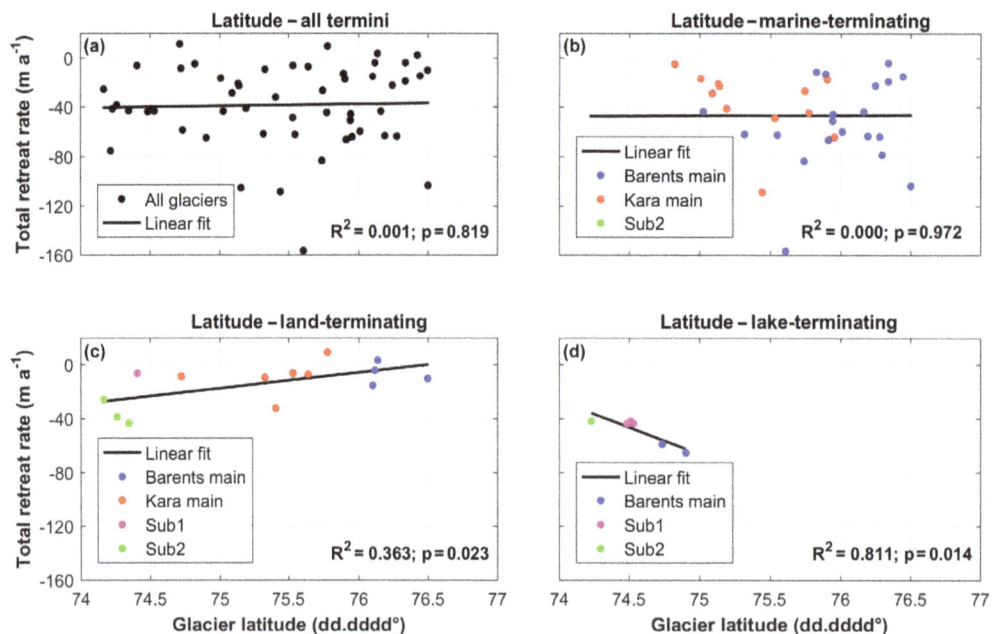

Figure 3. Linear regression of total retreat rate (1986–2015) versus glacier latitude. Latitude was regressed against total glacier retreat rate for (a) all outlet glaciers in the study sample; (b) marine-terminating glaciers only; (c) land-terminating glaciers only; and (d) lake-terminating glaciers only. In all cases, the linear regression line is shown, as are the associated R^2 and p values. The R^2 value indicates how well the line describes the data, and the p value indicates the significance of the regression coefficients, i.e. the likelihood that the predictor and response variable are unrelated.

Table 2. Wilcoxon test results, used to assess significant differences in retreat rates between each manually identified time interval (1976–1986, 1986–2000, 2000–2013, 2013, 2015). Retreat rate data were tested separately for each terminus type, and marine-terminating glaciers were further sub-divided by coast. Following convention, p values of < 0.05 are considered significant and are highlighted in bold.

	Barents Sea marine-terminating	Kara Sea marine-terminating	Land-terminating	Lake-terminating
76–86/86–00	0.440	0.538	0.982	0.486
76–86/00–13	**> 0.001**	**0.018**	0.085	0.686
76–86/13–15	**0.008**	0.497	0.945	0.686
86–00/00–13	**0.001**	**0.008**	0.223	0.886
86–00/13–15	**0.001**	0.935	0.909	0.886
00–13/13–15	**> 0.001**	**0.009**	0.597	0.686

$(-22.7 \, \text{m} \, \text{a}^{-1})$ and were very similar to values in 1973/76–1986 $(-27.2 \, \text{m} \, \text{a}^{-1})$ and 1986–2000 $(-22.4 \, \text{m} \, \text{a}^{-1})$ (Fig. 4). On both the Barents Sea and Kara Sea coasts, the temporal pattern of marine-terminating outlet glacier retreat showed large variability, both between individual glaciers and over time (Fig. 5).

Following our initial analysis, we used change-point analysis to further assess the temporal patterns of glacier retreat, by identifying the timing of significant breaks in the data. On the Barents Sea coast, five glaciers underwent a significant change in retreat rate from the early 1990s onwards (Fig. 6). Of these, retreat rates on four glaciers (MAK, TAI2, VEL, and VIZ; see Fig. 1 for glacier locations and names) subsequently increased, whereas retreat was slower on INO

between 1989 and 2006. The most widespread step change on the Barents Sea coast occurred in the early 2000s, after which nine glaciers retreated more rapidly (Fig. 6). A second widespread change in glacier retreat rates occurred in the mid-2000s, which was also the second change-point for four glaciers (Fig. 6). Of these eight glaciers, only VOE retreated more slowly after the mid-2000s change-point. On the Kara Sea coast, we see a broadly similar temporal pattern, with two glaciers showing a significant change in retreat rate from the early 1990s and again in 2005 and 2007 (Fig. 6). In the case of MG, retreat rates were higher after each breakpoint, whereas for SHU1 retreat rates were lower between the 1990s and mid-2000s. Four glaciers began to retreat more rapidly from 2000 onwards, and five other glaciers showed a

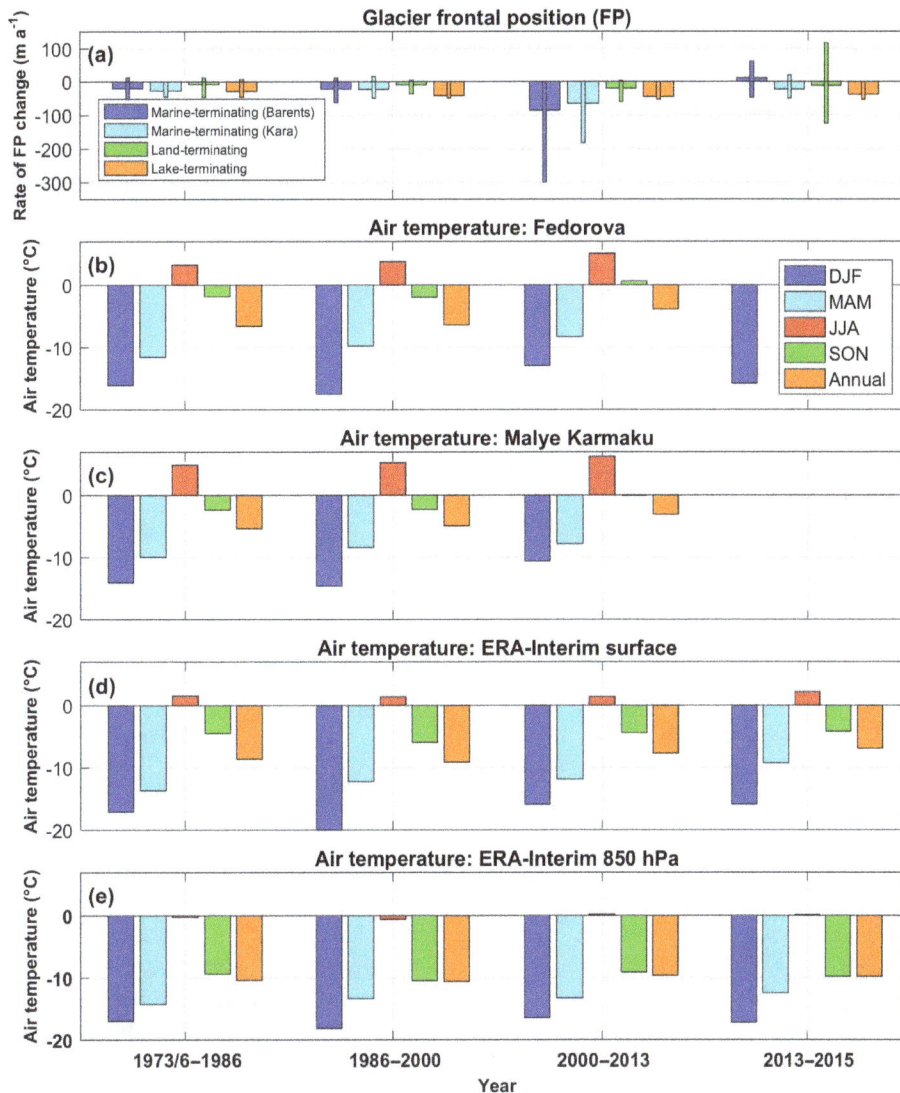

Figure 4. Mean retreat rates for Novaya Zemlya outlet glaciers, and mean air temperatures at E. K. Fedorova (WMO ID: 20946) and Malye Karmaku (WMO ID: 20744) (Fig. 1). Data are split into four time periods, based on manually identified breaks in the glacier retreat data: 1973/76–1986, 1986–2000, 2000–2013, and 2013–2015. **(a)** Retreat rates were calculated separately for different terminus types, and marine-terminating glaciers were further sub-divided into those terminating into the Barents Sea versus the Kara Sea. Wide bars represent mean values, and thin bars represent the total range (i.e. minimum and maximum values) within each category. **(b–e)** Mean seasonal air temperatures (December–February, March–May, June–August, and September–November) and mean annual air temperatures for E. K. Fedorova **(b)**, Malye Karmaku **(c)**, ERA-Interim surface **(d)**, and ERA-Interim 850 hPa pressure level **(e)**. Note that only mean values for E. K. Fedorova in January–March are calculated for 2013–2015, due to data availability.

significant change in retreat rates beginning between 2005 and 2010 (Fig. 6), with VER being the only glacier to show a reduction in retreat rates after this change (Fig. 6). Focusing on lake-terminating glaciers, a significant change in retreat rates began between 2006 and 2008 on all but one glacier, which began to retreat more rapidly from 2004 onwards (Fig. 6).

3.3 Climatic controls

At E. K. Fedorova, mean annual air temperatures were significantly warmer in 2000–2012 ($-3.9\,°C$) than in 1976–1985 ($-6.5\,°C$) or 1986–1999 ($-6.4\,°C$) (Fig. 4; Table 3). Looking at seasonal patterns, air temperatures were significantly higher during spring, summer, and autumn in 2000–2012 than in 1976–1985 (Fig. 4; Table 3). Similarly, air temperatures in 2002–2012 were significantly higher in summer, autumn, and than in 1986–1999 (Fig. 4; Table 3). Sum-

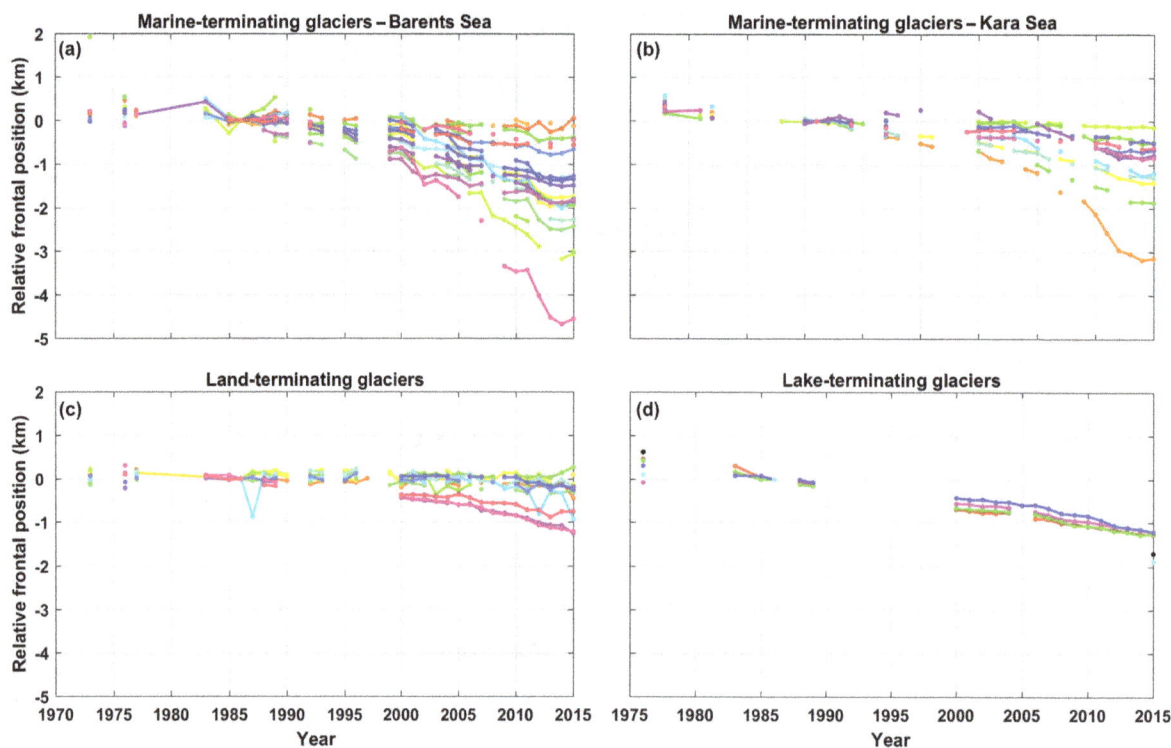

Figure 5. Relative glacier frontal position over time, from 1973 to 2015, for (a) marine-terminating outlet glaciers on the Barents Sea coast; (b) marine-terminating outlet glaciers on the Kara Sea coast; (c) land-terminating outlet glaciers; and (d) land-terminating outlet glaciers. Within each plot, frontal positions for each glacier are distinguished by different colours.

mer air temperatures averaged 5.1 °C in 2000–2012, compared to 3.8 °C in 1986–1999 and 3.3 °C in 1976–1985 (Fig. 4). Warming was particularly marked in winter, increasing from −16.1 °C (1976–1985) and −17.5 °C (1986–1999) to −12.9 °C in 2000–2012 (Fig. 4). Winter air temperatures then reduced to −15.9 °C for the period 2013–2015 (Fig. 4), although this change was not statistically significant (Table 3). A similar change in mean annual air temperatures was evident on Malye Karmakuly, where temperatures were significantly higher in 2000–2012 (−3.1 °C) than in 1976–1985 (−5.4 °C) or 1986–1999 (−5.0 °C) (Table 3; Fig. 4). In all seasons, air temperatures were significantly higher in 2000–2012 than in 1976–1985 (Table 3), with the largest absolute increases occurring in winter (Fig. 4). However, only autumn air temperatures were significantly warmer in 2000–2012 than in 1986–1999 (Fig. 4; Table 3). No significant differences in air temperatures were observed between 1976–1985 and 1986–1999 at either station (Table 3).

In the ERA-Interim reanalysis data, mean annual air temperatures increased significantly between 1986–1999 and 2000–2012 at both the surface and 850 hPa pressure level (Table 3). Winter (surface) and autumn (850 hPa) temperatures also warmed significantly between these time intervals (Table 3). Surface air temperatures were significantly warmer in 2013–2015 than in 1986–1999, in winter and annually (Table 3). No significant differences in air temper-

atures were observed at either height between 2000–2012 and 2013–2015 for any season (Table 3). Surface air temperatures were comparable between 2000–2012 and 2013–2015 in winter and autumn, and somewhat warmer in spring (+ 2.6 °C) and summer (+0.7 °C) in 2013–2015 (Fig. 4). At 850 m height, winter (−0.7 °C) and autumn temperatures were slightly cooler (−0.7 °C), and summer temperatures were warmer (+0.8 °C) in 2013–2015 than in 2000–2012 (Fig. 4). At the regional scale, warmer surface air temperatures penetrate further into the Barents Sea and the southern Kara Sea with each time step (Supplement Fig. S1). We observed a similar, although less marked, northward progression of the isotherms at 850 hPa level (Supplement Fig. S1).

On the Barents Sea coast, sea ice concentrations during all seasons were significantly lower in 2000–2012 than in 1976–1985 or 1986–1999, as was the number of ice-free months (Fig. 7; Table 4). Between 1976–1985 and 2000–2012, mean winter sea ice concentrations reduced from 68 to 35 %, mean spring values declined from 59 to 28 %, and mean autumn averages fell from 27 to 7 % (Fig. 7). Mean summer sea ice concentrations reduced slightly, from 12 to 5 % (Fig. 7). Over the same time interval, the number of ice-free months increased from 3.0 to 6.9 (Fig. 7). Summer sea ice concentrations on the Barents Sea coast reduced significantly between 2000–2012 and 2013–2015, but no significant change was observed in any other month, nor in the number of ice-free

Table 3. *P* values for Wilcoxon tests for significant differences in mean seasonal and mean annual air temperatures, for the periods 1976–1985, 1986–1999, 2000–2013, and 2013–2015. Following convention, *p* values of < 0.05 are considered significant and are highlighted in bold.

Station	Time interval	Season				
		DJF	MAM	JJA	SON	Annual
E. K. Fedorova	13–15/86–99	0.432				
E. K. Fedorova	13–15/76–85	0.937				
E. K. Fedorova	00–12/13–15	0.287				
E. K. Fedorova	00–12/86–99	**0.011**	0.643	**0.043**	**0.008**	**0.013**
E. K. Fedorova	00–12/76–85	0.186	**0.035**	**0.045**	**0.003**	**0.003**
E. K. Fedorova	86–99/76–85	0.188	0.089	0.704	0.495	0.828
Malye Karmakuly	13–15/86–99					
Malye Karmakuly	13–15/76–85					
Malye Karmakuly	00–12/13–15		–	–	–	–
Malye Karmakuly	00–12/86–99	0.017	0.840	0.056	**0.007**	0.017
Malye Karmakuly	00–12/76–85	**0.038**	**0.041**	**0.045**	**0.004**	**> 0.001**
Malye Karmakuly	86–99/76–85	0.623	0.086	0.5977	0.673	0.212
ERA-Interim (surface)	13–15/86–99	**0.032**	0.156	0.197	0.156	**0.006**
ERA-Interim (surface)	13–15/76–85	0.714	0.083	0.517	0.833	0.117
ERA-Interim (surface)	00–12/13–15	0.900	0.189	0.364	0.593	0.239
ERA-Interim (surface)	00–12/86–99	**0.006**	0.942	0.981	0.062	**0.044**
ERA-Interim (surface)	00–12/76–85	0.765	0.579	0.526	0.874	0.267
ERA-Interim (surface)	86–99/76–85	0.127	0.233	0.970	0.192	0.794
ERA-Interim (850 hPa)	13–15/86–99	0.591	0.509	0.432	0.500	0.206
ERA-Interim (850 hPa)	13–15/76–85	0.548	0.383	0.833	0.733	0.383
ERA-Interim (850 hPa)	00–12/13–15	0.521	0.611	0.782	0.511	0.900
ERA-Interim (850 hPa)	00–12/86–99	0.062	0.752	0.058	**0.041**	**0.004**
ERA-Interim (850 hPa)	00–12/76–85	0.831	0.303	0.939	0.751	0.132
ERA-Interim (850 hPa)	86–99/76–85	0.149	0.433	0.433	0.146	0.576

Table 4. *P* values for Wilcoxon tests for significant differences in mean seasonal sea ice concentrations and the number of ice-free months, for the periods 1976–1985, 1986–1999, and 2000–2013. Following convention, *p* values of < 0.05 are considered significant and are highlighted in bold.

Coast	Time interval	Season				Ice-free
		JFM	AMJ	JAS	OND	months
Barents	13–15/86–99	**0.003**	**0.012**	**0.003**	**0.003**	**0.003**
Barents	13–15/76–85	0.067	**0.017**	**0.017**	**0.017**	**0.017**
Barents	00–12/13–15	0.704	0.296	**0.039**	0.057	0.086
Barents	00–12/86–99	**0.002**	**0.009**	**0.019**	**> 0.001**	**0.001**
Barents	00–12/76–85	**0.006**	**0.002**	**0.002**	**0.001**	**0.002**
Barents	86–99/76–85	0.279	0.080	0.218	0.179	0.213
Kara	13–15/86–99	0.677	0.677	0.244	0.591	0.088
Kara	13–15/76–85	1	0.667	**0.017**	0.267	0.067
Kara	00–12/13–15	0.082	0.057	0.921	0.082	0.561
Kara	00–12/86–99	**> 0.001**	**> 0.001**	**> 0.001**	**> 0.001**	**0.037**
Kara	00–12/76–85	**> 0.001**	**> 0.001**	**> 0.001**	**> 0.001**	**0.011**
Kara	86–99/76–85	**0.003**	**0.034**	**0.028**	**0.001**	0.300

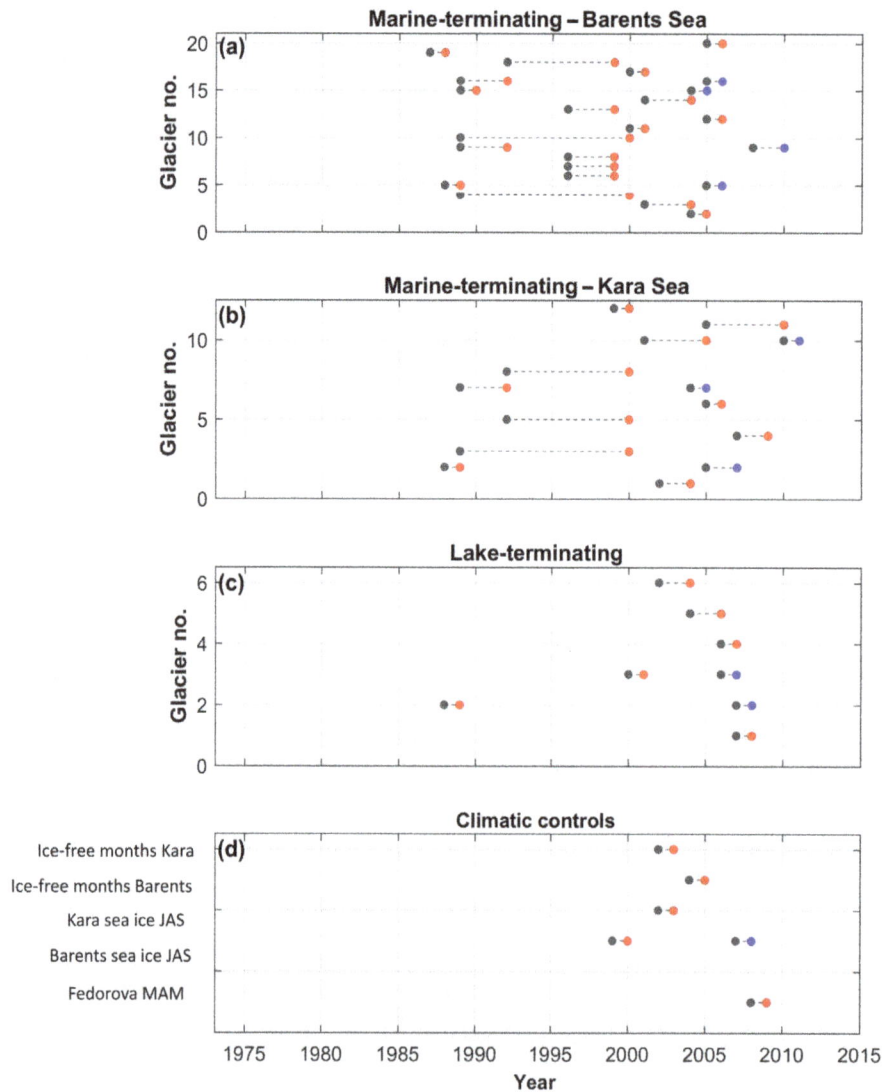

Figure 6. Results of the change-point analysis for glacier retreat rates and climatic controls. Red dots indicate the start of a significantly different period in the time series data, and grey dots represent the end of the previous period, with grey dashed lines connecting the two. This is done to account for missing data: we know that the change-point occurred between the grey and the red dot, and that the new phase of behaviour occurred from the red dot onwards, but not the exact timing of the change. Blue dots show the start of a second significant change in the time series. Frontal-position data were analysed separately for marine-terminating outlets on the Barents Sea coast (a), Kara Sea coast (b), and lake-terminating glaciers (c). (d) Change-point results for seasonal means in air temperatures and sea ice, and the number of ice-free months. Only climatic variables that demonstrated change-points are shown.

months (Fig. 7; Table 4). With exception of winter, sea ice concentrations were significantly lower in 2013–2015 than in 1976–1985 or 1986–1999 (Fig.4; Table 4). As on the Barents Sea coast, sea ice concentrations on the Kara Sea were significantly lower in all seasons in 2000–2012 than in 1976–1985 or 1986–1999 (Fig. 7; Table 4). Summer mean sea ice concentrations declined from 25 % in 1976–1985 to 13 % in 2000–2012 (Fig. 7). Over the same time interval, autumn mean concentrations reduced from 56 to 33 %, spring values declined from 87 to 73 %, and winter values decreased from 87 to 79 % (Fig. 7). The number of ice-free months also

reduced from 1.6 (1976–1985) to 3.0 (2000–2012) (Fig. 7). No significant differences were apparent between seasonal sea ice concentrations and the number of ice-free months in 2013–2015 and any other time period, with the exception of summer sea ice concentrations between 1976–1985 and 2013–2015 (Table 4).

Focusing on the change-point analysis, we see a significant change in air temperatures at E. K. Fedorova from 2008 onwards, after which air temperatures increased markedly (Fig. 6). On the Barents Sea coast, we observe significant breaks in summer sea ice concentrations at 2000 and 2008:

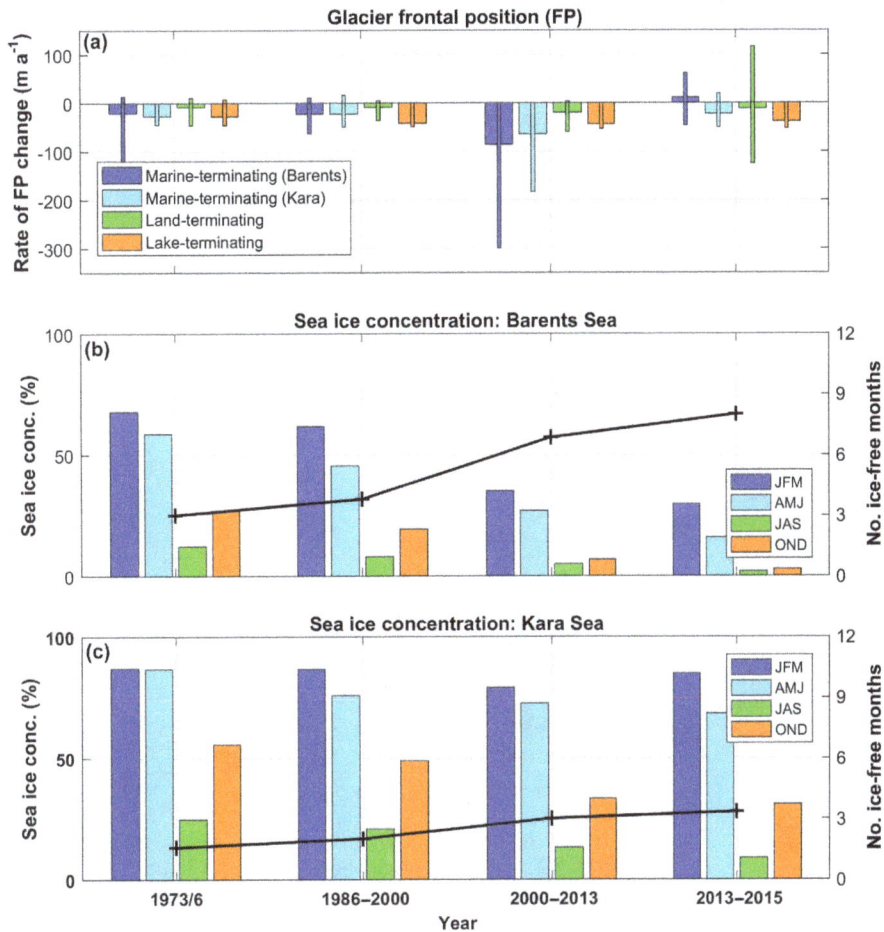

Figure 7. Mean retreat rates for Novaya Zemlya outlet glaciers, and seasonal mean sea ice concentrations and number of ice-free months for the Barents Sea and Kara Sea coasts. Data are split into four time periods, based on manually identified breaks in the glacier retreat data: 1973/76–1986, 1986–2000, 2000–2013, and 2013–2015. **(a)** Same as Fig. 4a. **(b, c)** Mean seasonal sea ice concentrations (January–March, April–June, July–September, and October–December) and number of ice-free months (thick black line) for the Barents Sea **(b)** and Kara Sea **(c)** coasts.

before 2000, summer sea ice showed a downward trend but large interannual variability; between 2000 and 2008, there was a slight upward trend and much lower variability; and from 2008 onwards, summer sea ice concentrations were much lower, showing both a downward trend and limited interannual variability (Supplement Fig. S2). From 2005 onwards, we observed much lower interannual variability in spring, summer, and autumn sea ice concentrations (Supplement Fig. S2). After 2005, summer sea ice concentrations on the Kara Sea coast showed much smaller interannual variability and had lower values (Supplement Fig. S3). The number of ice-free months increased significantly on both the Kara Sea (from 2003) and Barents Sea (from 2005) (Fig. 6).

Between 1970 and 1989, the summer and annual NAO index were largely positive, with a few years of negative values (Fig. 8a). From 1989 to 1994, values were all positive, followed by strongly negative values in 1995 (Fig. 8a). Subsequently, the summer and annual NAO index remained

weakly negative between 1999 and 2012, with values becoming increasingly negative in the final 5 years of this period (Fig. 8a). In 2013, the NAO index became strongly positive, particularly during summer, and values were also positive in 2015 and 2016 (Fig. 8a). The AO index follows an overall similar pattern to the NAO until ∼2000, although shifts are less distinct: the index is generally negative until 1988, followed by 5 years of more positive values. In the 2000s, the AO index fluctuates between positive and negative, and more negative summer values are observed in 2009, 2011, 2014, and 2015 (Fig. 8b). The AMO was generally negative from 1970 to 2000, although values fluctuated and were positive around 1990 (Fig. 8c). Subsequently, the AMO entered a positive phase from 2000 onwards (Fig. 8c).

At the broad spatial scale, data indicate that surface ocean temperatures have warmed in the Barents Sea over time (Fig. 9). Warming was particularly marked in the area extending approximately 100 km offshore of the Barents Sea

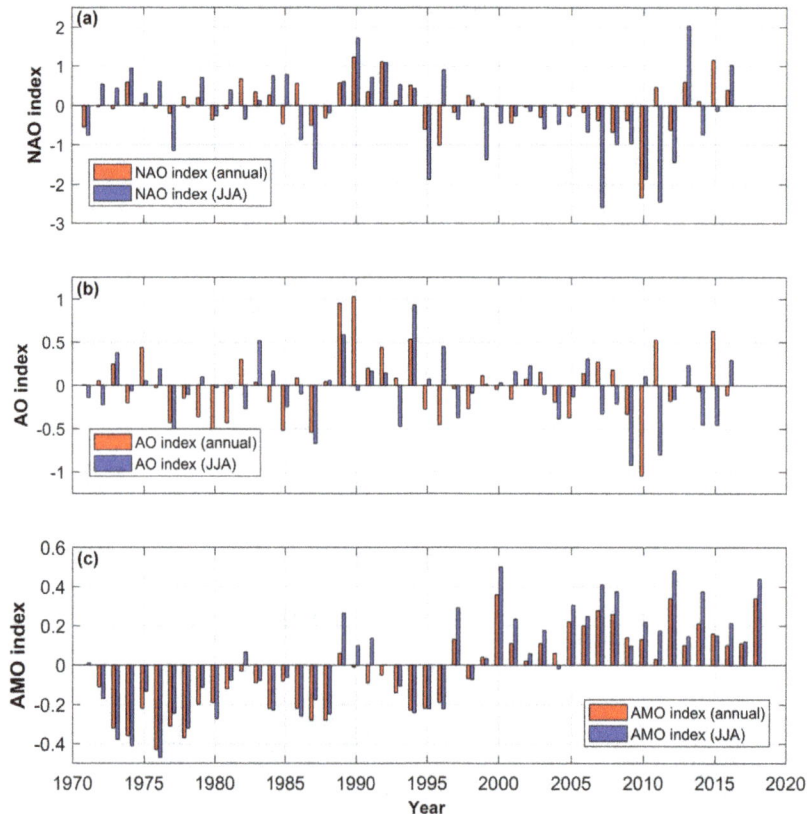

Figure 8. Time series of **(a)** North Atlantic Oscillation (NAO); **(b)** Arctic Oscillation (AO); and **(c)** Atlantic Multidecadal Oscillation (AMO) for 1970 to 2016. In each case, mean annual and mean summer values are shown.

coast and south of 76° N. Here, temperatures ranged between 2 and 4 °C in 1971–1980 and reached up to 7 °C by 2001–2012 (Fig. 9), although it should be noted that data are much sparser for the latter period. The Kara Sea also warmed over the study period, with temperatures increasing from 0–2 °C in 1971–1980 to 4–5 °C in 2001–2012 (Fig. 9). Although input data are comparatively sparse for 2001–2012, it appears that ocean temperatures have warmed in both the Barents and Kara seas at each time step, suggesting there may be a broad-scale warming trend in the region. At 100 m depth, the data suggest that warmer ocean water extends substantially during the study period, on both the Barents Sea and Kara Sea coasts (Fig. 9).

3.4 Glacier surging

During the study period, we observed three glaciers surging: ANU, MAS, and SER (Fig. 1). These were excluded from the analysis of glacier retreat rates and are discussed separately here. ANU has previously been identified as possibly surge type, based on the presence of looped moraine (Grant et al., 2009). Here, we identify an active surge phase, on the basis of a number of characteristics identified from satellite imagery and following the classification of Grant et al. (2009): rapid frontal advance, heavy crevassing, and dig-

itate terminus. High flow speeds are also evident close to the terminus (Melkonian et al., 2016), which is consistent with the active phase of surging. Our results show that advance began in 2008 and was ongoing in 2015, with the glacier advancing 683 m during this period (Fig. 10a). MAS was previously confirmed as surge type (Grant et al., 2009), and our data suggest that its active phase persisted between 1989 and 2007 (Fig. 10a). The imagery indicates that surging on MAS originated from the eastern limb of the glacier, which may be partially fed by the neighbouring glacier (Fig. 10b–f). The exact timing of this tributary surge is uncertain, but imagery from 1985 (Fig. 10c) shows limited evidence of surging, whereas a number of surge indicators are clearly visible by 1988, including looped moraines and rapid advance (Fig. 10d), suggesting it began in the late 1980s. The tributary glacier then advanced into the eastern margin of the main outlet of MAS, causing it to advance, and produced heavy crevassing on the eastern portion of its terminus (Fig. 10d and e). The main terminus of MAS reached its maximum extent for the study period in 2007, and the tributary continued advancing from the 1980s until 2007 (Fig. 10f). The role of the tributary glacier in triggering the surge is consistent with the lack of signs of surge type behaviour on the western margin of MAS and considerable visible displacement of

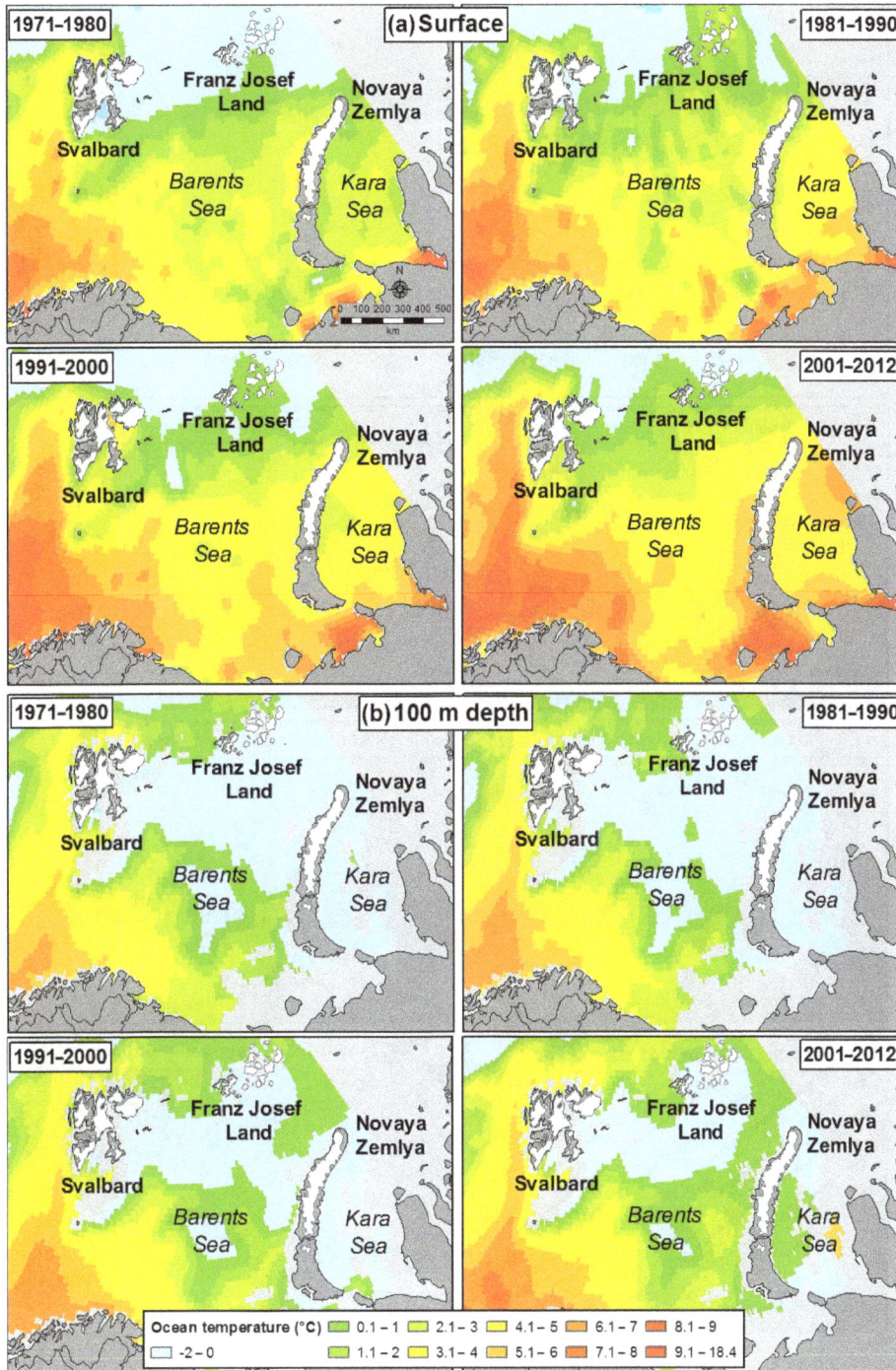

Figure 9. Ocean temperatures from the "Climatological Atlas of the Nordic Seas and Northern North Atlantic" (Korablev et al., 2014) at **(a)** the surface and **(b)** 100 m depth for the following time intervals: 1970–1981, 1981–1990, 1991–2000, and 2001–2012. These intervals were chosen to match as closely as possible with the glacier frontal-position data and other data sets. Note that data coverage was substantially lower for 2001–2012 than for other time periods. Further details on data coverage are available here: https://www.nodc.noaa.gov/OC5/nordic-seas/.

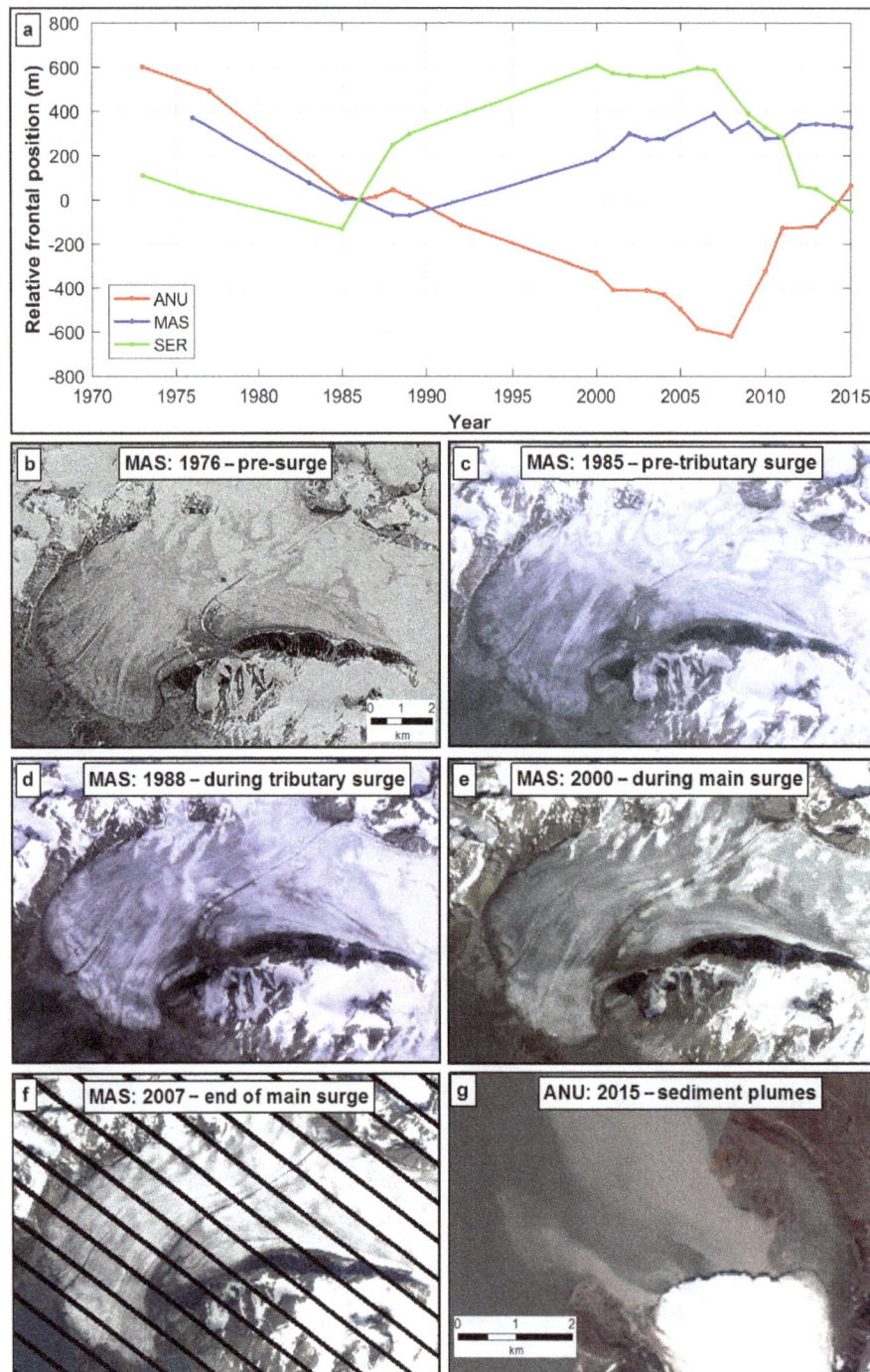

Figure 10. Glaciers identified as surging during the study period, based on the surge criteria compiled by Grant et al. (2009). **(a)** Glacier frontal position (relative to 1986) for glaciers identified as surge type: Anuchina (ANU), Mashigina (MAS), and Serp i Molot (SER). **(b)** Pre-surge imagery of MAS. Imagery source: Hexagon, 22 July 1976. **(c)** Tributary prior to the appearance of obvious surge-type features. Imagery source: Landsat 5, 26 July 1985. **(d)** MAS during the surge of its tributary. Imagery source: Landsat 5, 13 August 1988. **(e)** MAS during the surge of the main glacier trunk. Imagery source: Landsat 7, 13 August 2000. **(f)** MAS at the end of main glacier the surge, showing the maximum observed extent of the main terminus. Imagery source: Landsat 7, 8 July 2007. **(g)** Sediment plumes emerging from the margin of ANU during its recent surge. Imagery source: Landsat 8, 31 July 2015.

ice and surface features on the eastern tributary (Fig. 10b–f). SER was also confirmed as a surge-type glacier by Grant et al. (2009), who suggested that glacier advance occurred between 1976/77 and 2001. Our results indicate that advance began somewhat later, sometime between July 1983 and July 1986, and ended before August 2000 (Fig. 10a).

4 Discussion

4.1 Spatial patterns of glacier retreat

Our results demonstrate that retreat rates on marine-terminating outlet glaciers ($-46.9\,\mathrm{m\,a^{-1}}$) were more than 3 times higher than those on land ($-13.8\,\mathrm{m\,a^{-1}}$) between 1986 and 2015 (Fig. 2). This is consistent with previous shorter-term studies from Greenland (Moon and Joughin, 2008; Sole et al., 2008) and Svalbard (Dowdeswell et al., 2008), which demonstrated an order-of-magnitude difference between marine- and land-terminating glaciers. It also confirms that the differences in retreat rates, relating to terminus type, observed between 1992 and 2010 on NVZ (Carr et al., 2014) persist at multi-decadal timescales. Recent results suggest that marine-terminating glacier retreat and/or ice tongue collapse can cause dynamic thinning in the RHA (Melkonian et al., 2016; Willis et al., 2015), meaning that these long-term differences in retreat rates may lead to substantially higher thinning rates in marine-terminating basins at multi-decadal timescales. The Russian high Arctic is forecast to be the third-largest source of ice volume loss by 2100 outside of the ice sheets (Radić and Hock, 2011). However, these estimates only account for surface mass balance, not ice dynamics, meaning that they may underestimate 21st-century ice loss for the RHA. Consequently, dynamic changes associated with marine-terminating outlet glacier retreat on NVZ need to be taken into account in order to accurately forecast its near-future ice loss and sea level rise contribution.

Our data showed no significant difference in total retreat rates for marine-terminating ($-46.9\,\mathrm{m\,a^{-1}}$) and lake-terminating glaciers ($-49.1\,\mathrm{m\,a^{-1}}$). This contrasts with results from Patagonia, which were obtained during a similar time period (mid-1980s to 2010/11) and showed that lake-terminating outlet glaciers retreated significantly more rapidly than those ending in the ocean (Sakakibara and Sugiyama, 2014). For example, marine-terminating outlets retreat at an average rate of $-37.8\,\mathrm{m\,a^{-1}}$ between 2000 and 2010/11, whereas lake-terminating glaciers receded at $-80.8\,\mathrm{m\,a^{-1}}$ (Sakakibara and Sugiyama, 2014). Lake-terminating glacier retreat on NVZ also differs from Patagonia in that retreat rates are remarkably consistent between individual glaciers and remained similar over time (Figs. 4 and 5). Conversely, frontal-position changes in Patagonia showed major spatial variations, and retreat rates on several lake-terminating glaciers changed substantially between the

two halves of the study period (mid-1980s–2000 and 2000–2010/11) (Sakakibara and Sugiyama, 2014).

One potential explanation for the common behaviour of the lake-terminating outlet glaciers on NVZ is that retreat may be dynamically controlled and sustained by a series of feedbacks once it has begun. As observed on large Greenlandic tidewater glaciers, initial retreat may bring the terminus close to floatation, leading to faster flow and thinning, which promote further increases in calving and retreat (e.g. Howat et al., 2007; Hughes, 1986; Joughin et al., 2004; Meier and Post, 1987; Nick et al., 2009). This has been suggested as a potential mechanism for the rapid recession for Upsala Glacier in Patagonia (Sakakibara and Sugiyama, 2014) and Yakutat Glacier, Alaska (Trüssel et al., 2013). However, rapid retreat was not observed on all lake-terminating glaciers in Patagonia (Sakakibara and Sugiyama, 2014), and the potential for these feedbacks to develop depends on basal topography (e.g. Carr et al., 2015; Porter et al., 2014; Rignot et al., 2016). Consequently, the basal topography would need to be similar for each of the NVZ glaciers to explain the very similar retreat patterns, which is not implausible but perhaps unlikely. Alternatively, it may be that the proglacial lakes act as a buffer for atmospheric warming, due the greater thermal conductivity of water relative to air, and so reduce variability in retreat rates. Furthermore, lake-terminating glaciers are not subject to variations in sea ice and ocean temperatures, which may account for their more consistent retreat rates, compared to marine-terminating glaciers (Figs. 4 and 5). In order to differentiate between these two explanations, data on lake temperature changes during the study period and lake bathymetry would be required. However, neither are currently available, and we highlight this as an important area for further research, given the rapid recession observed on these lake-terminating glaciers.

For the period between 1986 and 2015, we find no significant difference in retreat rates between the Barents Sea and Kara Sea coasts (Fig. 2). This is contrary to the results of a previous shorter-term study, which showed that retreat rates on the Barents Sea coast were significantly higher than on the Kara Sea between 1992 and 2010 (Carr et al., 2014) and the higher thinning rates observed on marine outlets on the Barents Sea coast (Melkonian et al., 2016). Furthermore, there are substantial differences in climatic and oceanic conditions on the two coasts (Figs. 4 and 7) (Pfirman et al., 1994; Politova et al., 2012; Przybylak and Wyszyński, 2016; Zeeberg and Forman, 2001), so we would expect to see significant differences in outlet glacier retreat rates. This indicates that longer-term glacier retreat rates on NVZ may relate to much broader, regional-scale climatic change, which is supported by the widespread recession of glaciers across the Arctic during the past 2 decades (e.g. Blaszczyk et al., 2009; Carr et al., 2014; Howat and Eddy, 2011; Jensen et al., 2016; Moon and Joughin, 2008). One potential overarching control on NVZ frontal positions is fluctuations in the NAO, which covaries with Northern Hemisphere air temperatures, Arctic sea ice, and North Atlantic Ocean temperatures (Hurrell, 1995; Hurrell et al., 2003; IPCC, 2013). More recent work has also recognized the influence of the AMO on oceanic and atmospheric conditions in the Barents Sea and broader North Atlantic (Drinkwater et al., 2013; Oziel et al., 2016). Our data suggest that the major phases of frontal-position change on NVZ correspond to changes in the NAO and AMO (Fig. 8; Sect. 4.2): rapid retreat between 2000 and 2013 coincides with a weakly negative NAO and positive AMO, following almost 3 decades characterized by a generally positive NAO and negative AMO (Fig. 8). As such, these large-scale changes may overwhelm smaller-scale spatial variations between the two coasts of NVZ when retreat is considered on multi-decadal timescales.

Marine-terminating outlet glacier retreat rates do not show a linear relationship with latitude, and there is considerable scatter when the two variables are regressed (Fig. 3). This may be due to the influence of fjord geometry on glacier response to climatic forcing (Carr et al., 2014) and the capacity for warmer ocean waters to access the calving fronts. In contrast, southerly land-terminating outlets retreat more rapidly than those in the north, which we attribute to the substantial latitudinal air temperature gradient on NVZ (Zeeberg and Forman, 2001). Conversely, lake-terminating glaciers retreat more rapidly at more northerly latitudes (Fig. 3), which we speculate may relate to the bathymetry and basal topography of individual glaciers, but data are not currently available to confirm this.

4.2 Temporal patterns

Our results show that retreat rates on marine-terminating outlet glaciers on NVZ were significantly higher between 2000 and 2013 than during the preceding 27 years (Fig. 4). At the same time, land-terminating outlets experienced much lower retreat rates and did not change significantly during the study period (Figs. 4 and 5). This is consistent with studies from elsewhere in the Arctic, which identified the 2000s as a period of elevated retreat on marine-terminating glaciers (e.g. Blaszczyk et al., 2009; Howat and Eddy, 2011; Jensen et al., 2016; Moon and Joughin, 2008) and increasing ice loss (e.g. Gardner et al., 2013; Lenaerts et al., 2013; Moholdt et al., 2012; Nuth et al., 2010; Shepherd et al., 2012). As discussed above, recent evidence suggests that glacier retreat in the Russian high Arctic can trigger substantial dynamic thinning and ice acceleration (Melkonian et al., 2016; Willis et al., 2015), but it not currently incorporated into predictions of 21st-century ice loss from the region (Radić and Hock, 2011). Consequently, the period of higher retreat rates during the 2000s may have a much longer-term impact on ice losses from NVZ, and this needs to be quantified and incorporated into forecasts of ice loss and sea level rise prediction.

Within the decadal patterns of glacier retreat, we observe clusters in the timing of significant changes in marine-terminating glacier retreat rates (Fig. 6). Specifically, we see

breaks in the frontal-position time series on both the Barents Sea and Kara Sea coasts in the early 1990s, ~ 2000, and the mid-2000s (Fig. 6). This demonstrates some synchronicity in changes in glacier behaviour around NVZ, although it is not ubiquitous (Fig. 6). The timing of these changes coincides with those observed in Greenland, where the onset of widespread retreat and acceleration in south-east Greenland began in ~ 2000 (e.g. Howat et al., 2008; Moon and Joughin, 2008; Seale et al., 2011) and occurred from the mid-2000s onwards in the north-west (e.g. Carr et al., 2013b; Howat and Eddy, 2011; Jensen et al., 2016; McFadden et al., 2011; Moon et al., 2012). Whilst these changes could be coincidental, they may also relate to broad, regional-scale changes observed in the North Atlantic region during the 2000s (Beszczynska-Möller et al., 2012; Hanna et al., 2013, 2012; Holliday et al., 2008; Sutherland et al., 2013). Data demonstrate that the NAO was weakly negative from the mid-1990s until 2012, in contrast to strongly positive conditions in the late 1980s and early 1990s, and the AMO was persistently positive from 2000 onwards, following 3 decades of overall positive conditions (Fig. 8). These changes coincide with increases in glacier retreat rates, sea ice decline, and atmospheric warming in NVZ between 2000 and 2013 (Figs. 4 and 7).

Between the 1950s and mid-1990s, positive phases of the NAO were associated with the influx of warm Atlantic water into the Barents Sea (Hurrell, 1995; Loeng, 1991) and increased penetration of Atlantic cyclones and air masses into the region, which lead to elevated air temperatures and precipitation (Zeeberg and Forman, 2001). Conversely, negative NAO phases were associated with cooler oceanic and atmospheric conditions in the Barents Sea (Zeeberg and Forman, 2001). During this period, therefore, the impact of the NAO was opposite in the Barents Sea and in western portions of the Atlantic-influenced Arctic (e.g. the Labrador Sea) (Drinkwater et al., 2013; Oziel et al., 2016). However, since the mid-1990s, changes in the Barents Sea and the western Atlantic Arctic have been in phase, and warming and sea ice reductions have been widespread across both regions (Drinkwater et al., 2013; Oziel et al., 2016). As such, increased glacier retreat rates on NVZ during the 2000s (Figs. 4 and 5) may have resulted from the switch to a weaker, and predominantly negative, NAO phase from the mid-1990s (Fig. 8), which would promote warmer air and ocean temperatures, and reduced sea ice, as we observe in our data (Figs. 4 and 7). Previous studies have suggested a 3–5-year lag between NAO shifts and changes in conditions on NVZ, due to the time required for Atlantic water to transit into the Barents Sea (Belkin et al., 1998; Zeeberg and Forman, 2001), which is consistent with the onset of retreat in ~ 2000 (Figs. 4 and 8). However, it has recently been suggested that the NAO's role may have reduced since the mid-1990s and that the AMO may be the dominant influence on warming in the North Atlantic (Drinkwater et al., 2013; Oziel et al., 2016). The AMO is thought to promote blocking of high-pressure systems by westerly winds, which changes the wind field (Häkkinen et al., 2011). This allows warm water to penetrate further into the Barents Sea and other Nordic Seas, leading to atmospheric and oceanic warming during periods with a weakly negative NAO (Häkkinen et al., 2011). As such, rapid retreat on NVZ between 2000 and 2013 may have resulted from the combined effects of a weaker, more negative NAO from the mid-1990s and a more positive AMO from 2000 onwards (Fig. 8). This suggests that synoptic climatic patterns may be an important control on glacier retreat rates on NVZ and that the recent relationship between the NAO and glacier change on NVZ contrasts with that observed during the 20th century (Zeeberg and Forman, 2001).

Following higher retreat rates in the 2000s, our data indicate that marine-terminating glacier retreat slowed from 2013 onwards on both the Barents Sea and Kara Sea coasts, with several glaciers beginning to re-advance (Figs. 4 and 5). Our data demonstrate that marine-terminating glaciers on NVZ have previously undergone a step-like pattern of retreat, with short (1–2 year) pauses in retreat (Fig. 5). Thus, it is unclear whether this reduction in retreat rates is another temporary pause, before continued retreat, or the beginning of a new phase of reduced retreat rates. One possible explanation for reduced retreat rates on both coasts of NVZ is the stronger NAO values observed from the late 2000s onwards: winter 2009/10 had the most negative NAO for 200 years (Delworth et al., 2016; Osborn, 2011), and values were strongly positive in 2013 (Fig. 8a). This is consistent with the 3–5-year lag required for NAO-related changes in Atlantic water inflow to reach NVZ (Zeeberg and Forman, 2001), and so we speculate that reduced glacier retreat rates from 2013 onwards (Figs. 4 and 5) may relate to an increase in the influence of the NAO, relative to the AMO, from the late 2000s (Fig. 8). Evidence indicates that the impact of the NAO in the Barents Sea is now in phase with the western North Atlantic (Drinkwater et al., 2013; Oziel et al., 2016), and so a more positive NAO could lead to cooler conditions on NVZ and hence glacier advance. However, the relationship between large-scale features, such as the NAO and AMO; ocean conditions; and glacier behaviour is complex (Drinkwater et al., 2013; Oziel et al., 2016), and the period of glacier advance/reduced retreat on NVZ has lasted only 2 years. Consequently, further monitoring is required to determine whether this represents a longer-term trend or a short-term change and to confirm its relationship to synoptic climatic patterns.

Despite the changes in the NAO and AMO, our data show no significant change in sea ice concentrations, nor the length of the ice-free season, between 2000–2012 and 2013–2015 on either the Barents Sea or Kara Sea coast (Table 4; Fig. 7). Likewise, we see no significant change in winter (January–March) air temperatures at E. K. Fedorova (Table 3; Fig. 4) nor in the ERA-Interim data during any season (Table 3; Fig. 4). Although not significant, we see summer warming of 0.7 °C (surface) and 0.8 °C (850 hPa pressure level) in the ERA-Interim data (Fig. 4), which is the opposite of what we

would expect if reductions in air temperatures and surface melt were driving the slowdown in retreat rates. As such, reduced retreat rates do not seem to be directly linked to short-term changes in sea ice or air temperatures. They are also unlikely to result from changes in surface mass balance, as the response time of NVZ glaciers is likely to be slow: they have long catchments (~ 40 km) and slow flow speeds (predominantly < 200 m a^{-1}; Melkonian et al., 2016) and are likely to be polythermal. Furthermore, thinning rates between 2012 and 2013/14 averaged 0.4 m a^{-1} across the ice cap and reached up to 5 m a^{-1} close to the glacier termini (Melkonian et al., 2016), meaning that even a positive surface mass balance is very unlikely to deliver sufficient ice quickly enough to promote advance and/or substantially lower retreat rates. Instead, this may be a response to oceanic changes, which we cannot detect from available data; it may reflect a lagged response to forcing; and/or it may relate to more localized, glacier-specific factors. We suggest that the latter is unlikely, given the widespread and synchronous nature of the observed reduction in retreat rates (Figs. 4 and 5). Future work should monitor retreat rates to determine whether reduced retreat is persistent or is a short-term interruption to overall glacier retreat and collect more extensive oceanic data to assess its impact on this change. Furthermore, detailed data are also required to determine how short-term frontal-position fluctuations relate to changes in ice velocities and/or surface elevation.

Although we observe some common behaviour, in terms of the approximate timing and general trend in retreat, there is still substantial variability in the magnitude of retreat between individual marine-terminating glaciers (Figs. 4 and 5). Furthermore, not all glaciers shared common change-points, and certain outlets showed a different temporal pattern of retreat to the majority of the study population (Figs. 4–6). For example, INO retreated more slowly between 1989 and 2006 than during the 1970s and 1980s. We attribute these differences to glacier-specific factors and, in particular, the fjord bathymetry and basal topography of individual glaciers. Previous studies have highlighted the impact of fjord width on retreat rates on NVZ (Carr et al., 2014) and basal topography on marine-terminating glacier behaviour elsewhere (e.g. Carr et al., 2015; Porter et al., 2014; Rignot et al., 2016). This may result from the influence of fjord geometry on the stresses acting on the glacier once it begins to retreat: as a fjord widens, lateral resistive stresses will reduce, and the ice must thin to conserve mass, making it more vulnerable to calving (Echelmeyer et al., 1994; Raymond, 1996; van der Veen, 1998a, b), whilst retreat into progressively deeper water can cause feedbacks to develop between thinning, floatation, and retreat (e.g. Joughin and Alley, 2011; Joughin et al., 2008; Schoof, 2007). Thus, retreat into a deeper and/or wider fjord may promote higher retreat rates on a given glacier, even with common climatic forcing. In addition, differences in fjord bathymetry may determine whether warmer Atlantic water can access the glacier front (Porter et al., 2014; Rignot et al., 2016), which could promote further variations between glaciers. This highlights the need to collect basal topographic data for NVZ outlet glaciers, which is currently very limited but a potentially key control on ice loss rates.

4.3 Climatic and oceanic controls

Our data demonstrate that air temperatures were very substantially warmer between 2000 and 2012 than during the preceding decades and that sea ice concentrations were also much lower on both the Barents Sea and Kara Sea coasts during this period (Figs. 4 and 7). This is consistent with the atmospheric warming reported across the Arctic during the 2000s (e.g. Carr et al., 2013a; Hanna et al., 2013, 2012; Mernild et al., 2013) and the well-documented decline in Arctic sea ice (Comiso et al., 2008; Kwok and Rothrock, 2009; Park et al., 2015). As such, the decadal patterns of marine-terminating outlet glacier retreat correspond to decadal-scale climatic change on NVZ (Figs. 4 and 7), and exceptional retreat during the 2000s coincided with significantly warmer air temperatures and lower sea ice concentrations (Tables 2 and 3). Interestingly, step changes in the air temperature and sea ice data identified by the change-point analysis did not correspond to significant changes in outlet glacier retreat rates (Fig. 6), suggesting that such changes may not substantially influence retreat rates or that the relationship may be more complex, e.g. due to lags in glacier response.

The much lower retreat rates on land-terminating outlets (Fig. 4) may indicate an oceanic driver for retreat rates on marine-terminating glaciers. Previous studies have identified sea ice loss as a potentially important control on NVZ retreat rates (Carr et al., 2014), which fits with observed correspondence between sea ice loss and retreat, but it is unclear whether the two variables simply co-vary or whether sea ice can drive ice loss, by extending the duration of seasonally high calving rates (e.g. Amundson et al., 2010; Miles et al., 2013; Moon et al., 2015). The available ocean data indicate that temperatures were substantially warmer during the 2000s (Fig. 9), which would provide a plausible mechanism for widespread retreat on both coasts of NVZ (Fig. 4). However, oceanic data for the 2000s is sparse in the Barents and Kara seas, compared to previous decades, so it is difficult to ascertain the magnitude and spatial distribution of warming and to link it directly with glacier retreat patterns. Lake-terminating glaciers are not affected by changes in sea ice or ocean temperatures but could be influenced by air temperatures. However, despite much higher air temperatures in the 2000s, mean retreat rates on lake-terminating outlet glaciers were similar for each decade of the study (Fig. 4), suggesting that the relationship is not straightforward. Instead, the presence of lakes may at least partly disconnect these glaciers from climatic forcing, by buffering the effects of air temperatures changes and/or by sustaining dynamic changes, follow-

ing initial retreat (Sakakibara and Sugiyama, 2014; Trüssel et al., 2013).

4.4 Glacier surging

During the study period, we identify three actively surging glaciers, based on various lines of glaciological and geomorphological evidence (Copland et al., 2003; Grant et al., 2009), including terminus advance (Fig. 10). Frontal advance persisted for 18 years on MAS and 15 years on SER, whilst ANU began to advance in 2008, and this continued until the end of the study period (Fig. 10a). This is comparatively long for surge-type glaciers, which usually undergo short active phases over time frames of months to years (Dowdeswell et al., 1991; Raymond, 1987). For comparison, surges on Tunabreen, Spitzbergen, last only ∼ 2 years (Sevestre et al., 2015), and Basin 3 on Austfonna underwent major changes in its dynamic behaviour in just a few years (Dunse et al., 2015). Surges elsewhere can occur even more rapidly: the entire surge cycle of Variegated Glacier in Alaska takes approximately 1–2 decades, and the active phase persists for only a few months (e.g. Bindschadler et al., 1977; Eisen et al., 2005; Kamb, 1987; Kamb et al., 1985; Raymond, 1987). Furthermore, the magnitude of advance on these three glaciers is in the order of a few hundred metres, which is smaller than advances associated with surges on Tunabreen (1.4 km) and Kongsvegen (2 km) (Sevestre et al., 2015) and much less than the many kilometres of advance observed on Alaskan surge-type glaciers, such as Variegated Glacier (Bindschadler et al., 1977; Eisen et al., 2005). Consequently, the active phase on NVZ appears to be long, in comparison to other regions, and terminus advance is more limited, which may provide insight into the mechanism(s) driving surging here and may indicate that these glaciers are located towards one end of the climatic envelope required for surging in the Arctic (Sevestre and Benn, 2015).

During the active phase of the NVZ surge glaciers, we observe large sediment plumes emanating from the glacier terminus (Fig. 10g), which indicates that at least part of the glacier bed is warm-based during the surge. Together with the comparatively long surge interval, this supports the idea that changes in thermal regime may drive glacier surging on NVZ, as hypothesized for certain Svalbard glaciers (Dunse et al., 2015; Murray et al., 2003; Sevestre et al., 2015). In addition, the surge of MAS appears to have been triggered by a tributary glacier surging into its lateral margin (Fig. 10b–f). This demonstrates an alternative mechanism for surging, aside from changes in the thermal regime and/or hydrology conditions of the glacier, which has not been widely observed but will depend strongly on the local glaciological and topographical setting of the glacier. The data presented here focus only on frontal advance and glaciological/geomorphological evidence, whereas information on ice velocities is also an important indicator of surging (Sevestre and Benn, 2015). Consequently, information on velocity and surface elevation changes are needed to further investigate the surge cycle and its possible controls on NVZ. This is important, as NVZ is thought to have conditions that are highly conducive to glacier surging (Sevestre and Benn, 2015) but has a long surge interval. We therefore want to ensure that we can disentangle surge behaviour and the impacts of climate change on NVZ.

5 Conclusions

At multi-decadal timescales, terminus type remains a major overarching determinant of outlet glacier retreat rates on NVZ. As observed elsewhere in the Arctic, land-terminating outlets retreated far more slowly than those ending in the ocean. However, we see no significant difference in retreat rates between ocean- and lake-terminating glaciers, which contrasts with findings in Patagonia. Retreat rates on lake-terminating glaciers were remarkably consistent between glaciers and over time, which may result from the buffering effect of lake temperature and/or the impact of lake bathymetry, which could facilitate rapid retreat that is largely independent of climate forcing, after an initial trigger. We cannot differentiate between these two scenarios with currently available data. Retreat rates on marine-terminating glaciers were exceptional between 2000 and 2013, compared to previous decades. However, retreat slowed on the vast majority of ocean-terminating glaciers from 2013 onwards, and several glaciers advanced, particularly on the Barents Sea coast. It is unclear whether this represents a temporary pause or a longer-term change, but it should be monitored in the future, given the potential for outlet glaciers to drive dynamic ice loss from NVZ. The onset of higher retreat rates coincides with a more negative, weaker phase of the NAO and a more positive AMO, whilst reduced retreat rates follow stronger NAO years. This suggests that synoptic atmospheric and oceanic patterns may influence NVZ glacier behaviour at decadal timescales. Marine-terminating glaciers showed some common patterns in terms of the onset of rapid retreat (1990s, ~ 2000 and mid 2000s) but showed substantial variation in the magnitude of retreat, which we attribute to glacier-specific factors. Glacier retreat corresponded with decadal-scale climate patterns: between 2000 and 2013, air temperatures were significantly warmer than the previous decades and sea ice concentrations were significantly lower. Available data indicate oceanic warming, which could potentially explain why retreat rates on marine-terminating glaciers far exceed those ending on land, but data are comparatively sparse from 2000 onwards, making their relationship to glacier retreat rate difficult to evaluate. The surge phase on NVZ glaciers appears to be comparatively long and warrants further investigation to separate its impact on ice dynamics from that of climate-induced change and to determine the potential mechanism(s) driving these long surges. Recent results suggest that outlet glaciers can trigger dynamic losses on NVZ, but these processes are not yet included in estimates of the region's contribution to sea level rise. As such, it is vital to determine the longer-term impacts of exceptional glacier retreat during the 2000s and to monitor the near-future behaviour of these outlets.

Competing interests. The authors declare that they have no conflict of interest.

Acknowledgements. The authors thank Xavier Fettweis, Robbert McKnabb, and one anonymous reviewer for their constructive comments that helped to improve the manuscript.

Edited by: Xavier Fettweis

References

Amundson, J. M., Fahnestock, M., Truffer, M., Brown, J., Lüthi, M. P., and Motyka, R. J.: Ice mélange dynamics and implications for terminus stability, Jakobshavn Isbræ, Greenland, J. Geophys. Res., 115, F01005, https://doi.org/10.1029/2009JF001405, 2010.

Belkin, I. M., Levitus, S., Antonov, J., and Malmberg, S.-A.: "Great salinity anomalies" in the North Atlantic, Prog. Oceanogr., 41, 1–68, 1998.

Benn, D. I., Warren, C. R., and Mottram, R. H.: Calving processes and the dynamics of calving glaciers, Earth Sci. Rev., 82, 143–179, 2007.

Beszczynska-Möller, A., Fahrbach, E., Schauer, U., and Hansen, E.: Variability in Atlantic water temperature and transport at the entrance to the Arctic Ocean, 1997–2010, ICES J. Mar. Sci., 69, 852–863, https://doi.org/10.1093/icesjms/fss056, 2012.

Bindschadler, R., Harrison, W. D., Raymond, C. F., and Crosson, R.: Geometry and dynamics of a surge-type glacier, J. Glaciol., 18, 181–194, 1977.

Blaszczyk, M., Jania, J. A., and Hagen, J. M.: Tidewater glaciers of Svalbard: Recent changes and estimates of calving fluxes, Pol. Polar Res., 30, 85–142, 2009.

Carr, J. R., Stokes, C. R., and Vieli, A.: Recent progress in understanding marine-terminating Arctic outlet glacier response to climatic and oceanic forcing: Twenty years of rapid change, Prog. Phys. Geog., 37, 435–466, 2013a.

Carr, J. R., Vieli, A., and Stokes, C. R.: Climatic, oceanic and topographic controls on marine-terminating outlet glacier behavior in north-west Greenland at seasonal to interannual timescales, J. Geophys. Res., 118, 1210–1226, 2013b.

Carr, J. R., Stokes, C., and Vieli, A.: Recent retreat of major outlet glaciers on Novaya Zemlya, Russian Arctic, influenced by fjord geometry and sea-ice conditions, J. Glaciol., 60, 155–170, 2014.

Carr, J. R., Vieli, A., Stokes, C., Jamieson, S., Palmer, S., Christoffersen, P., Dowdeswell, J., Nick, F., Blankenship, D., and Young, D.: Basal topographic controls on rapid retreat of Humboldt Glacier, northern Greenland, J. Glaciol., 61, 137–150, 2015.

Chizov, O. P., Koryakin, V. S., Davidovich, N. V., Kanevsky, Z. M., Singer, E. M., Bazheva, V. Y., Bazhev, A. B., and Khmelevskoy, I. F.: Glaciation of the Novaya Zemlya, in: Glaciology IX section of IGY Program 18, 338 pp., Nauka, Moscow, Russia, 1968 (in Russian, with English summary).

Comiso, J. C., Parkinson, C. L., Gersten, R., and Stock, L.: Accelerated decline in the Arctic sea ice cover, Geophys. Res. Lett., 35, L01703, https://doi.org/10.1029/2007GL031972, 2008.

Copland, L., Sharp, M. J., and Dowdeswell, J. A.: The distribution and flow characteristics of surge-type glaciers in the Canadian High Arctic, Ann. Glaciol., 36, 73–81, 2003.

Delworth, T. L., Zeng, F., Vecchi, G. A., Yang, X., Zhang, L., and Zhang, R.: The North Atlantic Oscillation as a driver of rapid climate change in the Northern Hemisphere, Nat. Geosci., 9, 509–512, 2016.

Dowdeswell, J. and Williams, M.: Surge-type glaciers in the Russian High Arctic identified from digital satellite imagery, J. Glaciol., 43, 489–494, 1997.

Dowdeswell, J., Benham, T. J., Strozzi, T., and Hagen, J. M.: Iceberg calving flux and mass balance of the Austfonna ice cap on Nordaustlandet, Svalbard, J. Geophys. Res., 113, F03022, https://doi.org/10.1029/2007JF000905, 2008.

Dowdeswell, J. A., Hamilton, G. S., and Hagen, J. O.: The duration of the active phase on surge-type glaciers: contrasts between Svalbard and other regions, J. Glaciol., 37, 388–400, 1991.

Drinkwater, K., Colbourne, E., Loeng, H., Sundby, S., and Kristiansen, T.: Comparison of the atmospheric forcing and oceanographic responses between the Labrador Sea and the Norwegian and Barents seas, Prog. Oceanogr., 114, 11–25, 2013.

Dunse, T., Schellenberger, T., Hagen, J. O., Kääb, A., Schuler, T. V., and Reijmer, C. H.: Glacier-surge mechanisms promoted by a hydro-thermodynamic feedback to summer melt, The Cryosphere, 9, 197–215, https://doi.org/10.5194/tc-9-197-2015, 2015.

Echelmeyer, K. A., Harrison, W. D., Larsen, C., and Mitchell, J. E.: The role of the margins in the dynamics of an active ice stream, J. Glaciol., 40, 527–538, 1994.

Eckley, I., Fearnhead, P., and Killick, R.: Analysis of Changepoint Models, in: Bayesian Time Series Models, edited by: Barber, D., Cemgil, T., and Chiappa, S., Cambridge University Press, Cambridge, UK, 2011.

Eisen, O., Harrison, W. D., Raymond, C. F., Echelmeyer, K. A., Bender, G. A., and Gorda, J. L. D.: Variegated Glacier, Alaska, USA: a century of surges, J. Glaciol., 51, 399–406, 2005.

Enderlin, E. M., Howat, I. M., Jeong, S., Noh, M.-J., van Angelen, J. H., and van den Broeke, M. R.: An improved mass budget for the Greenland ice sheet, Geophys. Res. Lett., 41, 866–872, https://doi.org/10.1002/2013GL059010, 2014.

Gardner, A., Moholdt, G., Wouters, B., Wolken, G. J., Burgess, D. O., Sharp, M. J., Cogley, J. G., Braun, C., and Labine, C.: Sharply increased mass loss from glaciers and ice caps in the Canadian Arctic Archipelago, Nature, 473, 357–360, 2011.

Gardner, A., Moholdt, G., Cogley, J. G., Wouters, B., Arendt, A. A., Wahr, J., Berthier, E., Hock, R., Pfeffer, W. T., Kaser, G., Ligtenberg, S. R. M., Bolch, T., Sharp, M. J., Hagen, J. O., van den Broeke, M. R., and Paul, F.: A Reconciled Estimate of Glacier Contributions to Sea Level Rise: 2003 to 2009, Science, 340, 852–857, 2013.

Grant, K. L., Stokes, C. R., and Evans, I. S.: Identification and characteristics of surge-type glaciers on Novaya Zemlya, Russian Arctic, J. Glaciol., 55, 960–972, 2009.

Häkkinen, S., Rhines, P. B., and Worthen, D. L.: Atmospheric Blocking and Atlantic Multidecadal Ocean Variability, Science, 334, 655–659, 2011.

Hanna, E., Mernild, S. H., Cappelen, J., and Steffen, K.: Recent warming in Greenland in a long-term instrumental (1881–2012) climatic context: I. Evaluation of surface air temperature records, Environ. Res. Lett., 7, 045404, https://doi.org/10.1088/1748-9326/7/4/045404, 2012.

Hanna, E., Jones, J. M., Cappelen, J., Mernild, S. H., Wood, L., Steffen, K., and Huybrechts, P.: The influence of North Atlantic atmospheric and oceanic forcing effects on 1900–2010 Greenland summer climate and ice melt/runoff, Int. J. Climatol., 33, 862–880, 2013.

Higgins, R., Leetmaa, A., Xue, Y., and Barnston, A.: Dominant factors influencing the seasonal predictability of U.S. precipitation and surface air temperature, J. Climate, 13, 3994–4017, 2000.

Holliday, N. P., Hughes, S. L., Bacon, S., Beszczynska-Möller, A., Hansen, B., Lavin, A., Loeng, H., Mork, K. A., Østerhus, S., Sherwin, T., and Walczowski, W.: Reversal of the 1960s to 1990s freshening trend in the northeast North Atlantic and Nordic Seas, Geophys. Res. Lett., 35, L03614, https://doi.org/10.1029/2007GL032675, 2008.

Howat, I. M. and Eddy, A.: Multi-decadal retreat of Greenland's marine-terminating glaciers, J. Glaciol., 57, 389–396, 2011.

Howat, I. M., Joughin, I., and Scambos, T. A.: Rapid changes in ice discharge from Greenland outlet glaciers, Science, 315, 1559–1561, 2007.

Howat, I. M., Joughin, I., Fahnestock, M., Smith, B. E., and Scambos, T.: Synchronous retreat and acceleration of southeast Greenland outlet glaciers 2000–2006; Ice dynamics and coupling to climate, J. Glaciol., 54, 1–14, 2008.

Hughes, T.: The Jakobshavns effect, Geophys. Res. Lett., 13, 46–48, 1986.

Hurrell, J. W.: Decadal trends in the North Atlantic Oscillation: Regional temepratures and precipitation, Science, 269, 676–679, 1995.

Hurrell, J. W., Kushnir, Y., Visbeck, M. M., and Ottersen, G. G.: An Overview of the North Atlantic Oscillation, in: The North Atlantic Oscillation: Climate Significance and Environmental Impact, Geophysical Monograph Series, edited by: Hurrell, J. W., Kushnir, Y., Ottersen, G. G., and Visbeck, M. M., Washington, D.C., USA, 2003.

IPCC: Climate Change 2013: The Physical Science Basis. Contribution of Working Group I to the Fifth Assessment Report of the Intergovernmental Panel on Climate Change, Cambridge University Press, Cambridge, UK and New York, NY, USA., 2013.

Jensen, T. S., Box, J. E., and Hvidberg, C. S.: A sensitivity study of annual area change for Greenland ice sheet marine terminating outlet glaciers: 1999–2013, J. Glaciol., 62, 72–81, 2016.

Joughin, I. and Alley, R. B.: Stability of the West Antarctic ice sheet in a warming world, Nat. Geosci., 4, 506–513, 2011.

Joughin, I., Abdalati, W., and Fahnestock, M.: Large fluctuations in speed on Greenland's Jakobshavn Isbræ glacier, Nature, 432, 608–610, 2004.

Joughin, I., Howat, I. M., Fahnestock, M., Smith, B., Krabill, W., Alley, R. B., Stern, H., and Truffer., M.: Continued evolution of Jakobshavn Isbrae following its rapid speedup, J. Geophys. Res., 113, F04006, https://doi.org/10.1029/2008JF001023, 2008.

Kamb, B.: Glacier surge mechanism based on linked cavity configuration of the basal water conduit system, J. Geophys. Res., 92, 9083–9100, 1987.

Kamb, B., Raymond, C. F., Harrison, W. D., Engelhardt, H., Echelmeyer, K. A., Humphrey, N., Brugman, M. M., and Pfeffer, T.: Glacier surge mechanism: 1982–1983 surge of Variegated Glacier, Alaska, Science, 227, 469–479, 1985.

Killick, R. and Eckley, I. A.: Changepoint: An R Package for Changepoint Analysis, J. Stat. Softw., 58, 1–19, 2015.

Killick, R., Fearnhead, P., and Eckley, I.: Optimal Detection of Changepoints With a Linear Computational Cost, J. Am. Stat. Assoc., 107, 1590–1598, 2012.

Killick, R., Beaulieu, C., and Taylor, S.: EnvCpt: Detection of Structural Changes in Climate and Environment Time Series, R package version 0.1.1, available at: https://www.rdocumentation.org/packages/EnvCpt/versions/0.1.1, 2016.

Korablev, A., Smirnov, A., Baranova, O. K., Seidov, D., Parsons, A. R.: Climatological Atlas of the Nordic Seas and Northern North Atlantic (NODC Accession 0118478), Version 3.3, National Oceanographic Data Center, NOAA, Dataset, https://doi.org/10.7289/V54B2Z78, 2014.

Koryakin, V. S.: Glaciers of the New Earth in the XX century and global warming, Nature, 1, 42–48, 2013.

Kotlyakov, V. M., Glazovskii, A. F., and Frolov, I. E.: Glaciation in the Arctic, Her. Russ. Acad. Sci.+, 80, 155–164, 2010.

Kwok, R. and Rothrock, D. A.: Decline in Arctic sea ice thickness from submarine and ICESat records: 1958–2008, Geophys. Res. Lett., 36, L15501, https://doi.org/10.1029/2009GL039035, 2009.

Lenaerts, J. T. M., van Angelen, J. H., van den Broeke, M. R., Gardner, A. S., Wouters, B., and van Meijgaard, E.: Irreversible mass loss of Canadian Arctic Archipelago glaciers, Geophys. Res. Lett., 40, 870–874, 2013.

Loeng, H.: Features of the physical oceanographic conditions of the Barents Sea, Polar Res., 10, 5–18, 1991.

Matsuo, K. and Heki, K.: Current ice loss in small glacier systems of the Arctic Islands (Iceland, Svalbard, and the Russian High Arctic) from satellite gravimetry, Terr. Atmos. Ocean. Sci., 24, 657–670, 2013.

McFadden, E. M., Howat, I. M., Joughin, I., Smith, B., and Ahn, Y.: Changes in the dynamics of marine terminating outlet glaciers in west Greenland (2000–2009), J. Geophys. Res., 116, F02022, https://doi.org/10.1029/2010JF001757, 2011.

McMillan, M., Leeson, A., Shepherd, A., Briggs, K., Armitage, T. W. K., Hogg, A., Kuipers Munneke, P., van den Broeke, M., Noël, B., van de Berg, W. J., Ligtenberg, S., Horwath, M., Groh, A., Muir, A., and Gilbert, L.: A high-resolution record of Greenland mass balance, Geophys. Res. Lett., 43, 7002–7010, 2016.

McNabb, R. W. and Hock, R.: Alaska tidewater glacier terminus positions, 1948–2012, J. Geophys. Res.-Earth, 119, 153–167, 2014.

Meier, M. and Post, A.: What are glacier surges?, Can. J. Earth Sci., 6, 807–817, 1969.

Meier, M. F. and Post, A.: Fast tidewater glaciers, J. Geophys. Res., 92, 9051–9058, 1987.

Meier, M. F., Dyurgerov, M. B., Rick, U. K., O'Neel, S., Pfeffer, W. T., Anderson, R. S., Anderson, S. P., and Glazovsky, A. F.: Glaciers Dominate Eustatic Sea-Level Rise in the 21st Century, Science, 317, 1064–1067, 2007.

Melkonian, A. K., Willis, M. J., Pritchard, M. E., and Stewart, A. J.: Recent changes in glacier velocities and thinning at Novaya Zemlya, Remote Sens. Environ., 174, 244–257, 2016.

Mernild, S. H., Hanna, E., Yde, J. C., Cappelen, J., and Malmros, J. K.: Coastal Greenland air temperature extremes and trends 1890–2010: annual and monthly analysis, Int. J. Climatol., 34, 1472–1487, https://doi.org/10.1002/joc.3777, 2013.

Miles, B. W. J., Stokes, C. R., Vieli, A., and Cox, N. J.: Rapid, climate-driven changes in outlet glaciers on the Pacific coast of East Antarctica, Nature, 500, 563–566, 2013.

Moholdt, G., Hagen, J. O., Eiken, T., and Schuler, T. V.: Geometric changes and mass balance of the Austfonna ice cap, Svalbard, The Cryosphere, 4, 21–34, https://doi.org/10.5194/tc-4-21-2010, 2010a.

Moholdt, G., Nuth, C., Hagen, J. O., and Kohler, J.: Recent elevation changes of Svalbard glaciers derived from ICESat laser altimetry, Remote Sens. Environ., 114, 2756–2767, 2010b.

Moholdt, G., Wouters, B., and Gardner, A. S.: Recent mass changes of glaciers in the Russian High Arctic, Geophys. Res. Lett., 39, L10502, https://doi.org/10.1029/2012GL051466, 2012.

Moon, T. and Joughin, I.: Changes in ice-front position on Greenland's outlet glaciers from 1992 to 2007, J. Geophys. Res., 113, F02022, https://doi.org/10.1029/2007JF000927, 2008.

Moon, T., Joughin, I., Smith, B. E., and Howat, I. M.: 21st-Century evolution of Greenland outlet glacier velocities, Science, 336, 576–578, 2012.

Moon, T., Joughin, I., and Smith, B. E.: Seasonal to multiyear variability of glacier surface velocity, terminus position, and sea ice/ice mélange in northwest Greenland, J. Geophys. Res.-Earth, 120, 818–833, 2015.

Murray, T., Strozzi, T., Luckman, A., Jiskoot, H., and Christakos, P.: Is there a single surge mechanism? Contrasts in dynamics between glacier surges in Svalbard and other regions, J. Geophys. Res.-Sol. Ea., 108, 2237, https://doi.org/10.1029/2002JB001906, 2003.

Nick, F. M., Vieli, A., Howat, I. M., and Joughin, I.: Large-scale changes in Greenland outlet glacier dynamics triggered at the terminus, Nat. Geosci., 2, 110–114, 2009.

Nuth, C., Kohler, J., Aas, H. F., Brandt, O., and Hagen, J. O.: Glacier geometry and elevation changes on Svalbard (1936–90): a baseline dataset, Ann. Glaciol., 46, 106–116, 2007.

Nuth, C., Moholdt, G., Kohler, J., Hagen, J. O., and Kääb, A.: Svalbard glacier elevation changes and contribution to sea level rise, J. Geophys. Res., 115, F01008, https://doi.org/10.1029/2008JF001223, 2010.

Osborn, T. J.: Winter 2009/2010 temperatures and a record-breaking North Atlantic Oscillation index, Weather, 66, 19–21, 2011.

Oziel, L., Sirven, J., and Gascard, J.-C.: The Barents Sea frontal zones and water masses variability (1980–2011), Ocean Sci., 12, 169–184, https://doi.org/10.5194/os-12-169-2016, 2016.

Park, D.-S. R., Lee, S., and Feldstein, S. B.: Attribution of the recent winter sea ice decline over the Atlantic sector of the Arctic Ocean, J. Climate, 28, 4027–4033, 2015.

Pavlov, V. K. and Pfirman, S. L.: Hydrographic structure and variability of the Kara Sea: Implications for pollutant distribution, Deep-Sea Res. Pt. II, 42, 1369–1390, 1995.

Pfirman, S. L., Bauch, D., and Gammelsrød, T.: The Northern Barents Sea: Water Mass Distribution and Modification, The Polar Oceans and Their Role in Shaping the Global Environment Geophysical Monograph, 85, 77–94, 1994.

Politova, N. V., Shevchenko, V. P., and Zernova, V. V.: Distribution, Composition, and Vertical Fluxes of Particulate Matter in Bays of Novaya Zemlya Archipelago, Vaigach Island at the End of Summer, Adv. Meteorol., 15, 259316, https://doi.org/10.1155/2012/259316, 2012.

Porter, D. F., Tinto, K. J., Boghosian, A., Cochran, J. R., Bell, R. E., Manizade, S. S., and Sonntag, J. G.: Bathymetric control of tidewater glacier mass loss in northwest Greenland, Earth Planet. Sc. Lett., 401, 40–46, 2014.

Price, S., Payne, A. J., Howat, I. M., and Smith, B.: Committed sea-level rise for the next century from Greenland ice sheet dynamics during the past decade, P. Natl. Acad. Sci. USA, 108, 8978–8983, 2011.

Pritchard, H. D., Arthern, R. J., Vaughan, D. G., and Edwards, L. A.: Extensive dynamic thinning on the margins of the Greenland and Antarctic ice sheets, Nature, 461, 971–975, 2009.

Przybylak, R. and Wyszyński, P.: Air temperature in Novaya Zemlya Archipelago and Vaygach Island from 1832 to 1920 in the light of early instrumental data, Int. J. Climatol., 37, 3491–3508, https://doi.org/10.1002/joc.4934, 2016.

Radić, V. and Hock, R.: Regionally differentiated contribution of mountain glaciers and ice caps to future sea-level rise, Nat. Geosci., 4, 91–94, 2011.

Radić, V., Bliss, A., Beedlow, A. C., Hock, R., Miles, E., and Cogley, J. G.: Regional and global projections of twenty-first century glacier mass changes in response to climate scenarios from global climate models, Clim. Dynam., 42, 37–58, 2014.

Raymond, C. F.: How do glaciers surge? A review, J. Geophys. Res., 92, 9121–9134, 1987.

Raymond, C. F.: Shear margins in glaciers and ice sheets, J. Glaciol., 42, 90–102, 1996.

Rignot, E., Fenty, I., Xu, Y., Cai, C., Velicogna, I., Cofaigh, C. Ó., Dowdeswell, J. A., Weinrebe, W., Catania, G., and Duncan, D.: Bathymetry data reveal glaciers vulnerable to ice-ocean interaction in Uummannaq and Vaigat glacial fjords, west Greenland, Geophys. Res. Lett., 43, 2667–2674, 2016.

Sakakibara, D. and Sugiyama, S.: Ice-front variations and speed changes of calving glaciers in the Southern Patagonia Ice field from 1984 to 2011, J. Geophys. Res., 119, 2541–2554, 2014.

Schoof, C.: Ice sheet grounding line dynamics: steady states, stability, and hysteresis, J. Geophys. Res., 112, F03S28, https://doi.org/10.1029/2006JF000664, 2007.

Seale, A., Christoffersen, P., Mugford, R., and O'Leary, M.: Ocean forcing of the Greenland Ice Sheet: Calving fronts and patterns of retreat identified by automatic satellite monitoring of eastern outlet glaciers, J. Geophys. Res., 116, F03013, https://doi.org/10.1029/2010JF001847, 2011.

Sevestre, H. and Benn, D. I.: Climatic and geometric controls on the global distribution of surge-type glaciers: implications for a unifying model of surging, J. Glaciol., 61, 646–662, 2015.

Sevestre, H., Benn, D. I., Hulton, N. R. J., and Bælum, K.: Thermal structure of Svalbard glaciers and implications for thermal switch models of glacier surging, J. Geophys. Res.-Earth, 120, 2220–2236, 2015.

Sharov, A. I.: Studying changes of ice coasts in the European Arctic, Geo-Mar. Lett., 25, 153–166, 2005.

Shepherd, A., Ivins, E. R., A, G., Barletta, V. R., Bentley, M. J., Bettadpur, S., Briggs, K. H., Bromwich, D. H., Forsberg, R., Galin, N., Horwath, M., Jacobs, S., Joughin, I., King, M. A., Lenaerts, J. T. M., Li, J., Ligtenberg, S. R. M., Luckman, A., Luthcke, S.

B., McMillan, M., Meister, R., Milne, G., Mouginot, J., Muir, A., Nicolas, J. P., Paden, J., Payne, A. J., Pritchard, H., Rignot, E., Rott, H., Sørensen, L. S., Scambos, T. A., Scheuchl, B., Schrama, E. J. O., Smith, B., Sundal, A. V., van Angelen, J. H., van de Berg, W. J., van den Broeke, M. R., Vaughan, D. G., Velicogna, I., Wahr, J., Whitehouse, P. L., Wingham, D. J., Yi, D., Young, D., and Zwally, H. J.: A Reconciled Estimate of Ice-Sheet Mass Balance, Science, 338, 1183–1189, 2012.

Shumsky, P. L.: Modern glaciation of the Soviet Arctic, Sovremennoe oledenenie Sovetskoy Arktiki, Moscow – Leningrad, Russia, 1949 (in Russian).

Sole, A., Payne, T., Bamber, J., Nienow, P., and Krabill, W.: Testing hypotheses of the cause of peripheral thinning of the Greenland Ice Sheet: is land-terminating ice thinning at anomalously high rates?, The Cryosphere, 2, 205–218, https://doi.org/10.5194/tc-2-205-2008, 2008.

Sutherland, D. A., Straneo, F., Stenson, G. B., Davidson, F., Hammill, M. O., and Rosing-Asvid, A.: Atlantic water variability on the SE Greenland shelf and its relationship to SST and bathymetry, J. Geophys. Res.-Oceans, 118, 847–855, https://doi.org/10.1029/2012JC008354, 2013.

Sutton, R. T. and Hodson, D. L.: Atlantic Ocean forcing of North American and European summer climate, Science, 309, 115–118, 2005.

Trüssel, B. L., Motyka, R. J., Truffer, M., and Larsen, C. F.: Rapid thinning of lake-calving Yakutat Glacier and the collapse of the Yakutat Icefield, southeast Alaska, USA, J. Glaciol., 59, 149–161, 2013.

van den Broeke, M., Bamber, J., Ettema, J., Rignot, E., Schrama, E., van de Berg, W. J., van Meijgaard, E., Velicogna, I., and Wouters, B.: Partitioning Recent Greenland Mass Loss, Science, 326, 984–986, 2009.

van der Veen, C. J.: Fracture mechanics approach to penetration of bottom crevasses on glaciers, Cold Reg. Sci. Technol., 27, 213–223, 1998a.

van der Veen, C. J.: Fracture mechanics approach to penetration of surface crevasses on glaciers, Cold Reg. Sci. Technol., 27, 31–47, 1998b.

Willis, M. J., Melkonian, A. K., and Pritchard, M. E.: Outlet glacier response to the 2012 collapse of the Matusevich Ice Shelf, Severnaya Zemlya, Russian Arctic, J. Geophys. Res.-Earth, 120, 2040–2055, 2015.

Zeeberg, J. and Forman, S. L.: Changes in glacier extent on north Novaya Zemlya in the Twentieth Century, The Holocene, 11, 161–175, 2001.

Zhao, M., Ramage, J., Semmens, K., and Obleitner, F.: Recent ice cap snowmelt in Russian High Arctic and anti-correlation with late summer sea ice extent, Environ. Res. Lett., 9, 045009, https://doi.org/10.1088/1748-9326/9/4/045009, 2014.

Zhou, S., Miller, A., Wang, J., and Angell, J.: Trends of NAO and AO and their associations with stratospheric processes, Geophys. Res. Lett., 28, 4107–4110, 2001.

8

Assimilation of snow cover and snow depth into a snow model to estimate snow water equivalent and snowmelt runoff in a Himalayan catchment

Emmy E. Stigter[1], Niko Wanders[2], Tuomo M. Saloranta[3], Joseph M. Shea[4,5], Marc F. P. Bierkens[1,6], and Walter W. Immerzeel[1]

[1]Department of Physical Geography, Utrecht University, Utrecht, the Netherlands
[2]Department of Civil and Environmental Engineering, Princeton University, Princeton, NJ, USA
[3]Norwegian Water Resources and Energy Directorate (NVE), Oslo, Norway
[4]International Centre for Integrated Mountain Development, Kathmandu, Nepal
[5]Centre for Hydrology, University of Saskatchewan, Saskatchewan, Canada
[6]Deltares, Utrecht, the Netherlands

Correspondence to: Emmy E. Stigter (e.e.stigter@uu.nl)

Abstract. Snow is an important component of water storage in the Himalayas. Previous snowmelt studies in the Himalayas have predominantly relied on remotely sensed snow cover. However, snow cover data provide no direct information on the actual amount of water stored in a snowpack, i.e., the snow water equivalent (SWE). Therefore, in this study remotely sensed snow cover was combined with in situ observations and a modified version of the seNorge snow model to estimate (climate sensitivity of) SWE and snowmelt runoff in the Langtang catchment in Nepal. Snow cover data from Landsat 8 and the MOD10A2 snow cover product were validated with in situ snow cover observations provided by surface temperature and snow depth measurements resulting in classification accuracies of 85.7 and 83.1 % respectively. Optimal model parameter values were obtained through data assimilation of MOD10A2 snow maps and snow depth measurements using an ensemble Kalman filter (EnKF). Independent validations of simulated snow depth and snow cover with observations show improvement after data assimilation compared to simulations without data assimilation. The approach of modeling snow depth in a Kalman filter framework allows for data-constrained estimation of snow depth rather than snow cover alone, and this has great potential for future studies in complex terrain, especially in the Himalayas. Climate sensitivity tests with the optimized snow model revealed that snowmelt runoff increases in winter and the early melt season (December to May) and decreases during the late melt season (June to September) as a result of the earlier onset of snowmelt due to increasing temperature. At high elevation a decrease in SWE due to higher air temperature is (partly) compensated by an increase in precipitation, which emphasizes the need for accurate predictions on the changes in the spatial distribution of precipitation along with changes in temperature.

1 Introduction

In the Himalayas a part of the precipitation is stored as snow and ice at high elevations. This water storage is affected by climate change resulting in changes in river discharge in downstream areas (Barnett et al., 2005; Bookhagen and Burbank, 2010; Immerzeel et al., 2009, 2010). The Himalayas and adjacent Tibetan Plateau are important water towers, and water generated here supports the water demands of more than 1.4 billion people through large rivers such as the Indus, Ganges, Brahmaputra, Yangtze and Yellow River (Immerzeel et al., 2010). So far, the main focus has been on the effect of climate change on the glaciers and the resulting runoff. However, snow is an important short-term water reservoir in the Himalayas, which is released seasonally, contributing to river discharge (Bookhagen and Burbank, 2010; Immerzeel et al., 2009). The contribution of snowmelt to total runoff is highest in the western part of the Himalayas and lowest in the eastern and central Himalayas (Bookhagen and Burbank, 2010; Lutz et al., 2014).

Although Himalayan snow storage is important for the water supply in large parts of Asia, in situ observations of snow depth are sparse throughout the region. Many studies benefit from the continuous snow cover data retrieved from satellite imagery to estimate snow cover dynamics or contribution of snowmelt to river discharge (Bookhagen and Burbank, 2010; Gurung et al., 2011; Immerzeel et al., 2009; Maskey et al., 2011; Wulf et al., 2016). Studies about snowmelt in the Himalayas have predominantly relied on remotely sensed snow cover and a modeled melt flux estimating melt runoff resulting from this snow cover (e.g., Bookhagen and Burbank, 2010; Immerzeel et al., 2009; Tahir et al., 2011; Wulf et al., 2016). However, this approach provides no or limited information on snow water equivalent (SWE), which is an important hydrologic measure as it indicates the actual amount of water stored in a snowpack. SWE can be reconstructed based on the integration of a simulated melt flux over the time period of remotely sensed observed snow cover. However, this method provides only information on the peak SWE value and introduces errors when snowfall occurs during the melt season (Durand et al., 2008; Molotch, 2009; Molotch and Margulis, 2008). Currently there is only limited reliable information available on SWE for the Himalayas (Lutz et al., 2015; Putkonen, 2004). SWE can be retrieved with passive microwave remote sensing, but the results are highly uncertain, especially for mountainous terrain and wet snow (Dong et al., 2005). In addition, the spatial resolution is coarse and therefore inappropriate for catchment scale studies in the Himalayas. Estimating both the spatial and temporal distribution of SWE and snowmelt is important for flood forecasting, hydropower and irrigation in downstream areas.

Selection of a suitable snow model is critical to correctly represent snow cover and SWE. Snow models of different complexity exist and can be roughly divided into physically based and temperature-index models. Several studies have compared snow models of different complexity and their performance. Physically based models typically outperform temperature-index models in snowpack runoff simulations on a sub-daily timescale (Avanzi et al., 2016; Magnusson et al., 2011; Warscher et al., 2013). However, physically based and temperature-index models have a similar ability to simulate daily snowpack runoff (Avanzi et al., 2016; Magnusson et al., 2015). Avanzi et al. (2016) showed that the use of a temperature-index model does not result in a significant loss of performance in the simulation of SWE and snow depth with respect to a physically based model. Even though physically based models outperform temperature-index models in some cases, temperature-index models are often preferred, as data requirements and computational demands are lower.

Especially in the Himalaya, data availability constrains the choice of a snow model.

Assimilation of remotely sensed snow cover and ground-based snow measurements has been proved to be an effective method to improve hydrological and snow model simulations (Andreadis and Lettenmaier, 2006; Clark et al., 2006; Leisenring and Moradkhani, 2011; Liu et al., 2013; Nagler et al., 2008; Saloranta, 2016). Although different data assimilation techniques exist, Kalman filter techniques are often selected, due to their relatively low computation demand. They estimate the most likely solution using an optimal combination of observations and model simulations. Especially in catchments with strong seasonal snow cover, assimilation of remotely sensed snow cover is expected to be most useful as a result of fast changing conditions in the melting season (Clark et al., 2006).

The aim of this study is to estimate SWE and snowmelt runoff in a Himalayan catchment by assimilating remotely sensed snow cover and in situ snow depth observations into a modified version of the seNorge snow model (Saloranta, 2012, 2014, 2016). Climate sensitivity tests are subsequently performed to investigate the change of SWE and snowmelt runoff as result of changing air temperature and precipitation. The approach of modeling snow depth allows us to validate the quantity of simulated snow rather than snow cover alone and is a new approach in Himalayan snow research.

2 Methods and data

2.1 Study area

The study area is the Langtang catchment, which is located in the central Himalayas approximately 100 km north of Kathmandu (Fig. 1). The catchment has a surface area of approximately 580 km^2 from the outlet near Syabru Besi upwards. The elevation ranges from 1406 m above sea level (a.s.l.) at the catchment outlet to 7234 m a.s.l. for Langtang Lirung, which is the highest peak in the catchment. The climate is monsoon dominated and 68–89 % of the annual precipitation falls during the monsoon (Immerzeel et al., 2014). Spatial patterns in precipitation are seasonally contrasting, and there is a strong interaction between the orography and precipitation patterns. At the synoptic scale, monsoon precipitation decreases from south to north, but at smaller scales local orographic effects associated with the aspect of the main valley ridges (Barros et al., 2004) determine the precipitation distribution. Numerical weather models suggest that monsoon precipitation mainly accumulates at the southwestern slopes near the catchment outlet at low elevation, while winter precipitation mainly accumulates along high-elevation southern–eastern slopes (Collier and Immerzeel, 2015). Winter westerly events can also provide significant snowfall. Snow cover has strong seasonality with extensive, but sometimes erratic, winter snow cover and retreat of the

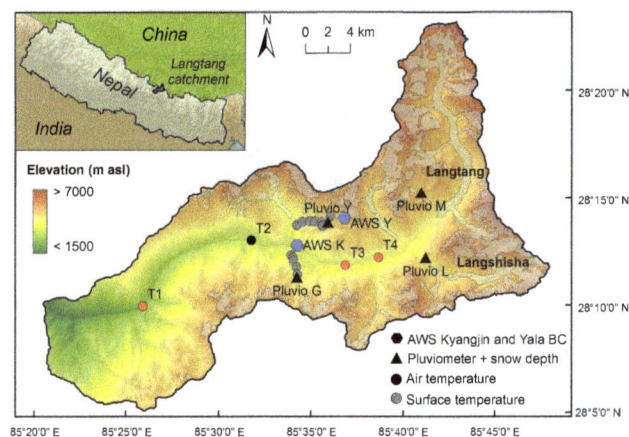

Figure 1. Study area with the locations of the in situ observations. Langtang and Langshisha refer to the two main glaciers in the upper Langtang valley.

snowline to higher elevations during spring and summer. For the upper part of the catchment (upstream of Kyangjin) it has been estimated that snowmelt contributes up to 40 % of total runoff (Ragettli et al., 2015).

2.2 Calibration and validation strategy

Remotely sensed snow cover, in situ observations and a modified version of the seNorge snow model were combined to estimate SWE and snowmelt runoff dynamics. The remotely sensed snow cover (Landsat 8 and MOD10A2 snow maps) was first validated with in situ snow cover observations provided by surface temperature and snow depth measurements. The snow model was used to simulate daily values of SWE and runoff and was forced by daily in situ meteorological observations of precipitation, temperature and incoming shortwave radiation. MOD10A2 snow cover and snow depth measurements were assimilated to obtain optimal model parameter values using an ensemble Kalman filter (EnKF; Evensen, 2003). The optimized parameters were used for a simulation without assimilation of the observations (open loop). Finally, the model outcome was validated with observed snow depth and Landsat 8 snow cover.

2.3 Datasets

2.3.1 Remotely sensed snow cover

MOD10A2

MOD10A2 is a Moderate Resolution Imaging Spectroradiometer (MODIS) snow cover product available at http://reverb.echo.nasa.gov/. The online sub-setting and reprojection utility was used to clip and project imagery for the Langtang catchment. MOD10A2 provides the 8-day maximum snow extent with a spatial resolution of ~ 500 m. If there is one snow observation within the 8-day period, then the

pixel is classified as snow. The 8-day maximum extent offered a good compromise between the temporal resolution and the interference of cloud cover. The snow mapping algorithm used is based on the normalized difference snow index (NDSI; Hall et al., 1995). The NDSI is a ratio of reflection in shortwave infrared (SWIR) and green light (GREEN) and takes advantage of the properties of snow – i.e., snow strongly reflects visible light and strongly absorbs SWIR – Eq. (1):

$$\text{NDSI} = \frac{\text{GREEN} - \text{SWIR}}{\text{GREEN} + \text{SWIR}}. \tag{1}$$

The NDSI is calculated with MODIS spectral bands 4 (0.545–0.565 μm) and 6 (1.628–1.652 μm). Pixels are classified as snow when the NDSI ≥ 0.4. Water and dark targets typically have high NDSI values, and, to prevent pixels from being incorrectly classified as snow, the reflection should exceed 10 and 11 % for spectral bands 2 (0.841–0.876 μm) and 4 respectively for a pixel to be classified as snow (Hall et al., 1995). A full description of the snow mapping algorithm is given by Hall et al. (2002).

Landsat 8

Landsat 8 imagery from 15 April 2013 to 5 November 2014 was downloaded from http://earthexplorer.usgs.gov/. Cloud-free scenes (10 out of 34), based on visual inspection, were used to derive daily snow maps with high spatial resolution (30 m). For each image digital numbers were converted to top of atmosphere reflectance. For Landsat 8 the NDSI was calculated with Eq. (1) with spectral bands 3 (0.53–0.59 μm) and 6 (1.57–1.65 μm). The chosen threshold value was equal to that used for the MOD10A2 snow cover product. The NDSI has proven to be a successful snow mapping algorithm for various sensors with a threshold value around 0.4 (Dankers and De Jong, 2004). Although the spectral bands have slightly different band widths and spectral positions, a threshold value of 0.4 gave satisfactory results when compared with in situ snow observations. In addition, the reflection in near-infrared light should exceed 11 % to prevent water from being incorrectly classified as snow (Dankers and De Jong, 2004). Therefore, a pixel is classified as snow when the NDSI value ≥ 0.4 and the reflectance in near-infrared light > 11 %.

2.3.2 In situ observations

Different types of snow and meteorological observations were available for the study period (January 2013–September 2014; Table 1, Fig. 1). Two transects of surface temperature measurements on a north- and south-facing slope provided information on snow cover. The 13 temperature sensors (Hobo Tidbits) were positioned on the surface and covered by a small cairn and recorded surface temperature with 10 min sampling intervals. Snow depths were measured with

Table 1. Overview of the in situ observations and their specifications. Locations are shown in Fig. 1.

Description	Code	Data availability (dd/mm/yy)	Latitude	Longitude	Elevation (m a.s.l.)	Observations*
Yala 1	Y1	06/05/13–03/05/14	28.22645	85.56878	4117	T_S
Yala 2	Y2	06/05/13–03/05/14	28.22897	85.57391	4214	T_S
Yala 3	Y3	06/05/13–03/05/14	28.2298	85.58051	4328	T_S
Yala 4	Y4	06/05/13–02/03/14	28.22932	85.58492	4441	T_S
Yala 5	Y5	06/05/13–03/05/14	28.22894	85.5908	4541	T_S
Yala 6	Y6	06/05/13–03/05/14	28.22635	85.5918	4656	T_S
Yala 7	Y7	06/05/13–02/03/14	28.22635	85.59246	4759	T_S
Yala 8	Y8	06/05/13–02/03/14	28.23342	85.59921	4960	T_S
Ganjala 1	G1	03/11/13–11/10/14	28.20305	85.56405	3908	T_S
Ganjala 2	G2	03/11/13–06/09/14	28.20155	85.56577	3998	T_S
Ganjala 3	G3	03/11/13–11/10/14	28.19899	85.56617	4094	T_S
Ganjala 4	G4	03/11/13–30/04/14	28.1938	85.56916	4201	T_S
Ganjala 5	G5	03/11/13–11/10/14	28.18831	85.57001	4300	T_S
Pluvio Yala	Pluvio Y	01/01/13–30/06/13 26/10/13–16/10/14	28.22900	85.59700	4831	T, SD
Pluvio Ganjala	Pluvio G	20/01/14–03/05/14	28.18625	85.56961	4361	SD
Pluvio Langshisha	Pluvio L	29/10/13–01/07/14	28.20265	85.68619	4452	SD
Pluvio Morimoto	Pluvio M	17/05/13–09/10/14	28.25296	85.68152	4919	T, SD
Lama Hotel	T1	01/01/13–07/10/14	28.16212	85.43073	2492	T
Langtang	T2	01/01/13–07/10/14	28.21398	85.52745	3557	T
Jathang	T3	01/01/13–07/10/14	28.1958	85.6132	3947	T
Numthang	T4	01/01/13–07/10/14	28.20213	85.64313	3983	T
AWS Kyangjin	AWS K	01/01/13–07/10/14	28.2108	85.5695	3862	T, SD, P, IR
AWS Yala base camp	AWS Y	01/01/13–07/10/14	28.23252	85.61208	5090	SD

* T_S: surface temperature, SD: snow depth, T: air temperature, P: precipitation, IR: incoming shortwave radiation.

sonic ranging sensors at four locations at 15 min intervals. Hourly measurements of snow depth were also made at the Kyangjin and Yala base camp automatic weather stations (AWS K and AWS Y; Fig. 1). Hourly means (or totals) of air temperature, liquid and solid precipitation, and incoming shortwave radiation were also recorded at AWS Kyangjin (Shea et al., 2015). Air temperature data were also acquired at several locations with 10 and 15 min recording intervals.

2.4 Model forcing

The snow model was forced with daily average and maximum air temperature, cumulative precipitation and average incoming shortwave radiation for the time period January 2013–September 2014. Hourly measurements of air temperature, precipitation and incoming shortwave radiation at AWS Kyangjin (Shea et al., 2015) were therefore aggregated to daily values. This study period was chosen based on availability of forcing data and observations. Daily temperature lapse rates were interpolated from the air temperature measurements throughout the catchment and used to extrapolate (average and maximum) daily air temperature observed at AWS Kyangjin (Fig. 1). The derived temperature lapse rates agree with the values found by Immerzeel et al. (2014). The daily observed precipitation and temperature lapse rates were

corrected in the modified seNorge snow model with the correction factors P and T_{lapse} respectively to account for the uncertainty related to undercatch and the derived temperature lapse rates (Table 2). Although temperature has a strong relation with altitude and can be accurately derived from multiple weather stations at different altitudes, small differences in the temperature lapse rate (e.g., $0.001\,°\mathrm{C\,m^{-1}}$) can result in temperature differences of up to several degrees at high altitude in Langtang due to the extreme topography (Immerzeel et al., 2014). Hence, there is a need to consider a potential correction on the temperature lapse rate. A correction is also applied to the daily observed precipitation as precipitation measurements are typically biased due to wind-induced undercatch, especially for solid precipitation (Wolff et al., 2015).

Collier and Immerzeel (2015) modeled the spatial distribution of precipitation in Langtang using an interactively coupled atmosphere and glacier mass balance model (Collier et al., 2013). Their study revealed seasonally contrasting spatial patterns of precipitation within the catchment. Monthly modeled precipitation fields from this study were therefore normalized and used to distribute the observed precipitation at AWS Kyangjin. Similarly, a radiation model (van Dam, 2001; Feiken, 2014) was used to extrapolate observed incoming shortwave radiation. The radiation model takes into

Table 2. Parameters in the snow model. Initial value indicates the uncalibrated parameter value and the value range indicates the range which is used for the sensitivity analysis. Sensitivity of snow depth (SD) and snow extent (SE) represents the difference between the 90th and 10th percentile of mean snow depth and snow extent resulting from the sensitivity analysis.

Parameter	Unit	Description	Initial value	Value range	Sensitivity SD (mm)	Sensitivity SE (km^2)
T_T	(°C)	Threshold temperature for onset of melt or refreezing	0[g]	−6 to 2[d, f, g]	157.3	57.25
F_{SR}	(m^2 mm W^{-2} d^{-1})	Melt factor dependent on incoming shortwave radiation	0.15[g]	0.13 to 0.19[d, g]	9.486	2.721
F_T	(mm °C^{-1} d^{-1})	Melt factor dependent on temperature	4.32[g]	2.54 to 5.19[d, g]	9.486	2.721
thr$_{snow}$	(°C)	Threshold for partitioning in rain or snow	0[g]	−1 to 1[e, g]	35.82	11.99
C_{rf}	(mm °C^{-1} d^{-1})	Degree-day refreezing factor	0.16[e]	0.08–0.40[e]	8.188	0.3248
a_{ini}	(–)	Decay of albedo deep snow (initial)	0.713[b]	–	–	–
α_u	(–)	Albedo of surface underlying snow (ground, ice)	0.15, 0.25[g]	–	–	–
a_1	(–)	Decay of albedo deep snow	0.112[b]	0.112 to 0.34[b, g]	56.39	7.279
a_2	(–)	Decay of albedo shallow snow	0.442[b]	0.3 to 0.5	0.2410	0.2818
a_3	(–)	Decay of albedo shallow snow (exponent)	0.058[b]	0.03 to 0.1	0.2001	0.2132
r_{max}	(–)	Maximum allowed fraction of liquid water in snowpack	0.1[e]	0.05 to 0.20[e]	31.66	0.3278
d^*	(cm)	Scaling length for smooth transition albedo from deep snow to shallow snow	2.4[b]	1 to 25	0.0012	0.0007
SS_1	(m)	Regression function of parameter snow holding depth dependence on slope angle	250[g]	200 to 300	10.86	2.033
SS_2	(–)	Regression function of parameter snow holding depth dependence on slope angle	0.172[g]	0.16 to 0.19	26.45	7.170
S_{min}	(°)	Minimum slope for avalanching to occur	25[a]	15 to 35	34.00	1.640
ρ_{av}	(kg L^{-1})	Density of avalanching snow	0.200[c]	–	–	–
ρ_{min}	(kg L^{-1})	Minimum density of new snow due to snowfall	0.050[e]	0.050 to 0.15[e]	–	–
a_{ns}	(–)	Coefficient for density of new snow	100[e]	–	–	–
η_0	(MN s m^{-2})	Coefficient related to viscosity of snow (at zero temperature and density)	7.6[e]	1 to 10[e]	75.75	–
C_5	(°C^{-1})	Coefficient for temperature effect on viscosity	0.1[e]	0.04 to 0.12[e]	10.44	–
C_6	(L kg^{-1})	Coefficient for density effect on viscosity	21[e]	15 to 35[e]	268.8	–
k_{comp}	(–)	Compaction factor	0.5[e]	–	–	–
P	(–)	Precipitation correction factor	1	0.6 to 1.4	320.1	14.17
T_{lapse}	(–)	Temperature lapse rate correction factor	1	0.9 to 1.1	116.0	24.63

[a] Bernhardt and Schulz (2010). [b] Brock et al. (2000). [c] Hopfinger (1983). [d] Pellicciotti et al. (2012). [e] Saloranta (2014). [f] Ragettli et al. (2013). [g] Ragettli et al. (2015).

account the aspect, slope, elevation and shading due to surrounding topography.

The model initial conditions for January 2013 (i.e., SWE and snow depth) were set by simulating year 2013 three times.

2.5 Modified seNorge model

The seNorge snow model (Saloranta, 2012, 2014, 2016) is a temperature-index model which requires only data of air temperature and precipitation. In addition, the seNorge snow model includes a compaction module that can be used to assimilate and validate snow depth rather than snow cover only. The low data requirements and the compaction module make the seNorge snow model suitable for application in this study.

The seNorge snow model was rewritten from its original code into the environmental modeling software PCRaster Python (Karssenberg et al., 2010) to allow spatiotemporal modeling of the SWE and runoff within the catchment. The snow is modeled as a single homogeneous layer with a spatial resolution of 100 m and a daily time step. The seNorge model was further improved by implementing a different melt algorithm, albedo decay and avalanching. These novel model components are described hereafter, and the model parameters used are given in Table 2.

2.5.1 Water balance and snowmelt

Precipitation in the model is partitioned as rain or snow based on an air temperature threshold thr$_{snow}$ (°C). The snowpack consists of a solid component and possibly a liquid component. Meltwater and rain can be stored within the snowpack until its water holding capacity is exceeded and has the possibility to refreeze within the snowpack. The original melt algorithm of the seNorge snow model is substituted by the enhanced temperature-index approach (Pellicciotti et al., 2005, 2008). When air temperature (T; °C) exceeds the temperature threshold for melt onset (T_T; °C), the potential melt (M_{pot}; mm d^{-1}) is calculated for each pixel by Eq. (2):

$$M_{pot} = T \cdot F_T + F_{SR} \cdot (1 - \alpha) \cdot R_{inc}, \tag{2}$$

where F_{SR} (m^2 mm W^{-2} d^{-1}) is a radiative melt factor, F_T (mm °C^{-1} d^{-1}) is a temperature melt factor, α (–) is the albedo of the snow cover and R_{inc} (W m^{-2}) is the incoming shortwave radiation. In case that the threshold temperature is negative, the potential melt can become negative when the radiation melt component is not positive enough to compensate for the negative temperature melt component. When the potential melt is negative it is set to zero to prevent negative values.

The simulated runoff in the seNorge snow model is the total runoff, i.e., the sum of snowmelt and rain. As the focus of this study is on snowmelt runoff it is necessary to split the runoff in snowmelt and rain runoff. Meltwater and rain fill up the snowpack until its water holding capacity is exceeded. The surplus is defined as snowmelt and rain runoff respectively. If both rain and snowmelt occur it is assumed that rain saturates the snowpack first. Rain falling on snow-free portions of the basin is included in the rain runoff totals.

2.5.2 Albedo decay

Decay of the albedo of snow is calculated with the algorithm developed by Brock et al. (2000) in which the albedo is a function of cumulative maximum daily air temperature T_{max} (°C). When T_{max} is above 0 °C the air temperature is summed as long as snow is present and no new snow has fallen. When T_{max} is below 0 °C the albedo remains constant. Albedo decay is calculated differently for deep snow (SWE $\geq$ 5 mm) and shallow snow (SWE < 5 mm). The albedo decay for deep snow is a logarithmic decay, whereas the decay for shallow snow is exponential. This results in a gradual decrease of the albedo for several weeks, which agrees with reality (Brock et al., 2000). When new snow falls the albedo is set to its initial value. In Langtang the observed albedo of fresh snow is 0.84 and the observed minimum precipitation rate to reset the snow albedo is 1 mm d^{-1} (Ragettli et al., 2015).

2.5.3 Avalanching

After snowfall events, avalanching occurs regularly on steep slopes in the catchment. Therefore, snow transport due to avalanching is considered to be an important process for redistribution of snow in the Langtang catchment (Ragettli et al., 2015). Snow avalanching is implemented in the model us-

ing the SnowSlide algorithm (Bernhardt and Schulz, 2010). For each cell a maximum snow holding depth SWE$_{max}$ (m), depending on slope S (°), is calculated using an exponential regression function following Eq. (3):

$$SWE_{max} = SS_1 + e^{-SS_2 \cdot S}, \tag{3}$$

where SS_1 and SS_2 are empirical coefficients. If SWE exceeds SWE$_{max}$ and the slope exceeds the minimum slope S_{min} for avalanching to occur, then snow is transported to the adjacent downstream cell. Snow can be transported through multiple cells within one time step.

As the snowpack is divided into an ice and liquid component, both the ice and liquid components should be transported downwards. Avalanches in the Langtang catchment mainly occur at high elevations where temperatures are low and (almost) no liquid water is present in the snowpack. It is therefore assumed that avalanches are dry avalanches and that no liquid water is present in the avalanching snow. When there is, in rare circumstances, liquid water present in avalanching snow, the liquid water is converted to the ice component to ensure water balance closure.

2.5.4 Compaction and density

The compaction module is described in detail in Saloranta (2014, 2016). In this module SWE is converted into snow depth. Change in snow depth occurs due to melt, new snow and viscous compaction. The change in snow depth due to new snow is adapted such that an increase in snow depth can occur due to both snowfall and deposition of avalanching snow. The increase in snow depth due to deposition of avalanching snow is calculated using a constant snow density for dry avalanches (200 kg m^{-3}; Hopfinger, 1983).

2.6 Data assimilation

2.6.1 Sensitivity analysis

In order to assess which model parameters to calibrate, a local sensitivity analysis was performed by varying the value of one parameter at a time while holding the values of other parameters fixed. This gives useful first order estimates for parameter sensitivity, although it cannot account for parameter interactions. Plausible parameter values were based on the literature (Table 2). The model was run in Monte Carlo (MC) mode with 100 realizations for each parameter. The values for the parameters were randomly chosen from a uniform distribution with defined minimum and maximum values for the parameters. The snow extent and snow depth were averaged over the study period and study area for the sensitivity analysis. The sensitivity of the modeled mean snow extent and mean snow depth were compared to the changes in parameter values. A pixel is determined to be snow covered in the model when the simulated SWE exceeds 1 mm. All the parameters were varied independently per run, except for the melt factors (F_T and F_{SR}), as these are known to be dependent on each other (Ragettli et al., 2015). Therefore, F_T and F_{SR} were varied simultaneously in the sensitivity analysis using a linear relation between these melt factors.

2.6.2   Parameter calibration

Using the ensemble Kalman filter (Evensen, 1994), data assimilation of snow extent and snow depth observations was used to calibrate model parameters using the framework developed by Wanders et al. (2013). Both the EnKF and particle filter (PF) have been used in several studies to assimilate snow observations into snow models (e.g., Charrois et al., 2016; Leisenring and Moradkhani, 2011; Liu et al., 2013; Magnusson et al., 2016). The EnKF and PF are similar in their approach (estimate the model uncertainty from the particle or ensemble spread). The EnKF can only be used for assimilation of continuous values and not for binary values (i.e., snow cover present or not). Therefore, it is necessary to assimilate snow extent (continuous values) into the model, which results in a partial loss of spatial information of snow cover. However, the EnKF has a higher efficiency when it deals with Gaussian data and related errors. The computational demand required for a PF exceeds the EnKF's computer requirements, due to the need to cover the entire (non-Gaussian) distribution. When the number of particles becomes too low, there is an additional risk of particle collapse, especially when one wants to take into account all the grid cells in the simulation with or without snow. This would require a total particle number exceeding the total number of grid cells in the domain, in combination with all the possible parameter combinations to avoid collapse of the filter. For a single site or small sites a PF would be a good alternative (e.g., Charrois et al., 2016; Magnusson et al., 2016), but, limited by the current available computational power, this is only feasible with an EnKF implementation. As we deal with continuous values, it is computationally efficient and allows for dual-state parameter estimations. The lower number of ensemble members compared to a PF allowed us to run multiple simulations over longer time periods, providing a better estimate of the potential of the EnKF improvements.

An advantage of the EnKF calibration framework is that it allows for the obtaining of an uncertainty estimate for the calibrated parameters. The EnKF obtains the simulation uncertainty by using an MC framework, where the spread in the ensemble members represents the combined uncertainty of parameters and input data. Unfortunately, the EnKF does not allow us to reduce and estimate the model structure uncertainty, since it relies on the assumption that the ensemble members are normally distributed. This assumption is no longer valid if multiple model schematizations are used. Therefore, it is assumed that the model is capable of accurately simulating the processes, when provided with the correct parameters. Besides the parameter and model uncertainty, there is uncertainty in the observations which are assimilated. The EnKF finds the optimal solution for the model states and parameters, based on the observations and modeled predicted values and their respective uncertainties. With sufficient observations the parameters will convert to a stable solution with an uncertainty estimate that is dependent on the observations error and the ability of the model to simulate the observations. It was found that 50 ensemble members are sufficient to obtain stable parameter solutions and correctly represent the parameter uncertainty.

The EnKF was applied for each time step that observations were available. The MOD10A2 snow extent was divided into six elevation zones. The snow extent per elevation zone was derived from the MOD10A2 snow cover and used for assimilation to include more information on spatial distribution of snow. The elevation zone breakpoints are at 3500, 4000, 4500, 5000 and 5500 m a.s.l. Snow maps with more than 30 % cloud cover and with obvious misclassification of snow were exempted from assimilation (3 snow maps out of 88). Only for cloud-free pixels, comparisons were made between modeled and observed snow extent. Two snow depth observation locations (Pluvio Langshisha and AWS Kyangjin; Fig. 1) were also assimilated.

The EnKF framework allows for the inclusion of an uncertainty in the assimilated observations. Point snow depth measurements have high uncertainties that are related to limited representativeness of point snow depth observations in complex terrain due to local influence of snow drift (Grünewald and Lehning, 2015). For the snow depth measurements a variance of 25 cm was chosen to represent the uncertainty of point snow depth measurements. The MOD10A2 snow extent was assigned an uncertainty based on the classification accuracy (fraction of correctly classified pixels) determined with the in situ snow observations (Sect. 3.1.2). The uncertainty is dependent on the snow extent (SE; m^2), i.e., an increase in uncertainty for an increase in snow extent. To prevent the uncertainty from becoming zero when there is no snow cover, the minimum variance for each zone was restricted to the average snow extent $\overline{SE}_{zone}$ (m^2) × the accuracy (–). Therefore, the variance σ^2 per elevation zone is defined following Eq. (4):

$$\sigma^2 = \max\left(\left(SE_{zone} \cdot accuracy\right)^2 \left(\overline{SE}_{zone} \cdot accuracy\right)^2\right). \quad (4)$$

The four most sensitive parameters (T_T, T_{lapse}, P and C_6) resulting from the sensitivity analysis were optimized based on the assimilation of snow depth and MOD10A2 snow extent. The first three parameters (T_T, T_{lapse} and P) influence both snow depth and snow extent and were optimized by assimilating MOD10A2 snow extent. The fourth parameter (C_6) is an empirical coefficient relating viscosity to snow density and only influences snow depth. C_6 was optimized by assimilating snow depth observations and taking into account the full uncertainty in the previously determined parameters. The two-step approach was chosen to restrict the degrees of freedom and to prevent unrealistic parameter estimates.

Table 3. Changes in temperature (ΔT) and precipitation (ΔP) for the climate sensitivity tests (same as Immerzeel et al., 2013).

Sensitivity test	ΔT (°C)	ΔP (%)
Dry, cold	1.5	−3.2
Dry, warm	2.4	−2.3
Wet, cold	1.3	12.4
Wet, warm	2.4	12.1

2.7 Climate sensitivity

Climate sensitivity tests were performed to investigate changes in SWE and snowmelt runoff as a result of temperature and precipitation changes. Climate sensitivity was tested by perturbing daily average air temperature, daily maximum air temperature and daily cumulative precipitation using a delta-change method. Immerzeel et al. (2013) extracted temperature and precipitation trends from all available CMIP5 simulations for the emission scenario RCP 4.5 for the Langtang catchment. They selected four models that ranged from dry to wet and from cold to warm. Four climate sensitivity tests were performed based on the projected changes in temperature and precipitation found by Immerzeel et al. (2013) (Table 3).

Figure 2 shows the monthly cumulative precipitation and the average daily maximum temperature per month measured at AWS Kyangjin for the study period. These data are also available for the time period 1988–2009 and are used to characterize the climatology of the catchment. Comparison of the measurements of the 1988–2009 period and the study period shows that the maximum temperature is similar for both time periods, whereas more variability exists in the cumulative precipitation. Especially in October, a large difference exists in cumulative precipitation, which is caused by a large precipitation event of approximately 100 mm during the study period.

3 Results and discussion

3.1 Validation of snow maps with in situ observations

3.1.1 In situ snow observations

Surface temperature is an indirect measure of presence of snow. Figure 3 shows observed surface temperature for two locations. Snow cover is distinguishable based on the low diurnal variability in surface temperature when snow is present due to the isolating effect of snow (Lundquist and Lott, 2008). An optimal threshold for distinguishing between snow and no snow was determined to be a 2 °C difference between daily minimum temperature and maximum temperature. The use of a larger temperature interval as threshold value was

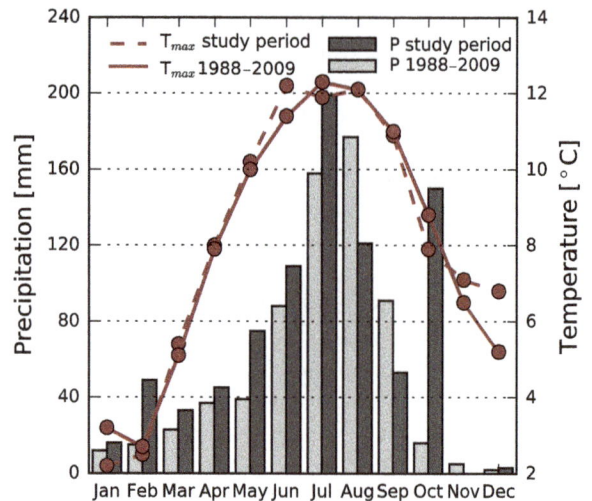

Figure 2. Comparison of maximum temperature ($T_{\max}$) and cumulative monthly precipitation (P) for the study period (January 2013–September 2014) and the 1988–2009 time series (based on measurements in Kyangjin). The average yearly cumulative precipitation is 853 and 663 mm for the study period and the 1988–2009 time series respectively.

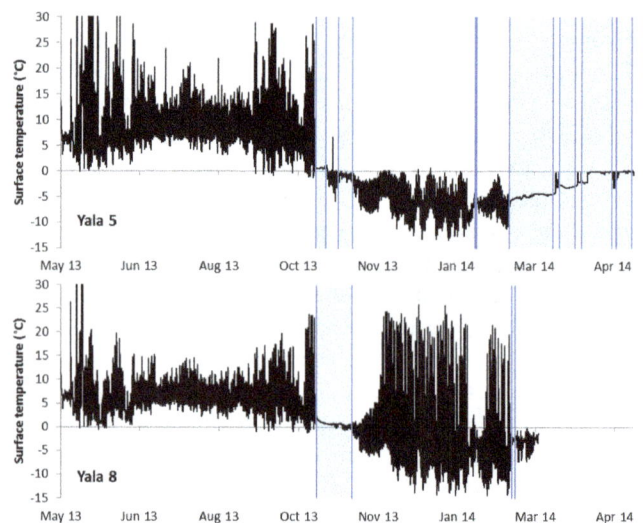

Figure 3. Observed surface temperature with 10 min interval at two locations (Table 1). The blue vertical lines indicate the start and end of the snow cover.

explored; however, as diurnal temperature variability is small during monsoon (Immerzeel et al., 2014) setting, the diurnal cycle temperature threshold above 2 °C may result in incorrect monsoon snow observations.

3.1.2 Remotely sensed snow cover

Both observed surface temperature and snow depth measurements were converted to daily and 8-day maximum binary snow cover values to validate Landsat 8 and MOD10A2

Table 4. Confusion matrices for comparison of Landsat 8 snow maps and MOD10A2 snow maps with in situ snow observations.

		MOD10A2		Landsat 8	
		Snow	No snow	Snow	No snow
In situ	Snow	83	31	20	3
	No Snow	75	438	18	106

Table 5. Confusion matrices for comparison of in situ snow observations provided by snow depth and surface temperature observations with remotely sensed snow maps (MOD10A2 and Landsat 8 combined).

		In situ snow depth		In situ surface temperature	
		Snow	No snow	Snow	No snow
Remotely sensed	Snow	52	16	51	77
	No Snow	17	80	17	464

Table 6. Parameter value range prior to calibration and after calibration. The standard deviation of posterior parameter values is based on the standard deviation of all members.

Parameter	Prior (min–max)	Posterior mean	Posterior standard deviation
T_T	−6 to 2	−8.18	1.66
T_{lapse}	0.9 to 1.10	1.10	0.01
P	0.6 to 1.4	1.31	0.02
C_6	15 to 35	16.07	0.52

snow cover respectively. We find that the classification accuracy of MOD10A2 and Landsat 8 snow maps based on all in situ snow observations is 83.1 and 85.7 % respectively. The classification accuracy is defined as the number of correctly classified pixels divided by the total number of pixels. Table 4 shows the confusion matrices. Misclassification can be a result of variability of snow conditions within a pixel and classification of ice clouds or high cirrus clouds as snow (Parajka and Blöschl, 2006). Large viewing angles, and consequently larger observation areas, may also result in misclassification (Dozier et al., 2008). MOD10A2 has a lower spatial resolution than Landsat 8 which likely causes the slightly lower accuracy for the MOD10A2 snow cover product (Hall et al., 2002). Visual inspection of MOD10A2 snow maps also revealed that some clouds are erroneously mapped as snow cover.

The accuracy of MODIS daily snow cover products are reported to be 95 % for mountainous Austria (Parajka and Blöschl, 2006) and 94.2 % for the upper Rio Grande basin (Klein and Barnett, 2003). The lower accuracy presented in this study is likely a result of the simplification of the 8-day composite product and more extreme relief and consequently larger spatial variability in snow cover. Besides classification errors, uncertainty in the in situ snow observations should be considered as well. For the in situ snow cover observations provided by surface temperature, there are relatively many observations for which snow is not observed in situ, while the MOD10A2 and Landsat 8 snow maps indicate that snow should be present (Table 5). This may be caused by the fact that a thin snow layer may not result in sufficient isolation to reduce the diurnal temperature fluctuations for observation as snow (Lundquist and Lott, 2008). This observation bias in the temperature-sensed snow cover data would indicate that MOD10A2 and Landsat 8 snow maps possibly have even higher accuracies than presented here based on this validation approach.

3.2 Model calibration

The results of the sensitivity of mean snow extent and mean snow depth to parameter variability are shown in Table 2. The sensitivity analysis shows that the threshold temperature for melt onset (T_T), precipitation bias (P), temperature lapse rate bias (T_{lapse}) and the coefficient for conversion for viscosity (C_6) are the most sensitive parameters. For the snow com-

paction parameters, snow depth is most sensitive for changes in C_6, which is in agreement with Saloranta (2014). The melt parameters F_{SR} and F_T influence melt directly but show small sensitivity, as these parameters are dependent on each other. A higher value for F_T coincides with a lower value for F_{SR} where the value of both parameters is climate zone dependent (Ragettli et al., 2015).

Only the four most sensitive parameters were chosen to be calibrated by the EnKF to limit the degrees of freedom and to prevent the absence of convergence in the solutions for the parameters. Table 6 shows the prior and posterior parameter distribution resulting from the assimilation of snow extent per zone and snow depth. The parameter values for T_{lapse}, P and C_6 show a narrow posterior distribution (i.e., small standard deviation) indicating that parameter uncertainty is small. T_{lapse} and P represent measurement uncertainties of the model inputs. After calibration the modeled precipitation is increased and the temperature lapse rate is slightly steeper (more negative) than derived. The calibrated value of T_T shows a large standard deviation indicating absence of convergence in parameter solutions. This can be either a result of insufficient data to determine the parameter value or insensitivity of the model to the parameter value. A negative value for T_T is plausible as melt can occur with air temperatures below 0 °C when incoming shortwave radiation is sufficient. Especially at low latitudes and high elevation, solar radiation is an important cause of snowmelt (Bookhagen and Burbank, 2010). T_T is reported to be as negative as −6 °C for Pyramid Station, Nepalese Himalayas (Pellicciotti

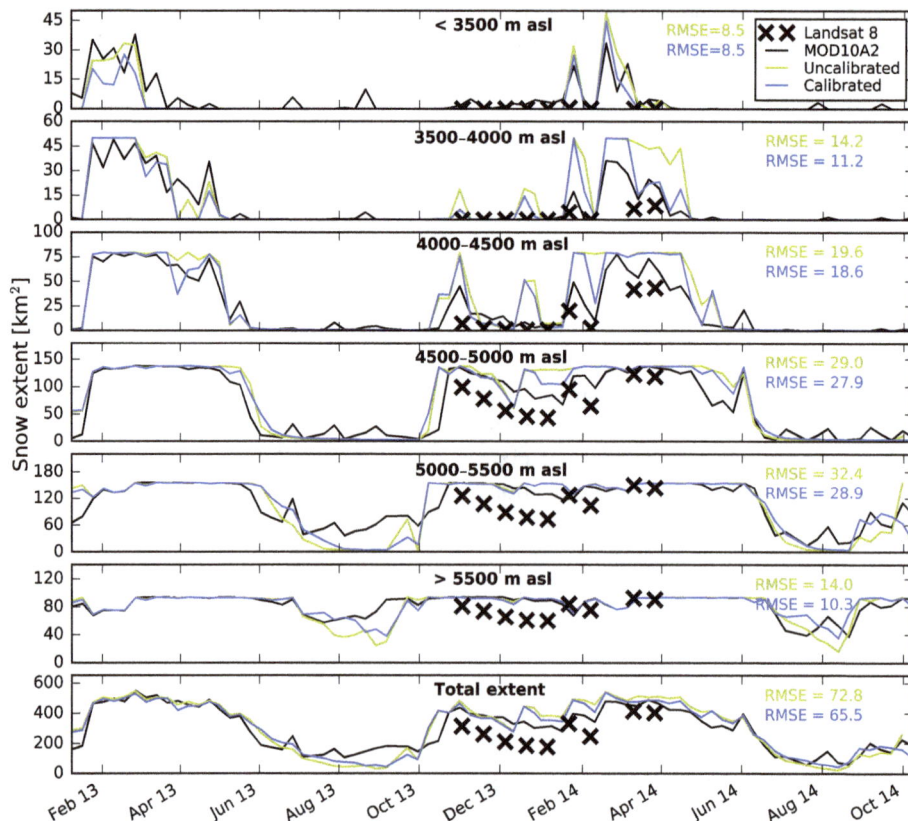

Figure 4. Modeled 8-day maximum snow extent before and after calibration (ensemble mean); Landsat 8 snow extent and MOD10A2 snow extent per elevation zone. The RMSE (km^2) is given per zone for the fit between modeled (before and after calibration) and MOD10A2 snow extent.

et al., 2012). Here T_T lies in a range which is even more negative than $-6\,°\mathrm{C}$. This is likely to be partly a result of the model structure. When T_T is negative the melt algorithm (Eq. 2) can give negative values. The temperature term in Eq. (2) becomes negative in case the air temperature is below zero degrees but higher than T_T. The reason for negative melt to occur in a few rare cases is a limitation of the EnKF calibration in combination with the enhanced temperature-index method. The EnKF does not allow us to constrain parameter ranges and this results in a relative low T_T, which may occasionally lead to negative melt when incoming shortwave radiation is low and the air temperature is above T_T. In those cases when negative melt occurs, it is capped to zero, and as a results the model is relatively insensitive for low temperatures close to the T_T and the EnKF does not converge into a parameter solution.

3.3 Model validation

3.3.1 Snow cover

Both the modeled and MOD10A2 snow extent show strong seasonality of snow cover in the catchment (Fig. 4). After calibration, modeled snow extent shows notable improvement in

elevation zone 3500–4000 m a.s.l. during the melt season in 2014. After calibration the threshold temperature for melt onset is lower, resulting in more and earlier onset of snowmelt. Consequently there is a decreased snow extent. The zones in the lower areas are expected to show most improvement, as this is the area where snow cover is ephemeral, and considerable improvements of the modeled snow extent in elevation zone 3500–4000 m a.s.l. are indeed observed (Fig. 4). The root mean square error (RMSE) decreased from 14.2 to 11.2 km^2 after calibration. The simulated snow extent agrees well with MOD10A2-observed snow cover for the higher elevation zones (> 4500 m a.s.l.). An exception is the snow extent in summer 2013 in the elevation zone 5000–5500 m a.s.l. The snow model underestimates the snow extent compared to the MOD10A2 snow extent. This discrepancy is possibly the result of (i) overestimation of simulated melt, (ii) an actual snow event that is simulated as rain by the model due to too-high air temperature or (iii) erroneous mapping of clouds as snow in the MOD10A2 snow cover.

The model classification accuracy of snow cover after calibration is 85.9 % based on pixel comparison between modeled 8-day maximum snow extent and MOD10A2 snow extent. The classification accuracy is the average classification accuracy over all members. There is only a slight increase of 0.2 % in accuracy after calibration; however, the performance was already high (85.7 %) before calibration. The classification accuracy is lower on steep slopes where avalanching is common, and as the snow extent in avalanching zones is highly dynamic, this is not well captured in the model. Calibration of parameters that influence avalanching might overcome this discrepancy to some degree; however, a more advanced approach to avalanche modeling may be required. In addition, the spatial resolution of the remotely sensed snow cover is likely to be insufficient to detect the avalanche dynamics. Other potential explanations for lower classification accuracies are uncertainties related to the simulated precipitation phase (rain or snow) and the simulated spatial distribution of precipitation based on Collier and Immerzeel (2015).

Landsat 8-derived snow extent is lower in winter than the modeled snow extent and the MOD10A2 snow extent (Fig. 4). Distinct differences between the Landsat 8 instantaneous snow cover observations and the MOD10A2 8-day maximum snow cover extents (Fig. 4) can be attributed to (i) the sensitivity of the Landsat 8 snow cover maps to misclassified snow pixels in the shaded area, (ii) the much higher spatial resolution of Landsat 8 (Hall et al., 2002), and (iii) the difference between an instantaneous image and an 8-day composite.

The model classification accuracy, based on pixel comparison with Landsat 8 snow maps, increased from 74.7 to 78.2 % after calibration. In Table 7 individual model classification accuracy is given based on comparison with each Landsat 8 snow map. Relative low accuracies occur in winter (especially on 20 December 2013 and 5 January 2014), and the model overestimates snow cover compared to the Landsat 8 snow maps (Fig. 4). The overestimation of snow cover by the model on 20 December 2013 is particularly large, and it can be explained by a small snow event (2.3 mm measured at Kyangjin) a few days before the acquisition. With below zero temperatures the model simulates a large snow cover extent, but based on a very small amount. Snow redistribution by wind, a patchy snow cover and/or sublimation may also explain the mismatch with the Landsat 8 snow cover in this particular case.

3.3.2 Snow depth

The observed and modeled snow depths at four locations are shown in Fig. 5. The simulated snow depth is given for the model simulations (i) without calibration, (ii) after calibration of snow extent, and (iii) after calibration of both snow extent and snow depth. After calibration with snow extent there is an increase in snow depth for Yala Pluvio and Yala BC for the entire snow season as result of increased simulated precipitation. For Langshisha and Kyangjin the snow depth mainly decreased after calibration with snow extent. These stations are at a lower elevation, and, since the threshold temperature for melt onset is lowered after calibration,

Table 7. Classification accuracy of modeled snow extent based on pixel comparison with Landsat 8 snow maps. Calibrated accuracies are averaged over all members and the standard deviation represents the standard deviation in individual member accuracies (after calibration).

Date (dd/mm/yy)	Accuracy uncalibrated	Accuracy calibrated	Standard deviation accuracy
	(%)		
02/11/13	80.96	84.41	0.12
18/11/13	78.43	79.15	0.11
04/12/13	77.41	77.10	0.05
20/12/13	54.97	60.38	0.08
05/01/14	63.46	67.07	0.07
20/01/14	74.30	81.33	0.04
06/02/14	65.55	73.24	0.05
10/03/14	84.94	89.67	0.05
26/03/14	87.03	86.90	0.04
11/04/14	80.29	82.92	0.05

this leads to reduced snow depth. At all locations the modeled snow depth decreased after calibration with both snow extent and snow depth due to lowering of the parameter relating snow density to snow depth. After calibration with both snow extent and snow depth, comparison of modeled and observed snow depth at Langshisha shows good agreement. Especially after calibration, the timing of the melt onset during spring is improved. For Yala Pluvio and Yala BC the agreement between modeled and observed snow depth is also good, though improvement of the timing of melt onset is limited. For Kyangjin the modeled snow depth does not agree as well with observed snow depth in spring 2013, but it improves in 2014. In spring the snow cover duration of snow events decreases after calibration and improves the fit with the observed snow depth.

Yala Pluvio and Yala BC are the only locations that serve as an independent validation of snow depth, as these stations are not used for the assimilation. The simulated melt onset in spring is later compared to what is observed. The diurnal variability of air temperature is high during the pre-monsoon season (March to mid-June; Immerzeel et al., 2014). Though daily average air temperatures are below zero, positive temperatures and snowmelt can occur in the afternoon above 5000 m a.s.l. (Shea et al., 2015; Ragettli et al., 2015). This can explain the difference between simulated and observed melt onset. Using an hourly time step might therefore improve the simulation of snowmelt in spring (Ragettli et al., 2015). While the timing of snowpack depletion at Yala Pluvio and Yala BC are offset from the observations, the modeled quantity of snow is in the same order of magnitude for both modeled and observed time series. Hence, there is no substantial overestimation or underestimation of snow depth. The RMSE between simulated and observed snow

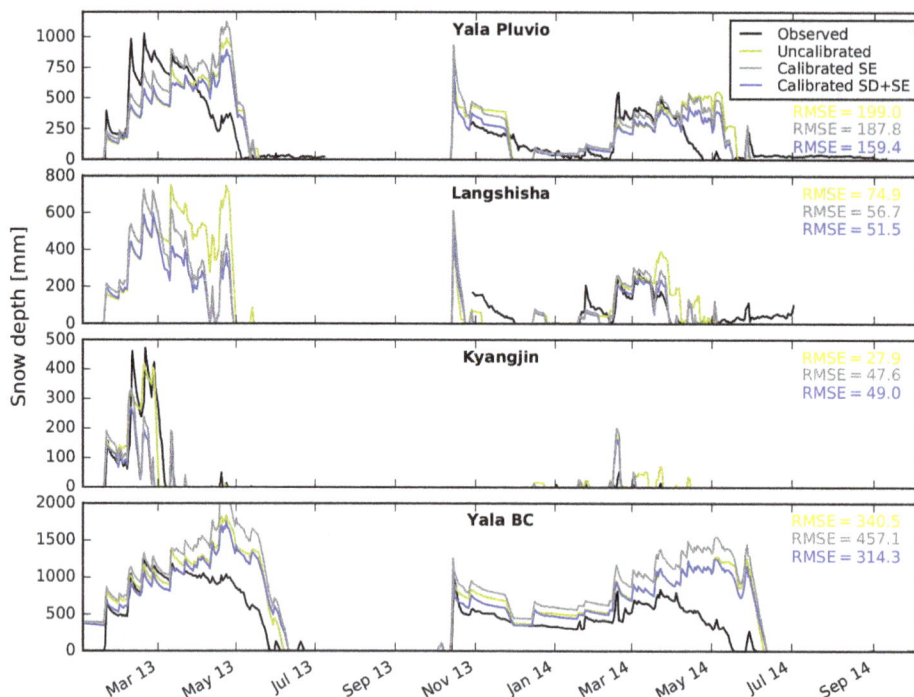

Figure 5. Observed snow depth and modeled snow depth (i) before calibration, (ii) after calibration of snow extent (SE), and (iii) after calibration of both snow extent and snow depth (SD + SE; ensemble mean) at three locations. The RMSE (mm) is given for the fit between modeled (before and after calibration) and observed snow depth.

depth decreases after calibration with both snow extent and snow depth compared to the uncalibrated simulation of snow depth and after calibration of only snow extent. This shows the benefit of assimilating both snow extent and snow depth into the snow model to obtain optimal parameter values.

While this study shows an approach in using snow depth observations for assimilation and validation, only four locations with snow depth observations were available. The number of available snow depth observations and the choice of different stations for assimilation might influence the results. Four snow depth observations are insufficient for systematic assimilation and independent validation. However, our approach is useful and is recommended for future studies in the Himalayas, in particular when more point observations of snow depth are available.

3.4 Climate sensitivity of SWE and snowmelt runoff

The cumulative basin-wide mean snowfall is 1222 mm for the simulation period. Nearly one-third (31.4 %) of the snowfall is transported to lower elevations due to avalanching, and 16.2 % of the snowfall is transported to elevations lower than 5000 m a.s.l. Transport of snow to lower elevations contributes to snowmelt runoff and has been estimated to be 4.5 % of the total water input for the upper part of the Langtang catchment (Ragettli et al., 2015).

The simulation of the SWE for the study period shows a pattern of increasing SWE with increasing elevation (Figs. 6

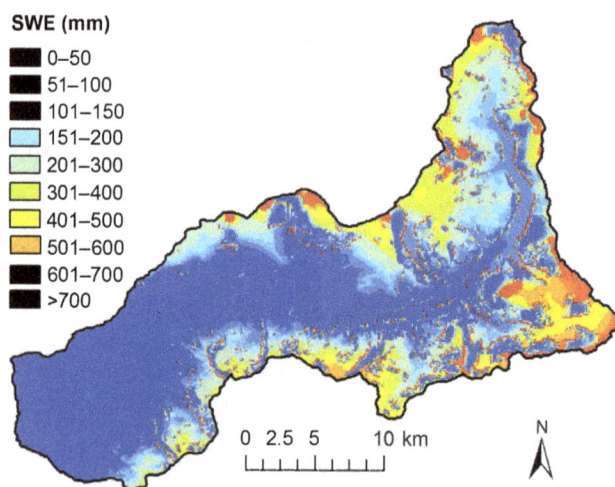

Figure 6. Spatial distribution of ensemble mean annual average snow water equivalent (SWE).

and 7). At higher elevation, air temperature is lower with more snow accumulation than melt, resulting in a higher gain in SWE over time. The glaciers Langtang and Langshisha are positioned at approximately the same elevation (Ragettli et al., 2015), though the SWE is considerably higher at the Langshisha glacier (Fig. 6) due to the precipitation distribution approach we use. Also, some areas at higher elevation

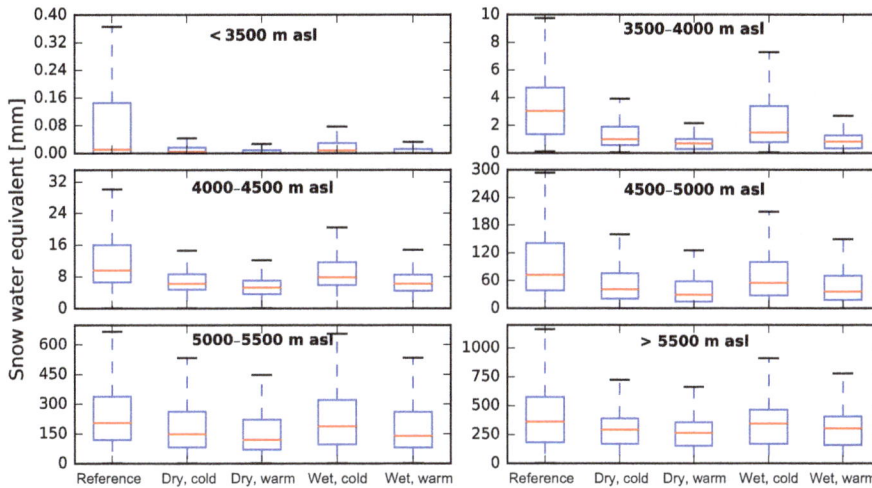

Figure 7. Boxplots of SWE per elevation zone averaged over the simulation period and all ensemble members for the study period (reference) and the four climate sensitivity tests (Table 3).

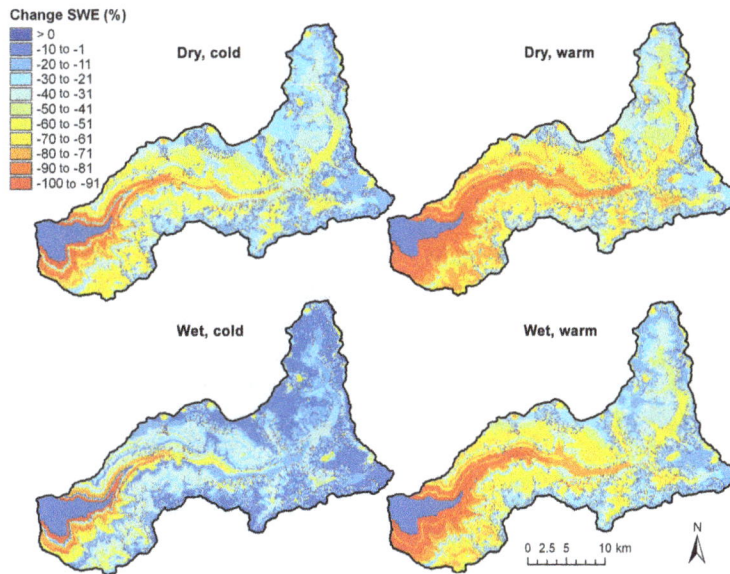

Figure 8. Change in SWE averaged over the simulation period and all members for each climate sensitivity test (Table 3).

show less SWE than surrounding areas at the same elevation. These areas represent the steep slopes in the catchment where avalanching occurs regularly. The transported snow accumulates below these steep slopes. The simulated avalanches are based on a simple model parameterization in which the snow is transported via single stream paths, resulting in a few pixels with extreme accumulation of SWE. This is mainly visible in the northeastern part of the catchment. Modeling the divergence of transported snow might improve the extreme accumulation simulated for some pixels.

For the climate sensitivity tests a delta-change method is used. This method has limitations as climate variability of future climate is not constant compared to the study period (Kobierska et al., 2013). In addition, Kobierska et al. (2013)

showed that changes in runoff due to climate change are predicted differently by a physically based snow model and a parameterized snow model for a glacierized catchment. Parameterized snow models (such as the modified seNorge snow model that is used in this study) are calibrated to fit the current climate and not future climate and might therefore be incapable of predicting future states of the snowpack. However, the scope of this study is to show the sensitivity of the SWE and snowmelt runoff to changes in air temperature and precipitation and not that of a full-fledged climate impact study. Therefore, the use of a parameterized snow model and the delta-change method is suitable in this case.

Figures 7 and 8 show the results of the absolute and relative change in SWE for different climate sensitivity tests.

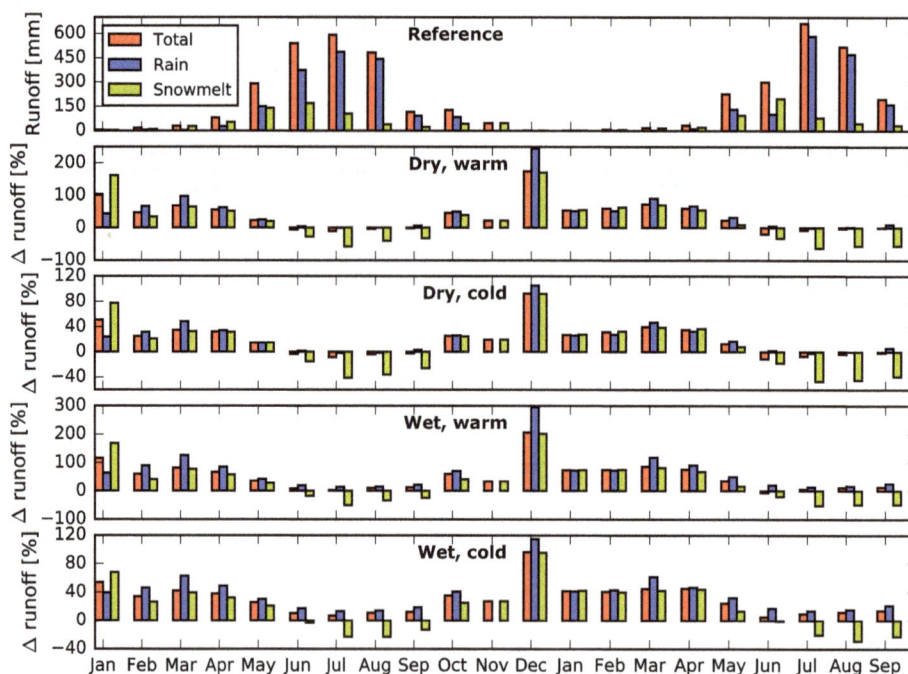

Figure 9. Modeled runoff at catchment outlet for the study period (January 2013–September 2014) and change in runoff compared to the study period for the climate sensitivity tests.

All climate sensitivity tests show a decrease in SWE, but the relative change is greatest at low elevations in the valley. We also observe a strong gradient of decreased relative change in SWE with increased elevation. An increase in temperature leads to an increase in melt and more precipitation in the form of rain instead of snow. Both processes result in decreased relative change of SWE with elevation. Near the catchment outlet there is an area with 100 % decrease in SWE, as precipitation will only fall as rain instead of snow.

A slight deviation from the elevational trend in SWE change occurs between 3000 and 4000 m a.s.l., which is a zone that could be sensitive to changes in the elevation at which snowfall occurs. The combination of snowfall at higher elevations due to higher temperature and the monthly differing spatial patterns in precipitation are likely to explain the banded patterns.

Changes in SWE and the spatial distribution of SWE will also be affected by changes in total precipitation. The influence of precipitation can be determined based on comparison of the two wet and dry climate sensitivity tests (Figs. 7 and 8). A decrease in precipitation results in decreased SWE as there is less snowfall. However, the increased precipitation for the wet/cold and wet/warm climate sensitivity tests (+12.1 and +12.4 % respectively) does not compensate for the temperature-related increase in melt and decrease in snowfall in the valley.

Reduced warming under the wet/cold climate sensitivity test results in a smaller decrease of SWE compared to the wet/warm climate sensitivity test, even in the valley. At

higher elevations, changes in SWE are weakly negative and in some areas positive. Snowpack sensitivity to temperature change decreases with elevation (Brown and Mote, 2009). The increased SWE under both wet climate sensitivity tests occurs in the southeastern part of the catchment where relatively large amounts of precipitation occur in winter (Collier and Immerzeel, 2015). Schmucki et al. (2015) showed similar results for the Alps. They showed that low- and mid-elevation stations are sensitive to temperature change but not to precipitation change. In contrast, at high-elevation stations an increase in precipitation partly compensates for an increase in temperature. The compensating effect of increased precipitation at high elevations is important for glacier systems and emphasizes the importance of accurate estimations of both change in precipitation and its spatial distribution.

The modeled snowmelt and rain runoff at the catchment outlet is greatest during the monsoon and lowest during winter (Fig. 9). Peak snowmelt and rain runoff occur in June and July respectively. The snowmelt season starts in March when temperatures and insolation are rising and continues until October. Snowmelt runoff contributes most to total runoff during pre-monsoon and early-monsoon (March–June), which is in agreement with Bookhagen and Burbank (2010). Validation of the simulated runoff with observed runoff was impossible, because (i) there were no reliable runoff data available for the study period, as there was no reliable rating curve, and (ii) the model focusses on rain and snowmelt runoff; however, glacier runoff and delay of runoff due to groundwater and glacier storage is not incorporated in the model structure.

The climate sensitivity of snowmelt and rain runoff is shown in Fig. 9. All climate sensitivity tests show an increase in snowmelt runoff from October to May. In contrast, snowmelt runoff decreases from June to September. Higher temperatures result in more snowmelt and less snowfall during winter and an early melt season which leads to a shift in the peak of snowmelt runoff. In other mountain regions similar changes in runoff patterns appear. Several studies in the Alps show that the peak in snowmelt runoff shifts from summer to late spring (Bavay et al., 2009, 2013; Kobierska et al., 2013). Immerzeel et al. (2009) showed that in the upper Indus Basin, the peak in snowmelt runoff appears 1 month earlier by 2071–2100 as result of an increase in temperature and precipitation. However, Immerzeel et al. (2012) showed that total snowmelt runoff remains more or less constant under positive temperature and precipitation trends in the upper part of the Langtang catchment. In their study snowmelt on glaciers is not defined as snowmelt runoff and is therefore a minor component of total runoff, leading to different results.

For the wet climate sensitivity tests, total runoff (i.e., the sum of snowmelt and rain runoff) increases throughout the year. The decrease in melt runoff during the late melt season is compensated by the increase in rain runoff as there is more precipitation. The future hydrology of the central Himalayas largely depends on precipitation changes, as it is dominated by rainfall runoff during the monsoon (Lutz et al., 2014). As we perturb the model with a percentage change in precipitation that is constant through the year, the absolute change in precipitation is greater in the monsoon than in winter. For climate sensitivity tests with decreased precipitation, total runoff from June to September decreases, but from October to May it increases as a result of increased snowmelt. Estimates of seasonal changes in precipitation are thus critical for determining whether rain and snowmelt runoff increases or decreases during monsoon.

4   Conclusions

Remotely sensed snow cover, in situ observations and a modified seNorge snow model were combined to estimate (climate sensitivity of) SWE and snowmelt runoff in the Langtang catchment. Validation of remotely sensed snow cover (Landsat 8 and MOD10A2 snow maps) shows high accuracies (85.7 and 83.1 % respectively) against in situ snow observations provided by surface temperature and snow depth measurements. Data assimilation of MOD10A2 snow cover and snow depth measurements using an EnKF proved to be successful for obtaining optimal model parameter values. Independent validations of simulated snow depth and snow cover against snow depth measurements and Landsat 8 snow cover show improvement after assimilation of snow depth and snow cover compared to results before data assimilation. The applied methodology of simultaneous assimilation of snow cover and snow depth allows for the calibration of

important snow parameters and validation of the snow depth rather than snow cover alone. This opens up new possibilities for future snow assessments and sensitivity studies in the Himalayas.

The spatial distribution of SWE averaged over the simulation period (January 2013–September 2014) shows a strong gradient of increasing SWE with increasing elevation. In addition, the SWE is considerably higher in the southeastern part of the catchment than the northeastern part of the catchment as a result of the spatial and temporal distribution of precipitation.

Finally the climate sensitivity study revealed that snowmelt runoff increases in winter and the early melt season (December–May) and decreases during the late melt season (June–September) as a result of the earlier onset of snowmelt due to increasing temperature. There is a strong relative decrease in SWE in the valley with increasing temperature due to more snowmelt and less precipitation as snow. At higher elevations an increase in precipitation partly compensates for increased melt due to higher temperatures. The compensating effect of precipitation emphasizes the importance and need for the accurate prediction of change in the spatial and temporal distribution of precipitation.

Competing interests. The authors declare that they have no conflict of interest.

Acknowledgements. This project was supported by funding from the European Research Council (ERC) under the European Union's Horizon 2020 research and innovation program (grant agreement no. 676819) and by the research programme VIDI with project number 016.161.308 financed by the Netherlands Organisation for Scientific Research (NWO). The authors thank Hendrik Wulf and two anonymous reviewers for their constructive comments that helped in improving the manuscript.

Edited by: Guillaume Chambon

References

Andreadis, K. M. and Lettenmaier, D. P.: Assimilating remotely sensed snow observations into a macroscale hydrology model, Adv. Water Resour., 29, 872–886, https://doi.org/10.1016/j.advwatres.2005.08.004, 2006.

Avanzi, F., Michele, C. De, Morin, S., Carmagnola, C. M., Ghezzi, A., and Lejeune, Y.: Model complexity and data requirements in snow hydrology: seeking a balance in practical applications, Hydrol. Process., 30, 2106–2118, https://doi.org/10.1002/hyp.10782, 2016.

Barnett, T. P., Adam, J. C., and Lettenmaier, D. P.: Potential impacts of a warming climate on water availability in snow-dominated regions, Nature, 438, 303–309, https://doi.org/10.1038/nature04141, 2005.

Barros, A. P., Kim, G., Williams, E., and Nesbitt, S. W.: Probing orographic controls in the Himalayas during the monsoon using satellite imagery, Nat. Hazards Earth Syst. Sci., 4, 29–51, https://doi.org/10.5194/nhess-4-29-2004, 2004.

Bavay, M., Lehning, M., Jonas, T., and Loewe, H.: Simulations of future snow cover and discharge in Alpine headwater catchments, Hydrol. Process., 23, 95–108, https://doi.org/10.1002/hyp.7195, 2009.

Bavay, M., Grünewald, T., and Lehning, M.: Response of snow cover and runoff to climate change in high Alpine catchments of Eastern Switzerland, Adv. Water Resour., 55, 4–16, https://doi.org/10.1016/j.advwatres.2012.12.009, 2013.

Bernhardt, M. and Schulz, K.: SnowSlide: A simple routine for calculating gravitational snow transport, Geophys. Res. Lett., 37, L11502, https://doi.org/10.1029/2010GL043086, 2010.

Bookhagen, B. and Burbank, D. W.: Toward a complete Himalayan hydrological budget: Spatiotemporal distribution of snowmelt and rainfall and their impact on river discharge, J. Geophys. Res., 115, F03019, https://doi.org/10.1029/2009JF001426, 2010.

Brock, B. W., Willis, I. C., and Sharp, M. J.: Measurement and parameterisation of albedo variations at Haut Glacier d'Arolla, Switzerland, J. Glaciol., 46, 675–688, 2000.

Brown, R. D. and Mote, P. W.: The response of Northern Hemisphere snow cover to a changing climate*, J. Climate, 22, 2124–2145, https://doi.org/10.1175/2008JCLI2665.1, 2009.

Charrois, L., Cosme, E., Dumont, M., Lafaysse, M., Morin, S., Libois, Q., and Picard, G.: On the assimilation of optical reflectances and snow depth observations into a detailed snowpack model, The Cryosphere, 10, 1021–1038, https://doi.org/10.5194/tc-10-1021-2016, 2016.

Clark, M. P., Slater, A. G., Barrett, A. P., Hay, L. E., Mccabe, G. J., Rajagopalan, B., and Leavesley, G. H.: Assimilation of snow covered area information into hydrologic and land-surface models and land-surface models, Adv. Water Resour., 29, 1209–1221, https://doi.org/10.1016/j.advwatres.2005.10.001, 2006.

Collier, E. and Immerzeel, W. W.: High-resolution modeling of atmospheric dynamics in the Nepalese Himalaya, J. Geophys. Res.-Atmos., 120, 9882–9896, https://doi.org/10.1002/2015JD023266, 2015.

Collier, E., Mölg, T., Maussion, F., Scherer, D., Mayer, C., and Bush, A. B. G.: High-resolution interactive modelling of the mountain glacier–atmosphere interface: an application over the Karakoram, The Cryosphere, 7, 779–795, https://doi.org/10.5194/tc-7-779-2013, 2013.

Dankers, R. and De Jong, S. M.: Monitoring snow-cover dynamics in Northern Fennoscandia with SPOT VEGETATION images, Int. J. Remote Sens., 25, 2933–2949, https://doi.org/10.1080/01431160310001618374, 2004.

Dong, J., Walker, J. P., and Houser, P. R.: Factors affecting remotely sensed snow water equivalent uncertainty, Remote Sens. Environ., 97, 68–82, https://doi.org/10.1016/j.rse.2005.04.010, 2005.

Dozier, J., Painter, T. H., Rittger, K., and Frew, J. E.: Time – space continuity of daily maps of fractional snow cover and albedo from MODIS, Adv. Water Resour., 31, 1515–1526, https://doi.org/10.1016/j.advwatres.2008.08.011, 2008.

Durand, M., Molotch, N. P., and Margulis, S. A.: Merging complementary remote sensing datasets in the context of snow water equivalent reconstruction, Remote Sens. Environ., 112, 1212–1225, https://doi.org/10.1016/j.rse.2007.08.010, 2008.

Evensen, G.: Sequential data assimilation with a nonlinear quasi-geostrophic model using Monte Carlo methods to forecast error statistics, J. Geophys. Res., 99, 143–162, 1994.

Evensen, G.: The Ensemble Kalman Filter?: theoretical formulation and practical implementation, Ocean Dynam., 53, 343–367, https://doi.org/10.1007/s10236-003-0036-9, 2003.

Feiken, H.: Dealing with biases: three archaeological approaches to the hidden landscapes of Italy, PhD thesis, Groningen University Library, Barkhuis, 2014.

Grünewald, T. and Lehning, M.: Are flat-field snow depth measurements representative? A comparison of selected index sites with areal snow depth measurements at the small catchment scale, Hydrol. Process., 29, 1717–1728, https://doi.org/10.1002/hyp.10295, 2015.

Gurung, D. R., Kulkarni, A. V., Giriraj, A., Aung, K. S., Shrestha, B., and Srinivasan, J.: Changes in seasonal snow cover in Hindu Kush-Himalayan region, The Cryosphere Discuss., 5, 755–777, https://doi.org/10.5194/tcd-5-755-2011, 2011.

Hall, D. K., Riggs, G. A., and Salomonson, V. V: Development of methods for mapping global snow cover using moderate resolution imaging spectroradiometer data, Remote Sens. Environ., 54, 127–140, https://doi.org/10.1016/0034-4257(95)00137-P, 1995.

Hall, D. K., Riggs, G. A., Salomonson, V. V., DiGirolamo, N. E., and Bayr, K. J.: MODIS snow-cover products, Remote Sens. Environ., 83, 181–194, https://doi.org/10.1016/S0034-4257(02)00095-0, 2002.

Hopfinger, E. J.: Snow Avalanche Motion and Related Phenomena, Annu. Rev. Fluid Mech., 15, 47–76, https://doi.org/10.1146/annurev.fl.15.010183.000403, 1983.

Immerzeel, W. W., Droogers, P., de Jong, S. M., and Bierkens, M. F. P.: Large-scale monitoring of snow cover and runoff simulation in Himalayan river basins using remote sensing, Remote Sens. Environ., 113, 40–49, https://doi.org/10.1016/j.rse.2008.08.010, 2009.

Immerzeel, W. W., van Beek, L. P. H. and Bierkens, M. F. P.: Climate change will affect the Asian water towers, Science, 328, 1382–1385, https://doi.org/10.1126/science.1183188, 2010.

Immerzeel, W. W., van Beek, L. P. H., Konz, M., Shrestha, A. B., and Bierkens, M. F. P.: Hydrological response to climate change in a glacierized catchment in the Himalayas, Climatic Change, 110, 721–736, https://doi.org/10.1007/s10584-011-0143-4, 2012.

Immerzeel, W. W., Pellicciotti, F., and Bierkens, M. F. P.: Rising river flows throughout the twenty-first century in two Himalayan glacierized watersheds, Nat. Geosci., 6, 742–745, https://doi.org/10.1038/ngeo1896, 2013. Hall

Immerzeel, W. W., Petersen, L., Ragettli, S., and Pellicciotti, F.: The importance of observed gradients of air temperature and precipitation for modeling runoff from a glacierized watershed in the Nepalese Himalaya, Water Resour. Res., 50, 2212–2226, https://doi.org/10.1002/2013WR014506, 2014.

Karssenberg, D., Schmitz, O., Salamon, P., de Jong, K., and Bierkens, M. F. P.: A software framework for construction of process-based stochastic spatio-temporal models and data assimilation, Environ. Model. Softw., 25, 489–502, https://doi.org/10.1016/j.envsoft.2009.10.004, 2010.

Klein, A. G. and Barnett, A. C.: Validation of daily MODIS snow cover maps of the Upper Rio Grande River Basin for the 2000-2001 snow year, Remote Sens. Environ., 86, 162–176, https://doi.org/10.1016/S0034-4257(03)00097-X, 2003.

Kobierska, F., Jonas, T., Zappa, M., Bavay, M., Magnusson, J., and Bernasconi, S. M.: Future runoff from a partly glacierized watershed in Central Switzerland?: A two-model approach, Adv. Water Resour., 55, 204–214, https://doi.org/10.1016/j.advwatres.2012.07.024, 2013.

Leisenring, M. and Moradkhani, H.: Snow water equivalent prediction using Bayesian data assimilation methods, Stoch. Env. Res. Risk A., 25, 253–270, https://doi.org/10.1007/s00477-010-0445-5, 2011.

Liu, Y., Peters-Lidard, C. D., Kumar, S., Foster, J. L., Shaw, M., Tian, Y., and Fall, G. M.: Assimilating satellite-based snow depth and snow cover products for improving snow predictions in Alaska, Adv. Water Resour., 54, 208–227, https://doi.org/10.1016/j.advwatres.2013.02.005, 2013.

Lundquist, J. D. and Lott, F.: Using inexpensive temperature sensors to monitor the duration and heterogeneity of snow-covered areas, Water Resour. Res., 44, W00D16, https://doi.org/10.1029/2008WR007035, 2008.

Lutz, A. F., Immerzeel, W. W., Shrestha, A. B., and Bierkens, M. F. P.: Consistent increase in High Asia's runoff due to increasing glacier melt and precipitation, Nature Climate Change, 4, 587–592, https://doi.org/10.1038/nclimate2237, 2014.

Lutz, A. F., Immerzeel, W. W., Litt, M., Bajracharya, S., and Shrestha, A. B.: Comprehensive Review of Climate Change and the Impacts on Cryosphere, Hydrological Regimes and Glacier Lakes, 2015.

Magnusson, J., Farinotti, D., Jonas, T., and Bavay, M.: Quantitative evaluation of different hydrological modelling approaches in a partly glacierized Swiss watershed, Hydrol. Process., 25, 2071–2084, https://doi.org/10.1002/hyp.7958, 2011.

Magnusson, J., Wever, N., Essery, R., Helbig, N., Winstral, A., and Jonas, T.: Evaluating snow models with varying process representations for hydrological applications: Snow model evaluation, Water Resour. Res., 51, 2707–2723, https://doi.org/10.1002/2014WR016498, 2015.

Magnusson, J., Winstral, A., Stordal, A. S., Essery, R., and Jonas, T.: Improving physically based snow simulations by assimilating snow depths using the particle filter, Water Resour. Res., 53, 1125–1143, https://doi.org/10.1002/2016WR019092, 2016.

Maskey, S., Uhlenbrook, S., and Ojha, S.: An analysis of snow cover changes in the Himalayan region using MODIS snow products and in-situ temperature data, Climatic Change, 108, 391–400, https://doi.org/10.1007/s10584-011-0181-y, 2011.

Molotch, N. P.: Reconstructing snow water equivalent in the Rio Grande headwaters using remotely sensed snow cover data and a spatially distributed snowmelt model, Hydrol. Process., 23, 1076–1089, https://doi.org/10.1002/hyp.7206, 2009.

Molotch, N. P. and Margulis, S. A.: Estimating the distribution of snow water equivalent using remotely sensed snow cover data and a spatially distributed snowmelt model: A multi-resolution, multi-sensor comparison, Adv. Water Resour., 31, 1503–1514, https://doi.org/10.1016/j.advwatres.2008.07.017, 2008.

Nagler, T., Rott, H., Malcher, P., and Müller, F.: Assimilation of meteorological and remote sensing data for snowmelt runoff forecasting, Remote Sens. Environ., 112, 1408–1420, https://doi.org/10.1016/j.rse.2007.07.006, 2008.

Parajka, J. and Blöschl, G.: Validation of MODIS snow cover images over Austria, Hydrol. Earth Syst. Sci., 10, 679–689, https://doi.org/10.5194/hess-10-679-2006, 2006.

Pellicciotti, F., Brock, B., Strasser, U., Burlando, P., Funk, M., and Corripio, J.: An enhanced temperature-index glacier melt model including the shortwave radiation balance: Development and testing for Haut Glacier d'Arolla, Switzerland, J. Glaciol., 51, 573–587, 2005.

Pellicciotti, F., Helbing, J., Rivera, A., Favier, V., Corripio, J., Araos, J., Sicart, J. E., and Carenzo, M.: A study of the energy balance and melt regime on Juncal Norte Glacier, semi-arid Andes of central Chile, using melt models of different complexity, Hydrol. Process., 22, 3980–3997, https://doi.org/10.1002/hyp.7085, 2008.

Pellicciotti, F., Buergi, C., Immerzeel, W. W., Konz, M., and Shrestha, A. B.: Challenges and Uncertainties in Hydrological Modeling of Remote Hindu Kush–Karakoram–Himalayan (HKH) Basins: Suggestions for Calibration Strategies, Mt. Res. Dev., 32, 39–50, https://doi.org/10.1659/MRD-JOURNAL-D-11-00092.1, 2012.

Putkonen, J. K.: Continuous Snow and Rain Data at 500 to 4400 m Altitude near Annapurna, Nepal, 1999–2001, Arct. Antarct. Alp. Res., 36, 244–248, 2004.

Ragettli, S., Pellicciotti, F., Bordoy, R., and Immerzeel, W. W.: Sources of uncertainty in modeling the glaciohydrological response of a Karakoram watershed to climate change, Water Resour. Res., 49, 6048–6066, https://doi.org/10.1002/wrcr.20450, 2013.

Ragettli, S., Pellicciotti, F., Immerzeel, W. W., Miles, E. S., Petersen, L., Heynen, M., Shea, J. M., Stumm, D., Joshi, S., and Shrestha, A.: Unraveling the hydrology of a Himalayan catchment through integration of high resolution in situ data and remote sensing with an advanced simulation model, Adv. Water Resour., 78, 94–111, https://doi.org/10.1016/j.advwatres.2015.01.013, 2015.

Saloranta, T. M.: Simulating snow maps for Norway: description and statistical evaluation of the seNorge snow model, The Cryosphere, 6, 1323–1337, https://doi.org/10.5194/tc-6-1323-2012, 2012.

Saloranta, T. M.: Simulating more accurate snow maps for Norway with MCMC parameter estimation method, The Cryosphere Discuss., 8, 1973–2003, https://doi.org/10.5194/tcd-8-1973-2014, 2014.

Saloranta, T. M.: Operational snow mapping with simplified data assimilation using the seNorge snow model, J. Hydrol., 538, 314–325, https://doi.org/10.1016/j.jhydrol.2016.03.061, 2016.

Schmucki, E., Marty, C., Fierz, C., and Lehning, M.: Simulations of 21st century snow response to climate change in Switzerland from a set of RCMs, Int. J. Climatol., 35, 3262–3273, https://doi.org/10.1002/joc.4205, 2015.

Shea, J. M., Wagnon, P., Immerzeel, W. W., Biron, R., Brun, F., Pellicciotti, F., Wagnon, P., Immerzeel, W. W., Biron, R., Brun, F., and Pellicciotti, F.: A comparative high-altitude meteorological analysis from three catchments in the

Nepalese Himalaya, Int. J. Water Resour. D., 31, 174–200, https://doi.org/10.1080/07900627.2015.1020417, 2015.

Tahir, A. A., Chevallier, P., Arnaud, Y., and Ahmad, B.: Snow cover dynamics and hydrological regime of the Hunza River basin, Karakoram Range, Northern Pakistan, Hydrol. Earth Syst. Sci., 15, 2275–2290, https://doi.org/10.5194/hess-15-2275-2011, 2011.

van Dam, O.: Forest filled with gaps: effects of gap size on water and nutrient cycling in tropical rain forest: a study in Guyana, PhD thesis, Tropenbos-Guyana Series, available at: https://dspace.library.uu.nl/handle/1874/532 (last access: 4 July 2017), 2001.

Wanders, N., De Jong, S. M., Roo, A., Bierkens, M. F. P., and Karssenberg, D.: The benefits of using remotely sensed soil moisture in parameter identification of large-scale hydrological models, Water Resour. Res., 50, 6874–6891, https://doi.org/10.1002/2013WR014639, 2013.

Warscher, M., Strasser, U., Kraller, G., Marke, T., Franz, H., and Kunstmann, H.: Performance of complex snow cover descriptions in a distributed hydrological model system: A case study for the high Alpine terrain of the Berchtesgaden Alps, Water Resour. Res., 49, 2619–2637, https://doi.org/10.1002/wrcr.20219, 2013.

Wolff, M. A., Isaksen, K., Petersen-Øverleir, A., Ødemark, K., Reitan, T., and Brækkan, R.: Derivation of a new continuous adjustment function for correcting wind-induced loss of solid precipitation: results of a Norwegian field study, Hydrol. Earth Syst. Sci., 19, 951–967, https://doi.org/10.5194/hess-19-951-2015, 2015.

Wulf, H., Bookhagen, B., and Scherler, D.: Differentiating between rain, snow, and glacier contributions to river discharge in the western Himalaya using remote-sensing data and distributed hydrological modeling, Adv. Water Resour., 88, 152–169, https://doi.org/10.1016/j.advwatres.2015.12.004, 2016.

9

Brief communication: Increasing shortwave absorption over the Arctic Ocean is not balanced by trends in the Antarctic

Christian Katlein[1], Stefan Hendricks[1], and Jeffrey Key[2]

[1] Alfred-Wegener-Institut Helmholtz-Zentrum für Polar- und Meeresforschung, 27570 Bremerhaven, Germany
[2] Center for Satellite Applications and Research, NOAA/NESDIS, Madison, Wisconsin, USA

Correspondence to: Christian Katlein (christian.katlein@awi.de)

Abstract. On the basis of a new, consistent, long-term observational satellite dataset we show that, despite the observed increase of sea ice extent in the Antarctic, absorption of solar shortwave radiation in the Southern Ocean poleward of 60° latitude is not decreasing. The observations hence show that the small increase in Antarctic sea ice extent does not compensate for the combined effect of retreating Arctic sea ice and changes in cloud cover, which both result in a total increase in solar shortwave energy deposited into the polar oceans.

1 Introduction

Changes in the Arctic and Antarctic cryosphere have been continuously monitored by different satellite programs since the 1970s. Arctic sea ice is becoming thinner (Haas et al., 2008) and younger (Maslanik et al., 2007) coupled with a decline in its extent (Serreze et al., 2007; Stroeve et al., 2012). This leads to a decrease in area-average sunlight reflection and thus to higher energy absorption in the Arctic Ocean (Perovich et al., 2011; Nicolaus et al., 2012). While some areas in Antarctica have also experienced a reduction of the sea ice cover, a modest overall gain of sea ice area has been observed in the Southern Hemisphere (Cavalieri et al., 1997; Stammerjohn et al., 2012), though there is some uncertainty in the magnitude of the trend (Eisenman et al., 2014). How these opposing trends relate to each other on a global scale is governed by a multitude of factors, such as the different latitudinal position of the ice cover and constraints by land masses, significant differences in the physical properties of the ice surface, and different forcing mechanisms from lower

latitudes (Meehl et al., 2016). A particular problem in relating Arctic and Antarctic trends and understanding their global impacts is the lack of a long-term, consistent observational dataset covering both poles.

The increased absorption of sunlight due to the loss of sea ice results in ocean warming, more ice loss, a decrease in albedo, and a further increase in absorbed sunlight. This is known as the ice–albedo feedback (Curry et al., 1995), a critical process in the global shortwave energy budget. Most of this added heat will be lost due to increased longwave emission during winter and will not be carried on into the next year. Especially in the Arctic, a longer sea ice melt season (Markus et al., 2009), thinner ice (Haas et al., 2008), and increased melt pond coverage (Rösel and Kaleschke, 2012) lead to increasing solar shortwave energy deposition in the ice–ocean system (Nicolaus et al., 2012), adding to the increase in absorption due to decreasing sea ice extent (Pistone et al., 2014). However, surface characteristics of Antarctic sea ice are less affected by global climate change (Allison et al., 1993; Brandt et al., 2005; Laine, 2008). Antarctic sea ice is mainly melting from below as the ice drifts away from the continent into warmer circumpolar waters, in contrast to the surface melting induced by melt ponding on Arctic sea ice. Therefore, sea ice extent losses in the Arctic are most pronounced during the Northern Hemisphere summer. In the Antarctic, the increasing extent of Antarctic sea ice is observed during the Southern Hemisphere winter, when the impact of sea ice cover on the shortwave energy balance is weaker.

Here we evaluate observations of the combined effect of different radiative processes in both hemispheres on the shortwave energy budget of the polar oceans. Our goal is to determine to what extent the increased absorption of solar shortwave energy caused by losses and other changes in Arctic summer sea ice are offset by potentially decreased absorption due to observed increases of sea ice extent in the Antarctic. The recently published Advanced Very High Resolution Radiometer (AVHRR) Polar Pathfinder – Extended (APP-x) dataset provides a novel tool to investigate this question on the global scale. It provides surface radiative properties and fluxes consistently derived twice daily (high and low sun) for both polar regions beginning in 1982 (Key et al., 2016). Its great advantage compared to earlier, individual products based on AVHRR data is that the integrated dataset inherently takes into account changes in cloud cover and albedo changes of various sources, allowing us to evaluate the actual shortwave energy deposition changes in the oceans poleward of 60° latitude. We calculated monthly averaged shortwave radiative fluxes into the ice–ocean system to estimate the partitioning of absorbed shortwave energy between sea ice and the unfrozen ocean surface.

2   Methods

The results presented here are based on version 1.0 of the AVHRR APP-x data. APP-x contains twice-daily data of many surface, cloud, and radiative properties retrieved at high-sun and low-sun times (04:00 and 14:00 local solar time for the Arctic; 02:00 and 14:00 for the Antarctic) from satellite data using a suite of algorithms and a radiative transfer model (Key et al., 2016). All variables are derived from the same set of satellite radiances, allowing an integrated view on the effects of sea ice changes. We use the variables ice thickness, surface albedo, cloud cover, and downwelling shortwave radiation at the surface. The APP-x record begins in 1982 and continues to the present day, though version 1.0 used in this study covers the period 1982–2014. Through validation studies, the APP-x variables used here have been determined to be of sufficient accuracy to be considered as climate data record variables. APP-x shortwave fluxes have been validated against observations from the SHEBA ice camp and the CERES satellite product (Riihelä et al., 2017). Details on the retrieval of the variables can be found in the Climate Algorithm Theoretical Basis Document (https://www1.ncdc.noaa.gov/pub/data/sds/cdr/CDRs/AVHRR_Extended_Polar_Pathfinder/AlgorithmDescription_01B-24b.pdf, Key and Wang, 2015). APP-x utilizes the Near-real-time Ice and Snow Extent (NISE) product from the National Snow and Ice Data Center (Boulder, Colorado, USA) for surface type identification.

The energy flux absorbed by the surface was calculated as

$$E = E_{\text{down}} \cdot (1 - \alpha) \qquad (1)$$

and multiplied by 12 h and the grid cell size to obtain the total amount of absorbed energy, where E_{down} inherently accounts for cloud cover. All grid cells with ice thickness greater than 0 were considered as ice covered. The APP-x data product does not contain a separate field for ice concentration, but ice thickness is only calculated for cells with an ice concentration > 15 % in the NISE product. Sea ice extent was calculated as the number of ice-covered grid cells multiplied by the cell size (25×25 km). This yields slightly larger numbers than comparable analyses on the direct basis of passive microwave sea ice concentration products with higher resolution, but the magnitude of changes proved to be unaffected.

Shortwave energy fluxes were calculated for the twice-daily data and averaged over each month to reduce the influence of retrieval errors and intermittent gaps in the data. For the calculation of total absorbed energy, twice-daily data were summed up to monthly values. For grid cells with invalid retrievals due to low light during winter, shortwave fluxes were set to 0, and the albedo was set to 1. Monthly data of average energy flux and total absorbed energy were then used for annual and long-term averages as well as for time series analysis. Data for the year 1994 were excluded from time series analysis due to a significant gap in the observations. Trends were calculated through linear regression using the MATLAB curve-fitting toolbox. All trends presented as significant have confidence levels above 95 %. The given error intervals for trends are 95 % confidence intervals.

3   Results

An analysis of the dataset revealed a decrease in September sea ice extent of $-0.126(\pm 0.03)$ million $\text{km}^2\,\text{year}^{-1}$ for the Northern Hemisphere and an insignificant increase of $0.012(\pm 0.02)$ million $\text{km}^2\,\text{year}^{-1}$ in the Southern Hemisphere summer in March (Fig. 1a). Antarctic sea ice extent increased $0.020(\pm 0.01)$ million $\text{km}^2\,\text{year}^{-1}$ in September during Southern Hemisphere winter, while winter sea ice loss in the Arctic is weaker in March with a loss of $-0.041(\pm 0.01)$ million $\text{km}^2\,\text{year}^{-1}$. Arctic sea ice extent losses are thus 5 times larger in magnitude than the small increases in ice area in the Antarctic, leading to a combined total sea ice loss of -0.106 million $(\pm 0.03)\,\text{km}^2\,\text{year}^{-1}$ in September and $-0.028(\pm 0.025)$ million $\text{km}^2\,\text{year}^{-1}$ in March. This reproduces the known global net loss of sea-ice-covered area during the last few decades (Stammerjohn and Smith, 1997; Stammerjohn et al., 2012; Stroeve et al., 2012). Thus, the slight gains in Antarctic sea ice area do not compensate for the areal loss of sea ice in the Arctic. Given the differences in the latitudinal distribution of Arctic and Antarctic sea ice, however, the impact of these changes on the absorption of solar radiation warrants further investigation.

The summer mean daytime albedo for ice-covered areas in the APP-x dataset was 0.34 for the Arctic (June–August) and 0.41 for the Antarctic (December–February), which compares well to earlier studies (Allison et al., 1993). The higher albedo of Antarctic sea ice may be caused mainly by a thicker

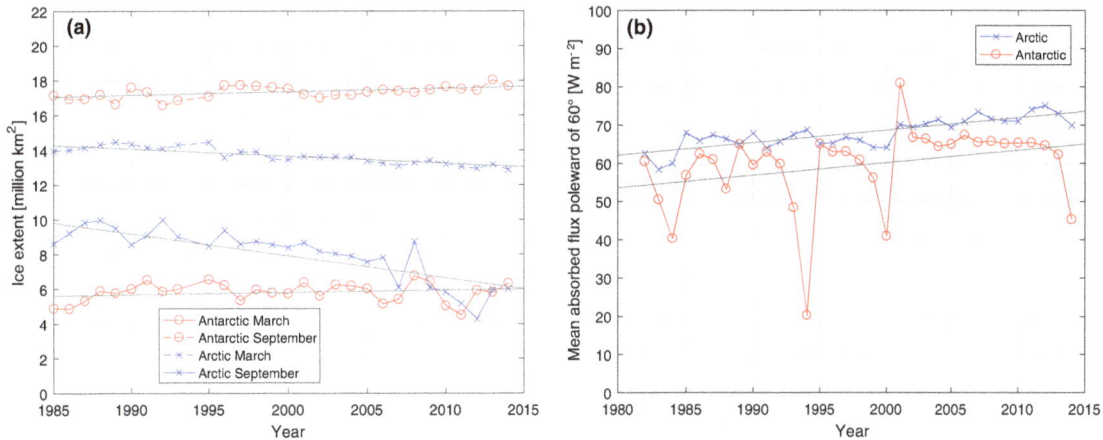

Figure 1. (a) Long-term trends of sea ice extent: temporal evolution of sea ice extent in Antarctica (red) and the Arctic (blue) for minimal (solid line) and maximal (dashed line) seasonal extent as derived from APP-x data. Grey lines indicate fitted linear trends. **(b)** Annual mean flux into ice–ocean system: shortwave radiative flux absorbed by the ice–ocean system poleward of 60° latitude in the Arctic (crosses) and Antarctic (circles).

snow cover with little surface melt and consequently the lack of meltwater pond formation. In accordance with the observed trend towards younger, predominantly seasonal Arctic sea ice with larger melt pond coverage, the APP-x dataset shows a decrease of the mean Arctic summer sea ice albedo, while in the Antarctic albedo trends show regional differences driven by the changes in ice concentration (Fig. 2). In this analysis of albedo trends, we only consider summer daytime albedos, as the albedos retrieved during wintertime are questionable due to low light levels and observation gaps. This does not significantly affect our analysis of energy fluxes, as the largest uncertainty in the albedo occurs with low fluxes, subsequently leading to a low energy flux uncertainty.

Antarctic sea ice exists mainly in the latitude zone between 55 and 77° S. In contrast, Arctic sea ice occupies the region between approximately 70 and 90° N. Due to its generally higher snow cover (Massom et al., 2001) and the 20 % higher albedo as well as its location at lower latitudes with higher shortwave insolation, the presence of sea ice does have a stronger impact on the local shortwave energy balance in the Antarctic. Mean annual shortwave energy uptake by the ice–ocean system poleward of 60° latitude was calculated twice daily from APP-x surface albedo and incoming solar radiation at the surface and averaged for each month. The use of these APP-x quantities inherently accounts for trends in cloud cover and surface albedo changes.

The mean annual shortwave energy flux into the ice–ocean system poleward of 60° accounts for 68 W m^{-2} in the Arctic, and 60 W m^{-2} in the Antarctic (Fig. 1b). Average Southern Hemisphere absorption shows high interannual variability throughout the satellite record, while the absorbed flux of the ice–ocean system in the Arctic clearly increased by 0.48 W m^{-2} yr^{-1}. While the trend in the absorption of solar

shortwave energy is more uniform across the Arctic Ocean due to uniform sea ice changes, there are large regional differences in the Antarctic (Fig. 2c, d). More solar energy is absorbed in the Bellingshausen and Amundsen seas, with smaller areas of decreasing energy absorption.

Combining the effects of cloud-cover-induced insolation changes, reduced sea ice extent, and a lower surface albedo, the total annual shortwave energy absorbed by the ice–ocean system north of 60° N increased at a rate of 1.7($\pm$0.5) $\times$ 10^{20} J yr^{-1}. In the Southern Hemisphere energy absorption by the ice–ocean system south of 60° S also increased at a very similar rate of 1.8($\pm$1.9) $\times$ 10^{20} J yr^{-1}, but due to the large interannual variability the trend is not statistically significant. Despite the increasing winter sea ice extent in the south, both hemispheres show a distinct increase in energy deposition in the ice–ocean system leading to ice melt and ocean warming. Increased ice extent in the Antarctic therefore does not decrease annual mean energy absorption as might be expected. An analysis of anomalies in sea ice extent, albedo, and shortwave energy deposition in the ice–ocean system shows that energy deposition is not directly correlated with the ice extent and albedo anomalies in general but can be offset by changes in cloud cover leading to increasing shortwave energy absorption in spite of albedo increases (Fig. 3).

In the Antarctic, the energy flux anomaly does not exhibit a strong relationship with ice extent anomalies overall (Fig. 3b, d), while in the Arctic anomalies in albedo and the resulting heat input into the ice–ocean system are much more closely related to the sea ice extent anomaly (Fig. 3a, c, e). Thus, in the context of the surface shortwave radiation balance, losses in Arctic sea ice and the resulting increase in solar energy absorption are not balanced by the moderate gains in sea ice extent in the Antarctic. On the global scale, changes

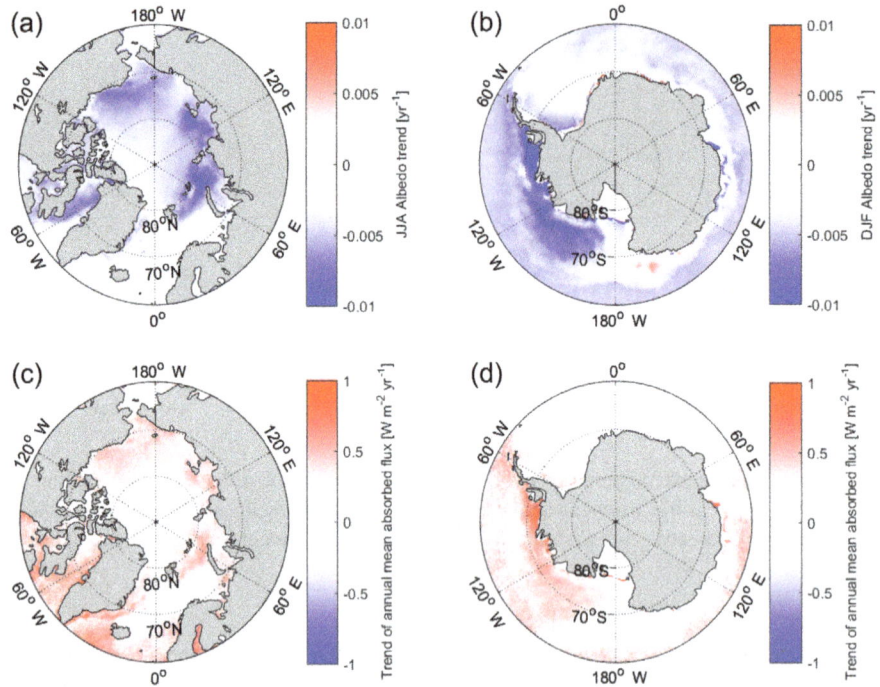

Figure 2. Spatial distribution of trends in surface albedo and overall energy deposition into the polar oceans: trends (yr^{-1}) of mean daytime summer sea ice albedo in the Arctic (**a**) and Antarctic (**b**) and trends ($W\,m^{-2}\,yr$) of shortwave energy flux (**c, d**) absorbed by the ice–ocean system. The latter inherently includes both changes in albedo and cloud cover due to the nature of the APP-x dataset.

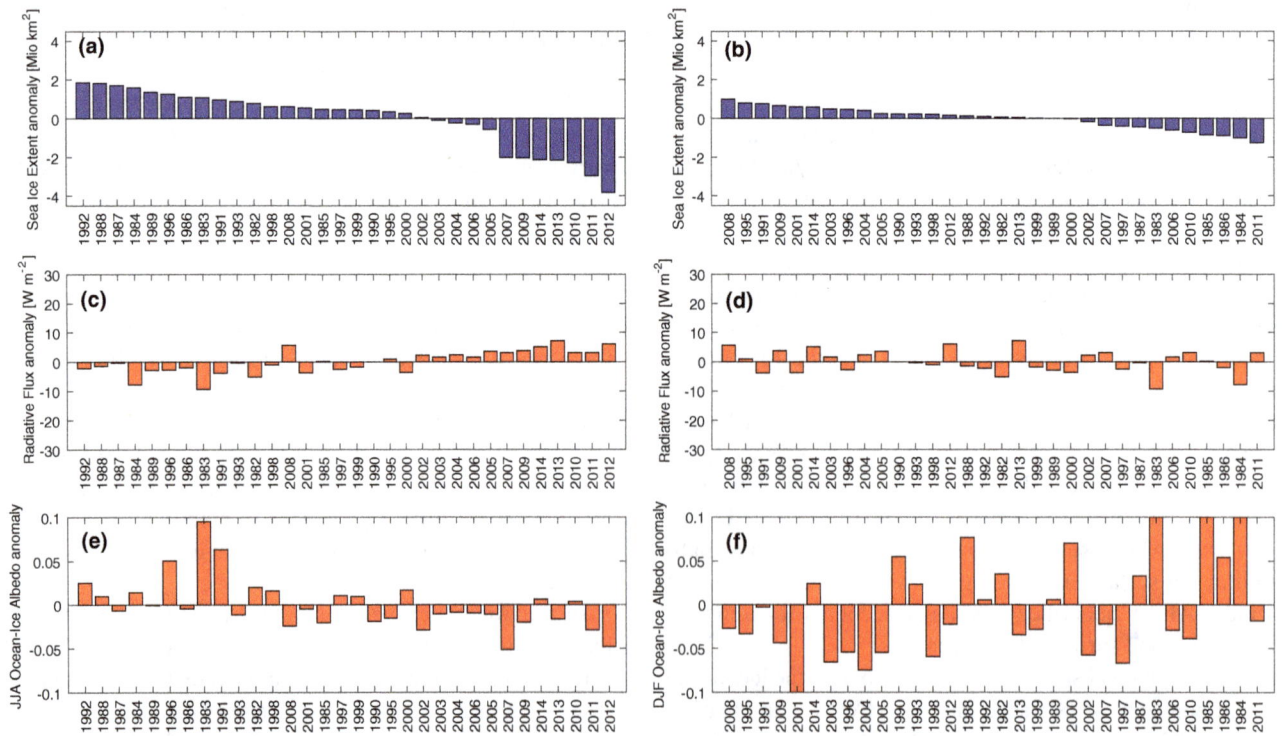

Figure 3. Anomaly of mean absorbed shortwave flux and surface albedo sorted from positive to negative sea ice extent anomaly: Arctic (**a, c, e**) and Antarctic (**b, d, f**) sea ice extent anomaly (**a, b**), annual anomaly of shortwave flux absorbed by the ice–ocean system poleward of 60° (**c, d**), and summer albedo anomaly (**e, f**).

in the shortwave energy partitioning in the polar oceans poleward of 60° latitude lead to a combined increased energy deposition of $3.5(\pm 2.1) \times 10^{20}\,\text{J yr}^{-1}$ comprised of positive energy absorption trends in both hemispheres despite moderate increases in Antarctic sea ice extent.

When extending our analysis from the oceans to all land and ocean areas poleward of 60° latitude, the result is similar with an increasing flux of $0.3\,\text{W m}^{-2}$ per year absorbed by the planet's surface poleward of 60°. This trend is somewhat weaker than over the ocean alone, as changes in land cover properties are not as pronounced as changes in ice extent and properties. Still, changing snow cover and prolonged melting seasons cause more absorption of sunlight and further heating of the climate system.

4 Conclusions

In conclusion, a consistent, long-term observational, satellite-based time series shows that changes in the shortwave energy budget caused by a decreasing Arctic sea ice cover are not balanced by the slight increases observed in Antarctic sea ice extent. Significant increases in Antarctic sea ice only occur during the Southern Hemisphere winter and thus have only a minor impact on the energy balance, while Arctic sea ice changes are accompanied by a spatially uniform decrease of sea ice albedo during the summer, further increasing the energy input to the northern polar ocean and thereby strengthening the ice–albedo feedback. This demonstrates the different roles that partitioning of solar shortwave radiation plays in the two very different sea ice zones of our planet. It is necessary to better understand the influence of the changes in different processes such as sea ice distribution, sea ice albedo, and cloud properties on the shortwave radiation budget in these two very different polar regions.

Author contributions. CK performed the calculations, prepared the figures, and wrote the text; SH had the initial idea for this study and contributed to the setup of the study; and JK contributed the APP-x data and provided insight into its use. All authors were involved in the writing of the manuscript.

Competing interests. The authors declare that they have no conflict of interest.

Acknowledgements. Christian Katlein is supported by the Helmholtz infrastructure initiative FRontiers in Arctic marine Monitoring (FRAM), and Stefan Hendricks received support from the European Space Agency. Stefan Hendricks was funded by the German Ministry of Economics and Technology (grant 50EE1008). We thank Xuanji Wang and Yinghui Liu of the Cooperative Institute for Meteorological Satellite Studies (CIMSS), University of Wisconsin–Madison, for their extensive work on the APP and APP-x climate data records, which was supported by the NOAA Climate Data Records Program. This study was funded by the Alfred-Wegener-Institut Helmholtz-Zentrum für Polar- und Meeresforschung. The views, opinions, and findings contained in this report are those of the authors and should not be construed as an official National Oceanic and Atmospheric Administration or U.S. Government position, policy, or decision.

Edited by: Dirk Notz

References

Allison, I., Brandt, R. E., and Warren, S. G.: East Antarctic sea ice: Albedo, thickness distribution, and snow cover, J. Geophys. Res.-Oceans, 98, 12417–12429, 1993.

Brandt, R. E., Warren, S. G., Worby, A. P., and Grenfell, T. C.: Surface Albedo of the Antarctic Sea Ice Zone, J. Climate, 18, 3606–3622, 2005.

Cavalieri, D. J., Gloersen, P., Parkinson, C. L., Comiso, J. C., and Zwally, H. J.: Observed Hemispheric Asymmetry in Global Sea Ice Changes, Science, 278, 1104–1106, 1997.

Curry, J. A., Schramm, J. L., and Ebert, E. E.: Sea Ice-Albedo Climate Feedback Mechanism., J. Climate, 8, 240–247, 1995.

Eisenman, I., Meier, W. N., and Norris, J. R.: A spurious jump in the satellite record: has Antarctic sea ice expansion been overestimated?, The Cryosphere, 8, 1289–1296, https://doi.org/10.5194/tc-8-1289-2014, 2014.

Haas, C., Pfaffling, A., Hendricks, S., Rabenstein, L., Etienne, J.-L., and Rigor, I.: Reduced ice thickness in Arctic Transpolar Drift favors rapid ice retreat, Geophys. Res. Lett., 35, L17501, https://doi.org/10.1029/2008GL034457, 2008.

Key, J. and Wang, X.: Climate Algorithm Theoretical Basis Document, Extended AVHRR Polar Pathfinder (APP-x), NOAA/NESDIS Center for Satellite Applications and Research and the National Climatic Data Center, 85 pp., 2015.

Key, J., Wang, X., Liu, Y., Dworak, R., and Letterly, A.: The AVHRR Polar Pathfinder Climate Data Records, Remote Sensing, 8, 197, https://doi.org/10.3390/rs8030167, 2016.

Laine, V.: Antarctic ice sheet and sea ice regional albedo and temperature change, 1981–2000, from AVHRR Polar Pathfinder data, Remote Sens. Environ., 112, 646–667, 2008.

Markus, T., Stroeve, J. C., and Miller, J.: Recent changes in Arctic sea ice melt onset, freezeup, and melt season length, J. Geophys. Res.-Oceans, 114, C12024, https://doi.org/10.1029/2009JC005436, 2009.

Maslanik, J. A., Fowler, C., Stroeve, J., Drobot, S., Zwally, J., Yi, D., and Emery, W.: A younger, thinner Arctic ice cover: Increased potential for rapid, extensive sea-ice loss, Geophys. Res. Lett., 34, L24501, https://doi.org/10.1029/2007GL032043, 2007.

Massom, R. A., Eicken, H., Hass, C., Jeffries, M. O., Drinkwater, M. R., Sturm, M., Worby, A. P., Wu, X., Lytle, V. I., Ushio, S.,

Morris, K., Reid, P. A., Warren, S. G., and Allison, I.: Snow on Antarctic sea ice, Rev. Geophys., 39, 413–445, 2001.

Meehl, G. A., Arblaster, J. M., Bitz, C. M., Chung, C. T. Y., and Teng, H.: Antarctic sea-ice expansion between 2000 and 2014 driven by tropical Pacific decadal climate variability, Nat. Geosci., 9, 590–595, https://doi.org/10.1038/ngeo2751, 2016.

Nicolaus, M., Katlein, C., Maslanik, J., and Hendricks, S.: Changes in Arctic sea ice result in increasing light transmittance and absorption, Geophys. Res. Lett., 39, L24501, https://doi.org/10.1029/2012GL053738, 2012.

Perovich, D. K., Jones, K. F., Light, B., Eicken, H., Markus, T., Stroeve, J., and Lindsay, R.: Solar partitioning in a changing Arctic sea-ice cover, Ann. Glaciol., 52, 192–196, 2011.

Pistone, K., Eisenman, I., and Ramanathan, V.: Observational determination of albedo decrease caused by vanishing Arctic sea ice, P. Natl. Acad. Sci. USA, 111, 3322–3326, 2014.

Riihelä, A., Key, J. R., Meirink, J. F., Kuipers Munneke, P., Palo, T., and Karlsson, K.-G.: An intercomparison and validation of satellite-based surface radiative energy flux estimates over the Arctic, J. Geophys. Res.-Atmos., 122, 4829–4848, 2017.

Rösel, A. and Kaleschke, L.: Exceptional melt pond occurrence in the years 2007 and 2011 on the Arctic sea ice revealed from MODIS satellite data, J. Geophys. Res., 117, C05018, https://doi.org/10.1029/2011JC007869, 2012.

Serreze, M. C., Holland, M. M., and Stroeve, J.: Perspectives on the Arctic's Shrinking Sea-Ice Cover, Science, 315, 1533–1536, 2007.

Stammerjohn, S., Massom, R., Rind, D., and Martinson, D.: Regions of rapid sea ice change: An interhemispheric seasonal comparison, Geophys. Res. Lett., 39, https://doi.org/10.1029/2012GL050874, 2012.

Stammerjohn, S. E. and Smith, R. C.: Opposing Southern Ocean Climate Patterns as Revealed by Trends in Regional Sea Ice Coverage, Climatic Change, 37, 617–639, 1997.

Stroeve, J. C., Serreze, M. C., Holland, M. M., Kay, J. E., Malanik, J., and Barrett, A. P.: The Arctic's rapidly shrinking sea ice cover: a research synthesis, Climatic Change, 110, 1005–1027, 2012.

10

A weekly Arctic sea-ice thickness data record from merged CryoSat-2 and SMOS satellite data

Robert Ricker[1,2], Stefan Hendricks[1], Lars Kaleschke[3], Xiangshan Tian-Kunze[3], Jennifer King[4], and Christian Haas[1,5]

[1]Alfred Wegener Institute, Helmholtz Centre for Polar and Marine Research, Bremerhaven, Bussestrasse 24, 27570 Bremerhaven, Germany
[2]Univ. Brest, CNRS, IRD, Ifremer, Laboratoire d'Oceanographie Physique et Spatiale (LOPS), IUEM, 29280 Brest, France
[3]Institute of Oceanography, University of Hamburg, Bundesstrasse 53, 20146 Hamburg, Germany
[4]Norwegian Polar Institute, Tromsø, Norway
[5]Department of Earth and Space Sciences and Engineering, York University, Toronto, ON, Canada

Correspondence to: Robert Ricker (Robert.Ricker@awi.de)

Abstract. Sea-ice thickness on a global scale is derived from different satellite sensors using independent retrieval methods. Due to the sensor and orbit characteristics, such satellite retrievals differ in spatial and temporal resolution as well as in the sensitivity to certain sea-ice types and thickness ranges. Satellite altimeters, such as CryoSat-2 (CS2), sense the height of the ice surface above the sea level, which can be converted into sea-ice thickness. Relative uncertainties associated with this method are large over thin ice regimes. Another retrieval method is based on the evaluation of surface brightness temperature (TB) in L-band microwave frequencies (1.4 GHz) with a thickness-dependent emission model, as measured by the Soil Moisture and Ocean Salinity (SMOS) satellite. While the radiometer-based method looses sensitivity for thick sea ice (> 1 m), relative uncertainties over thin ice are significantly smaller than for the altimetry-based retrievals. In addition, the SMOS product provides global sea-ice coverage on a daily basis unlike the altimeter data. This study presents the first merged product of complementary weekly Arctic sea-ice thickness data records from the CS2 altimeter and SMOS radiometer. We use two merging approaches: a weighted mean (WM) and an optimal interpolation (OI) scheme. While the weighted mean leaves gaps between CS2 orbits, OI is used to produce weekly Arctic-wide sea-ice thickness fields. The benefit of the data merging is shown by a comparison with airborne electromagnetic (AEM) induction sounding measurements. When compared to airborne thickness data in the Barents Sea, the merged product has a root mean square deviation (RMSD) of about 0.7 m less than the CS2 product and therefore demonstrates the capability to enhance the CS2 product in thin ice regimes. However, in mixed first-year (FYI) and multiyear (MYI) ice regimes as in the Beaufort Sea, the CS2 retrieval shows the lowest bias.

1 Introduction

Sea ice affects many climate-related processes, such as heat transfer between ocean and atmosphere or ocean circulation, but also marine operations (Meier et al., 2014). For decades, the variability and changes of the ice-covered region have been routinely observed by satellite remote sensing of sea-ice extent and area. However, the thickness of sea ice is a crucial parameter for the ice mass balance and is more difficult to observe. Recent satellite altimeter missions such as ICE-Sat or CryoSat-2 (CS2) demonstrated the capability to provide Arctic sea-ice thickness and volume estimates (Kwok et al., 2009; Laxon et al., 2013). They are used to measure freeboard (Fb), the height of the ice or snow surface above the water level, which can be converted into sea-ice thickness assuming hydrostatic equilibrium. CS2 was launched in 2010 and was primarily designed to measure the thickness of thick, perennial ice, but can also be used to retrieve first-year ice (FYI) thickness (Laxon et al., 2013). Nonetheless, the retrieval method shows considerable uncertainties over thin ice

Table 1. Summary of properties of input and output sea-ice thickness products in this study, including CryoSat-2 (CS2), SMOS, the weighted mean (WM) and the OI product (CS2SMOS).

Product	Temporal res.	Spatial res.	Coverage	Notes and applicability
CS2 (monthly)	1 month	25 km	Arctic wide	For studies of multiyear ice and thick first-year ice (> 1 m), high uncertainties for thin ice and in the marginal ice zone, constraints in regions where a snow climatology is inadequate
CS2 (weekly)	1 week	25 km	Gaps between orbits, sparse at lower latitudes	For studies of multiyear ice and thick first-year ice (> 1 m) where measurements are available, high uncertainties for thin ice (< 1 m) and in the marginal ice zone, constraints in regions where a snow climatology is inadequate
SMOS	1 day	12.5 km	Arctic wide	For studies of thin ice (< 1 m)
WM	1 week	25 km	Gaps between CS2 orbits	For studies of multiyear ice and of thin ice, where measurements are available
CS2SMOS	1 week	25 km	Arctic wide	For Arctic-wide studies on the entire thickness range, uses optimal interpolation

regimes and certainly in the marginal ice zones (Wingham et al., 2006; Ricker et al., 2014). On the other hand, the Soil Moisture and Ocean Salinity (SMOS) mission, launched in 2009, provides brightness temperature (TB) observations at microwave frequencies (L band), which can be exploited for thin ice thickness retrieval (Kaleschke et al., 2012).

Kaleschke et al. (2010) and Kaleschke et al. (2015) demonstrated the complementary nature of the relative uncertainties of CS2 and SMOS ice thickness retrieval methods. The CryoSat-2 sea-ice thickness product relies on accurate measurements of the height of the sea-ice surface above the water level, and therefore relative uncertainties are larger over thin ice (< 1 m). In contrast, the SMOS sea-ice thickness retrieval relies on the sensitivity of the brightness temperature to sea-ice thickness. While accuracy is high over thin ice, sensitivity gets lost over thick ice (> 1 m). Moreover, both sensor concepts have significantly different swath widths and revisit times and therefore provide different update rates of sea-ice thickness observations. Kaleschke et al. (2015) suggest that due to their different spatiotemporal sampling and resolution, and because of the complementary uncertainty due to the fundamental difference of the radiometric and altimetric measurement principle, a combination of both products has the capability to reduce uncertainties in relation to the individual products.

The spatial and interannual variability of sea-ice thickness is driven by dynamics and thermodynamics (Zhang et al., 2000; Kwok and Cunningham, 2016). For an accurate description of the Arctic sea-ice thickness distribution, it is necessary that thick and deformed ice as well as thin ice regimes

are represented adequately. Moreover, particularly the formation of new thin ice during the freeze-up characterizes a large area of the ice cover in autumn. In order to detect changes and interannual variabilities in such areas, accurate thin ice thickness estimates with high temporal and spatial resolution are required.

Wang et al. (2016) evaluate six different sea-ice thickness products, including SMOS and CS2, and find that all satellite products as well as the Pan-Arctic Ice-Ocean Modeling and Assimilation System (PIOMAS) overestimate the thickness of thin ice compared to airborne laser altimetry retrievals of NASA's Operation IceBridge. The smallest bias of 0.26 m over thin ice has been found when using the SMOS product.

Considering the complementarity of CS2 and SMOS retrievals and the need for a better representation of thin ice regimes in global-scale sea-ice thickness data products, the goal of this study is to provide a merged product of CS2 and SMOS sea-ice thickness retrievals, which has the capability to provide Arctic sea-ice thickness distributions over the entire thickness range with reduced uncertainties. We also aim for a weekly update rate of the merged product. This ensures that we obtain a sufficient coverage of CS2 observations over perennial sea ice, while, at the same time, we benefit from the daily update rates of SMOS observations in order to capture ice growth rates in thin ice regions during the freeze-up. We apply two different merging schemes. The first is represented by a weighted mean (WM), based on the individual uncertainties, which only provides estimates at grid cells where weekly observations are available. The second approach uses an optimal interpolation (OI) scheme for

Arctic-wide estimates. Table 1 summarizes the input thickness products and the merged products, their temporal and spatial resolution, and coverage and applicability depending on study purposes. In order to assess the improvement of the merged products, we use airborne sea-ice thickness data and compare them with co-located data of the merged products.

This paper is outlined as follows: in Sect. 2, we first present the individual sea-ice thickness products derived from CS2 and SMOS measurements, including a detailed description of input data and highlighting the complementarity of both thickness products. Then, we present methods to merge both sea-ice thickness data sets, based on a weighted mean and an optimal interpolation approach. In Sect. 3, the merged products are evaluated by a comparison with input products and by a cross-validation experiment. In Sect. 4, the merged products are evaluated using airborne electromagnetic (AEM) thickness sounding measurements. Finally, conclusions are drawn in Sect. 5.

2 Data and methods

This section is structured as follows: first, the input data (Sect. 2.1) are presented, and then the merging of weekly CS2 and SMOS data by applying a weighted mean based on the individual uncertainties with the product referred to as WM is described (Sect. 2.2). Finally, the merging of weekly CS2 and SMOS data by applying an OI scheme with the product referred to as CS2SMOS is explained (Sect. 2.3).

2.1 Input data

We use the Alfred Wegener Institute CS2 product (processor version 1.2; Ricker et al., 2014; Hendricks et al., 2016) and the SMOS sea-ice thickness retrieval from the University of Hamburg (processor version 3.1; Tian-Kunze et al., 2014; Kaleschke et al., 2016) as input ice thickness data. Auxiliary data of ice concentration and ice type were obtained from the Ocean and Sea Ice Satellite Application Facility (OSI SAF).

2.1.1 CryoSat-2 weekly sea-ice thickness retrieval

In the first step we use CS2 SIRAL level-1b orbit data files that are provided by ESA. They contain geolocation information and time of the Doppler beam formed radar echoes. SIRAL is operated in two different modes over sea ice. The synthetic aperture radar (SAR) mode covers major parts of the ice-covered area, while the interferometric mode (SIN) is applied mostly in coastal areas. Both modes serve for retrieving ice thickness but must be processed separately, as we discard the phase information of SIN waveforms (Kurtz et al., 2014).

The radar echoes (waveforms) are processed for each CS2 orbit according to Hendricks et al. (2016) and Ricker et al. (2014). A 50 % threshold-first-maximum retracker (Ricker et al., 2014; Helm et al., 2014) is used to obtain ellipsoidal

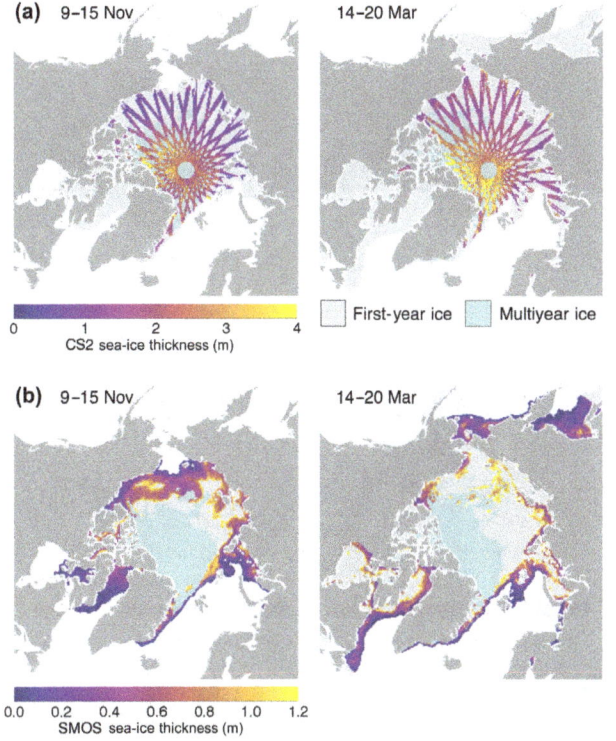

Figure 1. Example of weekly input data grids for November 2015 and March 2016. **(a)** Gridded weekly CryoSat-2 retrievals. **(b)** Gridded weekly mean SMOS retrievals derived from daily data. SMOS data are rejected over multiyear ice and when uncertainties are more than 1 m. The background fields indicate first-year and multiyear ice coverage.

surface elevations (L), which are corrected for geophysical perturbations like tides and atmospheric effects (Ricker et al., 2016). Geoid undulations and the mean sea-surface height (MSS) are removed by subtracting the Danish Technical University version 2015 (DTU15) MSS height:

$$L_{\text{MSS}} = L - \text{MSS}. \tag{1}$$

Ice and water are spatially separated by the pulse peakiness of the CryoSat waveforms. This is based on the fact that radar returns from surfaces that contain open water leads, i.e., openings in the ice pack, appear as specular echoes and can be separated from diffuse echoes that contain reflections from sea ice only (Laxon et al., 2003). The lead elevations are used to derive the instantaneous sea-surface height anomaly (SSHA) by interpolation. Finally, the SSHA is subtracted from the ice surface elevations to retrieve the freeboard (Fb):

$$\text{Fb} = L_{\text{MSS}} - \text{SSHA}. \tag{2}$$

Fb is corrected for a lower wave propagation speed inside the snow layer and can be converted into sea-ice thickness (Z) by assuming hydrostatic equilibrium (Laxon et al., 2003):

$$Z_{\text{cs2}} = \text{Fb} \cdot \frac{\rho_{\text{W}}}{\rho_{\text{W}} - \rho_{\text{I}}} + S \cdot \frac{\rho_{\text{S}}}{\rho_{\text{W}} - \rho_{\text{I}}}, \tag{3}$$

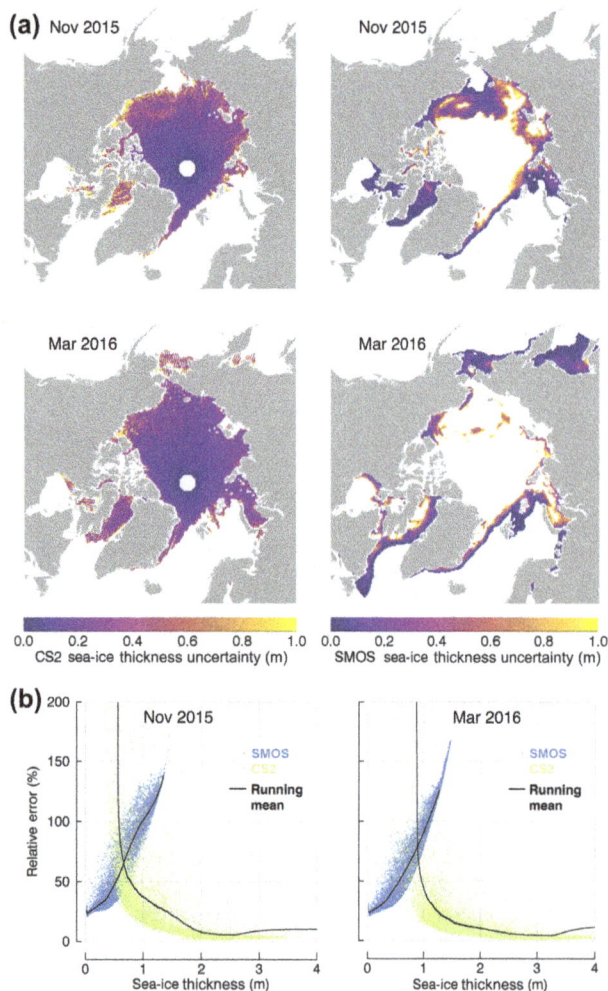

Figure 2. (a) Typical monthly sea-ice thickness uncertainty maps of the CryoSat-2 and SMOS retrievals from November 2015 and March 2016. The SMOS thickness uncertainty is masked where uncertainty is > 1 m. **(b)** Relative uncertainties from November 2015 and March 2016.

and $1024\,\mathrm{kg\,m^{-3}}$ for the sea water density. Z is calculated for each individual CS2 measurement along each orbit. All these retrievals are averaged on a 25 km Equal-Area Scalable Earth Grid version 2.0 (EASE2; Brodzik et al., 2012) within one calendar week (Fig. 1a).

CS2 sea-ice thickness uncertainties can be separated into observational uncertainties and systematic or bias uncertainties (Ricker et al., 2014). While observational uncertainties of individual measurements can be reduced due to spatial averaging, biases remain. The observational uncertainties of ice thickness retrievals from individual measurements contain uncertainties caused by speckle noise, sea-surface height estimation and densities of ice and snow (Ricker et al., 2014). They can easily reach values of > 1 m but will be reduced to the range of centimeters by spatial averaging. Figure 2a shows typical CS2 observational uncertainty maps for autumn and spring, mainly ranging between 0.1 and 1 m. Here, data points are averaged on a 25 km grid. The latitudinal dependency results from the denser orbit coverage towards the pole. In the marginal ice zones, when ice concentration decreases, many openings in the sea-ice cover can lead to an underrepresentation of sea ice. Moreover, when the sea-ice cover is characterized by many openings, so-called *snagging* leads to increased uncertainties in the range measurements (Armitage and Davidson, 2014). Biases mainly occur due to waveform processing and the lack of representation of interannual variability in the W99 snow climatology (Ricker et al., 2014).

2.1.2 SMOS weekly sea-ice thickness retrieval

Thin sea-ice thickness has been retrieved from the 1.4 GHz (L-band) brightness temperatures measured by SMOS for the winter seasons (15 October–15 April) from 2010 to present (Mecklenburg et al., 2016). The retrieval method consists of a thermodynamic sea-ice model and a one-ice-layer radiative transfer model (Tian-Kunze et al., 2014). The resulting plane layer thickness is multiplied by a correction factor assuming a log-normal thickness distribution. The algorithm has been used for the operational production of an SMOS-based sea-ice thickness data set from 2010 on (Tian-Kunze et al., 2014). In this study we use the most up-to-date version (v3.1) of the ice thickness data set, which has been produced operationally since October 2016. The v3.1 data for the previous winter seasons had been reprocessed using the same algorithm.

The v3.1 SMOS ice thickness data are based on v620 L1C brightness temperature data. Brightness temperatures used in the algorithm are the daily mean intensities averaged over incidence angles from 0 to 40°. The intensity is the average of horizontally and vertically polarized brightness temperatures, equal to $0.5\,(\mathrm{TB_h + TB_v})$. Over sea ice, the intensity is almost independent of incidence angle. By using the whole incidence angle range of 0–40°, we can reduce the brightness temperature uncertainty to about 0.5 K.

where S is the snow depth and ρ_S, ρ_I and ρ_W are the densities of snow, sea ice and sea water. S and ρ_S are represented by the modified Warren snow climatology (W99; Warren et al., 1999), meaning that S is reduced by 50 % over first-year ice to accommodate the recent change towards a seasonal Arctic ice cover (Kurtz and Farrell, 2011). FYI and multiyear ice (MYI) are separated by adopting the daily OSI SAF ice type product (Eastwood, 2012). We exclude CS2 measurements over the Hudson Bay and Baffin Bay as they are not located within the domain of the W99 climatology, referred to the area, which is constrained by in situ measurements from Soviet drifting stations and airborne landings from the 1950s to 1990 (Warren et al., 1999). In areas where no observations are available, the W99 polynomial fit is not reliable, being based only on extrapolation. We use ice densities of 916.7 and $882.0\,\mathrm{kg\,m^{-3}}$ for FYI and MYI (Alexandrov et al., 2010)

SMOS measurements are strongly influenced by radio frequency interference (RFI), especially in the first 2 years after its launch. In the previous processor RFI-contaminated snapshots have been discarded using a threshold value of 300 K, applied either to TB_h or TB_v. The new quality flags given in the v620 L1C data have been implemented to identify the data contaminated by RFI, by sun or by geometric effects to improve the quality of the radiometric data used for version 3.1.

To estimate the bulk ice temperature (T_{ice}) and bulk ice salinity (S_{ice}), which are the important input parameters in the radiation model, we need surface air temperature and sea-surface salinity (SSS) data as a boundary condition. The 2 m surface air temperature is extracted from JRA-25 atmospheric reanalysis (Onogi et al., 2007). SSS data used in the retrieval results from an integration of the MIT General Circulation Model (Marshall et al., 1997), including interannually varying surface forcing. From the daily surface salinity outputs from the model for the years 2002–2009, a weekly climatology was produced (Tian-Kunze et al., 2014).

Brightness temperatures over sea ice are simulated with the sea-ice radiation model adapted from Menashi et al. (1993), Kaleschke et al. (2010) and Kaleschke et al. (2012). The TB depends on the dielectric properties of the ice layer, which are a function of brine volume (Vant et al., 1978). The brine volume is a function of S_{ice} and T_{ice} (Cox and Weeks, 1983). For a thin ice layer, the ice temperature gradient within the ice can be assumed to be linear. The penetration depth of L band in the sea ice depends on the ice temperature and ice salinity. The retrieval algorithm works only under cold conditions. For the cold and less saline ice, the maximum retrievable ice thickness from SMOS can be up to 1.5 m.

The SMOS uncertainty given in the v3.1 product is estimated based on the uncertainty in the input parameters in the thermodynamic and radiation model as well as in the thickness distribution function (Tian-Kunze et al., 2014). At present, the estimation was carried out for each parameter – brightness temperature, ice temperature and ice salinity respectively, by keeping the other parameters constant. The uncertainty given in the product is then the sum of uncertainties caused by each parameter. In v3.1, we also varied the sigma in the log-normal ice thickness distribution function, which is used to convert plane layer ice thickness into heterogenous layer mean ice thickness in the retrieval. The average ice thickness uncertainty caused by the distribution function is less than 10 cm. This uncertainty is then added to the overall uncertainties caused by the brightness temperature, ice temperature and ice salinity. Errors caused by the assumptions about fluxes and snow thickness have not yet been included. The 100 % ice coverage assumption made in the retrieval can cause underestimation of ice thickness if the condition is not met.

For the merging, daily SMOS retrievals are averaged weekly and are projected on an EASE2 25 km grid to be co-

located with the CS2 retrievals. Here, we only allow SMOS thickness values with a corresponding uncertainty < 1 m, which corresponds to a maximum theoretical thickness of about 1.1 m. Furthermore we expect strong biases for the SMOS ice thickness in thicker MYI regimes. Therefore, we use the OSI SAF ice type product (Eastwood, 2012) to discard any SMOS grid cells that are indicated as MYI. The weekly composites are shown in Fig. 1b.

2.1.3 Complementarity of CryoSat-2 and SMOS sea-ice thickness products

The two main factors that drive the complementarity between the CryoSat-2 and SMOS sea-ice thickness products are the data coverage on the one hand and the sea-ice thickness uncertainties on the other hand.

Figure 2 shows typical uncertainty maps and the relative uncertainties of CS2 and SMOS monthly mean thickness retrievals from November 2015 and March 2016. While with SMOS relative uncertainties are lowest for thin ice (< 1 m), CS2 relative thickness uncertainties are smaller over thick ice and rise asymptotically towards thinner ice less than 1 m thick. This is due to the fact that CS2 thickness estimates over thin ice rely on the retrieval of small surface elevations slightly higher than sea level, while freeboard of thicker ice is much larger (Ricker et al., 2014). As a consequence, the relative uncertainty increases over thin ice, as measurement uncertainties do not decrease over thinner ice. Note that the CS2 uncertainties shown here represent observational uncertainties only. Systematic errors as associated with the usage of a snow climatology or due to variable snow penetration will increase the uncertainty of altimetry-based thicknesses (Ricker et al., 2014, 2015; Kwok, 2014; Armitage and Ridout, 2015).

Due to the different update rates of sea-ice thickness observations, CS2 grids are usually based on data composites from 1 month, while SMOS-based retrievals provide daily complete coverage of the ice-covered ocean up to about 85° N. Figure 1 compares weekly means of CS2 and SMOS for November 2015 and March 2016. While valid SMOS ice thickness estimates are found mostly in the marginal ice zones, the CS2 ice thickness retrieval covers major parts of the Arctic MYI. In November, during the freeze-up, SMOS retrievals cover major parts of the Beaufort Sea, Chuckchi Sea and East Siberian Sea. Towards spring, due to continued ice growth in these regions, the regions with SMOS retrievals retreat southwards, covering major parts of the Bering Sea and the Sea of Okhotsk (Fig. 1b).

Figure 3 illustrates the number of valid grid cells of the weekly means as shown in Fig. 1. The number of grid cells with co-located SMOS and CS2 estimates is less than 2000, while the number of grid cells that contain thickness estimates from CS2 or SMOS only is about 5000, highlighting the complementary data coverage of both sensors.

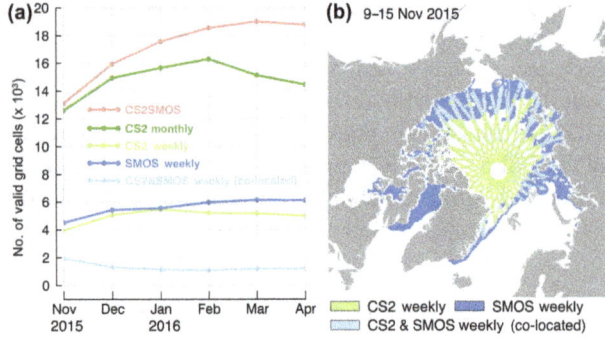

Figure 3. (a) Numbers of valid 25 km grid cells each month from November 2015 to April 2016. Here, "valid" grid cells are grid cells that contain a valid thickness estimate. **(b)** Spatial distribution of valid weekly thickness retrievals by CryoSat-2 and SMOS.

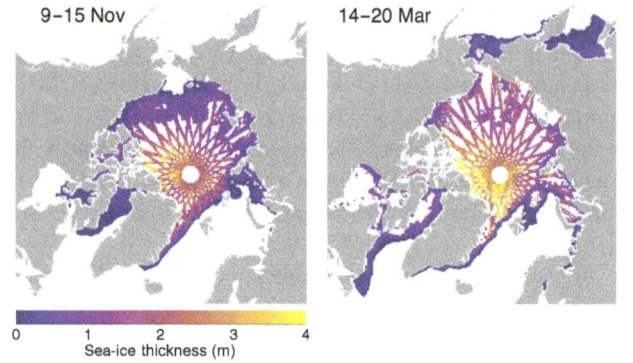

Figure 4. Weighted means of CryoSat-2 and SMOS weekly means during the target week, produced from fields shown in Fig. 1.

2.1.4 OSI SAF ice concentration and type

We use the OSI SAF sea-ice concentration (OSI-401-b) and type (OSI-403-b) products (Eastwood, 2012) in order to identify grid cells that contain $\geq 15\%$ sea ice and to classify them as FYI or MYI. The products are delivered daily, projected on a 10 km polar stereographic grid. To combine these data with the CS2 and SMOS thickness grids, we calculate weekly means that are projected on the EASE2 25 km grid (Brodzik et al., 2012) to be co-located with the thickness retrievals. The original ice type product contains grid cells that are flagged as *ambiguous*. We apply an inverse distance interpolation to those grid cells to obtain FYI or MYI flags for all ice-covered grid cells, because it is needed for further processing steps.

2.2 Weighted mean

We compute the weighted mean sea-ice thickness $\overline{Z}$ using weekly CS2 and SMOS ice thickness grids during the target week:

$$\overline{Z} = \frac{Z_{cs2}/\sigma_{cs2}^2 + Z_{smos}/\sigma_{smos}^2}{1/\sigma_{cs2}^2 + 1/\sigma_{smos}^2}, \tag{4}$$

where σ represents the observational uncertainty of the individual products. Figure 4 shows the weighted means for weeks in November 2015 and March 2016. In contrast to the OI approach, presented in the next section, the weighted mean only provides thickness estimates where observations are available during the target week, leaving data gaps in the CS2 domain. In the following we refer to the weekly weighted mean product as WM.

2.3 Optimal interpolation

To achieve complete spatial coverage, we use an OI scheme similar to Böhme and Send (2005) and McIntosh (1990) that enables the merging of data sets from diverse sources

on a predefined, so-called analysis grid. The input data are weighted based on their individual uncertainties and the modeled spatial covariances. OI minimizes the total error of observations and provides ideal weighting for the observations at each grid cell in the least square sense. In this section we present the processing methods, on which our OI approach is based. Figure 5 shows the processing scheme, which will be described in more detail in the following.

The OI scheme is used to get an objective estimate of values at observed or unobserved locations. The basic equation is

$$Z_a = Z_b + K[Z_o - H(Z_b)], \tag{5}$$

where the vector Z_a is the analysis field, i.e., each element represents a grid cell of the merged CS2SMOS ice thickness retrieval to be produced. Z_b is a background field vector, and Z_o is the vector that contains all SMOS and CS2 observations. Here we use already gridded, weekly mean CS2 and SMOS thickness estimates as observations, as shown in Fig. 1 and as described above. Using gridded data as observations reduces their observational uncertainties and provides equally distributed observations, which improves the performance of the OI. In addition, gridding of raw data reduces the number of available observations used for the OI, increasing the efficiency of the OI routine. We assume that the observations are static, i.e., remain temporally coherent within a week and do not change due to ice deformation and motion. Therefore, we neglect any temporal correlations. H is an operator that transforms the background field into the observation space. To be more specific, this is realized by an inverse distance interpolation method. K represents a weight matrix and is derived from error covariances. We aim to retrieve weekly analysis fields, based on calendar weeks from Monday to Sunday. Wet and warm snow or ice prevent the retrieval of summer sea-ice thickness estimates from CS2 or SMOS. Hence, the CS2SMOS product is limited to the period from end-of-October to April.

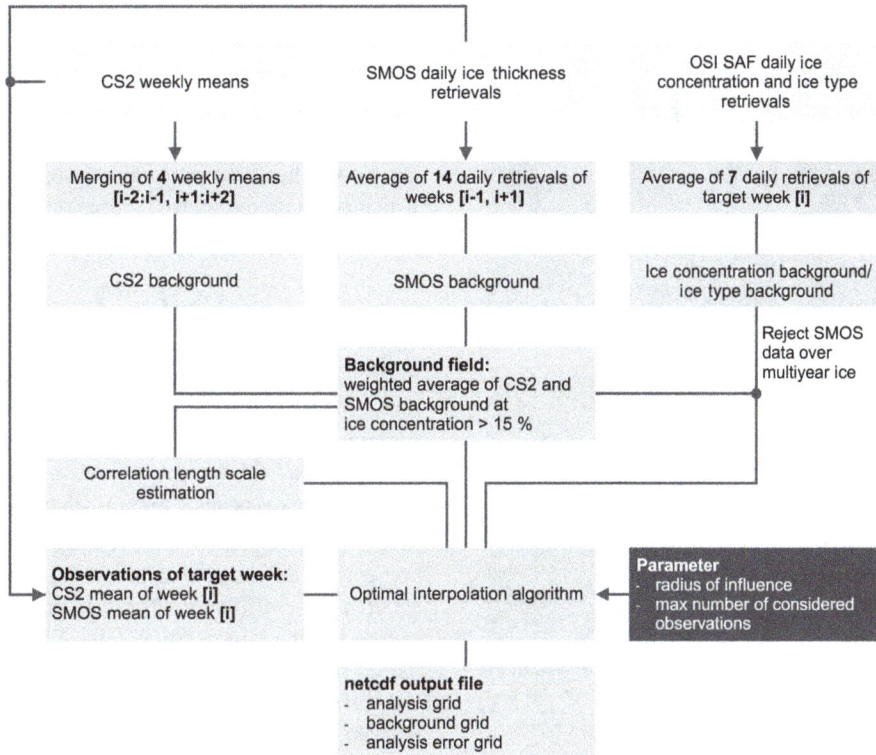

Figure 5. Optimal interpolation processing scheme. Week [i] represents the target week. The cycle is repeated for each week.

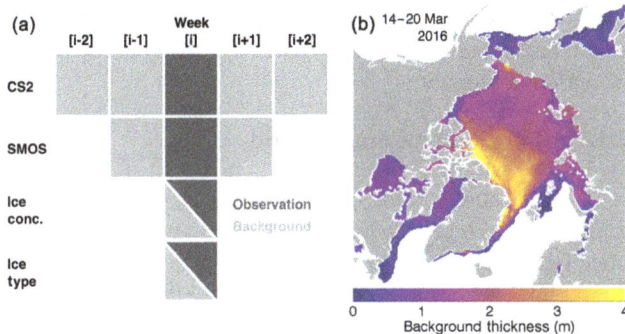

Figure 6. (a) The scheme illustrates how the background field and the observation field are generated from weekly input grids. Week [i] represents the target week. **(b)** Typical interpolated and low-pass-filtered background field as it is used for the optimal interpolation.

2.3.1 The background field

The weekly CS2 ice thickness composite possesses large gaps resulting from the limited orbital coverage (Fig. 1a). But for the OI approach, an Arctic-wide coverage is required for the background field. Therefore, we use a composite of retrievals from adjacent weeks to create a background field with nearly complete coverage for the Central Arctic at a certain target week (Fig. 6a). Here we combine data from the 2 weeks before and after the target week. Therefore, in con-

trast to CS2 near real-time sea-ice thickness retrievals (Tilling et al., 2016), products can only be released 2 weeks after data acquisition. In order to ensure independence between the observations and background field, CS2 data from the target week are not included in the background field. For the same reason, we use an SMOS weekly mean from 1 week before and after the target week. The initial background field is computed by a weighted mean using Eq. (4). Gaps in the weighted average are interpolated by using a nearest neighbor scheme. In order to reduce noise, the background field is low-pass filtered with a smoothing radius of 25 km before it is applied in the OI algorithm (Fig. 6b).

Since we use CS2 and SMOS retrievals for the background field beyond the target week and because the SMOS composite contains artifacts in coastal regions, we additionally use a weekly mean of the daily OSI SAF ice concentration product to determine the ice coverage during the target week. Here, we apply a threshold of 15 % and only grid cells that exceed this value will be considered as ice covered, which corresponds to the ice extent products provided by OSI SAF and the National Snow and Ice Data Center (NSIDC).

2.3.2 Correlation length scale estimation

The correlation length scale ξ controls the impact of a data point on the analysis grid point depending on their distance. Considering the grid resolution of 25 km, correlation length

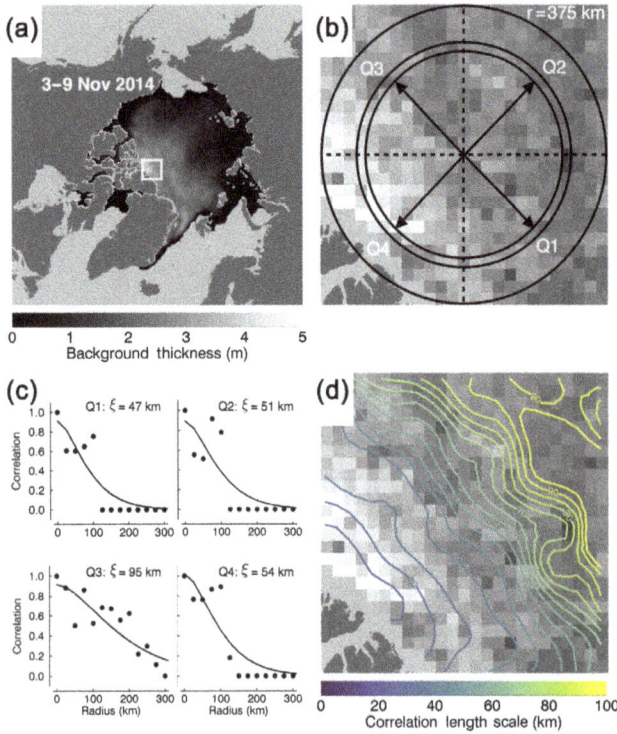

Figure 7. Scheme for the estimation of the correlation length scale ξ for a single grid cell for the target week 3–9 November 2014. **(a)** Background field with indicated area of interest (white box). **(b)** Adjacent ice thickness grid cells within a radius of 375 km are binned into annuli of distance and four quadrants. **(c)** Binned thickness estimates are used to calculate the structure function of each quadrant. ξ is estimated by fitting an exponential function. **(d)** Contour map of estimated correlation length scales for the considered area.

here is used in the sense of large-scale thickness gradients. For example, the correlation length scale estimate is large in the center of a certain ice type regime with similar ice thickness (i.e., level FYI). On the other hand, we expect a low ξ value at locations with strong thickness gradients, where distant observations are not representative of local conditions. Figure 7 illustrates the estimation of ξ for a certain grid cell Z'_0 in the Lincoln Sea during a week in November. In order to estimate ξ, we consider the unfiltered background field Z_b (Fig. 7a) and define a structure function ϵ^2. The structure function can be used to assess the change of ice thickness with distance and is related to the normalized auto correlation function $R(d, Q)$ as follows (Böhme and Send, 2005):

$$\epsilon^2(d, Q) = \overline{(Z'_0 - Z'_{Q,d})^2} = 2\overline{\sigma^2_{Z'}} - 2\overline{\sigma^2_{Z'}}R(d, Q),$$

$$R(d, Q) = 1 - \frac{\epsilon^2(d, Q)}{2\overline{\sigma^2_{Z'}}}. \tag{6}$$

Quadrants Q are defined to accommodate the anisotropy of the spatial ice thickness distribution (Fig. 7b). $\epsilon^2(d, Q)$ represents the square differences between ice thickness of the

grid cell and the ice thickness of the grid cells of binned 25 km distances d in a quadrant Q. $Z'_{Q,d}$ is the background thickness, binned according to d and Q. Figure 7b illustrates the annuli of distance and the four quadrants. $\overline{\sigma^2_{Z'}}$ values are the corresponding mean variances of a certain quadrant. With Eq. (6) we then obtain the auto correlation function $R(d, Q)$, which is computed up to a radius of 750 km (30 bins). In the next step, we fit a function of the form

$$C(d, \xi) = \left(1 + \frac{d}{\xi}\right)\exp\left(\frac{-d}{\xi}\right) \tag{7}$$

to $R(d, Q)$, using a least squares scheme, and obtain an estimate for ξ. Figure 7c shows the calculated auto correlation function $R(d, Q)$ and the functional fit (Eq. 7). A stronger decay of $R(d, Q)$ occurs with rising deviation between Z_0 and the thickness at a certain distance in a certain quadrant. $R(d, Q)$ can also become negative if $\epsilon^2(d, Q)/2\overline{\sigma^2_{Z'}}$ becomes > 1. In order to improve the fitting performance, we set $R(d, Q) = 0$ if $R(d, Q)$ becomes < 0. Furthermore, ξ is rejected if the computation fails. Finally, we average the ξ values from the four quadrants, as we do not use anisotropic weighting in the OI. In order to remove outliers and noise, the derived ξ grid is low-pass filtered with a smoothing radius of 25 km. Grid cells with failed computation are interpolated by a nearest neighbor scheme afterwards. Figure 7d shows the spatial correlation length scales ξ for 3–9 November 2014. It highlights the sensitivity to changing thickness gradients as ξ decreases towards the coast of the Canadian Archipelago, where higher sea-ice thickness gradients likely occur due to increased deformation.

2.3.3 Retrieving the analysis grid

In order to minimize the error covariances, the background error covariance matrix **B** in the observation space is multiplied by the inverted total error covariance matrix, leading to the optimal weight matrix **K** (McIntosh, 1990; Böhme and Send, 2005):

$$\mathbf{K} = \mathbf{B}\mathbf{H}^T(\mathbf{R} + \mathbf{H}\mathbf{B}\mathbf{H}^T)^{-1}, \tag{8}$$

where **R** is the error covariance matrix of the observations. In order to reduce computation expense we assume the following:

1. We neglect correlations of observation errors, which means that **R** is a matrix with nonzero elements only on the diagonal. These variances are represented by the respective SMOS and CS2 product uncertainties.

2. We assume that the influence of observations that are located far away from the analysis grid point can be neglected. Therefore, instead of computing the entire covariance matrix, we only consider observations within a radius of influence. This radius is set to 250 km to gather just enough observations in regions with large gaps, for

example over MYI between two CS2 orbits where valid SMOS observations are not available.

3. To further reduce computation expense we limit the number of matched observations to 120, meaning that, in the case of more matches, only the 120 closest observations are considered.

4. We generally assume that all observations are unbiased.

For practical reasons, we apply an iterative computation instead of applying the general matrix formulation in Eqs. (5) and (8). We iteratively calculate each element $z_{a_{m,n}}$ of the analysis field. Vector elements $(bh^T)_i$ and matrix elements $(hbh^T)_{i,j}$ are estimated using the correlation function in Eq. (7),

$$(bh^T)_i = \left(1 + \frac{d(x_{0_i}, x_{a_{m,n}})}{\xi_{m,n}}\right) \exp\left(\frac{-d(x_{0_i}, x_{a_{m,n}})}{\xi_{m,n}}\right),$$

$$(hbh^T)_{i,j} = \left(1 + \frac{d(x_{0_i}, x_{0_j})}{\xi_{m,n}}\right) \exp\left(\frac{-d(x_{0_i}, x_{0_j})}{\xi_{m,n}}\right), \quad (9)$$

with the Euclidian distance function:

$$d(x, y) = \|x - y\|. \quad (10)$$

Here, x_{0_i} and x_{0_j} represent the locations of the matched observations within the radius of influence. $x_{a_{m,n}}$ refers to the location of the analysis grid cell. As a consequence of Eq. (9), the impact of a data point decreases with increasing distance.

Computing $\mathbf{BH}^T$ and $\mathbf{HBH}^T$ allows the computation of the K weights that minimize the error covariances. When the analysis field is calculated iteratively, K will be a vector, containing the corresponding weights for the matched observations within the radius of influence, while in the general OI formulation $\mathbf{K}$ is a matrix. Thus, we retrieve the second part of Eq. (5), which is called *innovation* the difference between the observation field and the background field. This procedure is accomplished iteratively for each grid cell of the analysis field. The corresponding analysis error covariances are derived by

$$\sigma^2_{Z_a} = (\mathbf{I} - \mathbf{KH})\mathbf{B}, \quad (11)$$

where $\mathbf{I}$ is the identity matrix. Since we consider variances exclusively, we only calculate the diagonal elements of $\sigma^2_{Z_a}$. Figure 8 illustrates how the analysis thickness is derived at a certain analysis grid point, considering distant grid cells with ice thickness estimates of CS2 and SMOS. The K weights decrease with increasing distance to the analysis grid point as a consequence of Eq. (9). In addition, the individual uncertainties affect the weighting according to Eq. (8). The considered grid cell is located at the boundary between the CS2 and SMOS domain. In the following, we use *domain* as the regions where CS2 or SMOS data predominate. SMOS ice thicknesses of about 1 m reveal higher uncertainties than corresponding CS2 estimates (Fig. 2), and hence the K weights

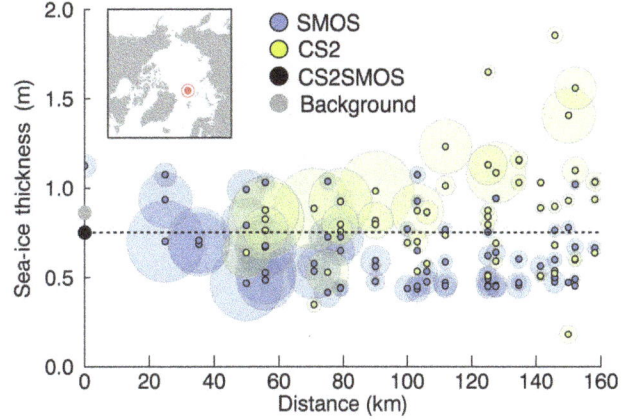

Figure 8. Example for CS2 and SMOS sea-ice thickness observations and their weighting to compose the CS2SMOS thickness estimate based on optimal interpolation at a grid cell in the Central Arctic first-year ice in November 2016. The x axis represents the distance of observations from the analysis grid cell. Normalized K weights are represented by the area of the circles.

of CS2 estimates exceed the SMOS weights for higher ice thicknesses. Figure 9 shows the innovation field, the merged CS2SMOS product and the analysis error field, which is the square root of the error variance (Eq. 11), for weeks in November 2015 and March 2016. The analysis error is a relative quantity with values between 0 and 1. It increases where the weekly CS2 retrieval leaves gaps and where valid SMOS observations are not available, for example at the North Pole or over MYI. In this case the analysis depends on the accuracy of the background field, leading to increased uncertainties.

3 Evaluation of the optimal interpolation

In this section, we aim to evaluate the CS2SMOS product derived from the OI scheme by a comparison with the individual satellite products. In addition, we carry out a cross-validation experiment by omission of random data to test the OI method.

3.1 Comparison with input products

Figure 10 illustrates the differences between CS2SMOS and the CS2 and SMOS retrievals from November 2015 to April 2016. The difference between CS2SMOS and SMOS weekly grids is shown in Fig. 10a, limited to grid cells with SMOS observations in the target week. Positive anomalies of up to 1 m occur mostly in the transition zone between the SMOS and the CS2 domain where the thick ice in the CS2 retrieval leads to an increase of ice thickness in these grid cells with respect to the SMOS data (Fig. 10a). However, the general pattern remains the same during the season. Subtracting the CS2 monthly mean sea-ice thickness from the CS2SMOS

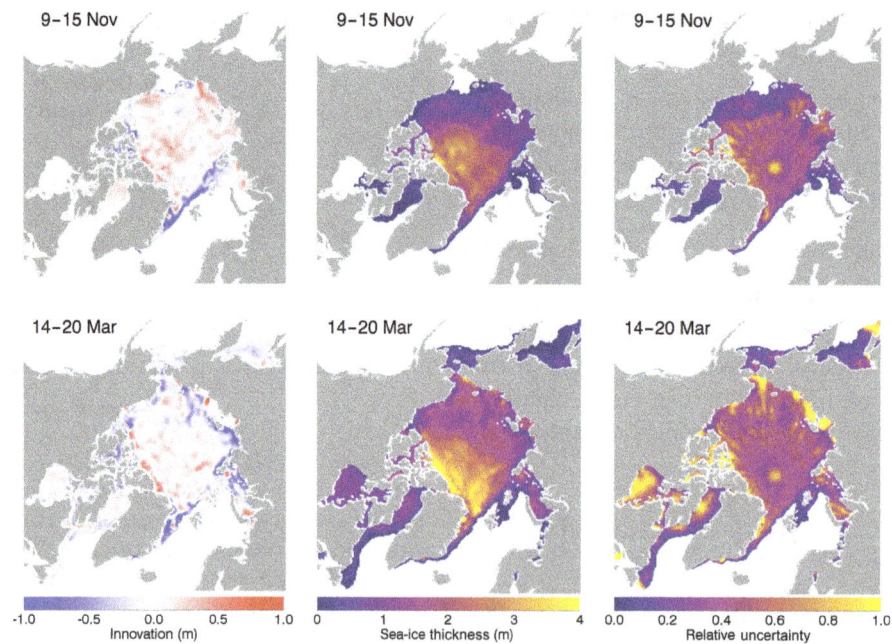

Figure 9. Optimal interpolation output grids for weeks in November 2015 and March 2016: the innovation field (left column) shows the difference between background field and the CS2SMOS ice thickness (center column). The right column shows the relative uncertainty associated with the optimal interpolation.

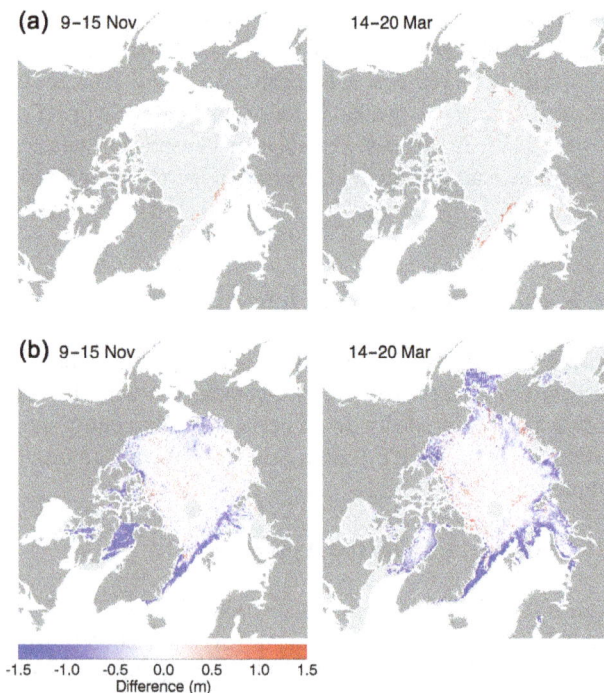

Figure 10. (a) Difference between CS2SMOS and weekly SMOS retrieval for weeks in November 2015 and March 2016. **(b)** Difference between CS2SMOS thickness for weeks in November 2015 and March 2016 and the corresponding monthly CryoSat-2 thickness retrieval.

product, represented by 1 week within each month, reveals substantial scattering between −1 and 1 m within the CS2 domain (Fig. 10b). This is mainly caused by the fact that the monthly retrieval is compared with the weekly product. During the different time spans, the regional sea-ice thickness distribution is subject to ice drift, convergence, and divergence, as well as thermodynamic ice growth. In addition, the OI algorithm evokes a low-pass filtering of the spatial thickness distribution due to the impact of distant grid cells, reducing the noise compared to the original CS2 product. Within the SMOS domain we find consistently negative anomalies, indicating a reduction of the CS2 ice thickness representation due to the impact of the coincident SMOS retrieval.

Figure 11a shows ice thickness distributions of monthly means of CS2 and weekly SMOS and CS2SMOS ice thickness retrievals for November 2015 and March 2016, illustrating the different thickness ranges of CS2 and SMOS retrievals. Table 2 presents the corresponding statistics for the entire winter season, including the mean and the standard deviation of each month or week respectively. The CS2 retrieval lacks sensitivity for thin ice (< 0.5 m) over the entire season. The gap in this thickness range can be closed by the SMOS retrieval. While the mean thickness of the CS2 retrieval consistently grows from 1.46 m in November to 1.90 m in April, the SMOS thickness mean remains at about 0.5 m after an increase from November to December. Due to the increasing uncertainties of the SMOS product towards thick ice, the distribution frequency steeply drops at about 1 m for each month. Therefore, the SMOS mean thickness

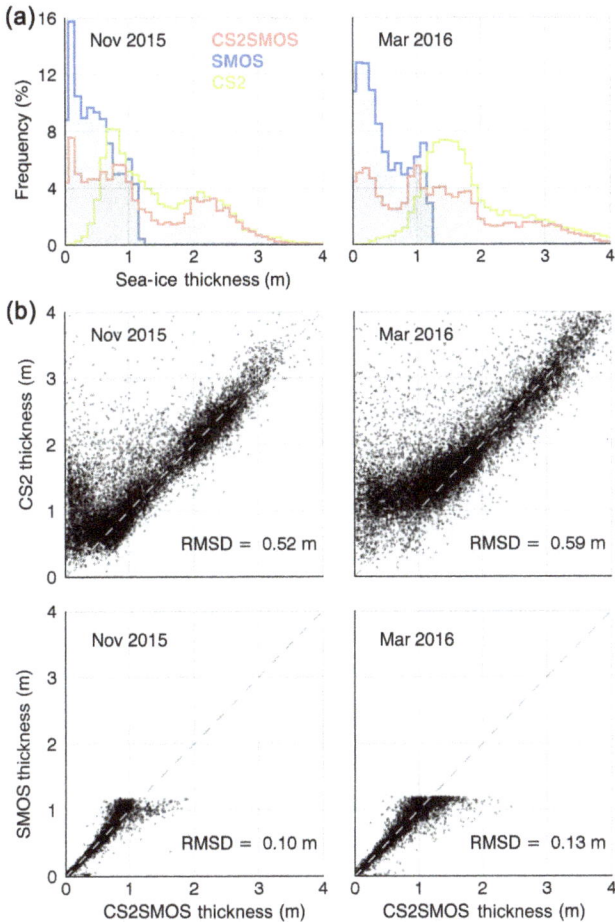

Figure 11. (a) Sea-ice thickness distributions of CryoSat-2, SMOS and CS2SMOS retrievals for November 2015 and March 2016. CS2SMOS is represented by 1 week in the middle of a month, while the CryoSat-2 and SMOS retrievals are monthly means. **(b)** Scatter diagrams illustrating the ice thickness differences between CS2SMOS and the individual satellite retrievals of CS2 and SMOS, for November 2015 and March 2016.

Table 2. Arctic-wide mean and standard deviation (SD) of the merged product (CS2SMOS), the individual CryoSat-2 (CS2) and Soil Moisture and Ocean Salinity (SMOS) retrievals for the winter season 2015–2016.

Mean (m)	Nov	Dec	Jan	Feb	Mar	Apr
CS2SMOS	1.16	1.19	1.23	1.29	1.34	1.35
CS2	1.46	1.53	1.65	1.66	1.83	1.90
SMOS	0.45	0.58	0.51	0.49	0.48	0.47
SD (m)						
CS2SMOS	0.88	0.81	0.81	0.92	0.97	0.99
CS2	0.76	0.76	0.72	0.73	0.75	0.78
SMOS	0.33	0.36	0.38	0.37	0.36	0.38

3.2 Cross-validation experiment

In order to test the robustness of the OI algorithm, we carry out a cross validation. We randomly remove grid cells of observations from the target week (see Figs. 5 and 6), with experiments for exclusion of 10 % (Fig. 12a), 25 % (Fig. 12b) and 50 % (Fig. 12c) of both CS2 and SMOS input grid cells. In the fourth case, all data contained in a box in the western Arctic are withdrawn (Fig. 12d). The box intentionally covers both the SMOS and the CS2 domain. After the data omission, the OI algorithm is applied using the reduced target week data set. The maps show the difference between the retrieved CS2SMOS sea-ice thickness and the withdrawn thickness data for each case. Compared to the SMOS domain, the ice thickness in the CS2 domain in the Central Arctic (Fig. 1) reveals a higher level of noise with deviations of up to 1 m. On the other hand, the SMOS domain shows a slightly negative shift of up to 10 cm in some areas. This can be explained by the different data coverages. We truncate the SMOS retrieval over thick ice, since the method does not apply for thick ice. On the other hand, the CS2 retrieval is used over the entire thickness range, but with higher uncertainties over thin ice. Therefore, CS2 thickness over thin ice is mostly reduced by the SMOS retrieval, while, in contrast, this is barely the case for SMOS data over thick ice, since it is cropped there. Hence, due to the optimal interpolation, there will be always a negative bias in the SMOS domain when doing the cross-validation experiment with the original input data from CS2 and SMOS.

The general pattern remains the same in all cases, independent of the fraction of data that are withdrawn in advance. The shape of the histograms of the differences indicates a normal distribution with similar standard deviations between 14 and 18 cm. The mean differences are −3 cm for the first three cases where data points have been withdrawn randomly and 1 cm where a box has been separated. The root mean square deviation (RMSD) is 23–25 cm for the first three cases and 17 cm for the last case. Here, the smaller RMSD is likely caused by the lack of thicker ice in the chosen box, which

is mostly affected by the boundary condition at about 1 m in conjunction with thermodynamic ice growth and the newly formed ice (< 0.1 m). The thickness distributions show the capability of the CS2SMOS product to combine the complementary ice thickness ranges. As a consequence, the standard deviation of the merged product ranges between 0.8 m (December) and 0.99 m (April) and therefore exceeds the standard deviations of the individual products that reach maximum values of 0.78 (CS2) and 0.38 (SMOS) in April. The scatter diagrams in Fig. 11b illustrate the thickness differences between CS2SMOS and the two individual products, with respect to the maps shown in Fig. 10. Using the SMOS data reduces the thickness in the CS2SMOS product below 1 m compared to the CS2 retrieval. The comparison between CS2SMOS and SMOS shows increasing scattering with rising thickness. As shown in Fig. 10, this originates from the transition zone between the CS2 and SMOS domain.

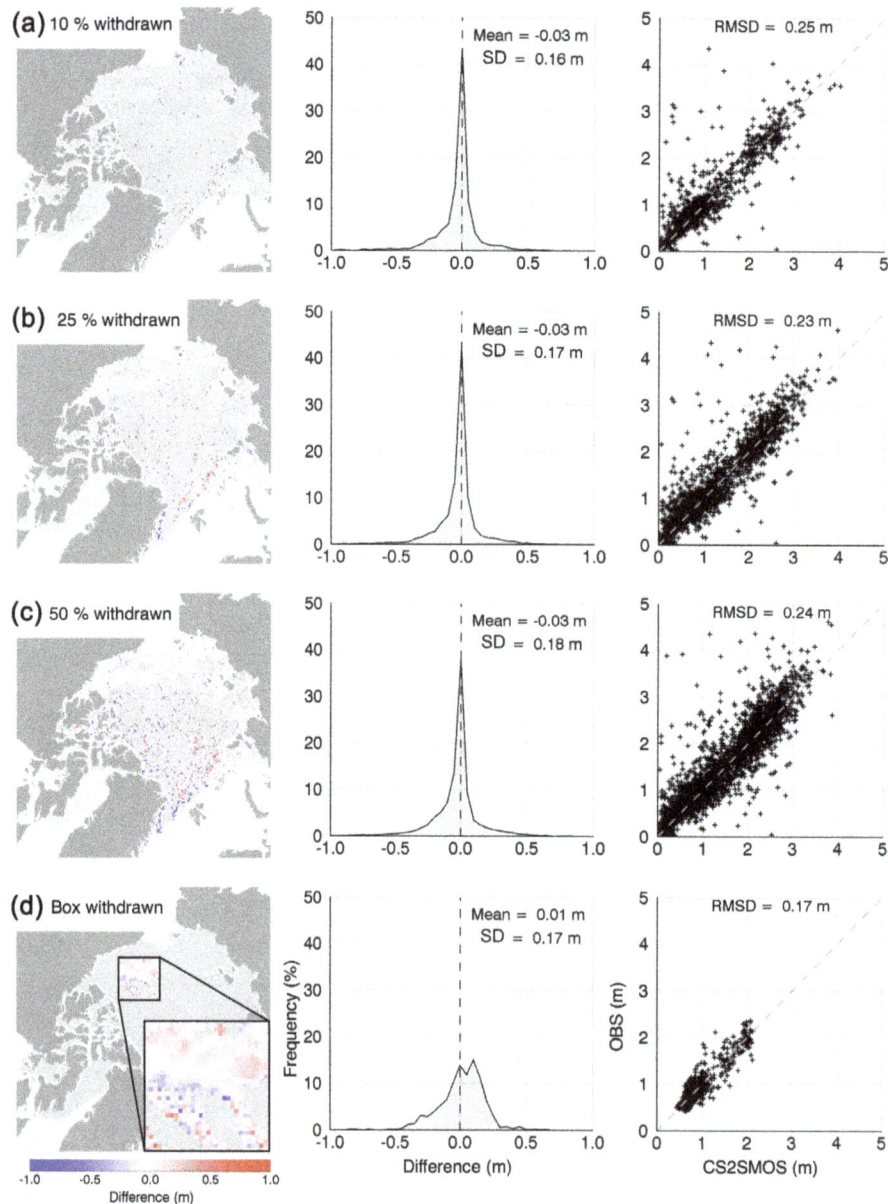

Figure 12. Cross-validation experiment for November 2015, showing the difference between CS2SMOS ice thickness, gridded CryoSat-2 and SMOS observations (OBS) that have been separated in advance as different fractions/areas of withdrawn data: **(a)** 10 %, **(b)** 25 %, **(c)** 50 % and **(d)** box. The maps show the withdrawn data subtracted from the CS2SMOS product. The histograms show the differences according to the maps, indicating the mean and standard deviation (SD) of the differences. Scatter diagrams indicate the root mean square deviation (RMSD).

does not contain sea ice thicker than about 2 m. This experiment demonstrates the performance of the applied algorithm. In particular, it shows that the background field mostly conserves the mean values even when co-located observations are missing.

4 Validation of the merged products with airborne EM

For validation of WM and CS2SMOS, we use sea-ice thickness measurements obtained during the SMOS-ice 2014 campaign east of the Spitsbergen archipelago and during the Canadian Arctic Sea Ice Mass Balance Observatory campaign in the Beaufort Sea in April 2016. Surveys have been carried out with an airborne electromagnetic induction thickness sounding device (EM-bird; Pfaffling et al., 2007; Haas et al., 2009; Hendricks, 2009) and are projected and averaged

Table 3. Statistics of the comparison of satellite retrievals with airborne EM thickness measurements (AEM), corresponding to Fig. 13. For each case we consider both the AEM modal thickness (AEM mode) and the AEM mean thickness (AEM mean). For the mean bias, AEM measurements are subtracted from the satellite retrievals. RMSD represents the root mean square deviation and r the Pearson correlation coefficient.

Beaufort Sea		RMSD (m)	Mean bias (m)	r
CS2SMOS	AEM mean	1.57	−0.86	0.48
	AEM mode	1.03	0.11	0.36
WM	AEM mean	1.49	−0.57	0.35
	AEM mode	1.13	0.30	0.26
SMOS	AEM mean	1.16	−0.38	0.37
	AEM mode	0.75	0.19	0.46
CS2	AEM mean	1.27	−0.17	0.52
	AEM mode	1.33	0.80	0.39
Barents Sea		**RMSD**	**Mean bias**	r
CS2SMOS	AEM mean	0.31	−0.25	0.61
	AEM mode	0.27	−0.11	0.56
WM	AEM mean	0.27	−0.17	0.73
	AEM mode	0.27	−0.05	0.63
SMOS	AEM mean	0.30	−0.24	0.7
	AEM mode	0.27	−0.11	0.67
CS2	AEM mean	0.97	0.82	−0.35
	AEM mode	1.11	0.95	−0.35

on a 25 km EASE2 grid as given by the satellite products. In addition to the mean AEM thickness in each grid cell, we also calculated the modal AEM thickness. The AEM data set represents total thickness, comprising snow and sea-ice thickness. Therefore, we add the climatological snow depth (modified W99) to the satellite products. Figure 13 shows the comparison between AEM ice thickness measurements and four satellite products at the two validation sites, the Beaufort Sea (Fig. 13a) and Barents Sea (Fig. 13b). The four satellite products are represented by CS2SMOS, WM, SMOS and CS2. The scatter diagrams illustrate the difference between the satellite products and the corresponding mean and modal AEM thickness. Statistics resulting from Fig. 13 are given in Table 3.

4.1 Beaufort Sea, April 2016

On 9 and 10 April, two AEM flights were carried out with a fixed wing DC3-T aircraft (Fig. 13a). The AEM measurements indicate high mean ice thickness variability ranging between 0.2 m and more than 5 m. Comparing the mean (2.2 m) and modal thickness (1.2 m) of the entire data set indicates substantial deformation. Thickness distribution and OSI SAF ice type data suggest two ice types. First-year ice, reaching a modal thickness of up to 1 m, and multiyear ice with a modal thickness ranging between 2 and 4 m.

The presence of two ice types and the drift along the Beaufort Gyre (Petty et al., 2016) make this region challenging for satellite observations, which are limited in spatial and temporal resolution. Especially scattered thick multiyear ice floes that drift along the Beaufort Gyre might not be captured by the OSI SAF ice type product, allowing for SMOS thickness estimates in MYI. Therefore, CS2SMOS, WM and SMOS underestimate the mean ice thickness by up to 0.86 m (CS2SMOS). On the other hand, the modal ice thickness is slightly overestimated by up to 0.3 m (WM). It is important to note that WM and SMOS do not provide a full data coverage. The SMOS data, for example, usually only cover first-year ice. This is also the reason why SMOS exhibits the smallest RMSD for mean and modal thickness (1.16 and 0.75 m). However, scatter diagrams show good agreement of AEM data and CS2SMOS, WM and SMOS retrievals within the first-year ice, up to about 1.2 m thickness (Fig. 13). CS2 shows the lowest bias (−0.17 m) for the mean ice thickness but the highest for the modal thickness. The scatter diagrams also indicate that CS2 is not able to capture high thickness gradients due to the presence of scattered heavily deformed multiyear ice, which is transported along with the Beaufort Gyre. As discussed above, the usage of SMOS data in CS2SMOS and WM leads to a stronger underestimation of mean ice thickness of deformed multiyear sea ice, compared to CS2. But it substantially improves the representation of first-year ice thickness. The comparison between WM and CS2SMOS shows that in areas where weekly observations are available, both retrievals show similar agreement with AEM measurements.

4.2 Barents Sea, March 2014

Between 19 and 26 March, eight AEM flights were carried out by a helicopter based on the Norwegian research vessel *Lance* (Fig. 13b; King et al., 2017). In contrast to the Beaufort Sea data, these data contain first-year ice only. Moreover, the degree of deformation is lower, indicated by only 0.1 m difference between mean and modal thickness of the entire data set. For CS2, the RMSD is 0.97 m for the AEM mean thickness and 1.11 m for the AEM modal thickness, indicating a slightly better representation of the mean thickness in the CS2 product. However, scattering is high and the mean bias of 0.82 m with respect to the mean AEM thickness suggests a strong bias towards thicker ice. Such errors might originate from erroneous sea-surface height interpolation along the CS2 orbits as well as from off-nadir lead ranging and retracker limitations (Ricker et al., 2014). The SMOS and CS2SMOS retrievals are almost identical for that region, which is caused in part by the better coverage of the SMOS retrieval in that region. In addition, this area is dominated by thin ice, leading to a higher weighting of the SMOS retrieval due to the lower uncertainties (Fig. 2). The scatter diagrams reveal a significantly better agreement of the AEM mean thickness measurements with the CS2SMOS, WM

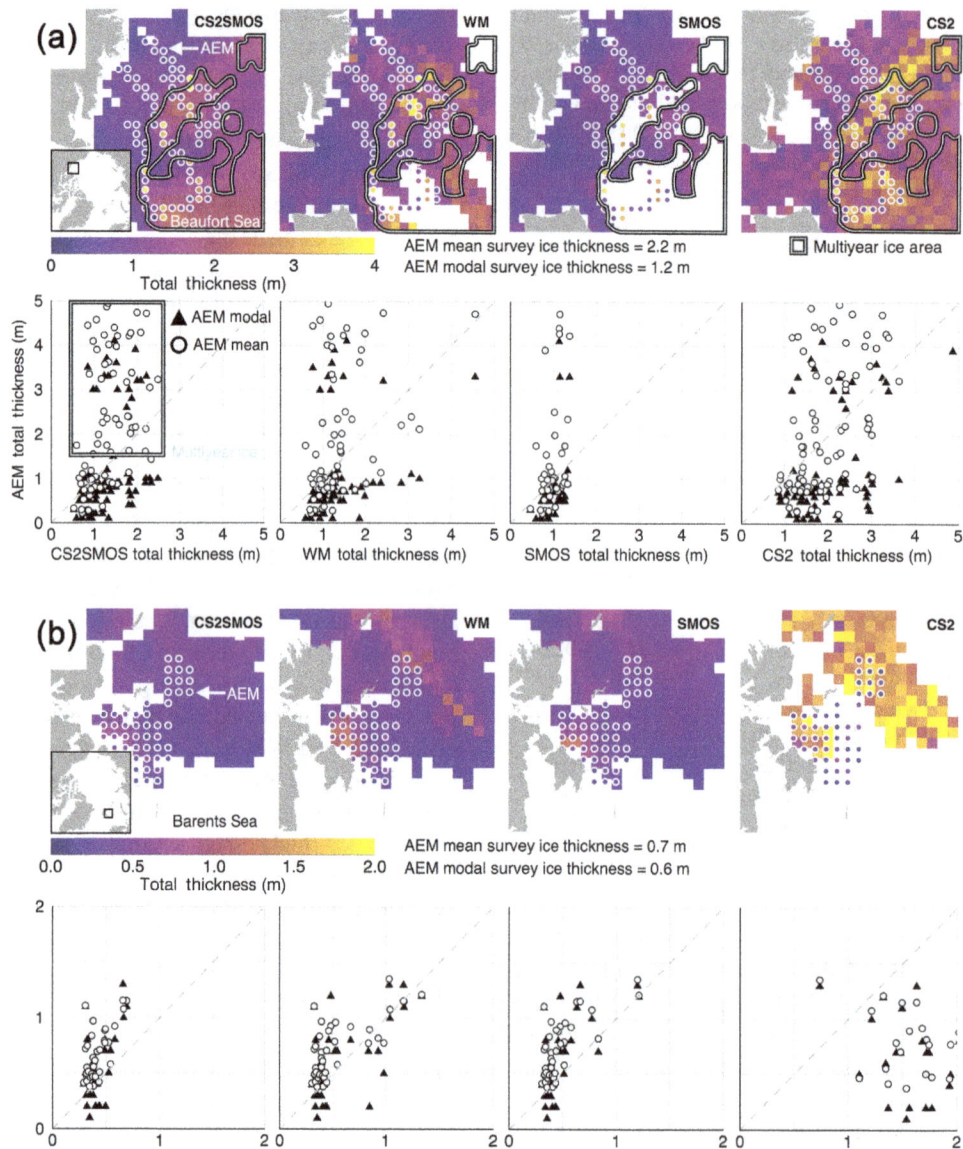

Figure 13. Comparison of satellite retrievals with airborne EM thickness measurements (AEM) over a mixed first-year and multiyear ice regime in the Beaufort Sea in April 2016 (**a**) and over thin ice in the Barents Sea east of Spitsbergen in March 2014 (**b**). AEM data are compared with the optimal interpolation product (CS2SMOS), the weighted mean (WM), the SMOS retrieval and the monthly CryoSat-2 thickness retrieval (CS2). AEM measurements are averaged on the 25 km EASE2 grid, providing mean and modal total thickness within a grid cell. AEM measurements in the scatter plots are capped at 5 m, while in (**a**) one mean AEM grid value exceeds the limit.

and SMOS retrievals (RMSD = 0.27–0.31 m, $r = 0.61$–0.73) than with the CS2 retrieval (RMSD = 0.97, $r = -0.35$). Hence, the reduction in RMSD considering CS2SMOS or WM compared to CS2 is roughly 0.7 m. The observed bias with respect to the mean AEM thickness is -0.25 m for CS2SMOS, -0.17 for WM and -0.24 m for SMOS, suggesting a bias towards thinner ice. The maps and scatter diagrams indicate that the CS2SMOS, WM and SMOS retrievals capture small thickness gradients visible in the AEM thickness data. This comparison provides evidence that using SMOS

data in areas with a thin ice regime will reduce the RMSD and the mean bias when compared to the CS2 product.

5 Conclusions

We presented methods to carry out the first joint data merging of CryoSat-2 sea-ice thickness fields and thin ice thickness estimates obtained from the L-band radiometer onboard the Soil Moisture and Ocean Salinity satellite. While CS2 lacks the capability to observe thin ice, SMOS is restricted to ice regimes thinner than about 1 m. We used two approaches for merging CS2 and SMOS ice thickness data: a weighted mean and an optimal interpolation scheme based on weekly CS2 and SMOS ice thickness grids. While the weighted mean product only provides estimates at grid cells where observations are available, the OI product (CS2SMOS) provides weekly Arctic-wide sea-ice thickness estimates with corresponding uncertainty estimates. We have shown that the merged products have the capability to allow for weekly thickness estimates that are sensitive to the entire thickness range, using the complementary sensitivity of the individual products to different thickness regimes. Moreover, the weekly merged products benefit from increased coverage at lower latitudes in conjunction with higher temporal resolution compared to the CS2 retrieval, which is important for observing ice growth during the freeze-up. In particular, the usage of the combined product will improve thickness retrievals in all areas with thin ice, which we have demonstrated using case studies from the Barents Sea during spring 2014 and Beaufort Sea during spring 2016. Comparisons with airborne electromagnetic thickness measurements reveal a reduction in root mean square deviation of about 0.7 m for CS2SMOS and WM, compared to the CS2 thickness retrieval in the Barents Sea. Moreover, the comparison shows that retrievals that use SMOS data seem to capture small thickness gradients in thin ice regimes, whereas the CS2 retrieval is very noisy. In the Barents Sea, the CS2 retrieval overestimates mean thin ice thickness by 0.8 m, while CS2SMOS, WM and SMOS underestimate it by about 0.2 m. The comparison with the AEM data has also revealed that WM represents a good estimate in regions where weekly data of SMOS and CS2 are available. For the observation of thicker multiyear ice (> 1 m) and mixed ice regimes as in the Beaufort Sea 2016, the CS2 product has the lowest bias, although limitations in capturing high thickness gradients due to heavily deformed ice exist. CS2SMOS, however, exclusively provides weekly ice thickness estimates covering the entire Arctic and combining CS2 and SMOS data. The OI approach used in this study can be adopted to merge sea-ice thickness or freeboard data sets derived from other satellite missions, such as the recently launched European Space Agency mission Sentinel-3, which carries a Ku-band radar altimeter similar to SIRAL onboard CS2.

Author contributions. RR developed the optimal interpolation algorithm and conducted the processing. SH processed the CryoSat-2 orbit files. LK and XT-K were responsible for the SMOS processing. JK processed the AEM data in the Barents Sea. Christian Haas processed the AEM data in the Beaufort Sea. RR wrote the paper, and all co-authors contributed to the discussion and gave input for writing.

Competing interests. The authors declare that they have no conflict of interest.

Acknowledgements. This work has been conducted in the framework of the European Space Agency project SMOS+ Sea Ice (contracts 4000101476/10/NL/CT and 4000112022/14/I-AM). Lars Kaleschke was responsible for the coordination and design of the SMOS+ Sea Ice project. Many thanks go to Matthias Drusch and the entire SMOS+ Sea Ice team. Moreover, this study is associated with the Deutsche Forschungsgemeintschaft (DFG EXC177) and the German Federal Ministry of Economics and Technology (grant 50EE1008). CryoSat-2/SMOS data from 2010 to 2016 are provided by http://www.meereisportal.de (grant REKLIM-2013-04). Jennifer King was funded by the Norwegian research council project "CORESAT" (NFR project number 222681). The helicopter work in the Barents Sea was supported by the Norwegian Polar Institute (NPI). Thanks also to the crews of RV *Lance* and Airlift helicopter and the engineer Marius Bratrein for their help collecting this data. The authors would like to thank the two anonymous referees for their constructive comments, which helped to improve the manuscript.

Edited by: Julienne Stroeve

References

Alexandrov, V., Sandven, S., Wahlin, J., and Johannessen, O. M.: The relation between sea ice thickness and freeboard in the Arctic, The Cryosphere, 4, 373-380, https://doi.org/10.5194/tc-4-373-2010, 2010.

Armitage, T. and Davidson, M.: Using the Interferometric Capabilities of the ESA CryoSat-2 Mission to Improve the Accuracy of Sea Ice Freeboard Retrievals, IEEE T. Geosci. Remote Sens., 52, 529–536, https://doi.org/10.1109/TGRS.2013.2242082, 2014.

Armitage, T. W. K. and Ridout, A. L.: Arctic sea ice freeboard from AltiKa and comparison with CryoSat-2 and Operation IceBridge, Geophys. Res. Lett., 42, 6724–6731, 2015GL064823, https://doi.org/10.1002/2015GL064823, 2015.

Böhme, L. and Send, U.: Objective analyses of hydrographic data for referencing profiling float salinities in highly variable environments, Deep Sea Res.-Pt. II, 52, 651–664, 2005.

Brodzik, M. J., Billingsley, B., Haran, T., Raup, B., and Savoie, M. H.: EASE-Grid 2.0: Incremental but Significant Improvements for Earth-Gridded Data Sets, ISPRS International Journal of Geo-Information, 1, 32–45, https://doi.org/10.3390/ijgi1010032, 2012.

Cox, G. F. and Weeks, W. F.: Equations for determining the gas and brine volumes in sea-ice samples, J. Glaciol., 29, 306–316, 1983.

Eastwood, S.: OSI SAF Sea Ice Product Manual, v3.8 edn., http://osisaf.met.no, 2012.

Haas, C., Lobach, J., Hendricks, S., Rabenstein, L., and Pfaffling, A.: Helicopter-borne measurements of sea ice thickness, using a small and lightweight, digital {EM} system, J. Appl. Geophys., 67, 234–241, https://doi.org/10.1016/j.jappgeo.2008.05.005, 2009.

Helm, V., Humbert, A., and Miller, H.: Elevation and elevation change of Greenland and Antarctica derived from CryoSat-

2, The Cryosphere, 8, 1539–1559, https://doi.org/10.5194/tc-8-1539-2014, 2014.

Hendricks, S.: Validierung von altimetrischen Meereisdickenmessungen mit einem helikopter-basierten elektronischen Induktionsverfahren, Ph.D. thesis, Universität Bremen, 2009.

Hendricks, S., Ricker, R., and Helm, V.: User Guide – AWI CryoSat-2 Sea Ice Thickness Data Product (v1.2), 2016.

Kaleschke, L., Maaß, N., Haas, C., Hendricks, S., Heygster, G., and Tonboe, R. T.: A sea-ice thickness retrieval model for 1.4 GHz radiometry and application to airborne measurements over low salinity sea-ice, The Cryosphere, 4, 583–592, https://doi.org/10.5194/tc-4-583-2010, 2010.

Kaleschke, L., Tian-Kunze, X., Maaß, N., Mäkynen, M., and Drusch, M.: Sea ice thickness retrieval from SMOS brightness temperatures during the Arctic freeze-up period, Geophys. Res. Lett., 39, L05501, https://doi.org/10.1029/2012GL050916, 2012.

Kaleschke, L., Tian-Kunze, X., Maas, N., Ricker, R., Hendricks, S., and Drusch, M.: Improved retrieval of sea ice thickness from SMOS and CryoSat-2, in: Geoscience and Remote Sensing Symposium (IGARSS), 2015 IEEE International, 5232–5235, IEEE, 2015.

Kaleschke, L., Tian-Kunze, X., Maaß, N., Beitsch, A., Wernecke, A., Miernecki, M., Müller, G., Fock, B. H., Gierisch, A. M., Schlünzen, K. H., Pohlmann, T., Dobrynin, M., Hendricks, S., Asseng, J., Gerdes, R., Jochmann, P., Reimer, N., Holfort, J., Melsheimer, C., Heygster, G., Spreen, G., Gerland, S., King, J., Skou, N., Søbjærg, S. S., Haas, C., Richter, F., and Casal, T.: {SMOS} sea ice product: Operational application and validation in the Barents Sea marginal ice zone, Remote Sensing of Environment, 180, 264–273, https://doi.org/10.1016/j.rse.2016.03.009, special Issue: ESA's Soil Moisture and Ocean Salinity Mission – Achievements and Applications, 2016.

King, J., Gerland, S., Spreen, G., and Bratrein, M.: Helicopterborne sea-ice thickness measurements from the 2014 IRO2 /ESA SMOSice campaign cruise in the Barents Sea region [Data set], Norwegian Polar Institute, https://doi.org/10.21334/npolar.2016.ee8f4f8d, 2016.

King, J., Spreen, G., Gerland, S., Haas, C., Hendricks, S., Kaleschke, L., and Wang, C.: Sea-ice thickness from field measurements in the northwestern Barents Sea, J. Geophys. Res.-Oceans, 122, 1497–1512, https://doi.org/10.1002/2016JC012199, 2017.

Kurtz, N. T. and Farrell, S. L.: Large-scale surveys of snow depth on Arctic sea ice from Operation IceBridge, Geophys. Res. Lett., 38, L20505, https://doi.org/10.1029/2011GL049216, 2011.

Kurtz, N. T., Galin, N., and Studinger, M.: An improved CryoSat-2 sea ice freeboard retrieval algorithm through the use of waveform fitting, The Cryosphere, 8, 1217–1237, https://doi.org/10.5194/tc-8-1217-2014, 2014.

Kwok, R.: Simulated effects of a snow layer on retrieval of CryoSat-2 sea ice freeboard, Geophys. Res. Lett., 41, 5014–5020, https://doi.org/10.1002/2014GL060993, 2014.

Kwok, R. and Cunningham, G. F.: Contributions of growth and deformation to monthly variability in sea ice thickness north of the coasts of Greenland and the Canadian Arctic Archipelago, Geophys. Res. Lett., 43, 8097–8105, https://doi.org/10.1002/2016GL069333, 2016.

Kwok, R., Cunningham, G. F., Wensnahan, M., Rigor, I., Zwally, H. J., and Yi, D.: Thinning and volume loss of the Arctic Ocean sea ice cover: 2003–2008, J. Geophys. Res., 114, C07005, https://doi.org/10.1029/2009JC005312, 2009.

Laxon, S., Peacock, N., and Smith, D.: High interannual variability of sea ice thickness in the Arctic region, Nature, 425, 947–950, 2003.

Laxon, S. W., Giles, K. A., Ridout, A. L., Wingham, D. J., Willatt, R., Cullen, R., Kwok, R., Schweiger, A., Zhang, J., Haas, C., Hendricks, S., Krishfield, R., Kurtz, N., Farrell, S., and Davidson, M.: CryoSat-2 estimates of Arctic sea ice thickness and volume, Geophys. Res. Lett., 40, 732–737, https://doi.org/10.1002/grl.50193, 2013.

Marshall, J., Adcroft, A., Hill, C., Perelman, L., and Heisey, C.: A finite-volume, incompressible Navier Stokes model for studies of the ocean on parallel computers, J. Geophys. Res.-Oceans, 102, 5753–5766, https://doi.org/10.1029/96JC02775, 1997.

McIntosh, P. C.: Oceanographic data interpolation: Objective analysis and splines, J. Geophys. Res., 95, 13529–13541, 1990.

Mecklenburg, S., Drusch, M., Kaleschke, L., Rodriguez-Fernandez, N., Reul, N., Kerr, Y., Font, J., Martin-Neira, M., Oliva, R., Daganzo-Eusebio, E., Grant, J., Sabia, R., Macelloni, G., Rautiainen, K., Fauste, J., de Rosnay, P., Munoz-Sabater, J., Verhoest, N., Lievens, H., Delwart, S., Crapolicchio, R., de la Fuente, A., and Kornberg, M.: ESA's Soil Moisture and Ocean Salinity mission: From science to operational applications, Remote Sensing of Environment, 180, 3–18, https://doi.org/10.1016/j.rse.2015.12.025, 2016.

Meier, W. N., Hovelsrud, G. K., van Oort, B. E., Key, J. R., Kovacs, K. M., Michel, C., Haas, C., Granskog, M. A., Gerland, S., Perovich, D. K., Makshtas, A., and Reist, J. D.: Arctic sea ice in transformation: A review of recent observed changes and impacts on biology and human activity, Rev. Geophys., 52, 185–217, https://doi.org/10.1002/2013RG000431, 2014.

Menashi, J. D., St. Germain, K. M., Swift, C. T., Comiso, J. C., and Lohanick, A. W.: Low-frequency passive-microwave observations of sea ice in the Weddell Sea, J. Geophys. Res.-Oceans, 98, 22569–22577, https://doi.org/10.1029/93JC02058, 1993.

Onogi, K., Tsutsui, J., Koide, H., Sakamoto, M., Kobayashi, S., Hatsushika, H., Matsumoto, T., Yamazaki, N., Kamahori, H., Takahashi, K., Kadokura, S., Wada, K., Kato, K., Oyama, R., Ose, T., Mannoji, N., and Taira, R.: The JRA-25 reanalysis, J. Meteorol. Soc. Jpn., 85, 369–432, 2007.

Petty, A. A., Hutchings, J. K., Richter-Menge, J. A., and Tschudi, M. A.: Sea ice circulation around the Beaufort Gyre: The changing role of wind forcing and the sea ice state, J. Geophys. Res.-Oceans, 121, 3278–3296, https://doi.org/10.1002/2015JC010903, 2016.

Pfaffling, A., Haas, C., and Reid, J. E.: Direct helicopter EM — Sea-ice thickness inversion assessed with synthetic and field data, Geophysics, 72, F127–F137, https://doi.org/10.1190/1.2732551, 2007.

Ricker, R., Hendricks, S., Helm, V., Skourup, H., and Davidson, M.: Sensitivity of CryoSat-2 Arctic sea-ice freeboard and thickness on radar-waveform interpretation, The Cryosphere, 8, 1607–1622, https://doi.org/10.5194/tc-8-1607-2014, 2014.

Ricker, R., Hendricks, S., Perovich, D. K., Helm, V., and Gerdes, R.: Impact of snow accumulation on CryoSat-2 range retrievals over Arctic sea ice: An observational approach with buoy data, Geophys. Res. Lett., 42, 4447–4455, 2015GL064081, https://doi.org/10.1002/2015GL064081, 2015.

Ricker, R., Hendricks, S., and Beckers, J. F.: The Impact of Geophysical Corrections on Sea-Ice Freeboard Retrieved from Satellite Altimetry, Remote Sensing, 8, 317, https://doi.org/10.3390/rs8040317, 2016.

Tian-Kunze, X., Kaleschke, L., Maaß, N., Mäkynen, M., Serra, N., Drusch, M., and Krumpen, T.: SMOS-derived thin sea ice thickness: algorithm baseline, product specifications and initial verification, The Cryosphere, 8, 997–1018, https://doi.org/10.5194/tc-8-997-2014, 2014.

Tilling, R. L., Ridout, A., and Shepherd, A.: Near-real-time Arctic sea ice thickness and volume from CryoSat-2, The Cryosphere, 10, 2003–2012, https://doi.org/10.5194/tc-10-2003-2016, 2016.

Vant, M., Ramseier, R., and Makios, V.: The complex-dielectric constant of sea ice at frequencies in the range 0.1–40 GHz, J. Appl. Phys., 49, 1264–1280, 1978.

Wang, X., Key, J., Kwok, R., and Zhang, J.: Comparison of Arctic Sea Ice Thickness from Satellites, Aircraft, and PIOMAS Data, Remote Sensing, 8, 713, https://doi.org/10.3390/rs8090713, 2016.

Warren, S. G., Rigor, I. G., Untersteiner, N., Radionov, V. F., Bryazgin, N. N., Aleksandrov, Y. I., and Colony, R.: Snow depth on Arctic sea ice, J. Climate, 12, 1814–1829, 1999.

Wingham, D., Francis, C., Baker, S., Bouzinac, C., Brockley, D., Cullen, R., de Chateau-Thierry, P., Laxon, S., Mallow, U., Mavrocordatos, C., Phalippou, L., Ratier, G., Rey, L., Rostan, F., Viau, P., and Wallis, D.: CryoSat: A mission to determine the fluctuations in Earth's land and marine ice fields, Adv. Space Res., 37, 841–871, https://doi.org/10.1016/j.asr.2005.07.027, 2006.

Zhang, J., Rothrock, D., and Steele, M.: Recent Changes in Arctic Sea Ice: The Interplay between Ice Dynamics and Thermodynamics, J. Climate, 13, 3099–3114, https://doi.org/10.1175/1520-0442(2000)013<3099:RCIASI>2.0.CO;2, 2000.

11

Combined diurnal variations of discharge and hydrochemistry of the Isunnguata Sermia outlet, Greenland Ice Sheet

Joseph Graly, Joel Harrington, and Neil Humphrey

Department of Geology and Geophysics, University of Wyoming, 1000 E. University Ave. Laramie, WY 82071, USA

Correspondence to: Joseph Graly (jgraly@iupui.edu)

Abstract. In order to examine daily cycles in meltwater routing and storage in the Isunnguata Sermia outlet of the Greenland Ice Sheet, variations in outlet stream discharge and in major element hydrochemistry were assessed over a 6-day period in July 2013. Over 4 days, discharge was assessed from hourly photography of the outlet from multiple vantages, including where midstream naled ice provided a natural gauge. pH, electrical conductivity, suspended sediment, and major element and anion chemistry were measured in samples of stream water collected every 3 h.

Photography and stream observations reveal that although river width and stage have only slight diurnal variation, there are large diurnal changes in discharge shown by the doubling in width of what we term the "active channel", which is characterized by large standing waves and fast flow. The concentration of dissolved solutes follows a sinusoidal diurnal cycle, except for large and variable increases in dissolved solutes during the stream's waning flow. Solute concentrations vary by $\sim 30\%$ between diurnal minima and maxima. Discharge maxima and minima lag temperature and surface melt by 3–7 h; diurnal solute concentration minima and maxima lag discharge by 3–6 h.

This phase shift between discharge and solute concentration suggests that during high flow, water is either encountering more rock material or is stored in longer contact with rock material. We suggest that expansion of a distributed subglacial hydrologic network into seldom accessed regions during high flow could account for these phenomena, and for a spike of partial silicate reaction products during waning flow, which itself suggests a pressure threshold-triggered release of stored water.

1 Introduction

Dissolved load in glacial outlet streams has long been employed as a metric for assessing water–rock interactions occurring beneath glaciers and ice sheets. Glacierized basins have comparable dissolved loads to non-glacial rivers, but are enriched in mobile cations and depleted in Si (Anderson et al., 1997). The chemistry of glacial water typically suggests that observed solute concentrations are reached due to the presence of reactive accessory minerals and fresh mineral surfaces in glacial sediments (Drever and Hurcomb, 1986). Dissolved load is therefore linked to physical erosion in subglacial environments (Anderson, 2005). Dissolved load is also indicative of the degree to which atmospheric gases have been sequestered by chemical processes in the subglacial environment (Hodson et al., 2000).

Diurnal variation of solute concentration is a potential indicator of meltwater routing and storage (e.g., Brown, 2002). Solute concentration is controlled by total water–rock contact over the water residence time in the subglacial environment and by the reactivity of minerals contacted by the water. In particular, two end member cases are expected: if dilution produces an inverse relationship between discharge and solute concentration, minimal changes in water–rock interaction over time are suggested, whereas if increased discharge is coupled to increased solute concentration, diurnal changes in the processes of water–rock interaction or storage are suggested.

Several studies of small alpine glaciers have found solute concentration and discharge to vary inversely, with rising discharges corresponding to falling concentrations of dissolved solutes (e.g., Collins, 1995; Hindshaw et al., 2011; Tranter et al., 1993; Tranter and Raiswell, 1991). Ions produced by saturation-limited reactions, such as calcite dissolution, can show increased load with discharge, but typically with diminished concentration per water volume (Mitchell and Brown, 2007). Elements that are limited by factors such as the rate of sorption/desorption will have constant flux levels and will only be diluted by increased water flow (Mitchell and Brown, 2007). Consequently, correlations between discharge and dissolved load are typically weak (Collins and MacDonald, 2004). These dilution relationships have been attributed to the dominance of channelized flow in alpine environments. In cases for which subglacial water is confined to fixed conduits, increased water flow will expand the size of the conduits and increase the speed of throughflow but will have a minimal impact on the area of water–bed contact (Nye, 1976; Röthlisberger, 1972).

Larger glacial systems have more complex water–rock interactions (e.g., Wadham et al., 2010), and have frequently demonstrated more complex hysteresis in the relationship between discharge and solute concentration. At the outlet of a large glacierized basin in SE Alaska, increases in dissolved load lag spikes in discharge by several days (Anderson et al., 2003). Anderson and others attribute this to storage of water in a distributed system only released during the waning stages of flow. In distributed or linked-cavity flow, increased discharge allows flowing water to spread out across the glacier's bed and thereby increase the area of water-bed contact (Humphrey, 1987; Kamb, 1987). Time series data from the Watson River, near Kangerlussuaq, West Greenland, show out-of-phase variation in discharge and solute concentration, with maximum daily solute concentrations occurring on the rising limb of the discharge hydrograph (Yde et al., 2014). However, on the scale of the melt season as a whole, Yde and others find a strong inverse correlation between discharge and solute concentration, which they attribute to conduits carrying a substantially higher portion of the meltwater flow than the distributed subglacial system. Lags between minimum discharge and peak solute concentrations have also been observed in karst dominated systems, such as Tsanfleuron, Swiss Alps, where flux from groundwater is hydrologically important (Zeng et al., 2012). Such lags are also observed in non-glacial streams and have long been understood to result from the mixing of groundwater, soil water, and surface runoff – each having a unique response time to rainfall events (e.g., Evans and Davies, 1998). The existence of a range of distributed and channelized flow mechanisms under larger ice bodies similarly suggests a range of response times to input from surface melt water.

Seven measurements of dissolved solute chemistry taken from samples collected over the course of 2 days in 2011 at the terminus of Insunnguata Sermia, a major land-terminating West Greenland outlet glacier, potentially show a direct relationship between solute concentration and discharge (Graly et al., 2014; Landowski, 2012). In this limited time series, solute concentration appears to peak at midafternoon, while discharge is high, and minimal in the early morning hours, with total variation of < 20 %. To investigate fur-

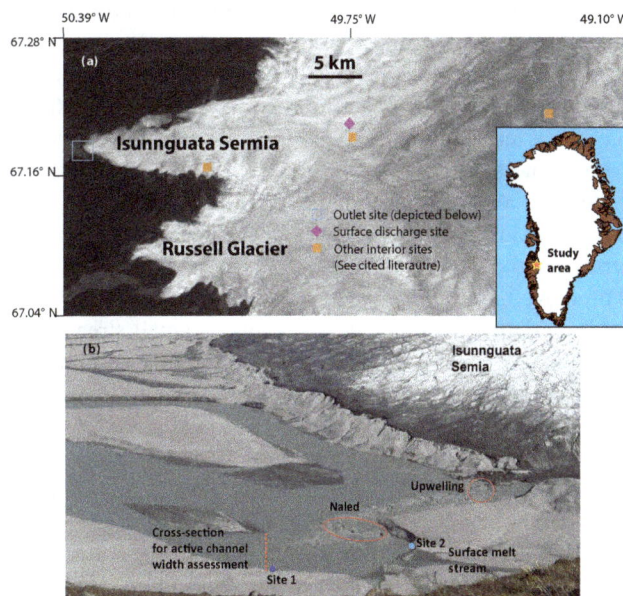

Figure 1. (a) Location of the study area on satellite imagery provided by polar geospatial data center. **(b)** Overhead photograph of the study area taken 21 July 2013 on an overlooking ridge, 400 m above and 900 m away from the stream. Important sampling and observational features are labeled. Samples were collected at site 1 from 10:00 on 16 July 2013 to 11:00 on 18 July 2013. Samples were collected at site 2 from 14:00 h on 18 July 2013 to 20:00 on 21 July 2013. Beginning at 10:00 on 18 July 2013, hourly photographs on the labeled naled and ∼ 100 m cross section were taken.

ther whether a direct relationship exists between discharge and solute concentration, we returned to the same site for a 6-day period of the summer of 2013, collecting eight samples per day for chemical analysis.

2 Field site

Water samples were collected from the terminus of the Isunnguata Sermia outlet of the Greenland Ice Sheet (Fig. 1). The outlet glacier occupies a deeply cut glacial valley, with surrounding hilltops > 400 m above sea level. Deep, glacially-carved trenches continue under the ice sheet for more than 20 km into the interior, where ice depths reach > 1000 m (Jezek et al., 2013). The Isunnugata Sermia outlet has a catchment that encompasses > 2400 km^2 of the ablation zone, making it one of the largest subglacial drainage basins in western Greenland (Palmer et al., 2011). The regional geology consists primarily of Paleoproterozoic gneisses and granitoids (van Gool et al., 2002), providing a silicate bedrock substrate for subglacial chemistry.

Water emerges from a single location on the south side of Isunnguata Sermia's terminus front ∼ 30 m above sea-level (Fig. 1) and traverses a broad, > 100 km long sandur to the fjord. Discharge of pressurized subglacial water creates a large upwelling capable of expelling water multiple meters into the air and, although no fully quantitative measurement could be made, peak discharge was estimated in the hundreds of $m^3 s^{-1}$, as is consistent with typical summer discharge in the nearby Watson River (Hasholt et al., 2013). Ice-cored moraines and frozen outwash shape the course of outlet waters. On the sandur, frozen outwash channels the water into a single thread, although the large sediment load creates rapidly changing channel and bed geometry. Near the terminus, the frozen outwash of the sandur is elevated ~ 2–4 m above the discharging stream. The main stream is also fed by minor ice surface melt streams. A small stream that runs along the south lateral margin of the glacier joins the main terminal outlet stream just below the primary upwelling site.

This work was performed as part of a wider program of coordinated studies of the Isunnguata Sermia terminus region. Hot water boreholes along a transect from the outlet to 40 km upstream have provided data regarding water pressure (Meierbachtol et al., 2013), ice temperature (Harrington et al., 2015), subglacial water chemistry (Graly et al., 2014), and mass balance between subglacial sediment and rock (Graly et al., 2016). More limited data sets from Isunnguata Sermia's terminal and lateral outlets were reported for the 2010, 2011, and 2012 seasons (Graly et al., 2014). The work reported here is based on samples and data collected over a 6-day period from 16 to 21 July 2013.

3 Methods

3.1 Discharge

Discharge measurements of the outlet were difficult to obtain. There is no exposed bedrock near the stream to act either as an elevation reference or to stabilize the river bed. Obtaining accurate cross profiles of the stream was prohibitively dangerous, with high flows, collapsing banks, and a considerable flux of mobile ice blocks. Attempts to install a stage pole were frustrated by considerable stage variation over time associated with cutting and filling of the stream bed. Once the stream opens out from its restriction by remnant glacier ice near the upwelling, stage is poorly correlated with discharge. The stream instead scours sediment during the rising limb and deposits it on the trailing limb of the daily hydrograph. It was decided that only relative discharge should be assessed. This was aided by repeat photography from two fixed locations.

Hourly stream photography began at 10:00 on 18 July and continuing through 20:00 on 21 July. From one vantage point, the central upwelling of subglacial water was photographed from the south, as it poured out from around the moraine. This vantage captured a ~ 1 m-high, mid-channel naled formed from freezing of outlet waters during winter months. The naled was variably covered or exposed as discharge varied and acted as a stream gauge in this respect.

This portion of the stream is restricted by frozen sediment and stream height is controlled by discharge.

A second vantage, from a rise above the south bank, captured a ~ 200 m long stretch of the outlet stream. In this portion of the stream, increased discharge caused scour and expansion of the stream's active channel. Photography allowed for an assessment of relative active channel width. Large waves and faster velocities are confined to this active channel. The distance between the upstream end of a persistent, midstream point bar and a distinct feature on the south shore was measured on each photograph (Fig. 1). The length of the portion of this transect characterized by large waves and flow features was also measured, allowing for the calculation of the fraction of the stream width contained by the active channel. On most of the photographs, the break between the large standing waves and the surrounding quiescent flow was unambiguous. The second vantage also allowed for an assessment of flow state and Froude number from the presence of features such as standing waves.

During the first 2 days of the sampling period, stream surface velocity was measured by having a person repeatedly run in pace with the movement of the stream surface along a 100 m stretch of the sandur. This was accomplished by observing visually consistent mobile features of the stream, such as lineations within waveforms and small pieces of floating ice. Measurements were taken during morning, afternoon, and evening stages to assess variation in velocity associated with high and low flow. During this earlier period, changes in the width of the active channel and volume of the water pouring over the naled were also observed (though without systematic photography from a consistent vantage).

3.2 Interior surface melt

In order to compare variation in terminus discharge to melt in the surface interior, we also consider discharge measurements from an interior ice sheet surface stream. The stream was gauged during June of 2012, so the data are not directly comparable to the measurements collected in 2013. However inasmuch as interior melt is primarily controlled by insolation, the stream's variation likely represents a typical pattern for the timing and scaling of diurnal summer surface melt fluctuations. Coincidentally, the progression of the Greenland Ice Sheet melt season was fairly comparable between June 2012 and July 2013, with ablation rates of 6–10 Gt day^{-1} during both periods (Langen et al., 2013). The supraglacial stream was gauged during a period in which bare ice was melting, so water retention in snow did not affect its hydrology.

The surface stream was located at 67.2° N and 49.8° E, ~ 25 km from the terminal outlet. Stream height was gauged with a calibrated pole drilled into ice. Surface velocity was measured by timing floating ice along a course of known distance. Cross-sectional area was directly measured in the region where the gauge was emplaced and calibrated to gauge height. Transect slope was measured by pole and automatic level. Six measurements of surface velocity were used to calculate an average Manning coefficient from the measured slope and hydraulic radius of the stream. Discharge was then calculated from change in gauge height. Stage height was measured every half hour or hour for a period from 11:30 on 18 June 2012 to 20:00 on 21 June 2012. During 18, 19, and 20 June, sunny weather predominated; 21 June had rainy, cooler weather.

3.3 Water sampling

Water sampling of Isunnguata Sermia's terminal outlet began at 08:00 local time on 16 July 2013 and continued in 3 h increments through 20:00 on 21 July 2013. Temperatures in the interior ablation zone measured at PROMICE KAN_M weather station stayed at ~ 0.5 positive degrees per day over 16–18 July and steadily rose to 1.3 positive degrees per day over 19–21 July. Samples were collected by lowering a liter Nalgene polypropylene bottle attached to an adjustable-length pole into discharging waters within 400 m of the subglacial upwelling. The bottle was dipped and rinsed in flowing stream water prior to final sample collection. Samples were initially collected from the south bank of the outlet stream, from the banks at the beginning of the outwash plain (Fig. 1; Site 1). During 17 July, the main course of the river shifted so that location had diminished flow and an emerging bank channeled waters from the lateral side stream to the location. Commencing at 14:00 on 18 July, sampling was relocated above the lateral side stream on the banks of the terminal moraine (Fig. 1; Site 2). The sampling location was not subsequently changed. Excepting periods in which the emerging bank channeled lateral stream water to the first location, both locations sampled water from the main subglacial outlet and should produce comparable results.

Upon collection, 125 mL of each water sample were pumped through 0.1 μm nylon filters, with filtered water and filter papers saved for laboratory analyses. A colorimetric alkalinity test, a conductivity measurement, and a pH measurement were performed on the remaining unfiltered sample. Alkalinity tests were performed with a Hach Model AL-AP alkalinity test kit. Results of field alkalinity tests were only accurate to 25 μM. Alkalinity was therefore also calculated by charge balance from the other measured ions. pH measurements and conductivity measurements were performed with Beckman–Coulter Φ460 multi-parameter meter. pH was measured using a low ionic strength probe, with a three-point calibration employed daily.

Subsequent water analyses were performed in the University of Wyoming Aqueous Geochemistry Lab. Concentrations of Si, Ca, Mg, Na, and K were measured on a Perkin Elmer Elan 6000 inductively coupled plasma quadrupole mass spectrometer (ICP-MS). Concentrations of SO_4^{2-}, Cl^-, NO_3^-, and Fl^- were measured on a Dionex ICS 500 ion chromatograph. Element and ion analyses were measured to-

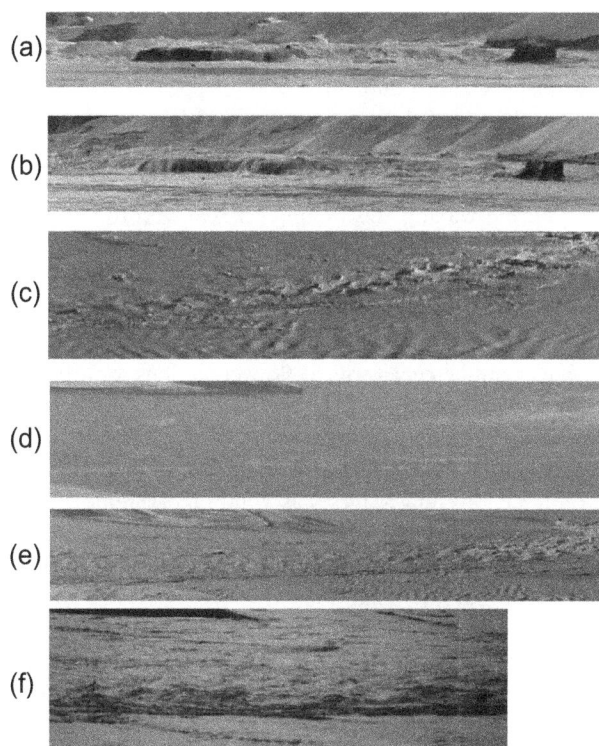

Figure 2. Photographs of typical flow patterns in the Isunnguata Sermia outlet. **(a)** Midstream naled exposed at during low flow (08:00). **(b)** Midstream naled covered during high flow (21:00). Panels **(c)**–**(f)** show images of flow captures from the overhead vantage during low (08:00), waxing (14:00), high (21:00) and waning (00:00) stages. Waxing and waning stages show different wave morphology but maintain standing wave features.

gether with procedural blanks, which were consistently measured below the lower limits of detection. The filter papers were dried overnight at 85 °C and weighed to assess suspended load.

4 Results

4.1 Discharge

Over the 4 days during which repeat photographic observations were made, photographs of the naled show consistent minima at 08:00, with the naled mostly exposed, and a small volume of water overtopping a portion of the ice body (Fig. 2). On 18 and 19 July, the naled was completely covered by flowing water from 19:00 to 00:00. On 20 July, it was covered from 16:00, and remained covered for the remainder of the study period.

Maximum discharge is harder to determine from observations of the naled alone. Once the naled is completely covered in water, visual interpretation of maximum flow is ambiguous. Some discrimination can be made based on the height of the covered naled feature compared to the surrounding waves and the angle at which the water pours over the naled (greater flows overtop the naled at a lower angle). From these features, maximum stream flow appeared to occur at 21:00 on 18 and 19 July, and 20:00 on 20 July.

Standing waves are observed at all flows (Fig. 2), although substantial differences in wave and surface morphology were noted during waxing and waning phases, with rougher water in waning flow and smoother water in waxing flow. The roughness change may represent a change in the sediment load of the river between the erosive waxing stage and the depositional waning stage. The persistence of standing waves implies near-critical-flow conditions, or a Froude number of approximately 1, for the entire study period. Measurements of stream velocity showed surface speeds of $2.86 \pm 0.12 \, \mathrm{m \, s^{-1}}$ (2σ, $n = 6$). Variation in velocity between morning and evening stages was within measurement error. The lack of relationship between stage and discharge and velocity has been noted before in sediment-laden glacial rivers (Humphrey and Raymond, 1994).

Based on calculations from a Froude number of 1 (i.e., stream velocity squared is equal to stream depth times acceleration from gravity) and assuming a total velocity within 20 % of surface velocity, stream depths of 0.5–0.9 m are suggested. These depth estimates were supported by observing ice blocks rolling or bouncing down the flow. The active channel's approximately constant stream velocity and persistent standing waves suggest a fairly constant stream depth. Wide areas of shallow slow water remained present during low flows and the total surface area of the stream remained approximately constant. Pole probing of these shallow areas suggests 10 to 20 cm depths. Because the active channel has an order of magnitude greater discharge per transect meter than the stream's marginal areas, we infer that the cross-sectional area of the active channel is the primary control on discharge. Rising discharge is accommodated by scouring on the margins of the active channel; falling discharge is accommodated by deposition.

Assessment of the active channel width from repeated photography shows substantial differences between morning hours ($\sim 05{:}00$–10:00), where 20–30 % of the stream is comprised of active channel characteristics, and late afternoon/evening hours ($\sim 18{:}00$–00:00) when > 40 % of the stream is comprised of active channel characteristics. These observations are generally consistent with assessments of the height of water pouring over the naled (Fig. 3). Though repeated photography from a consistent vantage was not performed during the first 2 days of the study, field observations and photographs from that period show similar changes in the active channel width and naled overflow.

4.2 Interior surface stream

The calculated Manning coefficient for the interior stream was 0.0117 ± 0.0018 (2σ). Its discharge varied by as much as an order of magnitude during the course of diurnal cy-

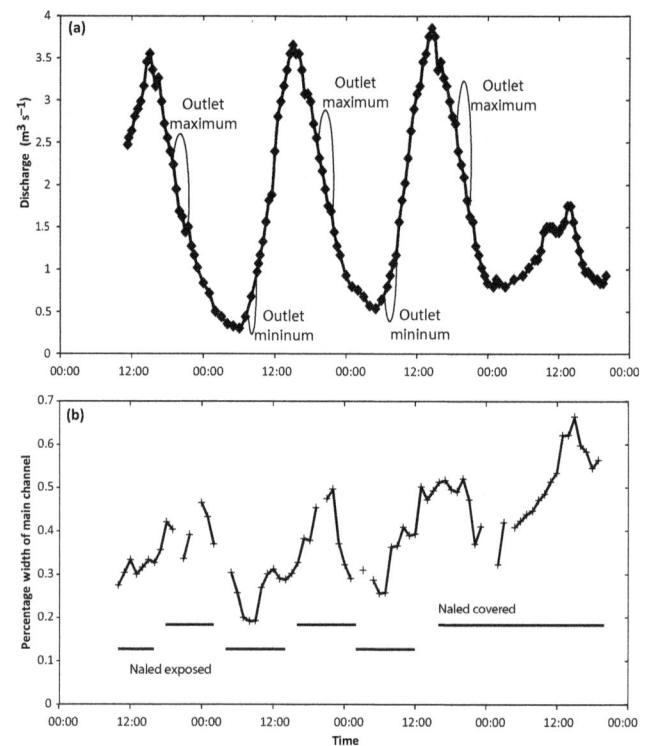

Figure 3. (a) Measured discharge in an interior surface stream over a 4-day period in 2012 compared to time ranges of maximum and minimum discharge as suggested by qualitative observation of flow volumes over midstream naled ice in the outlet of Isunnguata Sermia during the study period. **(b)** Assessment of percentage of distance between a point bar and the shore that is characterized by large waves suggestive of deep, fast flow. Periods of time in which the midstream naled is exposed and covered are included for comparison.

cles, with low values as small as $0.3 \, \mathrm{m^3 \, s^{-1}}$ and high values greater than $3.5 \, \mathrm{m^3 \, s^{-1}}$ (Fig. 3). Minimum stage heights consistently occurred around 04:00. Maximum stage heights consistently occurred at 14:00 or 15:00. These data contrast with our observations of water pouring over the naled. The naled minimum occurred approximately 4 h later than minimum of surface melt. The naled maximum occurred approximately 6 h later than maximum surface melt. This delay is representative of the integration of the travel time delays from the entire glacier catchment.

4.3 Water analyses

The sampled waters were generally chemically dilute, with 292 ± 50 micromoles per liter dissolved solutes (Table 1). Ca was the dominant cation, followed by Na, K, and Mg (Fig. 4). Mg abundances were an order of magnitude lower than the other major cations. Dissolved Si occurred at comparable abundance to Na. Standard deviations of the mass spectrometer measurements were < 1 % of the measured value. Bicarbonate (measured as alkalinity) was the dominant an-

Table 1. Field and laboratory measurements (results in micromolarity unless otherwise noted).

Sample	Date and time	pH	Electrical conductivity (μS)	Suspended sediment (gl^{-1})	Field alkalinity	Calculated alkalinity	Ca	K	Mg	Na	Si	Fl$^-$	Cl$^-$	NO$_3^-$	SO$_4^{2-}$
GL13-1	16 7 2013 8:10	8.6	9.7	1.84	150	141.2	63.1	23.2	6.4	38.1	38.7	0.7	18.2	0.8	19.7
GL13-2	16 7 2013 11:00	8.8	11.0	1.92	175	146.5	66.0	25.7	6.3	38.8	44.7	2.9	16.7	0.8	21.2
GL13-3	16 7 2013 14:00	8.7	11.5	NA	150	133.9	61.9	23.1	6.4	40.2	33.0	0.7	20.6	1.4	21.7
GL13-4	16 7 2013 17:00	8.7	9.4	NA	150	128.8	56.3	21.6	6.9	33.5	27.3	0.7	14.7	0.8	18.3
GL13-5	16 7 2013 20:00	8.7	10.0	3.28	150	114.8	50.3	15.6	4.9	28.6	27.8	0.6	7.5	0.5	15.6
GL13-6	16 7 2013 23:00	8.8	11.2	2.24	150	129.1	55.1	24.3	5.6	40.9	33.8	0.7	23.4	1.7	15.9
GL13-7	17 7 2013 2:00	8.7	8.8	2.80	150	124.0	56.0	18.8	5.5	29.7	33.0	0.6	11.5	<0.5	17.6
GL13-8	17 7 2013 5:00	8.8	10.3	1.60	175	135.9	59.7	27.0	5.9	38.9	34.3	<0.4	22.6	<0.5	19.4
GL13-9	17 7 2013 8:00	8.9	13.0	2.64	200	160.1	77.1	24.7	6.8	42.1	38.9	0.7	22.3	3.9	23.8
GL13-10	17 7 2013 11:00	8.5	12.7	2.40	150	127.4	61.8	22.1	6.5	33.6	34.5	0.7	15.2	1.2	23.8
GL13-11	17 7 2013 14:00	8.0	7.3	1.04	100	60.7	31.9	9.6	4.7	16.0	23.3	0.5	6.6	<0.5	15.6
GL13-13	17 7 2013 17:00	8.3	6.5	1.68	100	75.3	36.0	15.1	4.4	20.1	23.1	0.5	11.9	<0.5	14.1
GL13-14	17 7 2013 20:00	8.6	8.2	1.36	100	82.4	38.7	15.4	4.8	19.2	25.9	0.4	8.8	<0.5	15.0
GL13-15	17 7 2013 23:00	8.8	9.1	3.52	150	129.4	51.2	19.2	5.1	31.5	29.7	1.3	13.5	<0.5	9.5
GL13-16	18 7 2013 2:00	8.7	8.0	3.20	150	124.0	52.5	20.3	5.2	31.4	29.9	0.7	12.5	0.9	14.6
GL13-17	18 7 2013 5:00	8.5	10.7	2.88	150	131.3	58.2	21.7	6.1	32.7	33.8	0.9	14.6	0.8	17.7
GL13-18	18 7 2013 8:00	8.9	12.7	3.52	150	132.2	58.3	26.5	5.8	33.7	33.5	0.9	12.6	<0.5	21.5
GL13-19	18 7 2013 11:00	8.5	9.8	2.56	150	104.8	50.3	21.5	5.8	30.1	30.9	0.7	18.4	<0.5	20.0
GL13-20	18 7 2013 14:00	8.5	11.0	3.60	175	127.9	56.0	26.3	5.5	32.8	30.8	2.0	14.2	0.8	18.6
GL13-21	18 7 2013 17:00	8.7	9.4	3.60	150	112.8	48.9	22.9	4.7	30.1	28.1	0.6	13.8	0.7	16.2
GL13-22	18 7 2013 20:00	8.6	7.6	4.00	125	106.6	45.9	20.2	4.4	29.8	27.2	0.6	15.0	1.0	13.7
GL13-23	18 7 2013 23:00	8.6	7.1	3.20	100	90.4	44.9	19.5	4.4	33.3	26.0	0.5	34.9	0.6	12.5
GL13-24	19 7 2013 2:00	8.7	14.5	3.84	125	119.5	58.1	23.1	5.0	30.8	32.6	0.7	29.4	1.1	14.6
GL13-25	19 7 2013 5:00	8.3	9.4	2.56	150	123.8	50.3	23.6	4.8	28.4	30.2	<0.4	8.5	<0.5	15.0
GL13-26	19 7 2013 8:00	8.6	10.2	2.80	150	125.8	56.1	27.1	4.9	38.4	28.6	0.8	24.9	1.3	17.4
GL13-27	19 7 2013 11:00	8.7	10.5	2.64	150	129.9	59.2	28.3	5.3	35.1	30.4	0.8	18.6	2.1	20.5
GL13-28	19 7 2013 14:00	8.5	11.9	2.24	150	119.1	52.0	23.1	4.8	31.0	28.3	0.7	8.7	1.1	19.1
GL13-29	19 7 2013 17:00	8.4	11.7	3.44	125	116.9	51.3	25.0	4.7	31.0	26.1	0.7	16.4	1.5	16.3
GL13-30	19 7 2013 20:00	8.6	7.7	3.84	100	106.8	45.0	23.9	4.3	29.6	24.8	0.7	15.1	1.2	14.2
GL13-31	19 7 2013 23:00	8.6	7.4	2.96	100	104.7	42.7	24.5	4.0	29.3	23.9	0.6	14.6	0.8	13.3
GL13-32	20 7 2013 2:00	8.5	15.5	3.12	200	204.6	64.6	42.3	6.2	71.0	27.8	0.6	6.6	8.5	17.2
GL13-33	20 7 2013 5:00	8.0	8.6	3.68	150	129.8	55.7	24.5	4.8	30.5	26.8	0.7	10.1	1.8	16.9
GL13-34	20 7 2013 8:00	8.4	9.7	3.20	150	124.1	54.2	22.6	5.0	33.4	28.5	0.8	8.5	<0.5	20.5
GL13-35	20 7 2013 11:00	8.3	11.5	2.16	150	127.0	55.0	28.4	5.0	33.3	29.6	0.8	8.9	1.3	21.8
GL13-36	20 7 2013 14:00	8.4	9.9	2.88	150	117.9	51.9	24.1	4.8	32.5	28.5	0.8	8.7	1.3	20.7
GL13-37	20 7 2013 17:00	8.5	10.0	1.84	150	111.8	48.2	22.3	4.6	31.3	27.6	1.3	10.9	1.2	17.0
GL13-38	20 7 2013 20:00	8.7	9.1	2.00	125	102.8	42.1	18.1	4.1	29.1	24.6	0.7	7.4	<0.5	14.4
GL13-39	20 7 2013 23:00	8.1	6.6	1.44	100	81.6	42.1	15.5	5.5	22.8	29.4	<0.4	10.5	<0.5	19.2
GL13-40	21 7 2013 2:00	8.8	7.7	1.36	150	127.8	48.7	25.0	4.8	41.0	28.6	0.8	12.3	1.3	15.4
GL13-41	21 7 2013 5:00	8.2	7.2	3.68	150	110.5	45.8	21.1	4.5	34.1	25.9	0.7	12.1	1.2	15.6
GL13-42	21 7 2013 8:00	8.7	8.2	3.68	150	110.0	47.6	18.8	4.5	31.8	27.0	0.7	8.6	1.3	17.0
GL13-43	21 7 2013 11:00	8.4	8.8	2.72	150	112.0	46.9	20.9	4.5	33.3	26.8	0.8	8.3	0.7	17.6
GL13-44	21 7 2013 14:00	8.1	10.2	3.36	150	114.6	47.5	21.5	4.5	30.7	26.8	<0.4	9.0	<0.5	16.3
GL13-45	21 7 2013 17:00	8.3	8.1	2.64	150	102.7	44.4	17.1	4.0	29.6	24.2	1.0	10.2	1.4	14.1
GL13-46	21 7 2013 20:00	8.6	7.4	3.92	150	101.7	40.9	18.3	4.0	27.8	24.8	<0.4	7.7	<0.5	13.3

ion. SO_4^{2-} and Cl^- are detected in all samples, but occur at a concentration that is an order of magnitude lower. Trace amounts of NO_3^- and Fl^- were detected in some samples, at values an order of magnitude below SO_4^{2-} and Cl^- concentrations (Fig. 4). On average, field alkalinity measurements exceeded the alkalinity estimates from charge balance by $25 \pm 14 \,\mu M$ (2σ). Some over-measurement in the field titration is expected, as the value is recorded at the level at which the color tracer disappears (and therefore is a maximum compared to previous drop). Charge imbalance may also result from absorption of H^+ particles to suspended sediment in unfiltered water. Field electrical conductivity measurements showed similar results to the sum of laboratory-measured inorganic ions ($p < 0.0001$; Fig. 4). Suspended sed-

iment concentration did not show a consistent correlation or anti-correlation with dissolved load (Fig. 4).

The relative abundances of cation species are comparable to measurements taken at the Isunnguata Sermia terminus in the summer of 2011 (Graly et al., 2014). The SO_4^{2-}/alkalinity ratios are diminished compared to those measured in 2011, but are comparable to those found in samples collected in 2010 and 2012. The concentrations of suspended sediment are similar to those observed at nearby Leverett Glacier during the summer of 2010 (Cowton et al., 2012).

When normalized to average concentration, the magnitude and timing of cation and silica concentration variation were highly consistent between species over time (Fig. 5). Covariation of all cation and Si species is statistically significant

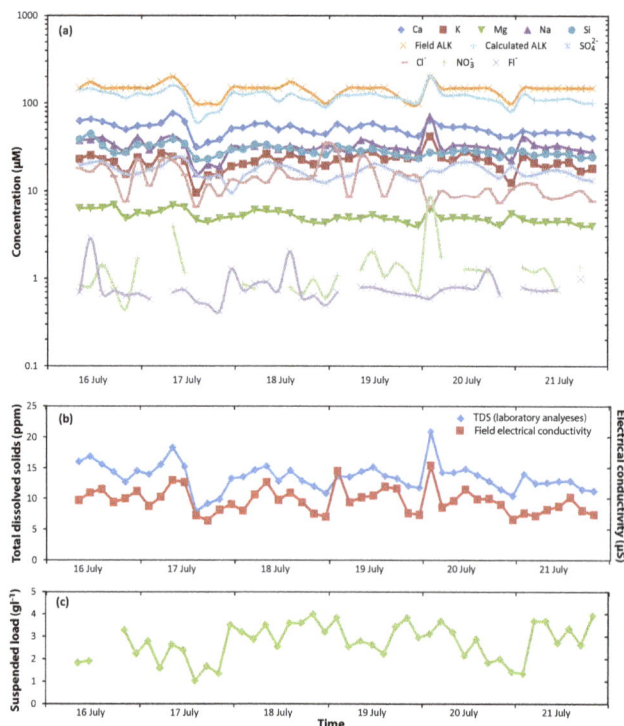

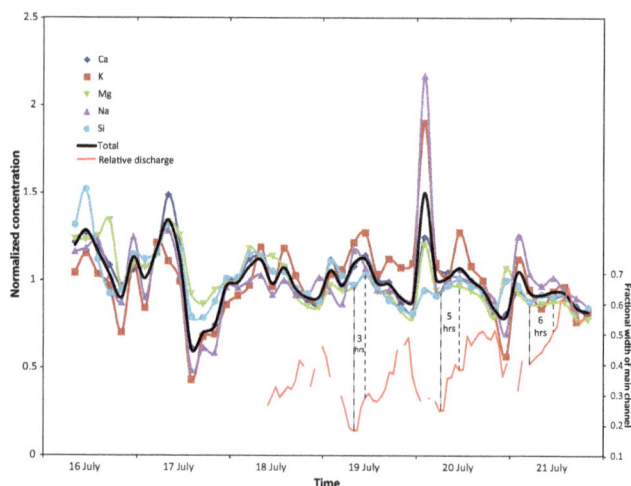

Figure 5. Concentration of dissolved cations and Si normalized to average concentration. Discharge from relative active channel width is shown for comparison. Lags between active channel width channel minima and solute concentration maxima are illustrated with dashed lines.

Figure 4. (a) Concentration of dissolved constituents in sampled waters over time, including laboratory measurements of cations and Si by ICP-MS, anions by ion chromatography, and field measurements of alkalinity (ALK). Alkalinity as calculated by charge balance is also depicted. **(b)** Total dissolved solids from the sum of the laboratory measurements and charge balance alkalinity (HCO_3) compared to field conductivity measurements. Covariation is statistically significant ($p < 0.0001$). **(c)** Dry weight of suspended sediment on filters.

with $p < 0.05$. Covariations of K-Mg, K-Si, and Na-Si have p-values ranging from 0.01 to 0.05; all others are < 0.0001. All cations and silica concentrations followed a diurnal pattern, with higher concentrations present during morning and early afternoon hours and significantly lower concentrations present during later afternoon and evening hours. In several of the studied cycles, large changes in total concentration were limited to the 20:00 and 23:00 samples, which are substantially lower than the other samples collected throughout the day.

The 11:00, 14:00, 17:00, and 20:00 samples from 17 July had substantially lower solute concentrations than would be otherwise suggested by diurnal fluctuations observed elsewhere in the record. This corresponds with the period during which an emerging bank partially separated site 1 from the main channel, allowing a surface-fed side stream to dilute the water.

At 02:00 on 20 July, there was a $> 60\%$ spike in total concentration of all cations. Similar, but smaller magnitude spikes were also present during the other four measured periods of waning flow. The largest of these spikes are substan-

tially more expressed in Na and K concentrations than in Ca, Mg, or Si (Fig. 5). During the final measured overnight spike, elevated concentrations of Mg and Si preceded those of Ca, Na, and K by a 3 h measuring period. The large variability in the magnitude of these spikes suggests that the 3 h sampling schedule was insufficiently frequent to characterize them entirely.

Anions generally followed similar patterns, but with greater variability (Fig. 4). In particular, Cl^- did not covary with other ions toward the end of the record; the 02:00 spike on 20 July coincided with a drop in Cl^- concentration. SO_4^{2-} concentrations increased very minimally during the waning flow spikes, and in one case declined. During the final spike, SO_4^{2-} rose during the 11:00 period, together with Mg and Si.

Excluding the spikes that occurred during waning flow, the highest concentration of dissolved solids occurred at 11:00 on 16, 19, 20, and 21 July (Fig. 5). On 17 and 18 July, the 11:00 sample was likely diluted by the side stream (which was a significant component of flow to site 1 during that period). Concentration minima were reached at 23:00 on 17 through 20 July. On 16 July, the minimum occurred in the 20:00 sample. The maximum solute concentration ranged between 22 to 49% larger than the minimum concentration, with an average daily value of $31 \pm 9\%$.

5 Discussion

5.1 Discharge and outlet stream observations

Observations from oblique photography suggest large diurnal changes in discharge. The width of the active channel, with deeper faster water, approximately doubles in the course of the day (Fig. 3b). An approximate doubling of discharge is also suggested by observations of the midstream naled. The naled is of comparable scale to the depth of the stream (both order of 1 m). Its exposure during low flow and burial during high flow suggests a change in stage comparable to its height. At the naled site, increased width of active channel flow is restricted by ice. Increases in flow height at the naled location are therefore approximately equivalent to increases in active channel width downstream.

During high flows, diurnal increases in discharge of up to 50 % of base value are observed in the Watson River at Kangerlussuaq, where a bridge over a narrow gorge has allowed for the construction of a reliable gauge (Hasholt et al., 2013). Contrastingly, Smith et al. (2015) found minimal diurnal variation in discharge at Isunnguata Sermia terminus. However, as Smith and others estimated discharge based solely on the surface area of the outlet stream water, their analysis missed the variation in the width of the active channel and height of its flow over static features that we present. Based on our limited observational record, it appears that changes in discharge at the Isunnguata Sermia terminus are similar or larger in magnitude to those recorded at the Watson River.

5.2 Diurnal changes in solute flux

The lag between relative discharge minima and maximum solute concentrations is similar to other glacial and nonglacial systems where waters of differing response times are merged into a single stream. Similar lags are observed when groundwater, soil water, and surface flow mix into a stream after a rainfall event (Evans and Davies, 1998). Substantial lags between discharge and solute flux are also observed where glacial melt mixes with groundwater in a karstic system (Zeng et al., 2012) and in the mixing of marginal melt streams with a subglacial pool in a polythermic setting (Skidmore and Sharp, 1999). Even in a small alpine system, observable chemical differences between the leading and lagging limbs of the discharge hydrograph have been attributed to a mixing of englacial and subglacial waters (Tranter and Raiswell, 1991).

In the context of the Greenland Ice Sheet, periods of inphase change between discharge and solute concentration are best explained by the flushing of a linked cavity or other distributed hydrological system as hydraulic pressure rises. Seasonal changes in ice velocity in this sector of the Greenland ice sheet have been linked to a combination of distributed and channelized subglacial flow (Bartholomew et al., 2011). Dye tracing of the hydrological connections between moulins and glacial outlets has also indicated a mixture of subglacial flow regimes (Chandler et al., 2013). Though the single upwelling structure of the terminus of Isunnguata Sermia implies locally channelized flow, observations of water pressures at interior sites (Meierbachtol et al., 2013) and hydrologic theory for low ice surface slopes (Werder et al., 2013) both suggest

that much of the catchment interior has a distributed flow system.

The sudden increases in solute concentration during waning flow suggest that discharge from subglacial regions with high concentrations of dissolved solutes is triggered when a threshold is reached. To our knowledge, the release of stored water during waning flow has only been previously documented on a multiday scale (Anderson et al., 2003), whereas here it occurred as part of the diurnal melt cycle. For slow-moving, distributed subglacial water to be both flushed by rising hydraulic pressures and released from storage by falling hydraulic pressures, multiple subglacial flow paths or mechanisms must be operating simultaneously.

The contrast in solute chemistry between the long-wavelength increases in solute concentration (in which, all major chemical constituents respond comparably) and the waning flow spikes (in which Na, K, and alkalinity dominate) suggests differing subglacial environments and mechanisms. Na- and K-dominated waters likely form in settings where water–rock interactions occur only over a limited time, such that cation exchange occurs on fresh feldspar and mica surfaces, but complete silicate dissolution and clay precipitation do not occur (Blum and Stillings, 1995; Graly et al., 2014). The lack of constituents derived from reactive accessory minerals such as pyrite (i.e., SO_4^{2-}) implies that the waters were reacting with sediment that has been depleted of accessory minerals. Such accessory mineral depletion can occur if sediment residence time in the subglacial system is sufficiently long (Graly et al., 2014). Sampling of sediment beneath ice boreholes has shown the greatest chemical depletion in portions of the ice sheet most influenced by distributed flow (Graly et al., 2016).

This variation in water chemistry suggests that the spike of chemical solutes comes from water that has temporarily entered regions of distributed flow as a part of a diurnal cycle. Modeling of subglacial water pressures suggests that near the ice sheet margin, water flows from conduits to the distributed cavity system at high conduit water pressures and back to conduits at low pressures (Meierbachtol et al., 2013). The spikes in solute concentration result from the crossing of a pressure threshold that allows water stored during high flow to suddenly enter the glacial outlet system. Solute concentration spikes might also be explained by creep closure of linked cavities that opened during high flow and expulsion of remaining solute-concentrated water. Anderson and others (2003) proposed a similar creep closure mechanism to explain increases in solute concentration during waning flow that occurred on a multiday scale in a mountain glacier setting. Following the Glen–Nye relation, the rate of creep closure of ice scales to approximately the third power of effective pressure (Cuffey and Paterson, 2010). Differences in the timing of these effects between ice sheets and mountain glaciers can therefore be explained by differences in ice thickness.

6   Conclusions

A semi-quantitative relative discharge record can be constructed through hourly photographic monitoring of the static and dynamic features of a large, sediment-laden glacial outlet stream. These assessments suggest large diurnal changes in discharge over the study period at the Isunnguata Sermia outlet of the Greenland Ice Sheet (cf. Smith et al., 2015). Simultaneously collected chemical measurements show substantially smaller fluctuation in dissolved load; thus this Greenland outlet glacier does not show the discharge-driven dilution of solute concentration that is common in smaller ice masses. Periods in which dissolved solute concentrations increase and decrease along with discharge, and abrupt and variable increases in solute concentration during waning flow imply that significant contributions to the solute load are made by changes to the routing and storage of meltwater in the subglacial system over the course of the day. In particular, these results indicate considerable diurnal exchange of water between the conduit and linked cavity drainage systems, as well as implying threshold pressure conditions for these exchanges.

Competing interests. The authors declare that they have no conflict of interest.

Acknowledgements. This work would not have been possible without funding from the Greenland Analogue Project (SKB, Posiva, NWMO) and NSF grant ARC-0909122. Janet Dewey assisted with laboratory analyses. Data from the Programme for Monitoring of the Greenland Ice Sheet (PROMICE) were provided by the Geological Survey of Denmark and Greenland (GEUS) at http://www.promice.dk. Thoughtful reviews by editor Rob Bingham and an anonymous referee greatly improved the manuscript.

Edited by: R. Bingham

References

Anderson, S. P.: Glaciers show direct linkage between erosion rate and chemical weathering fluxes, Geomorphology, 67, 147–157, 2005.

Anderson, S. P., Drever, J. I., and Humphrey, N. F.: Chemical weathering in glacial environments, Geology, 25, 399-402, 1997.

Anderson, S. P., Longacre, S. A., and Kraal, E. R.: Patterns of water chemistry and discharge in the glacier-fed Kennicott River, Alaska: Evidence of subglacial water storage cycles, Chem. Geol., 202, 297–312, 2003.

Bartholomew, I. D., Nienow, P., Sole, A., Mair, D., Cowton, T., King, M. A., and Palmer, S.: Seasonal variations in Greenland Ice Sheet motion: Inland extent and behaviour at higher elevations, Earth Planet. Sc. Lett., 307, 271–278, 2011.

Blum, A. E. and Stillings, L. L.: Feldspar dissolution kinetics, in: Chemical Weathering Rates of Silicate Minerals, edited by: White, A. F. and Brantley, S. L., Mineralogical Soc Amer, Chantilly, 291–351, 1995.

Brown, G. H.: Glacier meltwater hydrochemistry, Appl. Geochem., 17, 855–883, 2002.

Chandler, D. M., Wadham, J. L., Lis, G. P., Cowton, T., Sole, A., Bartholomew, I., Telling, J., Nienow, P., Bagshaw, E. B., Mair, D., Vinen, S., and Hubbard, A.: Evolution of the subglacial drainage system beneath the Greenland Ice Sheet revealed by tracers, Nat. Geosci., 6, 195–198, 2013.

Collins, D.: Dissolution kinetics, transit times through subglacial hydrological pathways and diurnal vatiations of solute content of meltwaters draining from an alpine glacier, Hydrol. Process., 9, 897–910, 1995.

Collins, D. N. and MacDonald, O. G.: Year-to-year variability of solute flux in meltwaters draining from a highly-glacierised basin, Nordic Hydrology, 35, 359–367, 2004.

Cowton, T., Nienow, P., Bartholomew, I., Sole, A., and Mair, D.: Rapid erosion beneath the Greenland ice sheet, Geology, 40, 343–346, 2012.

Cuffey, K. M. and Paterson, W. S. B.: The Physics of Glaciers, 4th Edn., Elsevier, Amsterdam, 2010.

Drever, J. I. and Hurcomb, D. R.: Neutralization of atmospheric acidity by chemical weathering in an alpine drainage basin in the North Cascade Mountains, Geology, 14, 221–224, 1986.

Evans, C. and Davies, T. D.: Causes of concentration/discharge hysteresis and its potential as a tool for analysis of episode hydrochemistry, Water Resour. Res., 34, 129–137, 1998.

Graly, J. A., Humphrey, N. F., Landowski, C. M., and Harper, J. T.: Chemical weathering under the Greenland Ice Sheet, Geology, 42, 551–554, 2014.

Graly, J. A., Humphrey, N. F., and Harper, J. T.: Chemical depletion of sediment under the Greenland Ice Sheet, Earth Surf. Proc. Land., 41, 1922–1936, 2016.

Harrington, J., Humphrey, N. F., and Harper, J. T.: Temperature distribution and thermal anomalies along a flowline of the Greenland Ice Sheet, Ann. Glaciol., 56, 98–104, 2015.

Hasholt, B., Mikkelsen, A. B., Nielsen, M. H., and Larsen, M. A. D.: Observations of runoff and sediment and dissolved loads from the Greenland Ice Sheet at Kangerlussuaq, West Greenland, 2007 to 2010, Z. Geomorphol., 57, 3–27, 2013.

Hindshaw, R. S., Tipper, E. T., Reynolds, B. C., Lemarchand, E., Wiederhold, J. G., Magnusson, J., Bernasconi, S. M., Kretzschmar, R., and Bourdon, B.: Hydrological control of stream water chemistry in a glacial catchment (Damma Glacier, Switzerland), Chem. Geol., 285, 215–230, 2011.

Hodson, A., Tranter, M., and Vatne, G.: Contemporary rate of chemical denudation and atmospheric CO_2 sequestration in glacier basins: An arctic perspective, Earth Surf. Proc. Land., 25, 1447–1471, 2000.

Humphrey, N. F.: Coupling between water pressure and basal sliding in a linked-cavity hydraulic system, The Physical Basis of Ice Sheet Modelling, IAHS Publ. No. 170, 105–118, 1987.

Humphrey, N. F. and Raymond, C. F.: Hydrology, erosion and sediment production in a surging glacier: Variegated Glacier, Alaska, 1982–83, J. Glaciol., 40, 539–552, 1994.

Jezek, K., Wu, X., Paden, J., and Leuschen, C.: Radar mapping of Isunnguata Sermia, Greenland, J. Glaciol., 59, 1135–1147, 2013.

Kamb, B.: Glacier surge mechansim based on linked cavity configuration of the basal water conduit system, J. Geophys. Res., 92, 9083–9100, 1987.

Landowski, C.: Geochemistry and subglacial hydrology of the West Greenland Ice Sheet, MS Thesis, Geology and Geophysics, University of Wyoming, 2012.

Langen, P. L., Ahlstrøm, A. P., Andersen, K. K., Andersen, S. B., Barletta, V., Box, J. E., Citterio, M., Colgan, W., Dybkjær, G., Fausto, R. S., Forsberg, R., Hansen, B., Hanson, S., Høyer, J. L., Sørensen, L. S., and Tonboe, R. T.: Polar Portal Season Report 2013, available at: http://polarportal.dk/en/nyheder/arkiv/2013-season-report/ (last access: 15 January 2017), 2013.

Meierbachtol, T., Harper, J., and Humphrey, N.: Basal drainage system response to increasing surface melt on the Greenland Ice Sheet, Science, 341, 777–779, 2013.

Mitchell, A. C. and Brown, G. H.: Diurnal hydrological – physicochemical controls and sampling methods for minor and trace elements in an Alpine glacial hydrological system, J. Hydrol., 332, 123–143, 2007.

Nye, J. F.: Water flow in glaciers: Jokulhlaups, tunnels, and veins, J. Glaciol., 17, 181–207, 1976.

Palmer, S., Shepherd, A., Nienow, P., and Joughin, I.: Seasonal speedup of the Greenland Ice Sheet linked to routing of surface water, Earth Planet. Sc. Lett., 302, 423–428, 2011.

Röthlisberger, H.: Water Pressure in Intra- and Subglacial Channels, J. Glaciol., 11, 177–203, 1972.

Skidmore, M. L. and Sharp, M. J.: Drainage system behaviour of a High-Arctic polythermal glacier, Ann. Glaciol., 28, 209–215, 1999.

Smith, L. C., Chu, V. W., Yang, K., Gleason, C. J., Pitcher, L. H., Rennermalm, A. K., Legleiter, C. J., Behar, A. E., Overstreet, B. T., Moustafa, S. E., Tedesco, M., Forster, R. R., LeWinter, A. L.,

Finnegan, D. C., Sheng, Y., and Balog, J.: (Efficient meltwater drainage through supraglacial streams and rivers on the southwest Greenland ice sheet, P. Natl. Acad. Sci. USA, 112, 1001–1006, 2015.

Tranter, M. and Raiswell, R.: The composition of the englacial and subglacial component in bulk meltwaters draining the Gornergletscher, Switzerland, J. Glaciol., 37, 59-66, 1991.

Tranter, M., Brown, G., Raiswell, R., Sharp, M., and Gurnell, A.: A conceptual model of solute aquisition by Alpine glacial meltwaters, J. Glaciol., 39, 573–581, 1993.

van Gool, J. A. M., Connelly, J. N., Marker, M., and Mengel, F. C.: The Nagssugtoqidian Orogen of West Greenland: Tectonic evolution and regional correlations from a West Greenland perspective, Can. J. Earth Sci., 39, 665–686, 2002.

Wadham, J. L., Tranter, M., Skidmore, M., Hodson, A. J., Priscu, J., Lyons, W. B., Sharp, M., Wynn, P., and Jackson, M.: Biogeochemical weathering under ice: Size matters, Global Biogeochem. Cy., 24, GB3025, doi:10.1029/2009GB003688, 2010.

Werder, M. A., Hewit, I. J., Schoof, C. G., and Flowers, G. E.: Modeling channelized and distributed subglacial drainage in two dimensions, J. Geophys. Res.-Earth, 118, 1–19, 2013.

Yde, J. C., Knudsen, N. T., Hasholt, B., and Mikkelsen, A. B.: Meltwater chemistry and solute export from a Greenland Ice Sheet catchment, Watson River, West Greenland, J. Hydrol., 519, 2165–2179, 2014.

Zeng, C., Gremaud, V., Zeng, H., Liu, Z., and Goldscheider, N.: Temperature-driven meltwater production and hydrochemical variations at a glaciated alpine karst aquifer: implication for the atmospheric CO_2 sink under global warming, Eviron. Earth Sci., 65, 2285–2297, 2012.

12

Dynamic response of an Arctic epishelf lake to seasonal and long-term forcing: implications for ice shelf thickness

Andrew K. Hamilton[1,2], **Bernard E. Laval**[1], **Derek R. Mueller**[2], **Warwick F. Vincent**[3], **and Luke Copland**[4]

[1]Department of Civil Engineering, University of British Columbia, Vancouver, British Columbia, Canada
[2]Department of Geography and Environmental Studies, Carleton University, Ottawa, Ontario, Canada
[3]Department of Biology and Centre for Northern Studies (CEN), Université Laval, Quebec City, Quebec, Canada
[4]Department of Geography, Environment, and Geomatics, University of Ottawa, Ottawa, Ontario, Canada

Correspondence to: Andrew K. Hamilton (andrew@madzu.com)

Abstract. Changes in the depth of the freshwater–seawater interface in epishelf lakes have been used to infer long-term changes in the minimum thickness of ice shelves; however, little is known about the dynamics of epishelf lakes and what other factors may influence their depth. Continuous observations collected between 2011 and 2014 in the Milne Fiord epishelf lake, in the Canadian Arctic, showed that the depth of the halocline varied seasonally by up to 3.3 m, which was comparable to interannual variability. The seasonal depth variation was controlled by the magnitude of surface meltwater inflow and the hydraulics of the inferred outflow pathway, a narrow basal channel in the Milne Ice Shelf. When seasonal variation and an episodic mixing of the halocline were accounted for, long-term records of depth indicated there was no significant change in thickness of ice along the basal channel from 1983 to 2004, followed by a period of steady thinning at $0.50\,\mathrm{m\,a^{-1}}$ between 2004 and 2011. Rapid thinning at $1.15\,\mathrm{m\,a^{-1}}$ then occurred from 2011 to 2014, corresponding to a period of warming regional air temperatures. Continued warming is expected to lead to the breakup of the ice shelf and the imminent loss of the last known epishelf lake in the Arctic.

1 Introduction

Polar aquatic ecosystems that depend on ice for their physical containment, for their surface covering, or as a source of freshwater are highly sensitive to climate conditions (White et al., 2007; Prowse et al., 2011; Wrona et al., 2016). Where ice dams freshwater that floats directly on seawater, such as for stamukhi lakes (ephemeral coastal lakes that form where river input is dammed behind sea ice ridges; Galand et al., 2008) or epishelf lakes (perennial lakes that form behind floating ice shelves), mass loss from the ice dam due to a shift in climate can lead to a thinning of the freshwater layer. This is particularly notable for epishelf lakes, which are known to persist for decades (Smith et al., 2006; Veillette et al., 2008) or even millennia (Doran et al., 2000; Antoniades et al., 2011) behind stable ice shelves. However, recent thinning and collapse of ice shelves in the Arctic due to climate warming (Vincent et al., 2001; Copland et al., 2007; Mueller et al., 2003, 2017a; England et al., 2008; White et al., 2015a) has resulted in physical changes of epishelf lakes in the region, including the complete loss of several lakes and a substantial reduction in the depth of the freshwater layer of the few that remained (Mueller et al., 2003, 2017b; Veillette et al., 2008; White et al., 2015b). Given their sensitivity to the state of the impounding ice shelf, observing changes to the water column structure of epishelf lakes can potentially provide a relatively simple way to monitor changes in the integrity of the impounding ice shelf (Mueller et al., 2003; Veillette et al., 2008).

Epishelf lakes form in ice-free areas adjacent to floating ice shelves and maintain a hydraulic connection to the sea (Gibson and Andersen, 2002). Epishelf lakes are numerous in Antarctica and are distributed around the margins of the continental ice sheet (Heywood, 1977; Gibson and Andersen, 2002; Laybourn-Parry et al., 2006; Smith et al., 2006) and were once relatively numerous along the northern coast

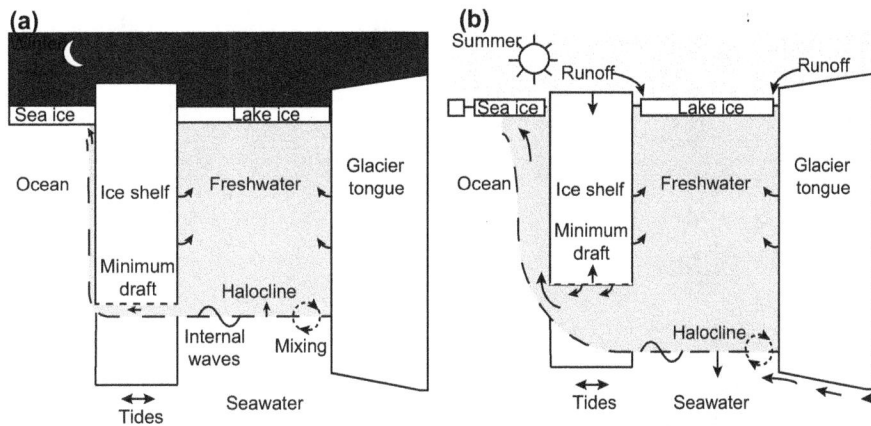

Figure 1. Schematic of the Milne Fiord epishelf lake showing processes that could influence the depth of the halocline in **(a)** winter and **(b)** summer. Potential influences include tidal flow, internal waves, mixing, deepening of the freshwater layer due to seasonal inflow, shoaling of the freshwater layer due to outflow beneath the ice shelf, changes in the minimum draft of the ice shelf, or melting along the outflow pathway changing the minimum draft of the ice shelf. Arrows indicating submarine melting of the walls of the ice shelf and glacier tongue and meltwater discharged at the base of the glacier tongue are included for completeness, although neither process will alter the depth of the epishelf lake. The ice walls are in hydrostatic equilibrium, so their melting will add volume to the freshwater layer but not change the depth. Meltwater runoff discharged at the bed of the glacier entrains seawater on its ascent, and observations suggest it spreads out below the epishelf lake halocline as a subsurface plume, which does not contribute to the thickness of the freshwater layer (Hamilton, 2016).

of Ellesmere Island, in the Canadian Arctic (Vincent et al., 2001; Veillette et al., 2008). Similar systems may exist elsewhere in the Arctic where ice shelves form, including Greenland, Franz Josef Land, and Severnaya Zemlya, but to date no epishelf lakes have been reported from these areas. The physical properties of epishelf lakes vary depending largely on the connection to the ocean below the ice shelf. At one extreme are those where the freshwater layer floats directly on seawater – such as Ablation Lake, West Antarctica (Heywood, 1977); Beaver Lake, East Antarctica (Wand et al., 2011); and Disraeli Fiord in the Canadian Arctic (Vincent et al., 2001) – while at the other extreme are epishelf lakes that are entirely fresh but have a restricted hydraulic connection to the ocean below the ice shelf (e.g., Schirmacher Oasis, East Antarctica; Bormann and Fritzsche, 1995), with lakes of varying degrees of stratification in between (e.g., southern Bunger Hills, East Antarctica; Gibson and Andersen, 2002).

For epishelf lakes where freshwater floats directly on seawater, the focus of this paper, snow and ice meltwater from the surrounding catchment is thought to accumulate until the thickness of the freshwater layer is equal to the minimum draft of the ice shelf (Hattersley-Smith, 1973; Jeffries and Krouse, 1984; Gibson and Andersen, 2002; Mueller et al., 2003). Excess freshwater inflow is then assumed to be exported below the base of the ice shelf to the open ocean. Where low tidal action and perennial ice cover limit mixing between the freshwater and saltwater layers, the halocline can be unusually sharp (with reported vertical salinity gradients of $\sim 40\,\mathrm{ppt\,m^{-1}}$; Heywood, 1977) and stable. The strong physical and chemical stratification creates a rare ecosystem where freshwater, brackish water, and saltwater biota can ex-

ist within a single water column (Laybourn-Parry et al., 2006; Veillette et al., 2011a). Thus changes to the physical structure of the water column can affect ecosystem functioning (Thaler et al., 2017).

Measuring changes in the depth of the halocline separating freshwater and seawater could provide a relatively straightforward means to infer changes in the minimum thickness of the adjacent ice shelf (Vincent et al., 2001; Mueller et al., 2003; Veillette et al., 2008). This has been undertaken extensively along the northern coast of Ellesmere Island, in the Canadian Arctic, where the climate has warmed at twice the global average (IPCC, 2013). For example, water column profiles collected in Disraeli Fiord behind the Ward Hunt Ice Shelf (WHIS), Ellesmere Island, showed that its epishelf lake thinned from a maximum depth of 63 m in 1954 to 33 m in 1999 (Veillette et al., 2008), suggesting a long-term thinning of the WHIS prior to its fracturing between 2001 and 2002, which resulted in the complete drainage of the epishelf lake (Mueller et al., 2003).

Questions remain, however, as to what extent changes in the depth and gradient of an epishelf lake halocline are related to changes in the ice shelf, and what other factors may be important (see Fig. 1 for a schematic of some of the processes discussed below). Vincent et al. (2001) suggested that the long-term thinning of the Disraeli Fiord epishelf lake could have been the result of preferential drainage of freshwater via a localized conduit at the base of the WHIS, not necessarily representative of changes in the mean thickness of the ice shelf. Veillette et al. (2008) suggested that observed interannual increases in the halocline depth of some Ellesmere Island epishelf lakes were not indicative of a thick-

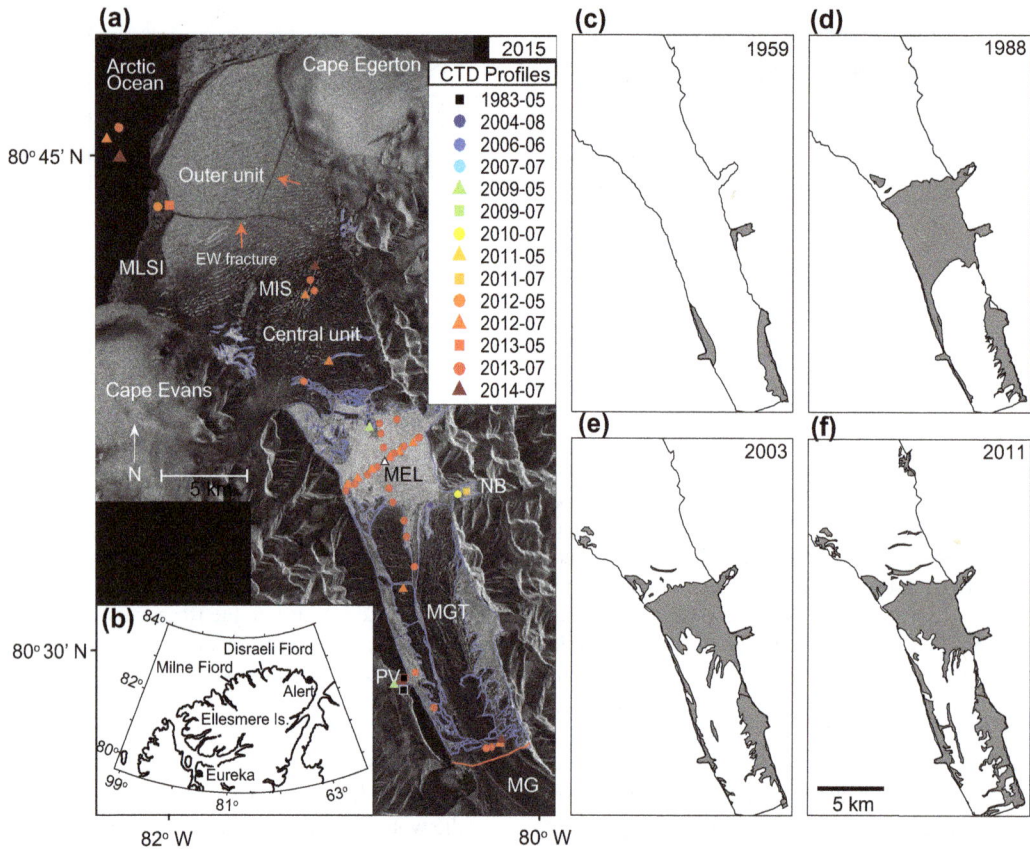

Figure 2. Map of Milne Fiord study area. **(a)** RADARSAT-2 image mosaic of Milne Fiord showing the extent of the Milne Fiord epishelf lake (MEL) in 2015, corresponding to the region of high backscatter (light grey) outlined in blue. The locations of CTD profiles collected during each field campaign are shown. The central and outer units of the Milne Ice Shelf (MIS), as well as two re-healed fractures (red arrows), are indicated, as well as the Milne Glacier (MG), grounding line (red line), Milne Glacier tongue (MGT), multiyear landfast sea ice (MLSI), the met station (white and black square), mooring (white and black triangle), and two small inlets unofficially named Purple Valley Bay (PV) and Neige Bay (NB). **(b)** Regional map of Ellesmere Island, Canada. The sequence of four panels on the right shows the increase in area of the MEL (grey) estimated from aerial and satellite imagery from **(c)** 1959, **(d)** 1988, **(e)** 2003, and **(f)** 2011, based on data from Mueller et al. (2017a). The coastline of Milne Fiord is outlined in black. White areas inside the coastline are glacier or ice shelf.

ening of the ice shelves but were more likely related to short-term vertical oscillations of the halocline due to tidal cycles, internal waves, or fjord circulation. Similarly, Smith et al. (2006) suggested the shoaling and weakening of the halocline in Ablation Lake, Antarctica, were not related to thinning of the George VI Ice Shelf but were more likely to reflect seasonal changes in the supply of fresh meltwater or variation due to changes in tides or the hydraulic connection to the sea below the ice shelf (e.g., Galton-Fenzi et al., 2012). Understanding of epishelf lake systems is, however, hampered by the fact that long-term records are based almost entirely on sparse conductivity–temperature–depth (CTD) profiles collected over intervals of years, or even decades, with little knowledge of the seasonal dynamics or spatial heterogeneity of these systems. Many studies have stressed the need for continuous monitoring to better understand the dynamics of epishelf lakes and their relationship to ice shelf thick-

ness (Vincent et al., 2001; Smith et al., 2006; Veillette et al., 2008).

In this study we aimed to identify factors controlling changes in the water column structure and the depth of the halocline of the last known epishelf lake in the Arctic, in Milne Fiord, Ellesmere Island, and to evaluate how these variations were related to changes in the spatial extent of the lake and to the state of the Milne Ice Shelf (MIS). We compiled archived CTD data collected in Milne Fiord since 1983, conducted extensive new CTD profiling between 2009 and 2014, and analyzed satellite data to monitor the long-term changes in the vertical structure and spatial extent of the lake. We deployed a mooring from 2011 to 2014 to continuously record changes in epishelf lake properties to investigate short-term variability, on timescales from tidal to seasonal, and determine the factors driving changes in the halocline. The complete epishelf lake halocline depth records are used to infer changes in the thickness and state of the Milne Ice

Shelf, and we discuss the implications of our findings on the interpretation of records from other ice shelf–epishelf lake systems.

Study site and background

Milne Fiord (82°35′ N, 80°35′ W) is 436 m deep and lies on the northern coast of Ellesmere Island adjacent to the Arctic Ocean (Fig. 2). The Milne Fiord epishelf lake (MEL) is dammed by the MIS, which spans 18 km across the mouth of the fjord. At the head of the fjord the Milne Glacier terminates in a 16 km long glacier tongue that forms the landward margin of the lake. The perennial freshwater ice cover of the MEL, which is approximately 1 m thick, extends throughout the inner fjord and appears as an area of high backscatter (lighter tones) in synthetic aperture radar (SAR) imagery, in contrast to the low backscatter (darker tones) of the MIS and the glacier tongue (Fig. 2a). From SAR imagery, Mortimer (2011) estimated the area of the epishelf lake as 52.5 km^2 in 2009, appearing to consist of a 6 km wide main basin between the inner edge of the MIS and the terminus of the Milne Glacier tongue (MGT), and two narrow arms extending 16 km along the sides of the glacier tongue to the Milne Glacier grounding line, although the actual spatial extent of the freshwater layer had not been confirmed with field observations.

The MEL was first discovered in 1983, when Jeffries (1985) collected water samples through 3.19 m of surface ice that revealed a 17.5 m deep freshwater layer separated from seawater by a sharp halocline only a few metres thick. The lake was not sampled again until 2004, when CTD profiles showed the freshwater layer had deepened to 18.3 m; by 2009 it had thinned to 14.3 m, which was assumed to correspond to the minimum draft of the MIS (Veillette et al., 2011b).

Surveys of the MIS have shown that its ice thickness varies from 94 to 8 m (Fig. 3; Mortimer et al., 2012; Hamilton, 2016). The topology of the ice shelf is quite variable owing to its complex origins and has been divided into three regions based on surface morphology and ice characteristics (Jeffries, 1986). The previously defined inner unit, which once abutted the Milne Glacier tongue, has been replaced with epishelf lake ice since the 1980s (Mortimer et al., 2012). The central unit currently forms the landward (southern) edge of the ice shelf and has an erratic surface morphology and highly variable thickness owing to substantial past input from tributary glaciers, and is apparent as a region of low backscatter (dark) in SAR imagery (Fig. 2a). The outer unit has a more uniform thickness (mean thickness ∼ 50 m) owing to its origin from marine ice accretion and snow accumulation, and it is distinguishable by a series of surface ridges and troughs running parallel to the coast, and relatively high backscatter in SAR imagery (Fig. 2a). The outer unit, however, is bisected by two re-healed fractures that have existed since at least 1950 (Hattersley-Smith et al., 1969). Ice thickness surveys indicate that the only ice thin enough to provide an

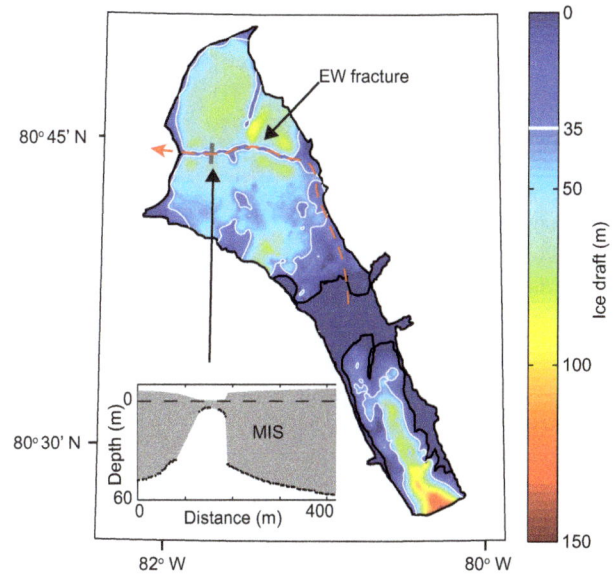

Figure 3. Map of the draft of ice in Milne Fiord indicating the likely drainage pathway of the MEL along a re-healed fracture in the MIS. The modelled ice draft is that presented in Hamilton (2016) from ground-based, aerial, and satellite ice thickness and surface elevation data assuming the ice is in hydrostatic equilibrium. The 35 m contour is highlighted to show the outflow restriction the MIS imposes across the mouth of the fjord for water above this level. The inset shows ice thickness across the basal channel from a 400 m ice-penetrating radar transect that crossed the east–west re-healed fracture in July 2013 (Hamilton, 2016).

outlet to the epishelf lake lies along these re-healed fractures (Fig. 3), although the drainage pathway of the lake has not been confirmed by observations. The MIS is in a state of negative mass balance (Mortimer et al., 2012) and is not expected to regenerate in the current climate (Copland et al., 2007; Veillette et al., 2011a; White et al., 2015a).

2   Methods

2.1   Area and volume

To track changes in the extent of the MEL, its area was estimated from optical imagery acquired by various aerial and satellite platforms from 1959 to 1988 and from 1992 onward from SAR imagery (Mueller et al., 2017b). The epishelf lake ice was discriminated from other surrounding ice types – including ice shelf, glacier ice, and marine ice – in optical imagery by its lack of surface topography, and in winter SAR imagery by its high backscatter signal (> −6 dB), produced by its lack of surface topography and freshwater underneath perennial lake ice (thin ice directly underlain by saltwater has a darker return; Veillette et al., 2008; White et al., 2015a). The epishelf lake was manually digitized in ArcGIS 10.2.2 at an image scale of ∼ 1 : 20 000, with a pixel size of 6.5 m,

Table 1. Milne Fiord epishelf lake depth, area, volume, and related observations.

CTD profiling dates	CTD profile location	No. of CTD profiles	MEL mean depth (±0.2 m)	PDDs to date (°C days) (yr total)	MEL PDD corr. depth (±0.2 m)	Area (km²)	Volume (km³)	Image source for area est.	Image acquisition date
1959/08/17	–	–	–	–	–	13.5	–	Aerial photo	1959/08/17
1963/08/29	–	–	–	–	–	13.5	–	Corona	1963/08/29
1983/05/25	PV	1	17.5[a]	0 (158)[c]	17.5	–	–	–	-
1988/08/08	–	–	–	–	–	67.3	–	SPOT-1	1988/08/08
								SPOT-1	1988/08/08
1992/01/29	–	–	–	–	–	60.6	–	ERS-1	1992/01/29
								ERS-1	1992/03/16
1998/01/13	–	–	–	–	–	61	–	RADARSAT-1	1998/01/13
2003/01/11	–	–	–	–	–	59.3	–	RADARSAT-1	2003/01/11
2004/08/06	NB	1	18.3[b]	98 (132)[c]	16.6	–	–	–	-
2006/06/03	NB	1	16.0[b]	4 (153)[c]	15.9	65	1.04	RADARSAT-1	2006/01/14
2007/07/13	NB	1	16.5[b]	78 (215)[c]	15.3	–	–	–	-
2009/05/29–05/30	PV	18	14.7	0 (274)	14.7	65.2	0.96	RADARSAT-2	2009/01/04
2009/07/04	NB	1	14.6	34 (274)	14.0	–	–	–	-
2010/07/09	NB	1	15.3	100 (185)	13.6	–	–	–	-
2011/05/10	MM	1	13.6	0 (278)	13.6	67.6	0.92	RADARSAT-2	2011/01/03
								RADARSAT-2	2011/02/28
2011/07/05	NB	1	14.4	94 (278)	12.8	–	–	–	-
2012/05/05–05/14	MM	3	9.5	0 (253)	9.5	64.4	0.61	RADARSAT-2	2012/02/03
								RADARSAT-2	2012/04/17
2012/06/28–07/09	MM, ML	23	10.6	50 (253)	9.8	–	–	–	-
2013/05/11–05/18	MM, ML	11	8.0	0 (92)	8.0	67	0.54	RADARSAT-2	2013/04/27
2013/07/04–07/22	MM, ML	46	8.1	10 (92)	7.9	–	–	–	-
2014/07/12–07/24	MM, ML	13	9.3	39 (110)	8.6	71.2	0.66	RADARSAT-2	2015/03/27

PV – Purple Valley Bay; NB – Neige Bay; MM – Milne Fiord mooring; ML – multiple locations inside and outside Milne Fiord. [a] Jeffries (1985). [b] Veillette et al. (2008).
[c] PDDs calculated from air temperatures estimated from the Eureka weather station, where $T_{Milne} = 0.87 \cdot T_{Eureka} - 3.26$.

and the area of the resultant polygons were calculated. Non-contiguous regions of lake ice in fractures in the MIS and between calved pieces of the MIS and Milne Glacier tongue were included in area estimates. Lake volume was estimated from area and depth, assuming a spatially uniform depth (see Sect. 3.3) and vertical shores. The volume estimated included the volume of lake surface ice. The depth used for volume calculations was estimated from hydrographic profiles (see Sect. 2.3) collected closest to the date of image acquisition (see Table 1).

2.2 Hydrography

Long-term changes in the structure of the epishelf lake were monitored using archived and newly acquired water column profiles. We present data from the first profile collected in 1983 using reversing thermometers and a Radiometer CDM80 Conductivity Meter (Jeffries, 1985). Near-annual CTD profiling commenced in 2004, with a directed and intensive sampling program from 2011 to 2014. CTD profiles were collected through drilled holes or natural leads in the ice, including fractures through the ice shelf and glacier

tongue, accessed by foot, snowmobile, or helicopter. Opportunistic profiles were collected in August 2004, June 2006, and July 2007 using a 1 Hz RBR XR-420 CTD. Subsequent profiles were collected in May and July of 2009; July 2010; May and July of 2011, 2012, and 2013; and July 2014 with a 6 Hz RBR XR-620 CTD. Profiles from May 2011 were collected using a 4 Hz Seabird SBE19+ CTD and from July 2011 using a 1 Hz Hydrolab HLX. The 2004 CTD profile was collected just north of the Milne Glacier tongue terminus, and from 2006 to 2010 profiles were collected in either Purple Valley Bay or Neige Bay (unofficial names), two shallow inlets on the west and east sides of the fjord, respectively (Fig. 2a). From 2011 onward multiple profiles were collected during each field campaign throughout the main fjord. Full depth profiles were usually collected to the bottom of the fjord, but here we focus on the upper 25 m of the water column. CTDs were calibrated once every 2 years after 2011; prior to this the CTDs were not regularly calibrated, so we interpret data prior to 2011 with this caveat on absolute accuracy. Profiles collected between 2004 and 2009 were previously published in Veillette et al. (2011b), although we have reprocessed all these data from raw conductivity and temperature measurements (where available) for consistency.

CTD data were processed in Matlab following a procedure that included correction for atmospheric pressure, application of a three-point low-pass filter in time to the raw pressure, temperature, and conductivity; alignment of conductivity and temperature with respect to pressure; a thermal cell mass correction (for the SBE19+ CTD data only); and loop editing (removal of pressure reversals) and bin-averaged to 0.2 m intervals. Derived variables were calculated using the International Thermodynamic Equation of Seawater 2010 (TEOS-10) Gibbs Seawater Oceanographic Matlab Toolbox (www.TEOS-10.org), with conservative temperature (Θ) and absolute salinity (S_A) reported. For freshwater lakes, the determination of salinity from measured temperature, conductivity, and pressure is dependent on the chemical composition of the water (Pawlowicz, 2008), data which we lack. However, the conductivity of two inflowing meltwater streams measured in 2012 and 2013 was low ($< 0.06 \, \text{mS cm}^{-1}$), suggesting that the source of ions to the lake (with conductivity generally $> 0.15 \, \text{mS cm}^{-1}$) was the underlying seawater and that the use of TEOS-10, which assumes seawater composition, was appropriate.

2.3 Lake depth from CTD profiles

In order to compare changes in the thickness of the freshwater layer, over time it was necessary to define the bottom of the epishelf lake, which was actually a continuum from freshwater to seawater. Previous studies have used the depth of the 3 ppt isohaline (Mueller et al., 2003; Veillette et al., 2008), or the depth of the halocline (which the authors qualitatively define as the zone of abrupt salinity change between freshwater and seawater; Veillette et al., 2011b). We formalized the definition of Veillette et al. (2011b) by defining the bottom of the lake (D_{EL}) as the depth of the stratification maximum as defined by the Brunt–Väisälä frequency:

$$D_{EL} = z(N^2_{max}),\tag{1}$$

where z is depth (positive downward) and N^2 is the Brunt–Väisälä frequency:

$$N^2 = \frac{g}{\rho}\frac{\partial\rho}{\partial z},\tag{2}$$

where g is gravitational acceleration and ρ is density of the water. Profiles of N^2 were averaged using a 12-point depth window before calculating the maximum. The advantage of this method is that the epishelf lake depth calculation is clearly defined, quantitative, and not dependent on the absolute salt content of the epishelf lake (which would affect the 3 ppt method). Due to the bottle sampling method used by Jeffries (1985) we could not accurately calculate N^2 for that profile, so we defined the bottom of the lake as the depth of the sample collected nearest the apparent stratification maximum. The depth of the lake as measured by a series

of 18 CTD profiles collected at a single location over 24 h in May 2009 varied by ± 0.2 m, so this was considered the depth measurement uncertainty from CTD profiles.

2.4 Current velocities

To understand the probability of shear-induced vertical mixing of the halocline, velocities of the upper water column were measured using an ice-anchored, downward-looking 300 kHz RDI acoustic Doppler current profiler (ADCP) at the mooring site in the centre of the MEL (Fig. 2a) over 4 days in May 2011, 7 days in July 2012, and 10 days in July 2013. The ADCP sampled at 2 min intervals, at 2 m bins, with 150 pings per ensemble, and the data were processed in Matlab.

2.5 Tidal height

Changes in water level were recorded using a bottom anchored RBR XR-620 CTD (accuracy is ± 0.37 dbar, drift 0.74 dbar a^{-1}) deployed at 355 m depth at the mooring site (Fig. 2a) from May 2011 to July 2012 sampling at 2 min intervals.

2.6 Meteorological time series

Meteorological conditions were recorded by a HOBO automated weather station (AWS) located at 10 m elevation in Purple Valley Bay (Fig. 2a). Only air temperature (at 1 and 2 m above ground) and shortwave solar radiation are reported here. Cumulative positive-degree days (PDDs), the daily integrated air temperatures above 0 °C, were calculated to provide a direct proxy for summer surface melting (Hock, 2003) and thus inflow to the lake. Prior to the AWS installation in 2009, we estimated summer air temperatures and PDDs in Milne Fiord from air temperature records at Eureka, Nunavut, 280 km to the south (http://climate.weather.gc.ca/historical_data/search_historic_data_e.html). Linear regression showed that daily mean air temperature in Milne Fiord, T_{Milne}, were related ($R^2 = 0.92$) to daily mean air temperature in Eureka by $T_{Milne} = 0.87 \cdot T_{Eureka} - 3.26$.

2.7 Mooring time series

Milne Fiord water properties were recorded continuously from May 2011 to July 2014 from a mooring deployed in the centre of the epishelf lake (Fig. 2a). The mooring was anchored to the epishelf lake ice and suspended down the water column. The mooring consisted of 20 RBR TR1050/60 temperature sensors, two RBR XR-420 freshwater conductivity–temperature (CT) sensors, two Seabird SBE37 CTs, and one RBR XR620 CTD from May 2011 to July 2012, and then was reduced to seven TR1060s, one XR420 CT, and one XR620 CTD for the remainder of the study. Calibrated accuracy for all sensors is ± 0.002 °C, ± 0.003 mS cm^{-1}, and ± 0.37 dbar, and nominal drift is ± 0.002 °C a^{-1}, ± 0.012 mS cm^{-1} a^{-1}, and ± 0.7 dbar a^{-1}. Some time series records were truncated for various reasons, including salinity going beyond the maximum sensor range (3 PSU) of the freshwater instruments (XR420s) or instrument malfunction. The mooring was serviced once or twice per year, and instruments were repositioned to track the halocline. Initially, the instruments were spaced every metre from the surface to 20 m depth, with increasing depth intervals below 25 m. Although the mooring instruments extended to the seabed at 355 m, in this paper we focus on the top 25 m of the water column. Instruments sampled at 30 to 120 s intervals. CTD profiles collected during mooring deployment and recovery were used to correct for instrument drift, which was within manufacturer specifications.

2.8 Lake depth from mooring time series

Seasonal changes in the depth of the epishelf lake were estimated from changes in salinity recorded by the conductivity sensor initially positioned within the halocline (at 13 m depth from May 2011 to May 2012). Salinity changes at a fixed depth over time were assumed to be driven by a vertical displacement of an otherwise unchanging initial vertical salinity gradient (obtained during CTD profiling). For example, an increase (decrease) in salinity was assumed to be caused by an upward (downward) displacement of the halocline. This method neglected other processes that could alter the salinity gradient, including horizontal advection of water masses or vertical mixing, and we discuss the impact of these in other sections.

We also estimated epishelf lake depth using temperature data from the more numerous and closely spaced thermistors on the mooring. The thermocline is coincident with the halocline because cold seawater below the halocline can circulate horizontally and continually remove heat at the base of the lake. Lake depth was therefore estimated from the depth of the isotherm corresponding to the average temperature at the depth of the N_{max}^2 measured by CTD profiles collected at the beginning and end of each mooring deployment. Temperature changes at a fixed depth were assumed to be caused by a vertical displacement of an otherwise unchanging thermocline (and halocline), and other processes that could alter the vertical distribution of heat – such as heating due to solar radiation, horizontal advection, and vertical heat flux or mixing across the halocline – were neglected. While these are broad assumptions, we note that all three methods used to determine lake depth (CTD profiling, the moored conductivity sensors, and moored thermistors) showed good agreement (with one notable exception detailed in Sect. 3.5.4), providing confidence in the methods.

3 Results

3.1 Stratification

The most conspicuous feature of all water column profiles collected in Milne Fiord was the presence of a several-metres-thick relatively warm, fresh (0–3 °C, $\sim 0.2\,\mathrm{g\,kg^{-1}}$) layer that was separated from cool, saline ($< -1\,°\mathrm{C}$, $\sim 30\,\mathrm{g\,kg^{-1}}$) water by a sharp halocline and thermocline (Fig. 4). The epishelf lake was apparent in the first water samples obtained in the fjord in 1983 (Jeffries, 1985), but it clearly thinned through time, and we investigate these changes in more detail below. Despite changes in the depth of the lake, several distinct layers in the epishelf lake and upper water column could be identified based on salinity and temperature characteristics (Fig. 2c).

In summer, a 1–2 m thick stratified layer with salinity $< 0.2\,\mathrm{g\,kg^{-1}}$ and temperature approaching the freshwater freezing point (0 °C) was present just below the surface ice–water interface. We termed this the surface melt layer given its origins from local melting of surface ice. Below the thin surface melt layer was a layer of nearly constant salinity (approximately $0.2\,\mathrm{g\,kg^{-1}}$), the mixed layer, extending from the base of the surface melt layer to the top of the halocline. The mixed layer was up to 8 m thick, with temperature between 0–2 °C; however it was not present in all years and usually only evident in summer; at other times the lake was weakly salinity stratified.

The strong salinity gradient below the mixed layer was divided into an upper and lower halocline. The upper halocline was the transition from the base of the mixed layer to the bottom of the lake (i.e., $z(N_{max}^2)$). A subsurface temperature maximum (up to 3 °C) was usually associated with the upper halocline. When the mixed layer was not present, the upper halocline extended to the base of the surface ice melt layer (if present) or to the ice–water interface. The gradient and thickness of the upper halocline varied among years, with a thicker and more gradual salinity gradient apparent prior to 2009 (e.g., in 2004 the upper halocline was 15 m thick and extended almost to the surface), while after 2009 the salinity gradient was thin and sharp (e.g., in June 2012 the upper halocline was < 3 m thick). The lower halocline was defined as extending below the $z(N_{max}^2)$ to the level at which properties within the fjord were equivalent to those at the same depth offshore (between 25 and 50 m). Temperatures in the lower halocline decreased rapidly with depth toward the freezing point of seawater ($< -1\,°\mathrm{C}$). The properties of the lower halocline varied among years and were likely dependent on local fjord processes, including interactions with ice and advection of subsurface glacial meltwater runoff, so the lower halocline was also referred to as fjord-modified water (Hamilton, 2016).

The salinity profiles showed a clear long-term thinning of the freshwater layer, from a maximum depth of 18.3 m in 2004 to a minimum of 8.0 m depth in 2013 (Fig. 4; Table 1).

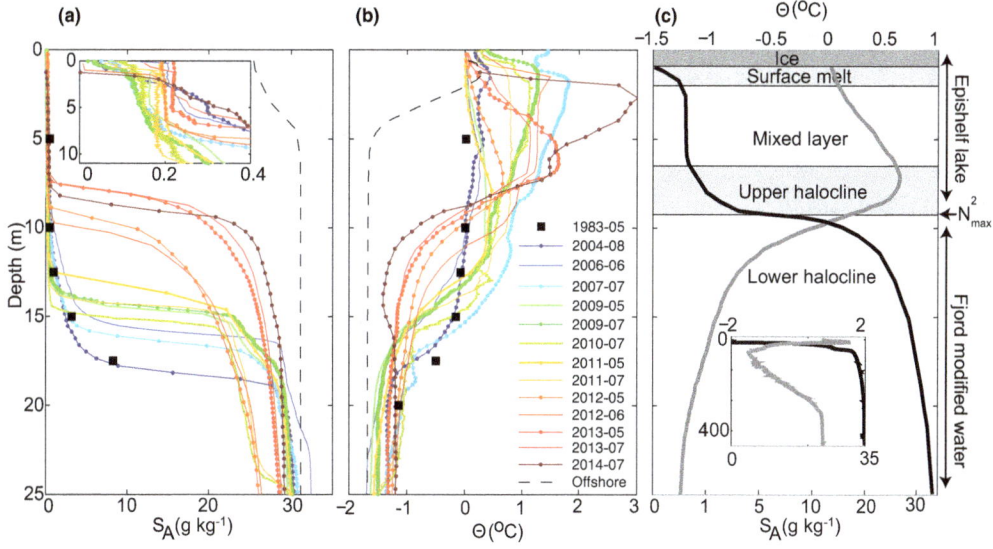

Figure 4. Changes in (a) salinity (S_A) and (b) temperature (Θ) properties of the upper 25 m of Milne Fiord from all field campaigns from 1983 to 2014. A single representative profile collected at the mooring site from each field campaign is shown when multiple profiles were collected. Inset in (a) shows a zoom-in of epishelf lake salinities. Dashed lines in (a) and (b) indicate a representative profile collected offshore of the MIS. (c) Idealized salinity (black line) and temperature (grey line) profiles showing the layers of the upper water column. Note the non-linear salinity scale. The epishelf lake is defined as extending from the surface to the buoyancy frequency maximum (N^2_{max}). The inset shows the full water column properties of the fjord to 440 m depth.

There was little change in the depth of the halocline between 1983 and 2004, then substantial thinning between 2004 and 2014. However, the CTD profiles indicate the magnitude and direction of change was not constant over time. For example, the halocline increased in depth between some years (e.g., 2006–2007, 2009–2010, and 2013–2014), while an apparent abrupt thinning occurred between 2011 and 2012. These changes are discussed in more detail in Sect. 3.5.3 and 3.5.4.

3.2 Current velocities

The relatively brief ADCP deployments (< 10 days) indicated a quiescent system with currents $< 2\,\mathrm{cm\,s^{-1}}$ in the upper 25 m of the water column (not shown). The currents were weakly baroclinic, with velocities near zero in the epishelf lake above the level of the halocline, increasing to $1–2\,\mathrm{cm\,s^{-1}}$ just below the halocline. The potential for velocity shear stress to generate vertical mixing in the water column was determined by calculating the gradient Richardson number, a ratio of stratification to velocity shear:

$$Ri = \frac{N^2}{\left(\frac{\partial u}{\partial z}\right)^2}, \tag{3}$$

where u is horizontal velocity ($\mathrm{m\,s^{-1}}$) and z is depth (m; positive z down). During all three periods of observation $Ri \gg 1$ across the halocline, indicating that stabilizing buoyancy forces dominate, and turbulent mixing was not expected.

3.3 Spatial extent

Although the depth of the halocline varied over time, the epishelf lake was present in all profiles collected landward (south) of the outer unit of the MIS in all years. Transects occupied over periods $< 24\,\mathrm{h}$ showed very little spatial heterogeneity in the depth of the halocline throughout the fjord. For example, $z(N^2_{max})$ varied by only ± 0.1 m in three profiles collected by helicopter on 29 June 2012 over 23 km from the grounding line of the Milne Glacier to a fracture in the central unit of the MIS (dark blue lines in Fig. 5a). Similarly, $z(N^2_{max})$ varied by only ± 0.05 m in six profiles collected across the 5.8 km width of the fjord (green lines in Fig. 5a). In addition, 21 profiles collected at multiple locations throughout the fjord in July 2013 showed $z(N^2_{max})$ varied by < 0.5 m over almost 3 weeks (Fig. 5b). We observed no evidence of spatial bias in the data that could inform the distribution of meltwater sources or their temporal variation, although the sparse sampling may not have been sufficient to capture such variation. Each of the profiling locations where the epishelf lake was present corresponded to a region of high backscatter in the RADARSAT-2 imagery acquired the winter prior, interpreted as freshwater lake ice, and provided verification of the remote-sensing method used to map the extent of the lake.

That the epishelf lake was observed even through fractures in the central unit of the MIS indicated that a network of

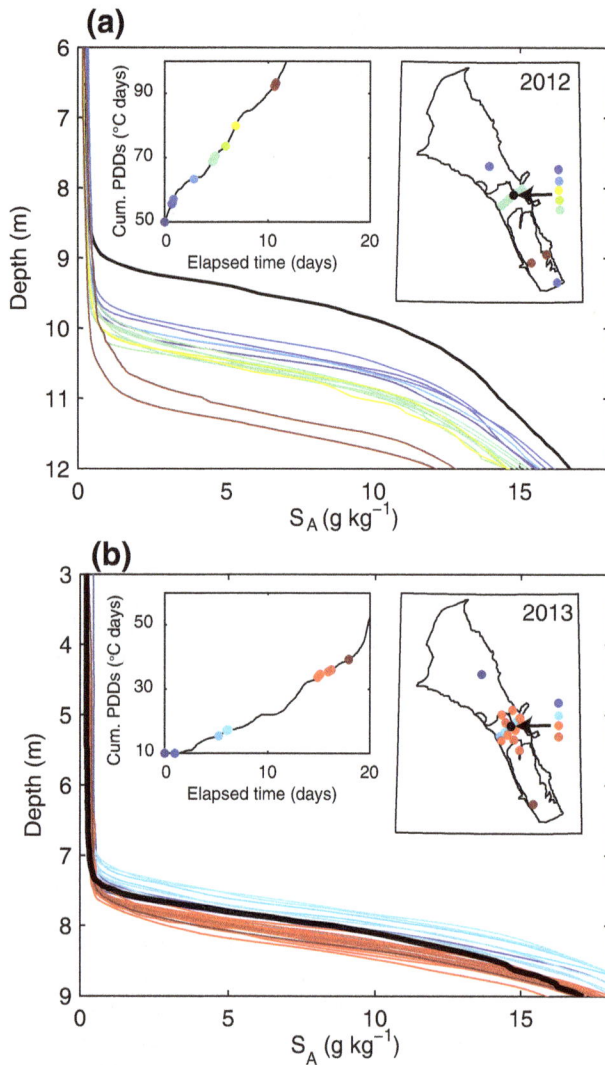

Figure 5. Salinity profiles showing seasonal changes in depth of the MEL halocline in **(a)** 2012 and **(b)** 2013. Profiles are coloured by time over the duration of each summer field campaign (10 days in June and July 2012 and 18 days in July 2013). A single profile collected in May of each year, prior to the onset of the melt season, is shown (black line). Note the different y axes range between **(a)** and **(b)**, although the increments are consistent. In **(a)** and **(b)** the left inset shows the PDDs (a proxy for the volume of surface meltwater inflow) accumulated during that field campaign, with the timing of profiles indicated (coloured circles). The right inset shows the profiling locations on a map of Milne Fiord (the MEL, MIS, and Milne Glacier tongue are outlined). Multiple profiles were collected at the mooring site (black circle) during each campaign.

fractures must allow water exchange under the central unit with the main basin of the epishelf lake. The fact that the epishelf lake was not present in profiles collected beyond the seaward edge of the MIS indicated that the ice dam for the

epishelf lake was likely located along the east–west-running fracture in the outer unit of the MIS (Fig. 3). Several attempts in 2012, 2013, and 2014 to profile through the re-healed fractures in the outer unit, to constrain the location of the ice dam and confirm the drainage pathway of the MEL, were unsuccessful. Investigation into the location of the ice dam and the drainage pathway of the MEL is ongoing.

3.4   MEL area and volume

Analysis of aerial and satellite imagery indicates that the area of the lake increased substantially over time, expanding from a few small ice-marginal lakes in 1959 to close to its present form by 1988, with an area of $71.2\,km^2$ as of 27 March 2015 (Fig. 2; Table 1). The largest change in area occurred sometime between 1959 and 1988, as the epishelf lake replaced the inner unit of the MIS. Increases in lake area after 1988 occurred due to retreat of the southern margin of the MIS, including calving of the MIS into the fjord, the creation of small satellite lakes in fractures of the MIS, and wastage along the margins of the Milne Glacier tongue. Some gains in area were partially offset by losses due to the advance of the terminus of the Milne Glacier tongue, which advanced 5.4 km down the fjord between 1950 and 2015.

The observed spatial uniformity of the depth of the lake (at a moment in time) allowed a straightforward determination of the volume of the lake from area and depth (Table 1). The earliest reliable estimate of the lake volume was $1.04\,km^3$ in 2006, after which the volume decreased substantially, reaching a minimum of $0.54\,km^3$ in 2013. The decrease in volume is largely due to the decrease in the depth of the lake; the area of the lake varied by only 10 % between 2006 and 2014, while the depth varied by over 50 % during this period. We note that measured surface ice thickness decreased from a maximum of 3.19 m in 1983 (Jeffries, 1985) to a minimum of 0.65 m in July 2010. However, changes in surface ice thickness did not affect estimated volumes as the surface ice was in hydrostatic equilibrium.

3.5   Temporal variability

3.5.1   Tidal

Water level records revealed that the tidal range in Milne Fiord was small, with a maximum range of 0.31 m between May 2011 and May 2012. The low tidal energy available for mixing in Milne Fiord has likely been an important factor in the long-term persistence of the epishelf lake halocline.

3.5.2   High frequency

Salinity sensors moored in the halocline showed evidence of high-frequency variations that were likely due to the passage of internal waves. Spectral analysis of the salinity time se-

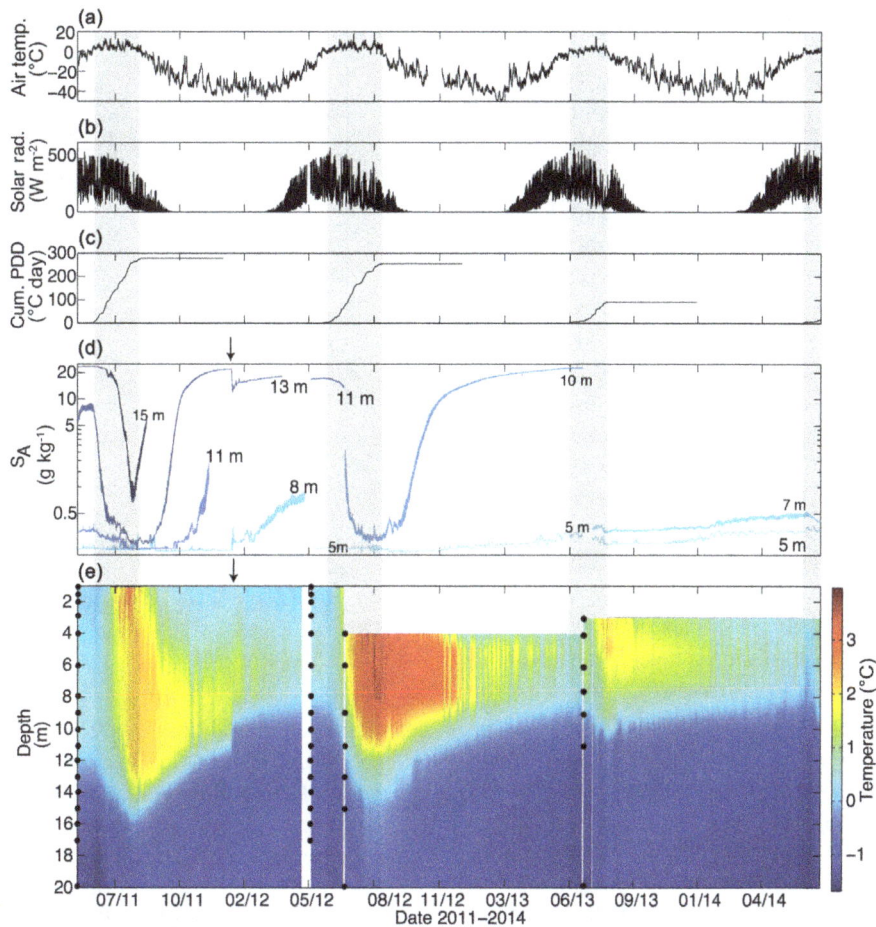

Figure 6. Time series of meteorological conditions and epishelf lake properties from May 2011 to July 2014 in Milne Fiord. **(a)** Air temperature. **(b)** Shortwave solar radiation. **(c)** Cumulative positive-degree days (PDDs). **(d)** Absolute salinity (S_A) from instruments moored at depths between 5 and 15 m. Note the logarithmic scale. **(e)** Temperature time series from thermistors moored between 1 and 20 m depth. Black circles indicate thermistor depths at each mooring deployment, and white areas indicate data gaps. Transparent grey regions mark the start and end of the surface melt season each year. Arrows in **(d)** and **(e)** indicate the timing of the mixing event shown in Fig. 8.

ries (not shown) revealed energy peaks at diurnal and semi-diurnal tidal periods in the halocline, as well as a strong non-tidal peak at 48 min and secondary peaks at 70 min and 5.7 h. However, at all of these periodicities salinity fluctuated by < 2 g kg^{-1}, equivalent to a vertical displacement of the halocline of < 0.2 m. The relatively low energy of the background internal wave field in Milne Fiord suggests they likely have a limited role in inducing mixing across the halocline and are primarily of concern as being a small source of error when determining epishelf lake depth from CTD profiles.

3.5.3   Seasonal

Records from the automated weather station over the period of the mooring deployment (Fig. 6) showed that, although the average air temperatures were well below zero (-18.8 °C from 15 May 2011 to 15 May 2014), the seasonal variation in air temperature was extreme, ranging from an hourly maximum of $+20.2$ °C in July 2012 to a minimum of -51.8 °C

in February 2013. The range in air temperature was driven, in part, by variation in solar radiation at this high latitude, which ranged from zero during the polar night (mid-October through February) to a daily maximum of ~ 650 W m^{-2} in late June during the period of 24 h of daylight between April and September. Summer melt conditions varied substantially among years, with cumulative positive-degree days of 278, 253, 92, and 110 in 2011, 2012, 2013, and 2014, respectively (note that for display purposes the meteorological record is truncated to match the mooring record in Fig. 6).

The time series of epishelf lake water temperature (Fig. 6e) revealed that the lake maintained a subsurface temperature maximum throughout the year, reaching a maximum of 2.5, 4.0, and 2.5 °C in 2011, 2012, and 2013, respectively. Temperatures decreased over winter but remained above freezing despite extremely low air temperatures. Below the epishelf lake there was far less seasonal variation in temperature, with the lower halocline remaining relatively cool (approximately −1 °C) all year.

The most striking features of the mooring time series are the substantial seasonal changes in the depth of the thermocline and halocline as revealed by records of salinity (Fig. 6d) and temperature (Fig. 6e). Sensors moored at depths corresponding to the thermo/halocline (referred to hereafter as simply the halocline; between 10 and 15 m from 2011 to 2013, and 8 to 10 m from 2013 to 2014) recorded a substantial freshening and warming from mid-June to mid-August each year. The magnitude of change varied year to year, with little change in summer of 2013. The commencement and duration of the deepening of the halocline corresponded to the onset and duration of the surface melt season, when air temperatures were above freezing (highlighted by the grey area each year in Fig. 6). Tracking the change in depth of the isotherm that corresponded to the $z(N_{max}^2)$ in May of each year indicated that the halocline deepened by 3.0, 3.3, and 1.0 m in summers 2011, 2012, and 2013, respectively.

After the epishelf lake reached its maximum depth at the cessation of the surface melt season each year, the salinity at the halocline quickly began to increase (Fig. 6d) while the temperature decreased (Fig. 6e), indicating shoaling of the halocline. We suggest this was the result of excess freshwater stored in the epishelf lake gradually draining under the ice shelf over winter (see Fig. 1). We note that the rate of shoaling each winter was not constant, as apparent by an abrupt change in depth of the thermocline that occurred in January 2012 (Fig. 6c), which is examined in more detail in the following section. The thickness of the freshwater layer reached a minimum in early June each year, with the depth of the halocline at this time consistently shallower than in June the previous year. Despite the substantial seasonal variation, the mooring records show the halocline shoaled on an interannual basis by 4.1 m between June 2011 and June 2012, 1.5 m between June 2012 and June 2013, and 1.0 m between June 2013 and June 2014.

The magnitude of deepening of the halocline during each summer appeared to be related to the intensity of the surface melt season. Changes in the depth of the epishelf lake were significantly correlated with the total number of PDDs accumulated over the season, having a ratio (Δz/PDD) of 0.017 m °C^{-1} day^{-1} ($n = 226$, $R^2 = 0.92$, $p = 0.005$; Fig. 7).

Further evidence of the linkage between the magnitude of deepening of the halocline and surface conditions is provided by time series of CTD profiles collected during the field campaigns in 2012 and 2013 (Fig. 5). In 2012, the halocline deepened a total of 1.9 m over 2 months, from 5 May to 9 July. In 2013, however, over a similar 2-month period (10 May to 19 July) the lake depth changed <0.5 m. Summer 2012 was unusually warm; a total of 93 PDDs had accumulated by the time the final profile was collected that year (9 July 2012). Summer 2013, however, was cool, and a total of only 38 PDDs had accumulated by the time the final profile

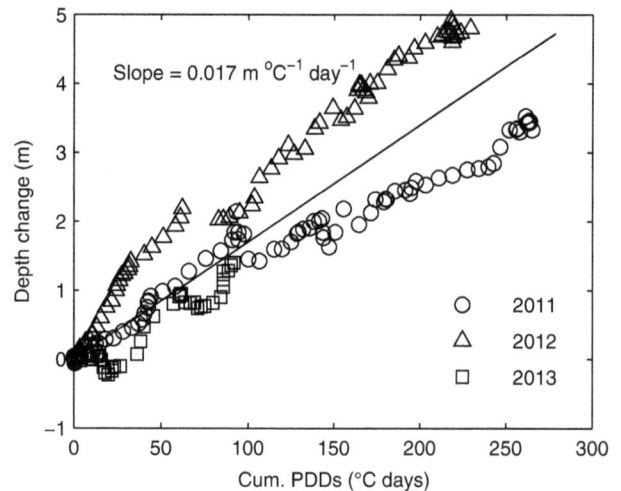

Figure 7. Correlation between the cumulative number of PDDs and the change in depth of the epishelf lake. Depth is estimated from the isotherm proxy during the melt season in 2011, 2012, and 2013.

was collected that year (22 July 2013), despite the 2013 field campaign ending almost 2 weeks later than the 2012 campaign. None of the field campaigns, however, spanned the duration of the entire summer melt season, so the mooring data provide the most complete record of seasonal change. These observations indicate that freshwater inflow to the lake from surface meltwater runoff increased the depth of the lake each year, the amount of deepening directly proportional to the intensity of surface melting.

3.5.4 Mixing event in January 2012

An abrupt change in temperature, salinity, and depth of the halocline occurred on 11 January 2012 at 06:00 UTC (Figs. 6 and 8). Over a duration of 18 h the salinity at 13 m depth dropped from 22 to 12 g kg^{-1} (Fig. 8a) and remained below 15 g kg^{-1} for the remainder of the winter (Fig. 6d). At the same time, the heat content of the upper 25 m of the water column was relatively steady (apart from some fluctuations during the actual event); the slow rate of heat loss was not substantially different from the long-term average over winter (Fig. 8b). During the event the upper portion of the thermocline (above 11 m depth) was displaced upwards 1.5 m, while isotherms in the lower portion of the thermocline (below 11 m depth) spread apart vertically (Fig. 8c). High-amplitude vertical fluctuations of isotherms persisted for several days following the event and were recorded from the uppermost thermistor (at 2 m depth) down to at least 50 m depth (not shown).

Profiles collected several months before and after the event showed a marked change in the depth of the epishelf lake and the gradient of the lower thermo/halocline (Fig. 8d and e). The changes were widespread and observed at several locations profiled throughout the fjord. These observations indi-

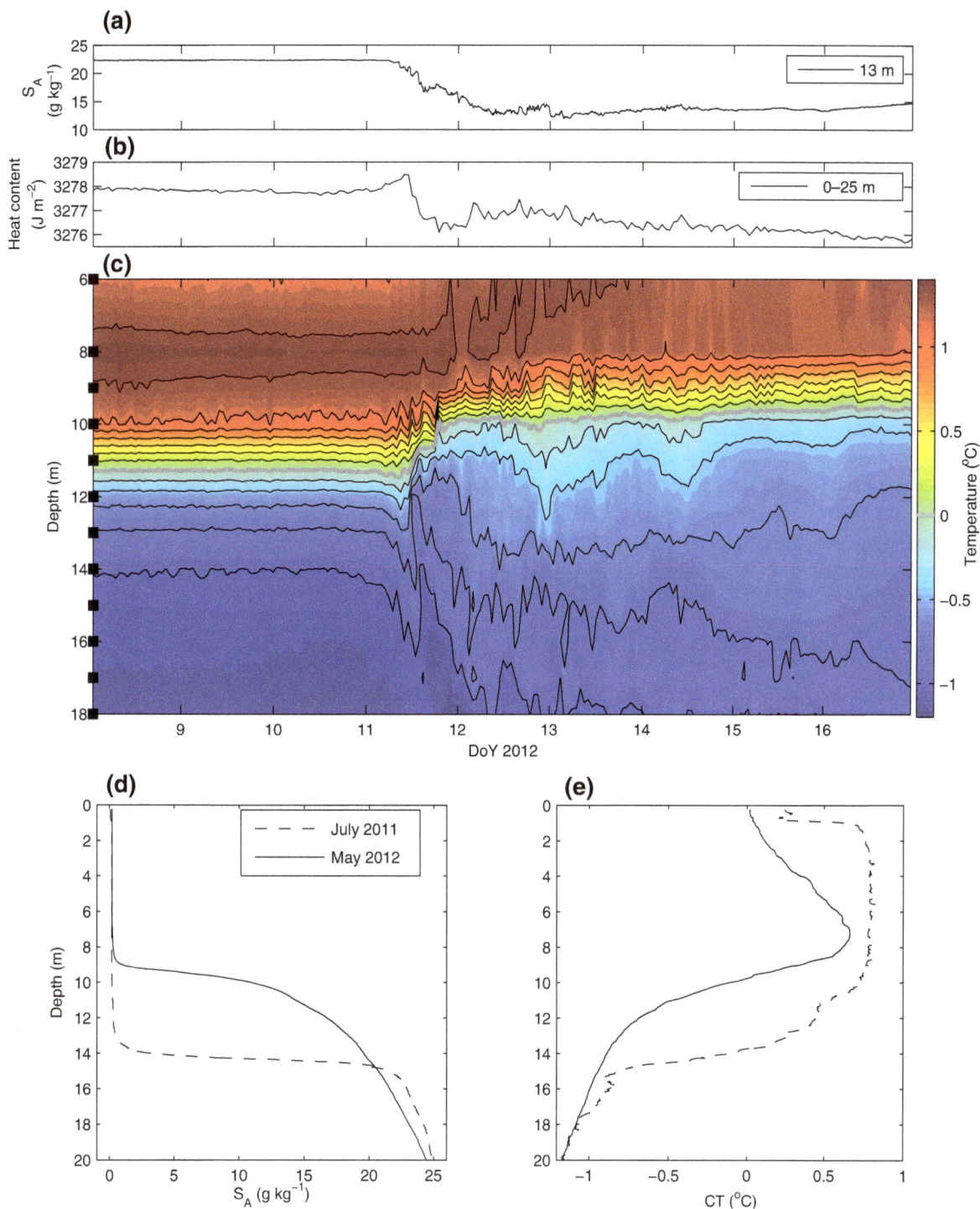

Figure 8. January 2012 MEL halocline mixing event. Time series of (**a**) salinity at 13 m depth, (**b**) heat content between 0 and 25 m, and (**c**) temperature between 6 and 18 m depth. The isotherm increment is 0.2 °C, and the 0 °C isotherm (grey) is highlighted as a proxy for the bottom of the epishelf lake. Black squares on y axis indicate the depths of thermistors. Profiles of (**d**) salinity and (**e**) temperature from field campaigns 6 months before (July 2011) and 4 months after (May 2012) the mixing event show the permanent change in the structure of the halocline (note that some of the change in the depth of the halocline is due to seasonal processes).

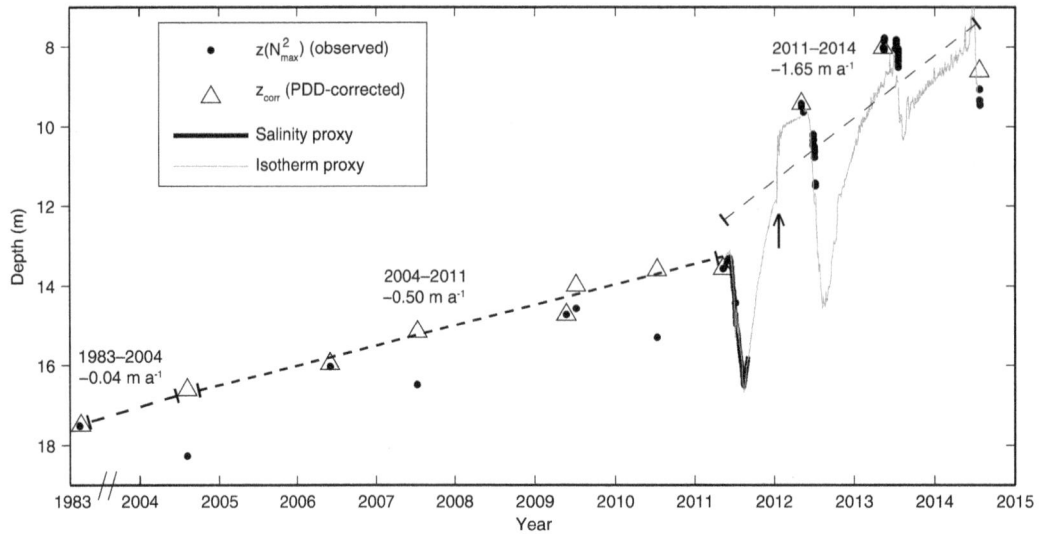

Figure 9. Interannual and seasonal changes in the depth of the MEL from 1983 to 2014. Shown are the observed lake depths from CTD profiling $\left(z(N_{max}^2)\right)$, PDD-corrected lake depths (z_{corr}; see Eq. 4) that account for the seasonal timing of the profile, and depths determined from moored salinity and temperature records from May 2011 to July 2014. Note the discontinuous x axis between 1983 and 2004. To reduce clutter, z_{corr} are shown only for the first profile collected during each field campaign after 2011. The arrow indicates the mixing event in January 2012. Dashed lines indicate a linear regression of z_{corr} for the periods 1983–2004, 2004–2011, and 2011–2014.

cated that a sudden mixing event occurred at the bottom of the epishelf lake that entrained warm, relatively fresh water from the lake downward, and cool, salty water upward. The result was an abrupt, irreversible thinning of the entire lake by 1.5 m.

3.5.5 Interannual

The complete record of lake depth inferred from CTD profiling between 1983 and 2014, as well as from the continuous mooring time series of temperature and salinity between 2011 and 2014 (Fig. 9), shows the depth of the lake decreased by a total of almost 10 m over this period, yet there was substantial seasonal variation. Lake depths derived from both the moored salinity and temperature records match and correspond to depths derived from CTD observations, providing confidence in the continuous multiyear isotherm proxy. The large degree of seasonal variation recorded by the mooring from 2011 to 2014 suggests that the depths of the lake determined from interannual CTD profiling prior to 2011 could have been influenced by the timing of profiling relative to the melt season. We have previously shown that the changes in depth of the lake during the summers of 2011, 2012, and 2013 were correlated with the cumulated PDDs that season. Therefore, assuming the relationship between lake depth and PDDs was constant over the full record, and because all profiles were collected just prior to or during the meltwater inflow period, we adjusted the observed lakes depths ($z(N_{max}^2)$) to account for differences in the timing of observations and the intensity of the melt season using the accumulated num-

ber of PDDs at the time of profiling. We calculated a PDD-corrected depth (z_{corr}) by

$$z_{corr} = z\left(N_{max}^2\right) - PDD(t)\frac{\Delta z}{PDD}, \qquad (4)$$

where PDD(t) is the number of PDDs accumulated from the start of the melt season until the time of profiling and $\Delta z / PDD$ is the ratio of seasonal changes in lake depth (Δz) to PDDs, taken as $0.017\,m\,°C^{-1}\,day^{-1}$ (Fig. 7). The PDD-corrected depth was interpreted as the theoretical minimum depth of the lake prior to the onset of meltwater inflow each year.

The PDD-corrected depths (triangles; Fig. 9; Table 1) suggested that, in contrast to the raw CTD observations, the lake actually thinned monotonically every year since 1983 (with the possible exception of 2013–2014, which is discussed further below), although the rate of thinning increased substantially over time. Between 1983 and 2004 the change in the depth of the lake was small, a total thinning of 0.9 m at an average rate of $0.04 \pm 1.2\,m\,a^{-1}$ (the large error due to uncertainty in the lake depth estimated from the 1983 water bottle samples). The lake then thinned steadily from 2004 to 2011 at an average rate of $0.50 \pm 0.05\,m\,a^{-1}$ ($n = 8$, $R^2 = 0.95$, $p < 0.001$). After this period there was a more dramatic phase of thinning, with the lake shoaling a total of 4.1 m from 2011 to 2012, with 1.5 m of that loss occurring during the January 2012 mixing event. This was followed by a further loss of 1.5 m from 2012 to 2013 and then an increase in depth by 0.6 m from 2013 to 2014 (although we note that the lake depth inferred from the continuous mooring record indicates the lake may have actually continued to thin between 2013 and 2014 and the PDD-corrected lake depth from CTD profiling may not have captured the actual minimum depth in 2014). The average rate of thinning for the period 2011–2014 was $1.65 \pm 0.74\,\mathrm{m\,a^{-1}}$ ($n = 4$, $R^2 = 0.71$, $p = 0.16$), over 3 times higher than in 2004–2011.

4   Discussion

Using a combination of data from CTD profiles, moorings, and remote sensing, we have documented changes in the water column structure, depth, and extent of the MEL since 1983. Our results have shown that the lake has existed in close to its current spatial extent for the last 3 decades and that the freshwater layer thinned substantially over the past decade. Our mooring records have shown that epishelf lakes are dynamic physical environments, undergoing depth changes of several metres in a single year owing, in part, to variation of meltwater inflow. In addition, epishelf lakes can be subject to abrupt events that can dramatically shift the halocline over very short (< 24 h) time periods. These results indicate that the utility of using changes in the depth of an epishelf lake to infer changes in the structure of the impounding ice shelf is dependent on an understanding of the freshwater budget and hydraulics of each system. In the following sections we first discuss possible causes of the January 2012 event given the magnitude of its effect on the depth of the halocline. Then we discuss the freshwater budget of the epishelf lake and what we can infer from our observations about changes in the thickness of the MIS and the implications of these findings for other related systems.

4.1   January 2012 event

The catastrophic drainage of the Disraeli Fiord epishelf lake due to the fracturing of the WHIS (Mueller et al., 2003) provides a motivation to consider the rapid shoaling of the MEL in January 2012 as a sudden drainage of the lake due to a fracturing of the MIS. However, first-order calculations indicate that this was not likely the case. Assuming that the 1.5 m depth change observed in the MEL was due to drainage of an equal amount of epishelf lake water under the MIS, then the total volume change across the $\sim 64.4\,\mathrm{km^2}$ lake was $9.8 \times 10^7\,\mathrm{m^3}$. For all of this water to drain out of the fjord through the ~ 10 m wide basal channel in the MIS over the observed 18 h event duration would have required a volume flux of $1.5 \times 10^3\,\mathrm{m^3\,s^{-1}}$, requiring outflow velocities of $> 100\,\mathrm{m\,s^{-1}}$. Even if the channel were an order of magnitude wider, the required velocities are still unrealistic, and we therefore consider it unlikely that the January 2012 event was related to a rapid drainage event.

In addition to shoaling of the thermocline (and by inference the halocline), the January 2012 event thickened the bottom half (i.e., water colder than $\sim 0\,^\circ\mathrm{C}$) of the thermo-

cline (Fig. 8), suggestive of mixing processes. A scale for the energy required to cause such mixing is given by the change in background potential energy of the water column through adiabatic redistribution of density (Winters et al., 1995). Lacking salinity profiles close to the timing of the event, we idealize the system as two layers, with warm freshwater above and cool saltwater below, and utilize the moored temperature time series to estimate the portion of the water column that mixed across the interface, represented by the change in depth of the thermocline. Assuming the mooring is representative of the entire epishelf lake and that complete mixing occurs, the change in background potential energy (ΔPE) is

$$\Delta\mathrm{PE} = \frac{1}{2}h_1 h_2 (\rho_2 - \rho_1) g A_\mathrm{L}, \qquad (5)$$

where h_1 is the thickness of the portion of the upper layer involved in mixing, h_2 is the thickness of the portion of the lower layer involved in mixing, ρ_1 is the density of the upper layer ($1000\,\mathrm{kg\,m^{-3}}$), ρ_2 is the density of the lower layer ($1025\,\mathrm{kg\,m^{-3}}$), g is gravitational acceleration ($9.81\,\mathrm{m\,s^{-2}}$), and A_L is $65\,\mathrm{km^2}$. From the temperature time series we chose a range of values for h_1 (0.5 and 1.5 m) and h_2 (2 and 7 m), which gives ΔPE of the order 10^8 to 10^{10} J, which we consider an upper bound due to the assumptions in the model.

We have considered several possible sources that could provide sufficient kinetic energy to induce mixing on this scale, including a tidal anomaly, tsunami, earthquake, iceberg capsize, and subglacial outburst flood, yet we lack convincing evidence that could identify a particular trigger as the cause. For example, the water level record shows no anomaly at this time that could indicate a large tidal excursion or tsunami occurred. A review of seismographic records from Ellesmere Island (http://earthquaketrack.com/r/ellesmere-island-nunavut-canada/recent) did not reveal any substantial earthquakes in January 2012. The increased production of icebergs in Milne Fiord from breakup of the glacier tongue and ice shelf suggests that turbulent energy released from calving or capsize of an iceberg could have generated mixing. Application of a scaling calculation (Eq. 4 in Burton et al., 2012) suggests the capsize of an iceberg with dimensions on the order of $50 \times 100 \times 150$ m, typical of Milne Fiord, could have generated sufficient energy for mixing; however a review of available satellite imagery does not provide any clear evidence of a substantial mid-winter capsizing event. Field observations of icebergs in the fjord up to 2015 indicate that most, particularly those calving from the ice shelf, are tabular, and only a few glacier-tongue-derived icebergs show evidence of capsize, perhaps explaining why halocline mixing events are rare (i.e., only one major mixing event observed during the multiyear mooring record). Likewise, we could find no clear evidence of the drainage of a supraglacial lake that might have triggered a sudden subglacial outburst flood, such as those observed in western Greenland by Kjeldsen et al. (2014). Although the cause of the mixing event remains uncertain, and we cannot rule out other mechanisms, such as the propagation of an offshore anomaly into the fjord, it seems plausible that the anomalously warm air temperatures in the summer of 2011 and the associated intense surface melting may have played a role in precipitating the mixing event. What our mooring records do clearly show is that continuous observations are the preferred method for tracking changes in epishelf lake depth, and, where they are lacking, the possibility of abrupt changes in epishelf lake depth that are unrelated to changes in ice shelf thickness must be considered when interpreting interannual observations.

4.2   Freshwater budget

The depth of an epishelf lake at a moment in time is determined by both seasonal and long-term factors, primarily the balance between inflow, outflow, the area of the lake, and the depth of the ice dam. To separate out the influence of changes in the depth of the ice dam from the other factors, it is necessary to consider the lake's freshwater budget. To do so, we can simplify an epishelf lake to an idealized form, by making the following assumptions: that the level of the ice dam is fixed in time and has a simple bathymetry with a horizontal bottom and vertical sidewalls, and that there is no mass or volume flux across the halocline, or through the ice walls (we revisit this assumption in Sect. 4.7). From conservation of volume the seasonal change in depth over time (dz/dt) can be expressed as

$$\frac{dz}{dt} = \frac{Q_{in} - Q_{out}}{A_L}, \tag{6}$$

where Q_{in} is the volumetric rate of inflow ($m^3 s^{-1}$) from surface runoff; Q_{out} is the volumetric rate of outflow ($m^3 s^{-1}$) below the ice shelf; and A_L is the surface area of the epishelf lake (m^2), which is assumed constant over a season. The rate of inflow will be dependent on the magnitude of surface runoff from the surrounding glacial catchment. The rate of outflow will depend on the internal hydraulics of freshwater drainage below the ice shelf. Interannual changes in the area of the epishelf lake will be determined by advective and thermodynamic changes in the position of the ice margins that make up the shoreline of the lake. Knowledge of each of these variables can then provide a means to isolate factors controlling depth changes that are, or are not, related to changes in the thickness of the ice dam. In the next sections we address what is known about these variables in the Milne Fiord setting.

4.3   Area

Changes in lake area could influence both the long-term changes in the depth of the lake and the seasonal response to inflow. On an interannual basis, if the volume of freshwater in the lake were conserved, an advective change in lake area could theoretically cause a change in depth (for example, an advance of the Milne Glacier terminus would reduce lake area and lead to an increase in lake depth). However, the largest contributor to the increase in area of the lake appears to have been the thermodynamic transformation of ice of the inner unit of the MIS to epishelf lake water through melting, a process that would have increased both the area and volume of the lake. This process would not have altered the depth of the lake because the melted ice was already in hydrostatic equilibrium. Therefore, the five-fold increase in area of the lake between 1959 and 1988 through melting of the ice shelf is not expected to have had a large influence on the interannual depth of the lake. The small 10 % fluctuations in lake area between 2004 and 2014 are also insufficient to account for the 50 % reduction in lake depth over this period, again indicating that area changes were not an important factor in driving the long-term shoaling of the lake.

On a seasonal basis, however, assuming the volume (and rate) of summer inflow was the same every year between 1959 and 1988, the five-fold increase in the area of the lake would have resulted in a five-fold decrease in the magnitude of summer deepening because the inflowing meltwater would have been distributed over a larger area. From 2004 to 2014, the 10 % interannual fluctuation in the area of the lake would have resulted in an equivalent 10 % variation in the magnitude of summer deepening, again assuming fixed inflow. Summer melt conditions, however, do vary substantially year to year; the annual cumulative PDDs varied by a factor of 3 between 2004 and 2014. This suggests that annual fluctuations in the volume (and rate) of meltwater inflow appear to be the primary factor determining the magnitude of seasonal deepening of the lake.

4.4   Inflows

We have shown that there is a positive correlation between the seasonal depth increase of the epishelf lake in summer and the cumulative number of PDDs, suggesting a direct relationship between freshwater inflow to the lake and surface melting of snow and ice, which is largely driven by air temperature. If, as a starting point, we assume that the observed increase in depth of the lake each year accounts for all runoff entering the fjord at the surface and neglect outflow, then using our estimated depth change rate of $0.017\,\mathrm{m\,PDD^{-1}}$ and the observed cumulative PDDs each year (ranging from 92 and 278 PDDs during the Milne Fiord weather station record from 2009 to 2014) equates to seasonal depth changes varying between 1.6 and 4.7 m. Combined with the average area of the lake ($65\,\mathrm{km^2}$), this suggests an seasonal increase in volume of 100–$300 \times 10^6\,\mathrm{m^3\,a^{-1}}$ over this period. Given that we have neglected outflow, this estimate is a lower bound on the change in volume of the lake and thus inflow from the fjord catchment.

The primary source of inflow to the lake is runoff from snow and glacier melt, and we can compare the seasonal volume change of the epishelf lake with an estimate of the total meltwater runoff from the Milne Fiord catchment. Lenaerts et al. (2013) estimated the average annual runoff from the $146\,000\,\mathrm{km}^2$ glaciated area of the Canadian Arctic Archipelago (CAA) from 2000 to 2011 was approximately $106\,\mathrm{Gt\,a}^{-1}$. Applying this estimate to the $1108\,\mathrm{km}^2$ glaciated area of the Milne Fiord catchment, which accounts for approximately 70 % of the total area of the catchment, suggests the annual meltwater inflow to Milne Fiord is at least $1.12 \times 10^9\,\mathrm{m}^3\,\mathrm{a}^{-1}$. If all of this meltwater flows into the $65\,\mathrm{km}^2$ MEL at the surface, the freshwater layer could theoretically deepen by 17 m each summer. In actual fact, though, an unknown portion of this meltwater inflow likely enters the fjord at the 150 m deep grounding line of the Milne Glacier (our observations suggest perhaps 10–28 % of the total runoff enters at the surface). Water column profiles indicate that the subglacial freshwater plume entrains a substantial amount of seawater on its ascent and spreads out in the lower halocline (Hamilton, 2016), likely not contributing substantially to the freshwater budget of the lake. Surface runoff therefore appears to be the main freshwater source to the epishelf lake. Although the exact proportions of surface versus subsurface inflow to Milne Fiord from meltwater runoff are unknown, our runoff estimate suggests there is ample inflow into Milne Fiord each summer to account for the observed seasonal depth increase of the epishelf lake. Furthermore, meltwater runoff from the CAA increased by 54 % in the periods 1971–2000 and 2000–2011 (Lenaerts et al., 2013), indicating that the long-term shoaling of the MEL cannot be explained by a decrease of the inflow term in the conservation of volume (Eq. 6).

We have argued that long-term changes in both inflow and the area of the lake are insufficient to account for the long-term shoaling of the lake. Therefore, understanding factors affecting the volumetric rate of outflow is key to understanding long-term changes in the depth of the lake. Next, we investigate the seasonal shoaling of the lake to better understand outflow hydraulics and use that information to inform what factors are driving the long-term shoaling.

4.5 Outflow

The mooring data showed that after inflow ceased each year the lake thinned over winter until the following melt season. The rate of the seasonal shoaling of the halocline was nonlinear, changing more rapidly at first and then gradually tapering off until the change with time was minimal just prior to the following melt season. The pattern indicated that the rate of shoaling depended on the relative depth difference between the halocline and the ice dam, suggesting an internal hydraulically controlled flow. Ice-penetrating radar ice thickness surveys collected between 2008 and 2015 (Mortimer et al., 2012; Hamilton, 2016) indicate that the outflow pathway under the MIS is likely along a narrow basal channel following the east–west re-healed fracture in the outer

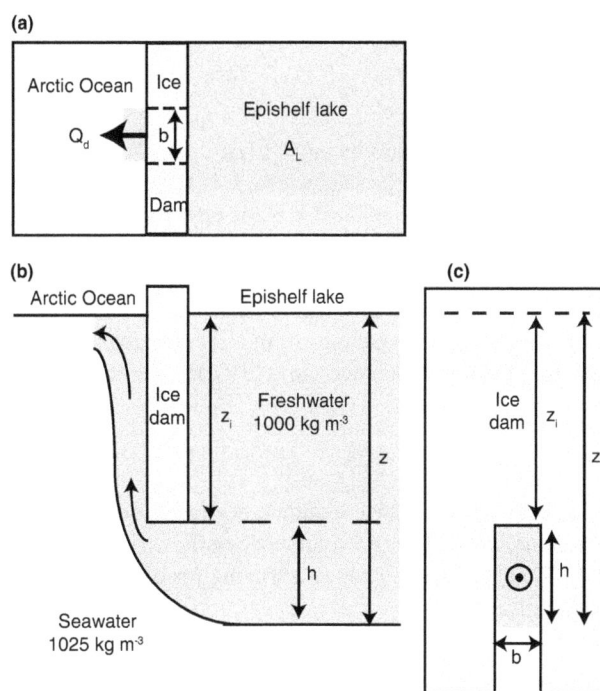

Figure 10. Schematic representation of epishelf lake outflow through a basal channel in the ice shelf dam in **(a)** plan, **(b)** elevation, and **(c)** cross-sectional views. The volumetric discharge is modelled using a modified form of the rectangular weir equation (Eq. 7).

unit. The measured ice thickness at a few locations along the mid-line of the basal channel is 8–11 m, suggesting the ice at the apex of the channel could be acting as a hydraulic control on outflow and thus determining the rate of change in the depth of the epishelf lake over winter. In this section we model the epishelf lake outflow as an internal hydraulically controlled flow through a weir and compare simulated changes in epishelf lake depth over winter to mooring observations.

An idealized schematic of epishelf lake outflow through a basal channel in the ice shelf is shown in Fig. 10, represented as a simple two-layer system, with freshwater overlying seawater ($\Delta\rho = 25\,\mathrm{kg\,m}^{-3}$). The ice dam acts as a hydraulic control, limiting two-way transport below the ice shelf. If the depth of the seawater layer is much greater than the depth of the freshwater layer, then the situation is analogous to single-layer flow through an inverted weir, but here the horizontal pressure gradient is supplied by the density difference between freshwater and seawater. For simplicity, if we assume a rectangular channel geometry and that the dimensions of the channel are constant in time (i.e., no melting or accretion on the ice walls and the depth of the ice dam is fixed), the volumetric outflow discharge (Q_{out}; $\mathrm{m}^3\,\mathrm{s}^{-1}$) can be estimated using a modified form of the Kindsvater–Carter rectangular weir equation (Kindsvater and Carter, 1959):

$$Q_{\text{out}} = \frac{2}{3}\sqrt{2g'}C_{\text{e}}bh^{\frac{3}{2}}, \tag{7}$$

where g' is reduced gravity ($g' = g(\Delta\rho/\rho)$), C_{e} is an empirically derived discharge coefficient, b is the effective width (m) of the outlet channel, and h is the effective depth (m) of the lake below the ice dam. b and h account for the effects of viscosity and wall friction and will therefore be somewhat smaller than the actual physical dimensions.

Assuming vertical sidewalls to the lake, the change in volume of the epishelf lake over time ($\mathrm{d}V/\mathrm{d}t$) is

$$\frac{\mathrm{d}V}{\mathrm{d}t} = A_{\text{L}}\frac{\mathrm{d}h}{\mathrm{d}t}. \tag{8}$$

During winter, inflow is assumed negligible, so the change in volume is equal to the volumetric outflow (i.e., $\mathrm{d}V/\mathrm{d}t = Q_{\text{d}}$). Equating Eqs. (7) and (8), solving for $\mathrm{d}h/\mathrm{d}t$, and integrating gives

$$h(t) = \left(\frac{1}{2}at + \frac{1}{\sqrt{h_0}}\right)^{-2}, \tag{9}$$

where

$$a = \frac{\frac{2}{3}\sqrt{2g'}C_{\text{e}}b}{A_{\text{L}}}, \tag{10}$$

and h_0 is the initial depth of the lake below the ice dam at $t = 0$.

From Eq. (9) we modelled the change in depth of the epishelf lake for 275 days each winter of the 3-year mooring record, from approximately September to May of 2011–2012, 2012–2013, and 2013–2014. We assume, for the time being, that all changes to the depth of the ice dam occurred during the melt season (which was not modelled), so the depth of the ice dam remained fixed during each run (although it differed between runs). The start time ($t = 0$) for each run was chosen as the date when air temperatures permanently fell below 0 °C for the winter, when meltwater inflow was assumed to cease. The initial depth of the lake ($z_{\text{MEL}}(t = 0)$) was estimated from the mooring temperature record. In the model, freshwater drains under the ice dam at a rate proportional to h until $z_{\text{MEL}}(t)$ shoals to the level of the ice dam, z_i (i.e., when $z_{\text{MEL}}(t) - z_i = h = 0$). The mooring records show that the lake was still shoaling when meltwater input commenced the following spring, suggesting the lake had not completely drained to the level of the ice dam. We therefore did not have a direct measure of z_i, which was required to estimate h_0. Instead, we estimated h_0 as the sum of the difference between the initial lake depth and the depth when $t = 240$ elapsed days (the maximum duration of the shortest continuous mooring record each year), plus some unknown offset c_h (i.e., $h_0 = z_{\text{MEL}}(t = 0) - z_{\text{MEL}}(t = 240 \text{ days}) + c_h$). The actual depth of the ice

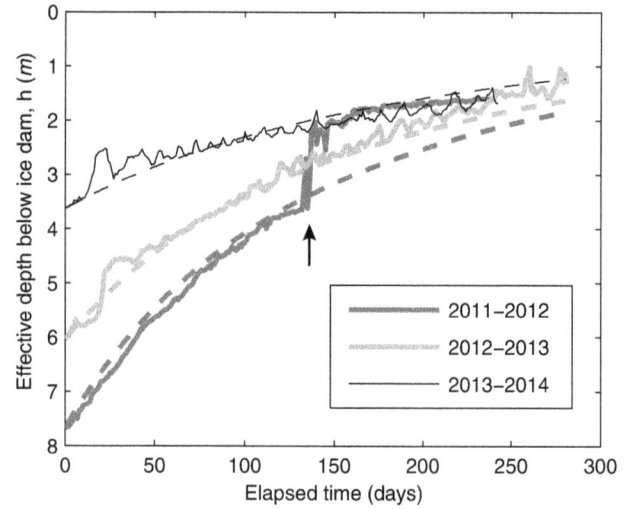

Figure 11. Change in the effective depth (h) of the epishelf lake below the ice dam over time during the winter of 2011–2012, 2012–2013, and 2013–2014. Elapsed time is measured from the end of the surface melt season each year. Observed depths (solid lines) are based on the isotherm proxy from the mooring record, while modelled depths (dashed lines) are based on a weir equation. Arrow indicates the mixing event that occurred in January 2012.

dam at $t = 240$ days is then $z_i(t = 240 \text{ days}) = z_{\text{MEL}}(t = 0) - h_0$. Note that to account for the abrupt 1.5 m shoaling of the halocline in January 2012 we added an equivalent 1.5 m to $z_{\text{MEL}}(t = 240 \text{ days})$ for that year. We found the best overall fit for all years was achieved when $c_h = 1.6$ m and $C_{\text{e}}b = 4.5$ m, so these values were held constant for all runs.

Modelled and observed changes to the depth of the lake relative to the ice dam are shown in Fig. 11. Note that the depth of the ice dam differed for each run, so lake depths are plotted relative to the level of the ice dam for that year. Despite h_0 varying by over a factor of 2 among the different years (from 7.5 m in 2011–2012 to 3.3 m in 2013–2014), the simple drainage model simulated the observed pattern of changes in the depth of the lake each winter well. The model could not account for the abrupt thinning of the lake in January 2012 (elapsed day 140 for 2011–2012) due to the mixing event, so observed and modelled values differ accordingly after this date. We note that the rate of shoaling of the lake slowed substantially after the mixing event, suggesting that the abrupt upward shift in the halocline moved it vertically closer to the level of the ice dam, thus reducing the rate of outflow. Overall, the results indicated that outflow drainage from the epishelf lake through the basal channel could generally be well simulated by weir outflow hydraulics.

To fully assess the model, we need to determine if the selected values for the parameters are physically realistic. For typical weirs the discharge coefficient, C_{e}, varies between 0.55 and 0.8 (ISO, 2008). If we assume that this range for C_{e}

is broadly appropriate for the MEL system, then the width of the channel, b, is between 5.6 and 8.2 m. This value is comparable to the estimated minimum width of the surface expression of the re-healed fracture at its narrowest point (2–8 m) from field observations and aerial imagery. Next, using the offset $c_h = 1.6$ m, we estimate the actual depth of the ice dam at the 240 day mark, approximately 1 May, varied from 9.4 m in 2012 to 7.5 m in 2014. This is broadly consistent with field measurements of 8 to 11 m thick ice measured along the re-healed fracture in July 2015. We conclude that the values chosen for the parameters are physically realistic and appropriate for this system, although further work is required to validate these.

The idealized model is useful in providing a simple explanation for the observed seasonal shoaling of the epishelf lake halocline over winter and suggests that the geometry of the outflow channel is a key factor in determining the depth of the lake, on both a seasonal and interannual basis. Importantly, the only variable that changed between runs was h_0, which was ultimately dependent on the depth of the ice dam that year. For the modelled depth changes to match the observed differences in the rate of shoaling each year (i.e., fastest shoaling in 2011–2012, slowest in 2013–2014) required that the depth of the ice dam differed, getting shallower each year, suggesting the ice dam is indeed thinning over time. This, admittedly, highly simplified model, however, does not identify what mechanism led to the change in the level of the ice dam over time, such as basal melting (or accretion) along the channel, mass loss or gain on the surface or bottom of the ice shelf, and the hydrostatic adjustment of the ice shelf from these processes. In the following section we address what the long-term changes in the depth of the epishelf lake indicate about changes in the thickness of the ice shelf.

4.6   Changes in thickness of the MIS

Over the long-term, the depth of the epishelf lake is determined by the maximum depth of the ice along the lake's outflow pathway, referred to as the depth of the ice dam. Assuming the ice shelf is free-floating, shoaling of the ice dam could be caused by three mechanisms: (1) mean surface ablation thinning the ice shelf and causing an upward shift of ice dam due to hydrostatic adjustment; (2) localized submarine melting of the ice along the basal channel due to warm epishelf lake outflow leading to thinning of the ice dam; or (3) submarine accretion of ice on the deeper portions of the ice shelf (not in the channel itself) thickening the ice shelf and causing an upward shift of the ice dam with hydrostatic adjustment (in this scenario lateral bridging stresses across the channel must allow ice in the channel to be raised out of hydrostatic equilibrium). Given that the Milne Ice Shelf is in a state of negative mass balance (Mortimer et al., 2012), the latter mechanism is unlikely, and the shoaling of the lake

is likely due to either surface ablation or melting along the basal channel, or some combination of the two.

The long-term record of change in lake depth (Fig. 9) suggests the thickness of the ice dam was relatively stable from 1983 to 2004 but then thinned dramatically by almost 10 m between 2004 and 2014. From 2004 to 2011, changes in the depth of the lake suggest a steady thinning of the ice dam at a rate of 0.50 ± 0.05 m a^{-1}. After this period, there was a more rapid phase of thinning, with more interannual variability. Between May 2011 and May 2012 the lake abruptly shoaled by 4.1 ± 0.4 m; however, 1.5 m of this thinning was due to the mixing event in January 2012, suggesting the actual thinning of the ice dam over this period was 2.6 ± 0.4 m. From May 2012 to May 2013 the change in lake depth suggests the ice dam thinned a further 1.5 ± 0.4 m. Following these 2 years of rapid thinning, the rate of ice loss appears to have slowed, or even reversed (indicating thickening of the ice dam), between 2013 and 2014. The annual average rate of shoaling of the lake between 2011 and 2014 was 1.65 m a^{-1}; however, if we account for the 1.5 m shoaling due to the January 2012 event that did not appear related to a change in ice thickness, then this suggests the ice dam thinned at a rate of 1.15 m a^{-1} over this period.

This pattern of accelerated ice loss is consistent with widespread rapid cryospheric loss along northern Ellesmere Island over the past decade. Since 2000, changes have included the breakup of the Ayles, Markham, Petersen, Serson, and Ward Hunt ice shelves (Copland et al., 2007; Mueller et al., 2008, 2017a; White et al., 2015a); the loss of multi-year landfast sea ice (Pope et al., 2012); and the thinning of epishelf lakes along this coast (Veillette et al., 2008), including the long-term thinning and catastrophic drainage of the Disraeli Fiord epishelf lake (Veillette et al., 2008; Mueller et al., 2003). The accelerated thinning of the MIS after 2004 also corresponds to a period of rapid loss of perennial ice cover of Ward Hunt Lake, which lies 115 km to the east of Milne Fiord (Paquette et al., 2015), and a sharp increase in mass loss from glaciers and ice caps in the Canadian Arctic (Gardner et al., 2011; Lenaerts et al., 2013). These widespread changes strongly suggest that regional climate warming is the driving factor behind the thinning of the MEL and MIS.

Mortimer et al. (2012) showed that the MIS was in a state of negative mass balance between 1981 and 2009, thinning at an average annual rate of 0.29 ± 0.1 m a^{-1} over this period, albeit with substantial spatial variability. The central unit of the MIS experienced a higher rate of thinning, while a repeat transect that crossed the east–west fracture indicates a low and spatially variable thinning rate of 0.08 ± 0.08 m a^{-1} for the outer unit. Using average annual surface mass loss of 0.07 m a^{-1} recorded at the nearby WHIS between 1989 and 2003 (Braun, 2017), Mortimer et al. (2012) inferred that basal melting was a key contributor to overall thinning of the MIS over that period. We suggest, however, that the low rate of thinning of the outer unit could be accounted for by surface ablation alone prior to 2009.

The ice dam for the epishelf lake appears to be located along the basal channel in the outer unit of the MIS, and thus thinning of this region would be reflected in the depth of the epishelf lake. The rate of thinning of the outer unit as measured by Mortimer et al. (2012) is of the same order as that inferred from the epishelf lake records over a similar period (0.1 ± 1 m a^{-1} for 1983–2009). Thus, assuming surface ablation at MIS was equivalent to that at WHIS, thinning of the ice dam due to surface ablation alone could have accounted for shoaling of the epishelf lake over the period 1983 to 2009. In fact, the higher-temporal-resolution lake depth observations suggest that the majority of ice dam thinning between the 1980s and 2009 actually occurred after 2004. Similarly, the majority of the surface mass loss between 1981 and 2009 on the WHIS actually occurred after 2002 (Braun, 2017), observations which preceded an even more rapid phase of ice loss.

Surface mass loss in Milne Fiord was recorded intermittently from 2009 to 2014 by a relatively sparse network of eight ablation stakes installed on the MIS and the Milne Glacier tongue, showing an annual area-averaged surface mass loss of 0.78 ± 0.64 m a^{-1} for this period (Hamilton, 2016). Accounting for hydrostatic adjustment of the floating ice shelf, this would have resulted in a shoaling of the ice dam by 0.69 ± 0.56 m a^{-1}, greater than the observed shoaling of the lake for 2004–2011 but less than that observed from 2011 to 2014. Thus, surface ablation of the ice shelf alone appears sufficient to account for the change in depth of the epishelf lake prior to 2011, yet other mechanisms are required to explain the higher rate of thinning inferred from the epishelf lake records in 2011 and 2012.

We suggest that the rapid increase in mass loss inferred from the lake record after 2011 was fundamentally due to anomalously warm air temperatures in summer 2011 and 2012, which had a two-fold effect. First, the thinning of the MIS from enhanced surface ablation led to a upward shift of the ice dam through hydrostatic adjustment. Second, the increased surface melting resulted in a large seasonal meltwater inflow to the lake and substantial deepening of the halocline, and it increased the outflow volume flux through the basal channel, causing increased flux of heat to the ice walls and ceiling of the channel, ultimately leading to enhanced melting and thinning of the ice dam. Therefore, although the long-term changes in the depth of the epishelf lake directly reflect the thickness of ice along the basal channel, they cannot be used to infer a mean thinning rate of the entire ice shelf. The long-term change in the depth of the epishelf lake, however, provides a useful indicator of the state of the ice shelf, in particular the weakening of the structural integrity of the outer unit by thinning along the basal channel.

After the enhanced melting and shoaling of the ice dam during the warm summers of 2011 and 2012, the depth of the epishelf lake remained relatively stable between 2013 and 2014. The summers of 2013 and 2014 were the coolest on record since the installation of the weather station in 2009, suggesting the significant reduction in surface ablation, meltwater production, and outflow through the basal channel combined meant that the ice dam did not substantially change in thickness over those years. Despite the short-term departure in 2013–2014 from the long-term thinning trend, however, it is likely that with continued regional climate warming the pattern of thinning of the lake and ice dam will continue. The warming trend continued in 2015 and 2016, with 2016 being the warmest summer on record at Milne Fiord (https://tinyurl.com/milnewx), and we expect thinning of the lake and ice dam was renewed.

Although the depth of the epishelf lake appears to be determined by changes in the thickness of the ice dam in the outer unit, the higher mean thinning rate of the central unit is an interesting phenomenon and warrants discussion. Unlike the outer unit, surface ablation alone does not appear sufficient to explain the high thinning rate of the central unit prior to 2009. While the spatial heterogeneity of surface ablation for the MIS is not well constrained, and differences in wind-blown snow deposition or surface albedo may be factors in the differential melt rates, it seems likely that the higher thinning rate of the central unit is due, in part, to enhanced submarine melting caused by the presence of the epishelf lake. The presence of the highly stratified epishelf lake results in an increased heat content within the fjord and heat that is available year-round to drive submarine melting above the halocline. We therefore suggest that the presence of the relatively warm water of the epishelf lake is in fact contributing to the thinning and breakup of the central unit.

4.7 Submarine melt

The volume of freshwater input from submarine melt around the perimeter of the epishelf lake, as well as a submarine melt rate, can be constrained by considering a rough heat balance of the epishelf lake over winter. To obtain an upper bound on the melt rate, we simplify the heat balance by assuming all heat in the epishelf lake is lost to melting of ice around the lake perimeter. This assumes vertical heat flux through surface ice and heat flux across the halocline are negligible (and the latter avoids complications of advective heat loss from epishelf lake outflow below the ice dam). We further assume ice temperature is at its melting point. The change of mooring temperatures over each winter (ΔT), from 15 August to the following 1 June from 2011 to 2014, between 2 and 8 m depth varies from 0.5 to 2.5 °C. This equates to an annualized rate of heat loss ($\Delta H / \Delta t$) on the order of 10^{15} J a^{-1} (where $\Delta H / \Delta t = \rho_{\mathrm{w}} c_{\mathrm{p}} \Delta T / \Delta t \, V_{\mathrm{w}}$, where water density ($\rho_{\mathrm{w}}$) is 1000 kg m^{-3}, the specific heat capacity of water (c_{p}) is 4.18×10^3 J kg^{-1} °C^{-1}, and the volume of the lake (V_{w}) between 2 and 8 m depth is 3.75×10^8 m^3). The volume flux of freshwater from submarine melting of perimeter ice (($\Delta H / \Delta t)/(L_{\mathrm{i}} \rho_{\mathrm{i}}$)) is therefore on the order of 10^6

to $10^7\,\mathrm{m}^3\,\mathrm{a}^{-1}$ (where the latent heat of fusion of ice (L_i) is $3.34 \times 10^5\,\mathrm{J\,kg}^{-1}$ and density of ice (ρ_i) is $900\,\mathrm{kg\,m}^{-3}$). We previously estimated that the annual volume of surface runoff into the lake was on the order of $10^8\,\mathrm{m}^3\,\mathrm{a}^{-1}$ (i.e., 10–28 % of the total meltwater runoff from the catchment). Thus, input from submarine melting around the lake perimeter is a relatively small proportion (1–10 %) of the total freshwater inflow to the lake and therefore has a relatively minor role in determining stratification, and we were justified in neglecting submarine melt from the freshwater budget.

The volume flux of freshwater from submarine melting can be used to calculate a horizontal melt rate of the ice walls around the lake perimeter using the surface area of the ice walls. The ice shelf and glacier tongue form a perimeter to the lake $\sim 40\,\mathrm{km}$ long, giving an ice wall surface area between 2 and 8 m depth on the order of $10^5\,\mathrm{m}^2$, resulting in an ice perimeter melt rate on the order of 10^1 to $10^2\,\mathrm{m\,a}^{-1}$. We consider this melt rate an upper bound given the assumptions in the heat loss calculation and because the actual perimeter of the lake is substantially longer if the network of fractures in the MIS and MGT are taken into account. Although crude, this estimate does appear to be broadly consistent with the average annual increase in area of the epishelf lake observed by remote sensing between 2011 and 2014 (Table 1) of $1.2 \pm 3.9 \times 10^6\,\mathrm{m}^2\,\mathrm{a}^{-1}$, which if averaged around the ice perimeter of the lake suggests a melt rate on the order of $10^1 \pm 10^1\,\mathrm{m\,a}^{-1}$. Thus, although submarine melting of the MIS and MGT driven by the relatively warm epishelf lake is a relatively small contributor to the freshwater budget and does not influence halocline depth, it does appear to be a plausible mechanism contributing to the areal expansion of the lake over the past few decades and warrants further investigation.

4.8 Implications

Our results have shown that the timing of CTD profiling is critical to prevent aliasing the long-term, interannual record of epishelf lake depth with seasonal changes. This is especially important if epishelf lake depth is to be used as a climate indicator, as suggested by Veillette et al. (2011b). In the absence of continuous records, the best long-term estimate for the lake depth is obtained from CTD profiling each year just prior to the initiation of the melt season, which for Milne Fiord would be 1 June. Observations at this time capture the annual minimum depth of the epishelf lake prior to inflow. Although the lake depth may not have reached equilibrium by early June, the depth measured at this time is arguably the most reliable indicator of the long-term state of the lake and likely the best indicator of the actual depth of the ice dam. Profiles collected during the melt season must account for the variations in summer meltwater inflow, and the method developed here based on PDD records provides an approach towards correcting for this variability.

The results from Milne Fiord have important implications for the interpretation of water column time series from other Arctic and Antarctic epishelf lakes. Veillette et al. (2008) observed interannual deepening of the epishelf lakes in Ayles and Markham fjords, Ellesmere Island, on the order of 2 m between 2006 and 2007; however, the timing of data collection suggests that the change was due to seasonal variation of meltwater inflow, as subsequent profiles were collected later in the melt season. Similarly, Smith et al. (2006) suggested that interannual thinning of the freshwater layer in two epishelf lakes at Ablation Point, Antarctica, between 1973 and 2001 could either indicate thinning of the George VI Ice Shelf or, alternatively, be due to seasonal processes. The 1973 observations from Smith et al. (2006) were taken in December, over a month later in the austral summer than those collected in November 2000, and we therefore suggest that seasonal processes were the more likely explanation. Continuous monitoring over a full seasonal cycle would, however, be required to test this hypothesis.

Veillette et al. (2008) presented a 5-decade-long record of CTD profiles from Disraeli Fiord that showed a change in the depth of the epishelf lake from 63 m in 1954 to 33 m in 1999, implying a 30 m change in thickness of the WHIS over this period (note that we estimated this change from the level of the halocline in Veillette et al., 2008, Fig. 5a, not from the level of the 3 ppt isohaline as shown in Veillette et al., 2008, Fig. 5b). However, we caution that a shift in the timing of observations relative to the melt season (from September in 1954 and 1960 to June in 1999) means seasonal signals could have biased the long-term record. If we assume outflow of the Disraeli Fiord epishelf lake was restricted to a narrow basal channel in the WHIS similar to that of the MIS, a likely scenario given the fracturing of the WHIS along a sinuous surface depression (Mueller et al., 2003), then we can estimate the seasonal change in depth of the Disraeli Fiord epishelf lake from the volume of meltwater inflow each summer. Lenaerts et al. (2013) estimated the average meltwater runoff from the CAA during the period 1971–2000 was $69\,\mathrm{Gt\,a}^{-1}$, which if applied to the glaciated area of Disraeli Fiord ($1385\,\mathrm{km}^2$) suggests $\sim 650 \times 10^6\,\mathrm{m}^3\,\mathrm{a}^{-1}$ of meltwater entered the fjord from glaciers (snowmelt from the $715\,\mathrm{km}^2$ unglaciated area of the catchment only accounts for $\sim 1 \times 10^6\,\mathrm{m}^3\,\mathrm{a}^{-1}$ assuming average annual snow accumulation of 0.15 m w.e.; Braun et al., 2004). If all the runoff enters Disraeli Fiord at the surface, this could have seasonally altered the depth of the $223\,\mathrm{km}^2$ epishelf lake (Keys, 1978) by $\sim 3\,\mathrm{m}$. If we assume maximal seasonal aliasing in the observations, then it is possible the actual change in depth was only 24 m, 20 % less than that inferred from the raw observations. Although the long-term record clearly indicates thinning of the WHIS along the outflow pathway prior to breakup, this example highlights that accounting for seasonal effects is important if the rate of change of ice shelf thickness is to be inferred from CTD profiles. The seasonal depth change will be most pronounced in epishelf lakes with a small surface area, large seasonal inflow, and a narrow outflow channel.

Meltwater runoff from the CAA is predicted to more than double by the end of the 21st century (Lenaerts et al., 2013). In the unlikely event that the MIS remains intact during this century, seasonal changes in the depth of the MEL could double in magnitude, further increasing the need to account for seasonal variation. However, given the long-term shoaling of the epishelf lake, expanding network of fractures in the MIS, and predicted climate warming, further weakening of the ice shelf will likely lead to a large-scale calving event and complete drainage of the epishelf lake in the near future.

5 Conclusions

Our detailed observations from Milne Fiord show that the depth of epishelf lakes can vary substantially owing to seasonal variations in meltwater inflow. This could potentially alias long-term records of lake depth from opportunistic CTD profiles unless seasonal variations are properly accounted for, which requires continuous monitoring and knowledge of various factors, including the inflow volume flux from the surrounding catchment, lake area and hypsometry, and outflow hydraulics under the ice shelf. Accounting for these factors is particularly important if records of epishelf lake depth are used as proxies for changes in ice shelf thickness. We emphasize even then that epishelf lake depth is only indicative of the thickness of the ice dam, that is, the minimum thickness of the ice along the outflow path of the epishelf lake, which for Milne Fiord appears to be a narrow basal channel in the MIS. Monitoring ice thickness along a basal channel is important, however, as it provides a critical indication of the structural integrity of the ice shelf. Observational verification of the outflow pathway would allow the basal channel hypothesis to be confirmed or refuted, but, more importantly, it would locate the thinnest and weakest fault line in the ice shelf and a likely site for future large-scale fracturing of the MIS.

The existence of epishelf lakes is highly sensitive to the interactions of the cryosphere with the atmosphere and hydrosphere. The warming climate in the Arctic and Antarctic has resulted in an increased freshwater flux to the coast, with the potential to increase the volume and depth of epishelf lakes. However, widespread climate-driven thinning and collapse of ice shelves are outpacing the effects of increased freshwater inflow, leading to accelerated thinning and loss of these cryospheric systems in the Arctic. At the current rate of thinning the Milne Fiord epishelf lake, the last known epishelf lake in the Arctic, could be lost within the next decade, although continued breakup of the Milne Ice Shelf suggests a catastrophic drainage could occur at any time.

Code and data availability. Matlab code used in the analysis, as well as in situ hydrographic and glaciological field data, can be obtained from the corresponding author. Meteorological data from the Automated Weather Station (AWS) in Milne Fiord can be obtained online at https://tinyurl.com/milnewx. Northern Ellesmere Island ice shelf extents can be downloaded from the Polar Data Catalogue (www.polardata.ca; CCIN: 12721).

Author contributions. We applied a sequence-determines-credit approach for the sequence of authors. AKH, BEL, and DRM conceived the study; AKH, DRM, WFV, and LC conducted the field work; AKH, BEL, and DRM analyzed the data; and AKH prepared the manuscript with contribution from all coauthors.

Competing interests. The authors declare that they have no conflict of interest.

Acknowledgements. We thank two anonymous reviewers for their constructive comments that helped improved this manuscript. This work was funded by grants from the Natural Sciences and Engineering Research Council (NSERC) of Canada; Canada Foundation for Innovation; Ontario Research Fund; University of Ottawa; and ArcticNet, a Network of Centres of Excellence of Canada. We thank the Polar Continental Shelf Program for providing excellent logistical support and Parks Canada for use of facilities. Andrew K. Hamilton was supported by graduate scholarships from NSERC and the Association of Canadian Universities for Northern Studies Garfield Weston Foundation, and awards from the Northern Scientific Training Program. We thank Humfrey Melling, Greg Lawrence, and Roger Pieters for equipment loans and helpful advice on the research and an early version of the manuscript, and Sam Brenner, Kelly Graves, Jill Rajewicz, Miriam Richer-McCallum, Denis Sarrazin, Adrienne White, and Nat Wilson for assistance in the field.

Edited by: Andreas Vieli

References

Antoniades, D., Francus, P., Pienitz, R., St-Onge, G., and Vincent, W. F.: Holocene dynamics of the Arctic's largest ice shelf, P. Natl. Acad. Sci. USA, 108, 18899–18904, https://doi.org/10.1073/pnas.1106378108, 2011.

Bormann, P. and Fritzsche, D.: The Schirmacher Oasis, Queen Maud Land, East Antarctica, and its surroundings, Justus Perthes Verlag, Gotha, Germany, 448 pp., 1995.

Braun, C.: The surface mass balance of the Ward Hunt Ice Shelf and Ward Hunt Ice Rise, Ellesmere Island, Nunavut, Canada, in: Arctic Ice Shelves and Ice Islands, edited by: Copland, L. and Mueller, D. R., Springer, Dordrecht, 149–183, https://doi.org/10.1007/978-94-024-1101-0_6, 2017.

Braun, C., Hardy, D. R., Bradley, R. S., and Sahanatien, V.: Surface mass balance of the Ward Hunt Ice Rise and Ward Hunt Ice Shelf,

Ellesmere Island, Nunavut, Canada, J. Geophys. Res.-Atmos., 109, D22110, https://doi.org/10.1029/2004JD004560, 2004.

Burton, J. C., Amundson, J. M., Abbot, D. S., Boghosian, A., Cathles, L. M., Correa-Legisos, S., Darnell, K. N., Guttenberg, N., Holland, D. M., and MacAyeal, D. R.: Laboratory investigations of iceberg capsize dynamics, energy dissipation and tsunamigenesis, J. Geophys. Res.-Earth, 117, F01007, https://doi.org/10.1029/2011JF002055, 2012.

Copland, L., Mueller, D. R., and Weir, L.: Rapid loss of the Ayles Ice Shelf, Ellesmere Island, Canada, Geophys. Res. Lett., 34, L21501, https://doi.org/10.1029/2007GL031809, 2007.

Doran, P. T., Wharton, Jr., R. A., Lyons, W. B., Des Marais, D. J., and Andersen, D. T.: Sedimentology and geochemistry of a perennially ice-covered epishelf lake in Bunger Hills Oasis, East Antarctica, Antarct. Sci., 12, 131–140, 2000.

England, J. H., Lakeman, T. R., Lemmen, D. S., Bednarski, J. M., Stewart, T. G., and Evans, D. J. A.: A millennial-scale record of Arctic Ocean sea ice variability and the demise of the Ellesmere Island ice shelves, Geophys. Res. Lett., 35, L19502, https://doi.org/10.1029/2008GL034470, 2008.

Galand, P. E., Lovejoy, C., Pouliot, J., Garneau, M.-E., Vincent, W. F., and others: Microbial community diversity and heterotrophic production in a coastal Arctic ecosystem: A stamukhi lake and its source waters, Limnol. Oceanogr., 53, 813–823, https://doi.org/10.4319/lo.2008.53.2.0813, 2008.

Galton-Fenzi, B. K., Hunter, J. R., Coleman, R., and Young, N.: A decade of change in the hydraulic connection between an Antarctic epishelf lake and the ocean, J. Glaciol., 58, 223–228, https://doi.org/10.3189/2012JoG10J206, 2012.

Gardner, A., Moholdt, G., Wouters, B., Wolken, G., Burgess, D., Sharp, M., Cogley, J., Braun, C., and Labine, C.: Sharply increased mass loss from glaciers and ice caps in the Canadian Arctic Archipelago, Nature, 473, 357–360, https://doi.org/10.1038/nature10089, 2011.

Gibson, J. and Andersen, D.: Physical structure of epishelf lakes of the southern Bunger Hills, East Antarctica, Antarct. Sci., 14, 253–261, https://doi.org/10.1017/S095410200200010X, 2002.

Hamilton, A. K.: Ice-ocean interactions in Milne Fiord, PhD Thesis, The University of British Columbia, Vancouver, Canada, available at: http://hdl.handle.net/2429/59051, 2016.

Hattersley-Smith, G.: Ice shelf and fiord ice problems in Disraeli Fiord, Northern Ellesmere Island, NWT, Tech. rep., Defence Research Board, Department of National Defence Canada, Ottawa, Canada, 12 pp., 1973.

Hattersley-Smith, G., Fuzesy, A., and Evans, S.: Glacier depths and in northern Ellesmere Island: airborne radio echo sounding in 1966, DREO Technical Note 69-6, Defense Research Establishment Ottawa, 55 pp., 1969.

Heywood, R. B.: A limnological survey of the Ablation Point area, Alexander Island, Antarctica, Phil. T. R. Soc. B, 279, 39–54, https://doi.org/10.1098/rstb.1977.0070, 1977.

Hock, R.: Temperature index melt modelling in mountain areas, J. Hydrol., 282, 104–115, https://doi.org/10.1016/S0022-1694(03)00257-9, 2003.

IPCC: Climate Change 2013: The Physical Science Basis. Contribution of Working Group I to the Fifth Assessment Report of the Intergovernmental Panel on Climate Change, Cambridge University Press, Cambridge, Uk, New York, NY, USA, https://doi.org/10.1017/CBO9781107415324, 2013.

ISO: International Organization of Standards, ISO 1438:2008(E), Hydrometry – Open channel flow measurement using thin-plate weirs, 2008.

Jeffries, M. and Krouse, H. R.: Arctic ice shelf growth, fiord oceanography and climate, Z. Gletscherkunde Glazialgeologie, 20, 147–153, 1984.

Jeffries, M. O.: Physical, chemical and isotopic investigations of Ward Hunt Ice Shelf and Milne Ice Shelf, Ellesmere Island, NWT, PhD Thesis, University of Calgary, Calgary, 358 pp., 1985.

Jeffries, M. O.: Glaciers and the morphology and structure of Milne Ice Shelf, Ellesmere Island, N.W.T., Canada, Arctic Alpine Res., 18, 397–405, https://doi.org/10.2307/1551089, 1986.

Keys, J. E.: Water regime of Disraeli Fjord, Ellesmere Island, Defense Research Establishment Ottawa, Ottawa, Tech. Rep. 792, 50 pp., 1978.

Kindsvater, C. and Carter, R.: Discharge characteristics of rectangular thin-plate weirs, T. Am. Soc. Civ. Eng., 124, 772–801, 1959.

Kjeldsen, K. K., Mortensen, J., Bendtsen, J., Petersen, D., Lennert, K., and Rysgaard, S.: Ice-dammed lake drainage cools and raises surface salinities in a tidewater outlet glacier fjord, west Greenland: Godthåbsfjord, J. Geophys. Res.-Earth, 119, 1310–1321, https://doi.org/10.1002/2013JF003034, 2014.

Laybourn-Parry, J., Madan, N. J., Marshall, W. A., Marchant, H. J., and Wright, S. W.: Carbon dynamics in an ultra-oligotrophic epishelf lake (Beaver Lake, Antarctica) in summer, Freshwater Biol., 51, 1116–1130, https://doi.org/10.1111/j.1365-2427.2006.01560.x, 2006.

Lenaerts, J., Angelen, J. H., Broeke, M. R., Gardner, A. S., Wouters, B., and Meijgaard, E.: Irreversible mass loss of Canadian Arctic Archipelago glaciers, Geophys. Res. Lett., 40, 870–874, https://doi.org/10.1002/grl.50214, 2013.

Mortimer, C. A.: Quantification of changes for the Milne Ice Shelf, Nunavut, Canada, 1950–2009, MSc thesis, University of Ottawa, Ottawa, 200 pp., 2011.

Mortimer, C. A., Copland, L., and Mueller, D. R.: Volume and area changes of the Milne Ice Shelf, Ellesmere Island, Nunavut, Canada, since 1950, J. Geophys. Res., 117, 1–12, https://doi.org/10.1029/2011JF002074, 2012.

Mueller, D. R., Vincent, W. F., and Jeffries, M. O.: Breakup of the largest Arctic ice shelf and associated loss of an epishelf lake, Geophys. Res. Lett., 30, 1–4, https://doi.org/10.1029/2003GL017931, 2003.

Mueller, D. R., Copland, L., Hamilton, A., and Stern, D.: Examining Arctic ice shelves prior to the 2008 breakup, EOS T. Am. Geophys. Un., 89, 502–503, https://doi.org/10.1029/2008EO490002, 2008.

Mueller, D. R., Copland, L., and Jeffries, M.: Changes in Canadian Arctic ice shelf extent since 1906, in: Arctic Ice Shelves and Ice Islands, edited by: Copland, L. and Mueller, D., 109–148, Springer, Dordrecht, https://doi.org/10.1007/978-94-024-1101-0_5, 2017a.

Mueller, D. R., Copland, L., and Jeffries, M. O.: Northern Ellesmere Island ice shelf and ice tongue extents, v. 1.0 (1906–2015), Nordicana, D28, https://doi.org/10.5885/45455XD-24C73A8A736446CC, 2017b.

Paquette, M., Fortier, D., Mueller, D. R., Sarrazin, D., and Vincent, W. F.: Rapid disappearance of perennial ice on Canada's most northern lake, Geophys. Res. Lett., 42, 1433–1440, https://doi.org/10.1002/2014GL062960, 2015.

Pawlowicz, R.: Calculating the conductivity of natural waters, Limnol. Oceanogr.-Meth., 6, 489–501, https://doi.org/10.4319/lom.2008.6.489, 2008.

Pope, S., Copland, L., and Mueller, D.: Loss of multi-year landfast sea ice from Yelverton Bay, Ellesmere Island, Nunavut, Canada, Arct. Antarct. Alp. Res., 44, 210–221, https://doi.org/10.1657/1938-4246-44.2.210, 2012.

Prowse, T., Alfredsen, K., Beltaos, S., Bonsal, B. R., Bowden, W. B., Duguay, C. R., Korhola, A., McNamara, J., Vincent, W. F., Vuglinsky, V., Walter Anthony, K. M., and Weyhenmeyer, G. A.: Effects of changes in Arctic lake and river ice, AMBIO, 40, 63–74, https://doi.org/10.1007/s13280-011-0217-6, 2011.

Smith, J. A., Hodgson, D. A., Bentley, M. J., Verleyen, E., Leng, M. J., and Roberts, S. J.: Limnology of two Antarctic epishelf lakes and their potential to record periods of ice shelf loss, J. Paleolimnol., 35, 373–394, https://doi.org/10.1007/s10933-005-1333-8, 2006.

Thaler, M., Vincent, W. F., Lionard, M., Hamilton, A. K., and Lovejoy, C.: Microbial community structure and interannual change in the last epishelf lake ecosystem in the North Polar Region, Front. Mar. Sci., 3, 275, https://doi.org/10.3389/fmars.2016.00275, 2017.

Veillette, J., Mueller, D. R., Antoniades, D., and Vincent, W. F.: Arctic epishelf lakes as sentinel ecosystems: Past, present and future, J. Geophys. Res., 113, 1–11, https://doi.org/10.1029/2008JG000730, 2008.

Veillette, J., Lovejoy, C., Potvin, M., Harding, T., Jungblut, A. D., Antoniades, D., Chenard, C., Suttle, C. A., and Vincent, W. F.: Milne Fiord epishelf lake: A coastal Arctic ecosystem vulnerable to climate change, Ecoscience, 18, 304–316, https://doi.org/10.2980/18-3-3443, 2011a.

Veillette, J., Martineau, M., Antoniades, D., Sarrazin, D., and Vincent, W.: Effects of loss of perennial lake ice on mixing and phytoplankton dynamics: insights from High Arctic Canada, Ann. Glaciol., 51, 56–70, 2011b.

Vincent, W., Gibson, J., and Jeffries, M.: Ice-shelf collapse, climate change, and habitat loss in the Canadian high Arctic, Polar Record, 37, 133–142, https://doi.org/10.1017/S0032247400026954, 2001.

Wand, U., Hermichen, W.-D., Br ü ggemann, E., Zierath, R., and Klokov, V. D.: Stable isotope and hydrogeochemical studies of Beaver Lake and Radok Lake, MacRobertson Land, East Antarctica, Isot. Environ. Healt. S., 47, 407–414, https://doi.org/10.1080/10256016.2011.630465, 2011.

White, A., Copland, L., Mueller, D., and Van Wychen, W.: Assessment of historical changes (1959–2012) and the causes of recent break-ups of the Petersen Ice Shelf, Nunavut, Canada, Ann. Glaciol., 56, 65–76, https://doi.org/10.3189/2015AoG69A687, 2015a.

White, A., Mueller, D., and Copland, L.: Reconstructing hydrographic change in Petersen Bay, Ellesmere Island, Canada, inferred from SAR imagery, Remote Sens. Environ., 165, 1–13, https://doi.org/10.1016/j.rse.2015.04.017, 2015b.

White, D., Hinzman, L., Alessa, L., Cassano, J., Chambers, M., Falkner, K., Francis, J., Gutowski, W., Holland, M., Holmes, R. M., Huntington, H., Kane, D., Kliskey, A., Lee, C., McClelland, J., Peterson, B., Rupp, T. S., Straneo, F., Steele, M., Woodgate, R., Yang, D., Yoshikawa, K., and Zhang, T.: The arctic freshwater system: Changes and impacts, J. Geophys. Res., 112, G04S54, https://doi.org/10.1029/2006JG000353, 2007.

Winters, K. B., Lombard, P. N., Riley, J. J., and D'Asaro, E. A.: Available potential energy and mixing in density-stratified fluids, J. Fluid Mech., 289, 115–128, https://doi.org/10.1017/S002211209500125X, 1995.

Wrona, F. J., Johansson, M., Culp, J. M., Jenkins, A., Mård, J., Myers-Smith, I. H., Prowse, T. D., Vincent, W. F., and Wookey, P. A.: Transitions in Arctic ecosystems: Ecological implications of a changing hydrological regime, J. Geophys. Res.-Biogeo., 121, 650–674, https://doi.org/10.1002/2015JG003133, 2016.

13

Impact of MODIS sensor calibration updates on Greenland Ice Sheet surface reflectance and albedo trends

Kimberly A. Casey[1,2,a], **Chris M. Polashenski**[1,3], **Justin Chen**[4], **and Marco Tedesco**[5,6]

[1]Thayer School of Engineering, Dartmouth College, Hanover, NH 03755, USA
[2]Cryospheric Sciences Lab, NASA Goddard Space Flight Center, Greenbelt, MD 20771, USA
[3]Cold Regions Research and Engineering Laboratory, Alaska Projects Office, US Army Corps of Engineers, Fairbanks, AK 99709, USA
[4]Department of Computer Science, Stanford University, Stanford, CA 94305, USA
[5]Lamont–Doherty Earth Observatory, Columbia University, NY 10964, USA
[6]NASA Goddard Institute for Space Studies, New York, NY 10025, USA
[a]now at: Land Remote Sensing Program, US Geological Survey, Reston, VA 20192, USA

Correspondence to: Kimberly A. Casey (kimberly.a.casey@nasa.gov)

Abstract. We evaluate Greenland Ice Sheet (GrIS) surface reflectance and albedo trends using the newly released Collection 6 (C6) MODIS (Moderate Resolution Imaging Spectroradiometer) products over the period 2001–2016. We find that the correction of MODIS sensor degradation provided in the new C6 data products reduces the magnitude of the surface reflectance and albedo decline trends obtained from previous MODIS data (i.e., Collection 5, C5). Collection 5 and 6 data product analysis over GrIS is characterized by surface (i.e., wet vs. dry) and elevation (i.e., 500–2000 m, 2000 m and greater) conditions over the summer season from 1 June to 31 August. Notably, the visible-wavelength declining reflectance trends identified in several bands of MODIS C5 data from previous studies are only slightly detected at reduced magnitude in the C6 versions over the dry snow area. Declining albedo in the wet snow and ice area remains over the MODIS record in the C6 product, albeit at a lower magnitude than obtained using C5 data. Further analyses of C6 spectral reflectance trends show both reflectance increases and decreases in select bands and regions, suggesting that several competing processes are contributing to Greenland Ice Sheet albedo change. Investigators using MODIS data for other ocean, atmosphere and/or land analyses are urged to consider similar re-examinations of trends previously established using C5 data.

1 Introduction

The Greenland Ice Sheet (GrIS) has experienced substantial mass loss during the past three decades, resulting in sizeable contribution to sea level rise (Krabill et al., 2000; Fettweis, 2007; van den Broeke et al., 2009; Rignot et al., 2011; Velicogna and Wahr, 2013; Enderlin et al., 2014). Partitioned estimates of GrIS mass loss have shown that surface melt contributes substantially to annual ice loss (Box, 2013; van den Broeke et al., 2016). Snow and ice surfaces become darker and have reduced visible-to-near-infrared (VNIR) reflectance and albedo due to the deposition of particulates, re-emergence of engrained particulates, biological activity, snow grain metamorphosis and the presence of melt (LaChapelle, 1969; Warren and Wiscombe, 1980; Kohshima et al., 1993; Painter et al., 2001; Takeuchi, 2009; Hodson et al., 2017). As surface melt is both a factor and result of surface darkening, the potential exists for a variety of melt–albedo feedbacks to further enhance shortwave absorption and accelerate melt, increasing mass loss and sea level rise contributions (Wiscombe and Warren, 1980; Tedesco et al., 2015). Moderate Resolution Imaging Spectroradiometer (MODIS) observations are among the best tools available to evaluate these albedo feedbacks. Several studies have used MODIS data to study the magnitude of ongoing albedo trends and assess their role in enhancing the impact of climate warming on the ice sheet (e.g., Tedesco et al., 2011,

2016; Box et al., 2012; Stroeve et al., 2013). A challenge in using long-term satellite records to detect change is maintaining consistent instrument performance. Recent literature has discussed MODIS Terra sensor calibration degradation and its impacts on apparent data trends (e.g., Franz et al., 2008; Wang et al., 2012; Lyapustin et al., 2014; Polashenski et al., 2015; Sayer et al., 2015). Specifically related to GrIS trends, a recent work by Polashenski et al. (2015) indicated that uncorrected sensor degradation in MODIS Collection 5 (C5) data, particularly on the Terra platform, was contributing significantly to an apparent albedo declining trend over the GrIS and that the albedo trend in large areas of the ice sheet which do not experience melt (the dry snow area) may disappear once Collection 6 (C6) calibrations are applied. This study compares GrIS surface reflectance, snow and land albedo (bands 1–7) trends obtained from MODIS C5 with those obtained from the newly released MODIS C6 data to identify and differentiate the impact of sensor calibration drift from actual surface changes. We present these comparisons by spectral band and spatial filters and discuss the implications that C6 calibration enhancements have on our understanding of GrIS albedo decline and the mechanisms driving this decline. The identified band 1–7, 459–2155 nm, and broadband albedo spatial and temporal patterns and trends provide insight toward the physical mechanisms likely dominating GrIS albedo decline, which is key to understanding the surface energy balance of the Greenland Ice Sheet.

2   Background

MODIS instruments operate onboard both the NASA Earth Observing System (EOS) Terra and Aqua satellites and collect Earth observations in 36 spectral bands ranging from 0.4–14.4 μm at spatial resolutions of 250–1000 m (Barnes et al., 1998). The MODIS data record is now in excess of 17 years for Terra and 15 years for Aqua; both sensors are operating well past their 6-year design life. MODIS calibrations are updated periodically to reflect new understanding of instrument changes, with the entire data record reprocessed as a new "Collection". Significant revisions in the calibration approach were initiated in C6 (Toller et al., 2013), resulting in a relatively large adjustment to end products. Before launch, both MODIS sensors were calibrated via laboratory light sources. Because the optics of the sensor (including reflecting mirrors and electronics) were expected to degrade over time, the MODIS sensor design included methods for post-launch onboard calibration. However, the onboard calibration did not sufficiently account for degradation of the calibrators due to the large degradation of the solar diffuser (detailed in Lyapustin et al., 2014). This led to a long-term drift in calibration, most pronounced on the Terra sensor, with the largest in the blue band (B3), and decreasing with increasing wavelength (Xiong and Barnes, 2006).

The discovery of systemic nonphysical trends in MODIS Terra products by science users (e.g., Kwiatkowska et al., 2008; Levy et al., 2010; Wang et al., 2012) motivated research into independent trend characterization for C6 calibration. Analysis of observations of remote desert targets that were assumed to be nearly invariant over the long term were used to constrain the long-term drift of the observations. These so-called pseudo-invariant targets such as deserts, deep convective clouds and high-elevation ice sheet investigations led to a new vicarious approach for calibration, which relies on collection of pseudo-invariant Earth site data continually to provide a reference over time. The approach permits calibration at multiple angles of incidence (AOIs) on the challenging-to-characterize scan mirror, rather than two angles of incidence available through the onboard solar diffuser and moon observations through the space view port (Sun et al., 2012; Lyapustin et al., 2014). The resulting calibration is of lower precision than a well-characterized mirror calibrated on a lunar standard but provides significant improvement to the C5 data (Lyapustin et al., 2014). Long-term degradation of the solar diffuser stability monitor and detectors will continue to be evaluated and addressed as the MODIS record proceeds (Toller et al., 2013).

The impact of C5 to C6 updates on higher-level MODIS products cannot be represented by a simple offset because the magnitude of calibration revision is dependent on mirror side, AOI and time. The correction reaches several percent (in absolute reflectance values) in the worst cases (i.e., B3, near the end of the C5 record, roughly 2013 to 2015). Lyapustin et al. (2014) report C5 to C6 adjustment of B3 top-of-atmosphere (TOA) reflectance reaches approximately 0.02 (from 0.225 to 0.245) at the Libya 4 pseudo-invariant site by Terra mission day 5000. Residual error after C6 calibrations (when compared to pseudo-invariant desert sites) is found to be on the order of several tenths of 1 % in TOA reflectance (Lyapustin et al., 2014). The residual error is now within the stated accuracy of the products but may still be large enough to impact end user's scientific results, particularly in products derived from band ratios such as aerosol or vegetation indices. Further possible improvements to sensor calibration have been discussed (Meister and Franz, 2014; Xiong et al., 2015), including those that address polarization correction on the Terra sensor (Lyapustin et al., 2014). A thorough description of the C5 calibration degradation can be found in Lyapustin et al. (2014), and further details on C6 sensor characterization can be found in Toller et al. (2013) and at the MODIS Characterization Support Team website at http://mcst.gsfc.nasa.gov/calibration/information. MODIS C6 updates included not only sensor calibration algorithms applied to Level 1 data, but also updates to algorithms used to derive higher-level data products (including products used in this study) as well as the aerosol retrieval and correction algorithms, the cloud and cloud shadow detection algorithms, and quality assurance bands (Levy et al., 2013; Platnick et al., 2015). Detailed documentation of modifications to the products used in this paper can be found in the MOD09 user guide at http://modis-sr.ltdri.org/guide/MOD09_UserGuide_v1.4.pdf, the MOD10 user guide at http://modis-snow-ice.gsfc.nasa.gov/uploads/C6_MODIS_Snow_User_Guide.pdf and the MCD43 documentation at https://www.umb.edu/spectralmass/terra_aqua_modis/v006. Our analysis investigates Terra and Aqua records from three product types, namely MOD09/MYD09, MOD10/MYD10, and MCD43 for the entire summer data record, though it does not directly compare MODIS C5 and C6 data with ground data. Several MODIS cryospheric calibration and validation investigations, some of which include in situ data, have been performed, such as Stroeve et al. (2005, 2013), Moody et al. (2007), Hall et al. (2008), Alexander et al. (2014), Wright et al. (2014), Polashenski et al. (2015), and Zhan and Davies (2016). Recently, Box et al. (2017) found Terra MOD10A1 albedo substantially improves in relative accuracy from C5 to C6, agreeing with the Greenland Climate Network and Programme for Monitoring of the Greenland Ice Sheet station data from mid-May through August for the majority of GrIS south of 80° N within 0.04 (unitless albedo from 0–1, see Box et al., 2017, Fig. 5b). Accuracy of in situ ice sheet automated weather station measurements remains challenging due to limitations of unattended stations and interference of factors including ice riming, high wind speeds, low temperatures, ablation, tilt and ice flow (van den Broeke et al., 2004; van As and Fausto, 2011). Considerable biases have been reported in GrIS automated weather station data (Stroeve et al., 2006, 2013; Wang et al., 2016; Ryan et al., 2017).

3 Methods

3.1 Processing MODIS data

We analyzed the MODIS Terra and Aqua 8-day surface reflectance products (MOD09A1/MYD09A1; Vermote, 2007, Vermote, 2015a for C6 Terra, Vermote, 2015b for C6 Aqua), Terra and Aqua daily snow cover and broadband albedo products (MOD10A1/MYD10A1; Hall et al., 2006a for C5 Terra; Hall et al., 2006b for C5 Aqua; Hall and Riggs, 2016a for C6 Terra; Hall and Riggs, 2016b for C6 Aqua), and combined Terra and Aqua platform daily land surface albedo products (MCD43A3; Schaaf and Wang, 2015) for both C5 and C6 collections over Greenland. The MOD09A1 and MYD09A1 land surface reflectance products from the Terra and Aqua platforms, respectively, provide surface reflectance in bands 1–7, corresponding to a wavelength range of 459–2155 nm (Table 1). Data are provided as an 8-day product, which contains the best available Level 2 gridded observation during the 8-day period, based on observation coverage, view angle, clouds and aerosol loading. The MOD10A1/MYD10A1 Terra/Aqua daily snow cover products include daily broadband snow albedo and quality as-

Table 1. MODIS sensor reflective band 1–7 characteristics.

MODIS band	Bandwidth (nm)
Band 1	620–670
Band 2	841–876
Band 3	459–479
Band 4	545–565
Band 5	1230–1250
Band 6	1628–1652
Band 7	2105–2155

surance observations, which are the focus of our processing, in addition to a snow cover index calculated from Level 1 radiance data. The MCD43A3 land surface albedo product provides both direct hemispherical reflectance (black-sky albedo, BSA) and bihemispherical reflectance (white-sky albedo, WSA) for bands 1–7 from a bidirectional reflectance distribution function (BRDF) inversion of all available observations during the 16-day moving window centered on the date of interest. Data are provided every 8 days in C5 and daily in C6.

Because the high latitude and low solar zenith angles over the GrIS are at the extreme of MODIS capabilities, we filtered the data to ensure use of only the highest quality retrievals for our analysis. MOD09A1/MYD09A1 (M*D09A1) data were filtered by using the band quality assurance layer where all four band quality flags were set to 0 for each band, and solar zenith angle observations were below 70°. MOD10A1/MYD10A1 (M*D10A1) daily snow cover data were filtered using methods similar to those in Box et al. (2012), where values outside the M*D10A1 scientific dataset albedo values are excluded and a filter removes pixels whose albedo is more than 2 standard deviations from the 11-day running median or which differ from the 11-day running mean by more than 0.04 (albedo is a dimensionless number on a scale of 0.0–1.0). Processing MCD43A3 datasets was carried out following methods described in Stroeve et al. (2013) Sect. 2.1. We use only data with the highest quality (inversion flag set to 0 in the MCD43A2 product), which represents data derived from time periods where sufficient cloud-free, high-quality observations are available for full BRDF inversion. We did not use data derived with the backup algorithm, even though evidence suggests it performs almost as well (Stroeve et al., 2013). During the summer season chosen for the processing timeframe (1 June–31 August), the majority of the data are acquired near solar noon, ensuring many high-quality inversions in the MCD43A3 data.

After quality filtering, data are processed to produce an annual mean albedo for dry and wet snow and ice areas of the ice sheet with methods of Polashenski et al. (2015). Wet snow and ice are GrIS areas that have experienced melt at any point during the current year prior to the date of interest. Dry snow is snow which has not at any point experienced

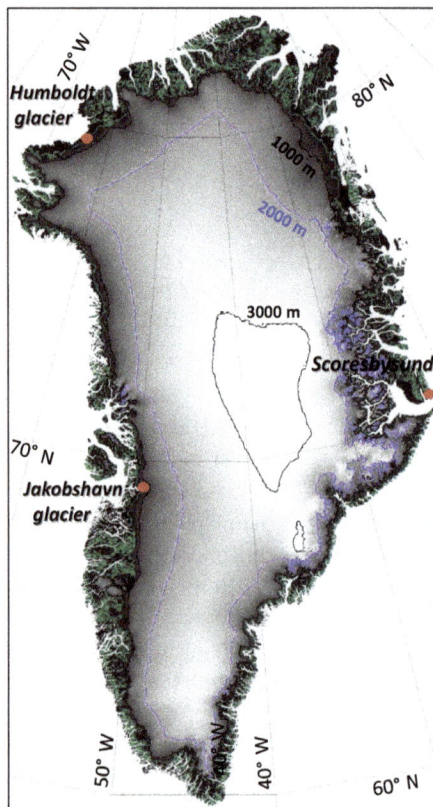

Figure 1. Location and topographic map of the Greenland Ice Sheet with 500 m (green), 1000 m (black), 2000 m (blue) and 3000 m (black) contour lines overlaid. Greenland surface elevation from Howat et al. (2014) is displayed, where darker greys indicate lower elevation (minimum at sea level, 0 m) and brighter greys and white indicate higher elevation (maximum at 3500 m).

melt. We do not return wet snow and ice to the dry snow category after it experiences melt until the following year when we are certain the melt surface has been buried, due to the residual impacts on albedo caused by melt occurrence. An elevation mask is applied using a GrIS digital elevation model (Howat et al., 2014) to group ice sheet areas at two different elevation bands (dry snow is snow above 2000 m, wet snow and ice is snow and ice within the elevations of 500 to 2000 m; Fig. 1). To these two elevation areas, high elevation and low elevation, we apply a melt mask generated from the regional climate model Modèle Atmosphérique Régional (MAR; Tedesco, 2014) to exclude pixels that did not match the predominant melt condition. MAR-indicated dry and wet snow and ice pixels were defined as those with no simulated melt occurring at any time during the summer (dry) and those showing one or more melt events (wet). The use of a dual filter, based on elevation and melt state, ensures that the areas discussed represent iconic surface types without contamination (e.g., for the dry snow area, the elevation cut off ensures non-melting bare ice at the margins of the ice sheet does not contribute). From the mosaicked, filtered and masked data,

the average of the daily mean of all remaining pixels from 1 June to 31 August is calculated for each year of the record. Though the time interval differs from the 15 May–15 July time interval chosen by Polashenski et al. (2015) to match the time when solar elevation angle is highest and albedo is most important to the GrIS, it is better aligned with prior studies and still captures the key behavior of the high-insolation period. We also conducted the same analysis for 15 May–15 July (not shown) and found very similar trend revisions and overall behavior of ice sheet albedo. Linear trends are calculated for display in map form using a least squares linear regression to the data at each pixel location.

All MODIS data product tiles were downloaded from the NASA US Geological Survey Land Processes Distributed Active Archive Center (LP DAAC) and National Snow and Ice Data Center (NSIDC). Data tiles were mosaicked and resampled via nearest-neighbor method to polar stereographic projection using the MODIS Reprojection Tool. The MODIS Reprojection Tool utilized to mosaic and resample MODIS tile data can be found at https://lpdaac.usgs.gov/tools/modis_reprojection_tool and the user guide at https://lpdaac.usgs.gov/sites/default/files/public/mrt41_usermanual_032811.pdf.

4 Results

4.1 Annual average summer M*D09A1 surface reflectance and M*D10A1 broadband albedo

Annual average summer (1 June–31 August) GrIS surface reflectance for dry and wet snow and ice areas from both Terra and Aqua data is presented in Figs. 2 and 3, respectively, as derived from the M*D09A1 and M*D10A1 C5 and C6 products. The discrepancy between the two data collections is indicated by the difference between dashed (C5) and solid (C6) lines of the same color. The adjustment from C5 to C6 is significantly greater for Terra than Aqua and greatest over the GrIS in the shortest wavelength bands, consistent with the sensor calibration degradation reported by Lyapustin et al. (2014). C6 reduces the discrepancy between Terra and Aqua data appreciably. Trends of the plotted data are quantified by linear regression in Tables 2 and 3 along with their statistical significance. Significant declining trends found in C5 Terra dry and wet snow and ice data with magnitude exceeding 0.01 decade^{-1} are no longer present in C6 data. Very small dry snow trends remain in C6 data, though not of strong statistical significance. Thus, these likely do not represent real changes on the surface. The C6 trend magnitude over the dry snow area is near the calibration drift of several tenths of a percent, and the trends show an incoherent pattern of albedo change. Specifically, B3, 459–479 nm, increases slightly, while B1, 620–670 nm, decreases. This could not be produced by expected physical mechanisms, for example, absorbing impurity concentration or grain size changes,

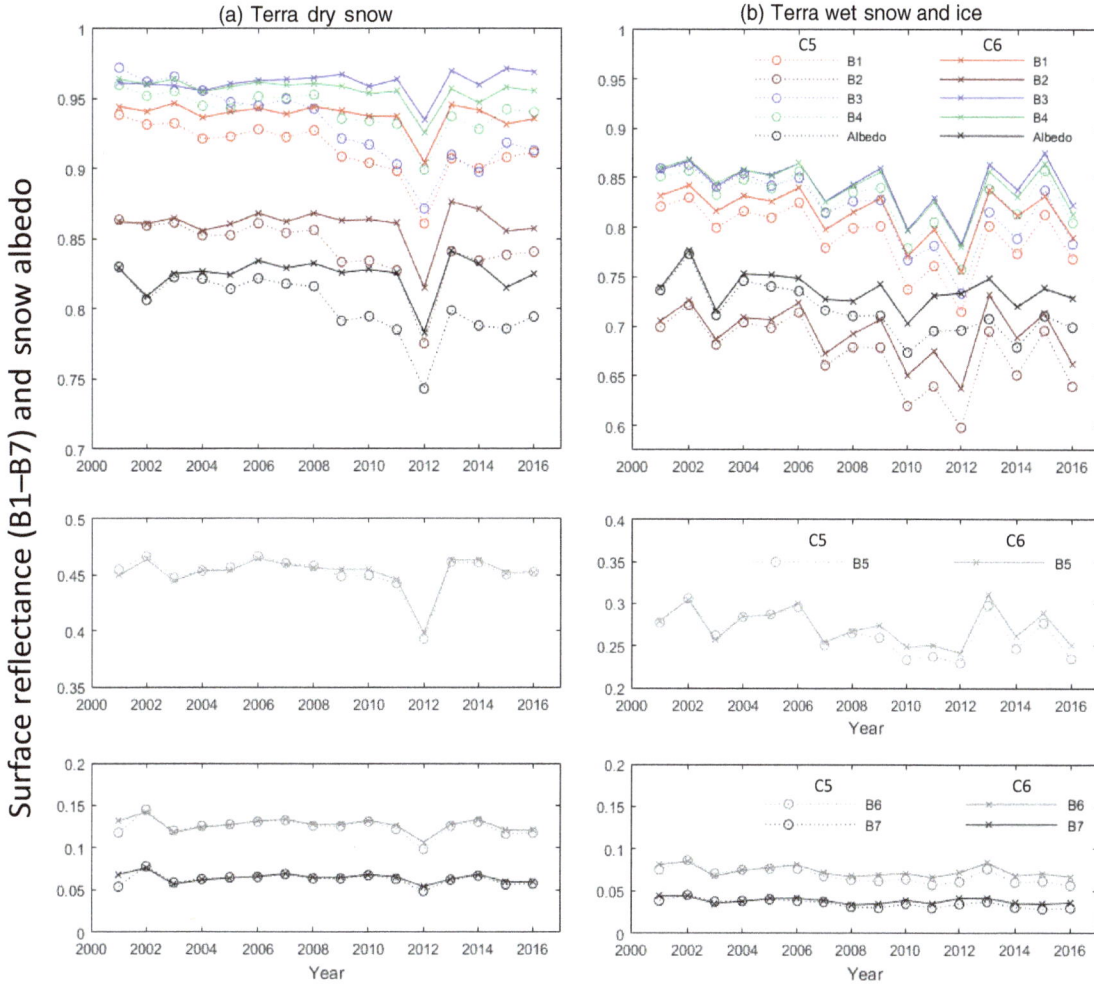

Figure 2. Terra MODIS GrIS C5 (dashed) and C6 (solid) average annual 1 June–31 August MOD09A1 surface reflectance (B1–B7) and MOD10A1 broadband snow albedo, denoted "albedo", for dry (**a**) and wet snow and ice (**b**). (Note the *y*-axis scale is different for dry vs. wet snow and ice for B1–4, albedo and B5).

which would both be expected to shift B1 and B3 in the same direction (see Warren and Wiscombe, 1980; Wiscombe and Warren, 1980, respectively). In wet snow and ice, significant trends seen in C5 Terra albedo nearly disappear in C6. Marginally, non-significant trends in wet snow and ice albedo remain across C6 visible bands, at magnitudes approximately one-third to one-half those of C5 data. We note that higher interannual variability (noise) in the wet snow and bare ice area limits trend significance at the $p \leq 0.05$ level, even though the absolute magnitude of trends (signal) is larger than in the dry snow area. The coherence across visible bands of marginally non-significant trends (nearly all bands decline for both sensors) and magnitude exceeding sensor accuracy suggest a physically real trend is likely, if not proven statistically by the satellite data products.

4.2 Annual average summer MCD43A3 albedo

Annual average summer (1 June–31 August) GrIS C5 and C6 MCD43A3 land surface direct hemispherical reflectances are presented in Fig. 4 for dry and wet snow and ice, showing results expected from a combination of Terra and Aqua data. Dashed lines represent C5 data, while solid lines represent C6. Revisions across the duration of the MODIS record are apparent between the MCD43A3 C5 and C6 data in both dry and wet snow and ice areas. Similar to M*D09A1 and M*D10A1 products, the MCD43A3 revisions result in a considerable decrease in trend magnitude. Statistically significant dry snow direct hemispherical reflectance declines, which had been apparent in C5 data, are reduced to magnitudes of or under $0.01\,\text{decade}^{-1}$ in C6. MCD43A3 direct hemispherical and bihemispherical reflectance trends (Table 3) remain significant or near-significant for B1 and B4, with incoherent patterns for other visible bands (B2, B3). As

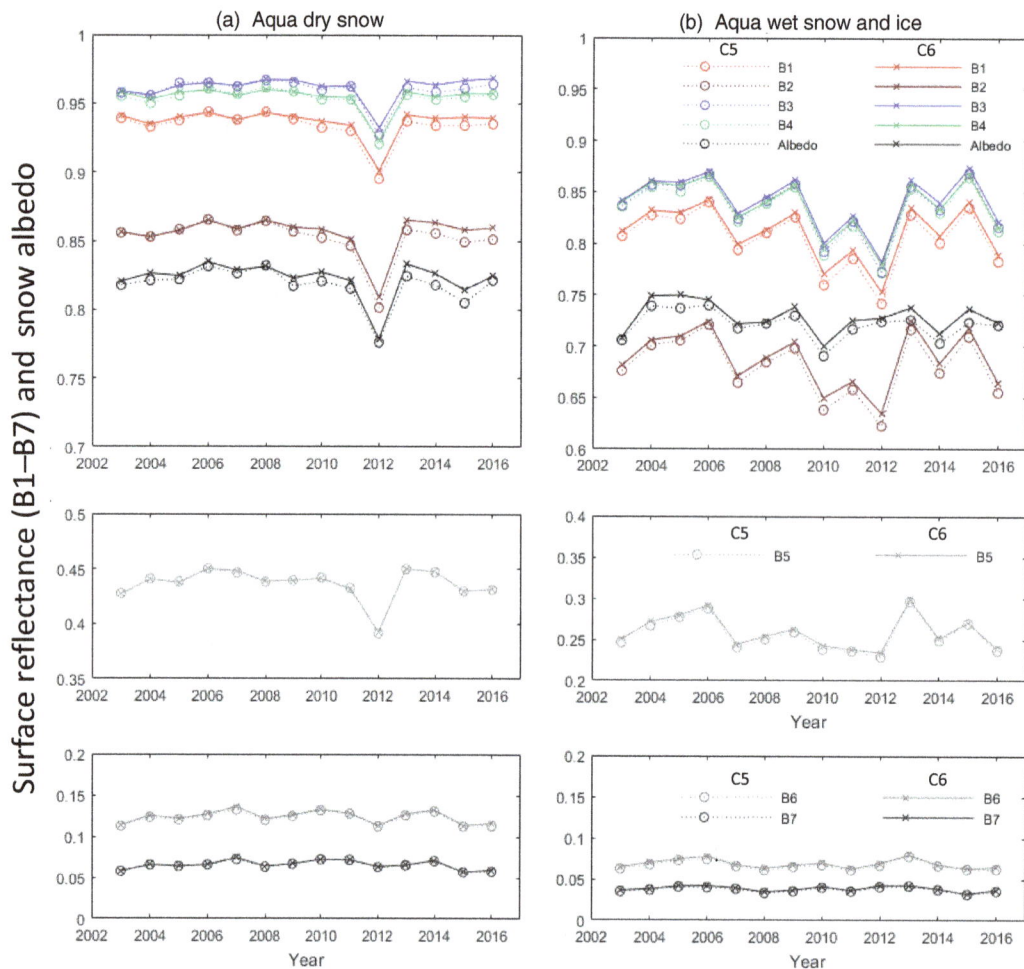

Figure 3. Aqua MODIS GrIS C5 (dashed) and C6 (solid) average annual 1 June–31 August MYD09A1 surface reflectance (B1–B7) and MYD10A1 (broadband snow albedo, denoted "albedo") for dry (**a**) and wet snow and ice (**b**). (Note the *y*-axis scale is different for dry vs. wet snow and ice for B1–4, albedo and B5).

discussed in Sect. 4.1, the spectral pattern, with conflicting trends in B2 and B3, indicates that these changes are likely not linked to physical processes. Wet snow and ice areas still exhibit coherent declining albedo trends after C6 revisions across B1 through B7, albeit of slightly lower magnitude than in C5. Like C5, individual band wet snow and ice trends have low (B2, B3) to marginal (B1) statistical significance due to large interannual variability in the wet snow and ice albedo (Fig. 4). Coherence across bands (as presented in Table 3 "BSA C6 wet snow and ice" and "WSA C6 wet snow and ice" trend columns), however, increases confidence in the trends. Impacts of C5 to C6 revision are very similar on WSA and BSA.

4.3 Spatial pattern of albedo trend

Maps of the decadal M*D10A1 broadband snow albedo trend over the entire MODIS record are shown in Fig. 5 for C5 (a, d) and C6 (b, c, e, f) data from both Terra and

Aqua sensors. Both sensors and both collections show similar spatial patterns, with the greatest albedo declines at low elevations of the ice sheet (Fig. 1), particularly on the western and southeastern margins. A statistically significant, ice-sheet-wide declining albedo trend in C5 Terra data is largely absent in C5 Aqua sensor data. The discrepancy between C5 Terra and Aqua data is not spatially dependent. C6 revisions change Terra trends upward by approximately 0.03 decade^{-1} across the ice sheet. Aqua revisions are smaller but also result in trends that are less negative in C6 than C5. Using revised C6 data, we mask the areas that have negligible trend (−0.01 to +0.01 decade^{-1}) in Fig. 5c, f. The magnitude of a "negligible" trend was determined by considering the errors of several tenths of a percent trend per decade remaining in C6 data collected to that obtained over pseudo-invariant desert sites (Lyapustin et al., 2014). Trends below this value must be considered below detection limit. The region of negligible trends covers nearly all of the dry snow and large portions of the upper reaches of the wet snow and ice, indicating

Table 2. Trends (decade^{-1}) and statistical significance of trends for M*D09A1 surface reflectance bands 1–7 and M*D10A1 broadband albedo. Statistically significant (where $p \leq 0.05$) trends are in bold, and marginally significant trends (where $p = 0.05$ to 0.1) are in italics.

Band	Terra C5 dry snow		Terra C6 dry snow		Aqua C5 dry snow		Aqua C6 dry snow	
	Trend	Significance	Trend	Significance	Trend	Significance	Trend	Significance
B1	**−0.027**	**0.003**	−0.007	0.166	−0.006	0.377	−0.002	0.761
B2	**−0.027**	**0.017**	−0.004	0.641	−0.010	0.289	−0.002	0.815
B3	**−0.051**	**0.000**	0.003	0.493	−0.001	0.902	0.004	0.509
B4	**−0.018**	**0.013**	*−0.009*	*0.073*	−0.002	0.718	−0.001	0.797
B5	−0.009	0.326	−0.005	0.600	−0.012	0.221	−0.012	0.232
B6	−0.008	0.166	−0.006	0.166	−0.007	0.211	−0.007	0.227
B7	−0.004	0.301	−0.004	0.162	−0.004	0.263	−0.004	0.292
Broadband albedo	**−0.032**	**0.003**	−0.002	0.808	−0.011	0.231	−0.008	0.419

Band	Terra C5 wet snow and ice		Terra C6 wet snow and ice		Aqua C5 wet snow and ice		Aqua C6 wet snow and ice	
	Trend	Significance	Trend	Significance	Trend	Significance	Trend	Significance
B1	**−0.037**	**0.029**	−0.023	0.102	−0.023	0.198	−0.020	0.222
B2	**−0.042**	**0.024**	−0.020	0.194	−0.026	0.166	−0.022	0.223
B3	**−0.054**	**0.003**	−0.017	0.227	−0.019	0.259	−0.018	0.281
B4	*−0.029*	*0.069*	*−0.025*	*0.073*	−0.020	0.237	−0.020	0.202
B5	**−0.027**	**0.043**	−0.014	0.257	−0.023	0.144	−0.024	0.115
B6	**−0.014**	**0.001**	*−0.006*	*0.067*	−0.008	0.102	*−0.009*	*0.087*
B7	**−0.008**	**0.001**	*−0.003*	*0.075*	−0.004	0.123	−0.005	0.109
Broadband albedo	**−0.040**	**0.001**	−0.013	0.166	−0.009	0.364	−0.007	0.495

that, even in some areas where it is well known that melt frequency is increasing (e.g., Fettweis et al., 2011; Box et al., 2012; Fausto et al., 2016), albedo trends are small enough to be challenging to confirm over the duration of the MODIS record. Trends with magnitude greater than sensor calibration limits of ~ 0.01 decade^{-1} are almost all negative. Significant areas of the southern ice sheet exhibit trends near -0.01 decade^{-1}, and a narrow band of substantial albedo decline, reaching -0.04 decade^{-1}, exists in areas around the periphery of the ice sheet. Based on MAR analysis, the positive albedo trends over northeastern Greenland are likely associated with a shift from no trend in accumulation to a trend of increasing accumulation (by 35 Gt yr^{-1}), starting in 2013. Direct hemispherical reflectance trends for select visible and near-infrared bands of MCD43A3 C6 data are presented in Fig. 6. Both positive and negative trends of statistical significance exist in all bands around the periphery of the ice sheet. These are discussed in detail below. Band 1 and band 3 data show spatially uniform trends that are generally under the sensor calibration accuracy of ~ 0.01 decade^{-1} across the interior of the ice sheet. Band 1 red visible light trends are positive and band 3 blue visible light trends are negative, a behavior inconsistent with albedo change caused by either light-absorbing impurity deposition or changes in surface grain properties, strongly suggesting these changes are not physically real. NIR bands 2 and 5 show trends in the upper elevations of the ice sheet that regionally exceed sensor calibration accuracy and significance, with notable spatial patterns. Positive albedo trends in these NIR bands dominate northeast

Greenland, while negative trends cover most of the remainder of the ice sheet. Band 5, in the near-infrared, is highly sensitive to grain size impacts. Figure 6d shows the opposing trends of albedo increasing primarily in the northeast and decreasing in the west and southern periphery. Figure 4 B5 shows modest interannual variability in the dry snow and more variability in the wet snow and ice areas, suggesting wet snow and ice albedo is worth investigating on seasonal and/or annual scales.

5 Discussion

5.1 Impact of C6 revision on scientific investigation of GrIS

The MODIS C5 to C6 revisions, and new surface reflectance and albedo trends over GrIS, have substantial ramifications for research seeking to evaluate the cause of enhanced surface melt, and hence mass loss, from the GrIS. Over the dry snow area, C5 Terra data and combined sensor data indicated a decadal trend of declining reflectance at a rate of up to several percent per decade, with high statistical significance (see Tables 2, 3; Fig. 5a). The albedo decline was greatest in short wavelength visible bands – a spectral signature consistent with enhanced dust deposition on the ice sheet (Dumont et al., 2014). The data, therefore, suggested that a snow albedo feedback initiated by dry snow processes resulted in increased GrIS melt. C6 revisions, particularly to short wavelength bands of Terra data, appreciably reduce these trends.

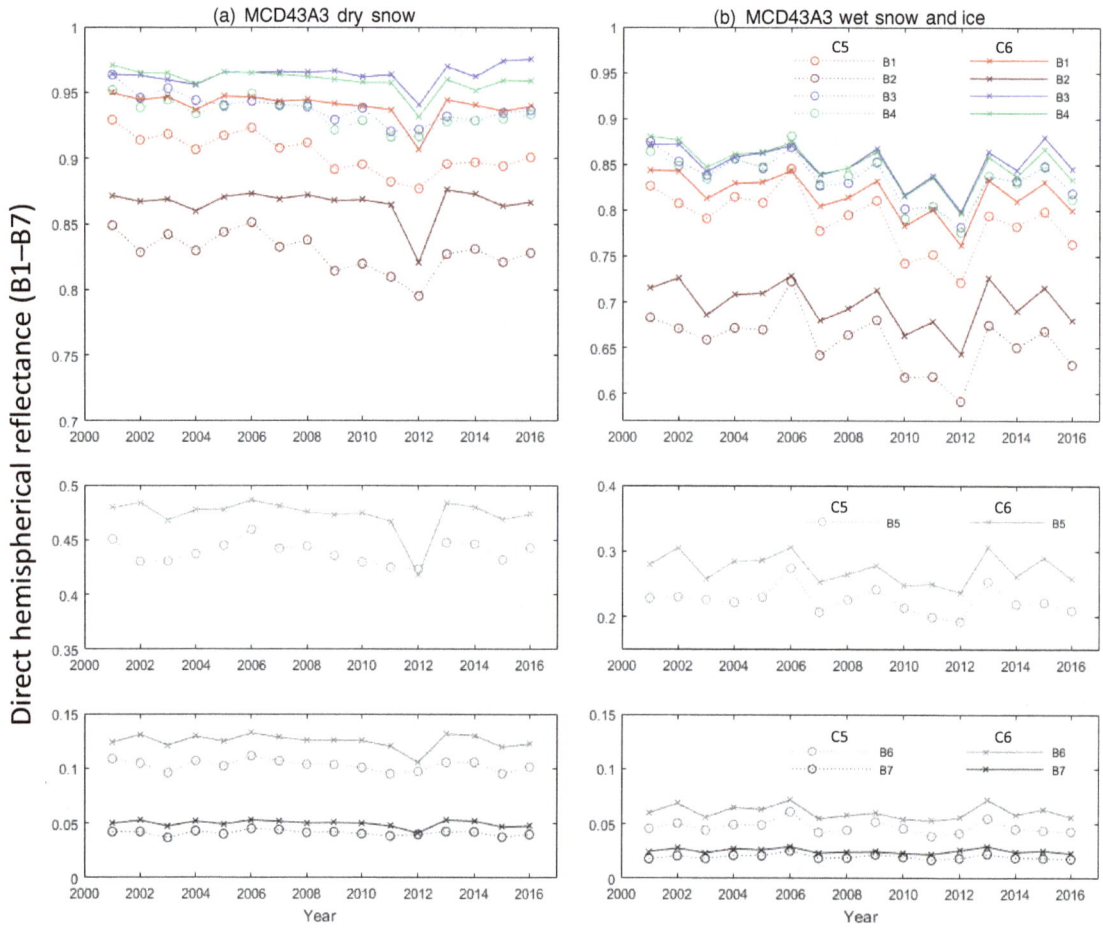

Figure 4. Greenland Ice Sheet average summer (1 June–31 August) MCD43A3 BSA C5 and C6 band 1–7 albedo for dry (**a**) and wet snow and ice (**b**). (Note the *y*-axis scale is different for dry vs. wet snow and ice for B1–4 and B5). Interestingly, there is a separation in C5 to C6 B1, B2, B4, B5 and B6 albedo values throughout the MODIS record.

C6 exhibits no statistically significant trends in visible nor NIR wavelength band surface reflectance from either sensor over the aggregated dry snow area exceeding the sensor calibration accuracy of ~ 0.01 decade^{-1}. Statistical significance of trends in MCD43 B4 C6 data is misleading. These trends, of a few tenths of a percent per decade, are within the range of a residual calibration error and their statistical significance may in fact reflect the ease with which a calibration error can be detected as a significant trend when superimposed on the relatively constant albedo of dry snow. Care in examination of the spatial and spectral variability and coherency between these trends (Figs. 5, 6) is recommended for future regional, in situ and/or process studies. Trends in C6 dry snow visible albedo (B1 and B3) are suspiciously consistent across the ice sheet, and opposite trends between these bands are inconsistent with expected mechanisms of albedo change. We conclude that dry snow visible band albedo is stable within MODIS' capabilities. This stable visible-wavelength albedo of the GrIS dry snow area supports conclusions in Polashenski et al. (2015), who found no in situ evidence of contin-

ual enhanced black carbon or dust deposition to support C5 trends. To note, forest fire events in North America and Asia have resulted in black carbon deposition to GrIS (e.g., Zennaro et al., 2014; Thomas et al., 2017), and such events have been predicted to increase during periods of drought in a warming climate (Soja et al., 2007; Flannigan et al., 2013). Deposition of black carbon and other absorbing impurities in the dry snow area is often buried by new snowfall; however, these impurities often have a stronger influence in reducing reflectance and albedo in wet snow and ice areas. In contrast to B1 and B3 decadal trends which are nearly stable in aggregate, NIR wavelength albedos (B2 and B5) show significant regional trends. Positive trends in NE Greenland offset negative trends across much of the remainder of the ice sheet in the aggregate dry snow data. The regional trends are statistically significant over large areas and exceed the expected magnitude of remaining calibration errors. The trends appear to indicate changing snow grain metamorphism. We speculate, based on the spectral change, that enhanced snowfall in the dry areas of NE Greenland is causing more rapid sur-

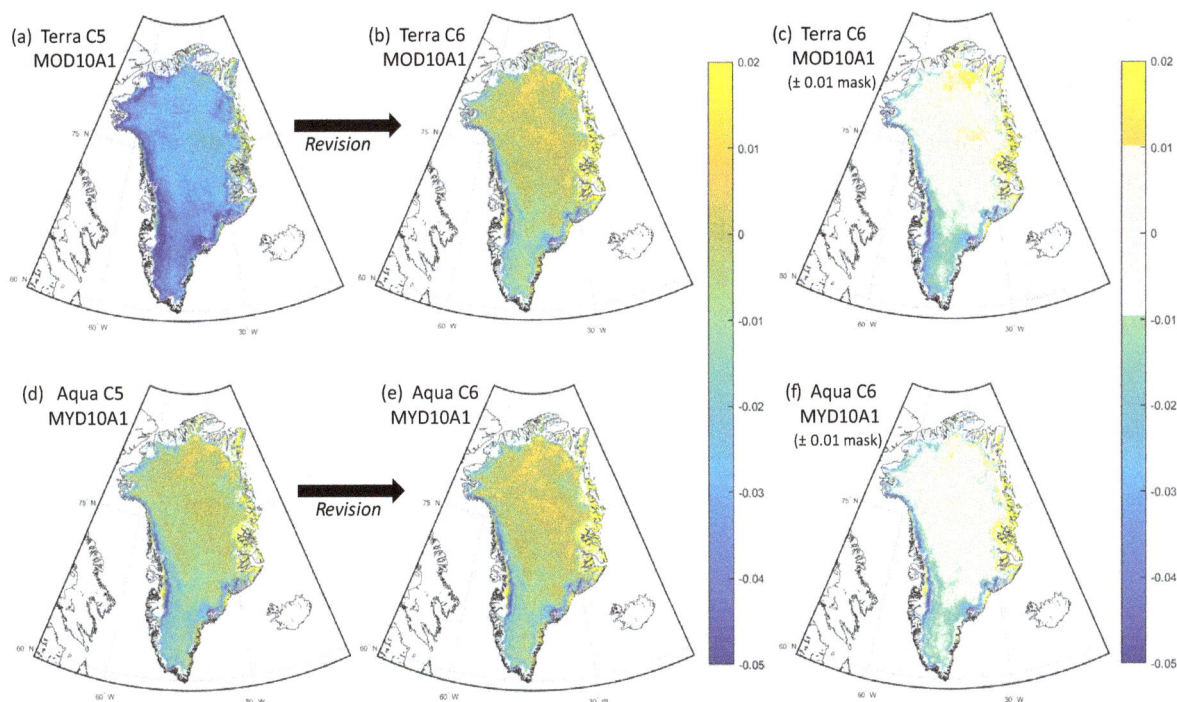

Figure 5. Maps **(a, b)** depict Terra MOD10A1 C5 and C6 2002–2016 decadal trend in broadband albedo. Maps **(d, e)** depict Aqua MYD10A1 C5 and C6 2003–2016 decadal trend in broadband albedo. Maps **(c, f)** show the same C6 decadal Terra and Aqua trends, respectively, with the sensor trends of ± 0.01 decade^{-1} masked out as white to visualize trends in albedo that are larger than expected sensor calibration uncertainty, including albedo decline (southeast and periphery GrIS) and albedo increase (northeast GrIS).

face burial and a lower age of surface grains, while increased temperatures and occasional melt are increasing grain size on southern and western areas of GrIS, where deposition was already relatively frequent. We find that the pattern does not appear to coincide well with the area of enhanced melt, as detected by passive microwave products (Tedesco et al., 2014) over the MODIS era (2000–present), indicating that this indeed appears to be more related to dry snow grain metamorphism and snowfall frequency than melt (see Fig. 7).

In wet snow and ice areas, the annual average broadband albedo and visible-wavelength reflectance shown for all products exhibits a downward trend for C5 and C6 data. The magnitude of these trends is reduced by approximately one half from C5 to C6 for MOD09A1 and by a small amount in MYD09A1 and MCD43A3. Statistical significance is harder to establish in the wet snow and ice areas due to much higher interannual variability (wet snow and ice, Figs. 2, 3, and 4). The C5 to C6 revisions to wet snow and ice albedo trends impact our understanding of GrIS surface energy and mass budgets and suggest a smaller role for albedo feedback in driving Greenland mass loss than previously indicated. The C5 to C6 revisions do not, however, demand a change in conclusions about reflectance and albedo declining trend existence in the wet snow and ice areas. Wet snow and ice areas on GrIS, in aggregate, still show coherent albedo decline across the nearly all visible and NIR wavelength bands from both

sensors in C6 data (Tables 2, 3). Reduction in reflectance and albedo of this magnitude (several tenths) has been shown to result in radiative forcing of several tens of W m^{-2} (e.g., 0.4 broadband albedo decline from dust on snow resulting in 80 W m^{-2} radiative forcing, Painter et al., 2007; 0.3 reduction in broadband visible albedo from black carbon and impurities on snow resulting in 70 W m^{-2} radiative forcing, Casey et al., 2017). The MODIS C6 reflectance and albedo data product results provide strong supporting evidence that enhanced melt processes (including melt-induced snow microstructure changes and melt-induced light-absorbing impurity accumulation changes) are creating an albedo feedback on the GrIS periphery.

The cause of these changes is important. We examined trends regionally along the margins of the ice sheet (Figs. 5 and 6) and find that trends can be supported by reasonable inferences about regional differences in snowfall frequency, melt duration and surface exposure of light-absorbing impurities on albedo control. In west Greenland, from Humboldt Glacier to Jakobshavn Glacier, substantial trends in NIR (B2, B5) albedo across a wide elevation range indicate increased presence of melting conditions. Trends in visible albedo are spatially correlated and most dominant at lower elevations where melt accumulation of impurities, exposure of bare ice and algal growth occurs. South of Jakobshavn on the western margin of the ice sheet, B1 and B3 visible albedo are sharply

Table 3. Trends (decade^{-1}) and statistical significance of trends for MCD43A3 directional hemispherical reflectance (or black-sky albedo, BSA) and bihemispherical reflectance (or white-sky albedo, WSA). Statistically significant (where $p \leq 0.05$) trends are in bold, and marginally significant trends (where $p = 0.05$ to 0.1) are in italics.

Band	BSA C5 dry snow		BSA C6 dry snow		WSA C5 dry snow		WSA C6 dry snow	
	Trend	Significance	Trend	Significance	Trend	Significance	Trend	Significance
B1	**−0.023**	**0.001**	*−0.009*	*0.096*	**−0.023**	**0.001**	*−0.009*	*0.093*
B2	**−0.017**	**0.026**	−0.005	0.499	**−0.017**	**0.030**	−0.005	0.517
B3	**−0.017**	**0.001**	0.004	0.343	**−0.017**	**0.001**	0.004	0.334
B4	**−0.014**	**0.006**	**−0.010**	**0.037**	**−0.014**	**0.006**	**−0.010**	**0.034**
B5	−0.003	0.596	−0.009	0.288	−0.003	0.652	−0.009	0.306
B6	−0.004	0.141	−0.004	0.317	−0.004	0.150	−0.003	0.345
B7	−0.001	0.344	−0.002	0.252	−0.001	0.376	−0.002	0.263

Band	BSA C5 wet snow and ice		BSA C6 wet snow and ice		WSA C5 wet snow and ice		WSA C6 wet snow and ice	
	Trend	Significance	Trend	Significance	Trend	Significance	Trend	Significance
B1	**−0.0369**	**0.0286**	*−0.0230*	*0.0645*	**−0.0357**	**0.0293**	*−0.0230*	*0.0644*
B2	*−0.0293*	*0.0842*	−0.0176	0.1979	−0.0269	0.1009	−0.0175	0.2004
B3	**−0.0292**	**0.0266**	−0.0133	0.2745	**−0.0282**	**0.0289**	−0.0133	0.2742
B4	**−0.0296**	**0.0412**	**−0.0254**	**0.0376**	**−0.0292**	**0.0397**	**−0.0255**	**0.0373**
B5	−0.0115	0.3073	−0.0135	0.2765	−0.0106	0.3469	−0.0136	0.2741
B6	−0.0031	0.3205	−0.0028	0.4178	−0.0030	0.3568	−0.0029	0.4038
B7	−0.0009	0.4445	−0.0010	0.4103	−0.0008	0.5199	−0.0010	0.3958

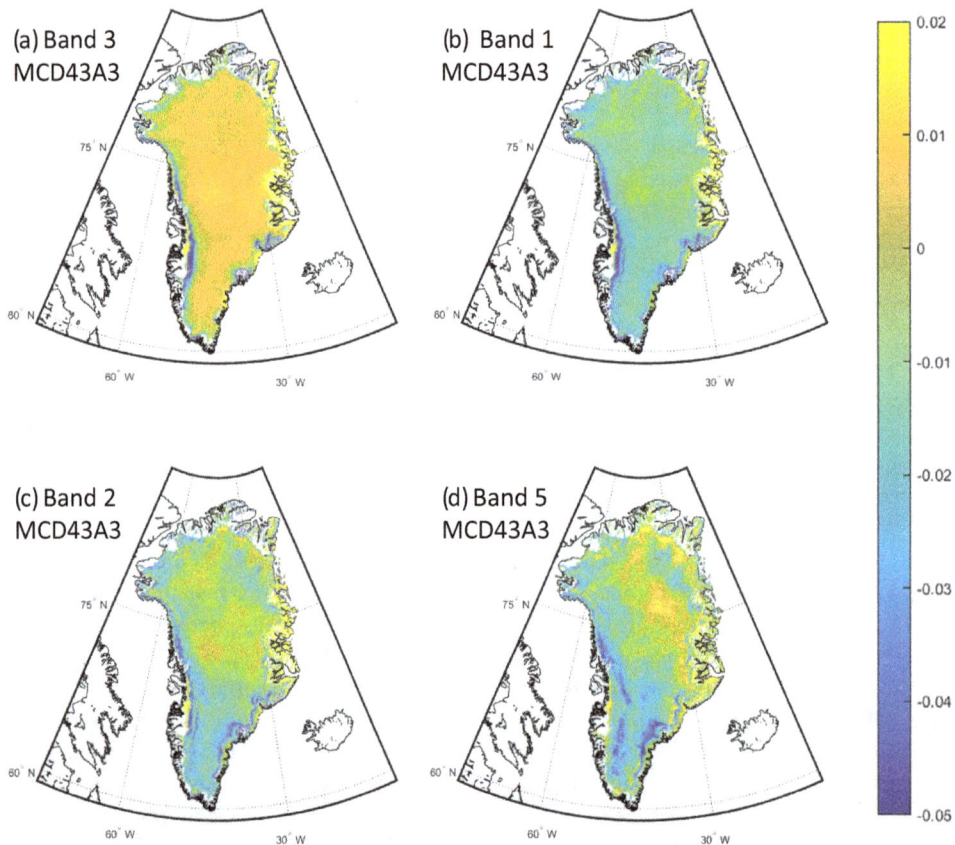

Figure 6. MCD43A3 C6 band 3 (459–479 nm), band 1 (620–670 nm), band 2 (841–876 nm) and band 5 (1230–1250 nm) 2003–2016 albedo trends. Band-specific maps show considerable spatial complexity in albedo trends.

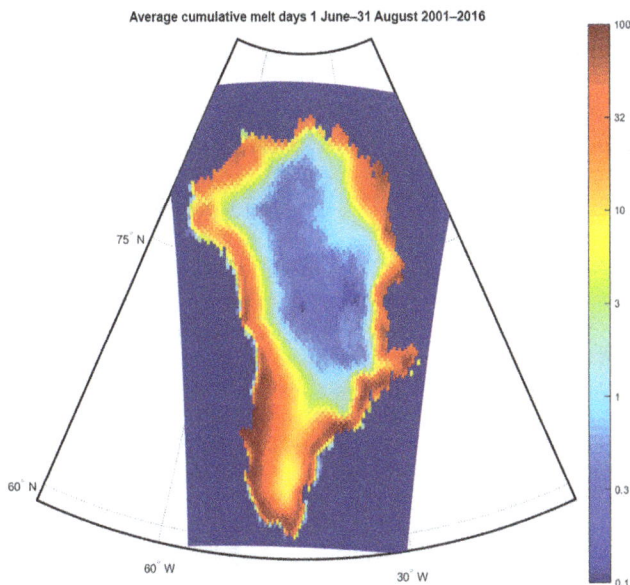

Figure 7. Average number of melt days experienced from 1 June to 31 August as determined from spaceborne passive microwave data (Tedesco, 2014). Note that the NE GrIS high-elevation areas show only a few melt days, almost entirely from 2012. The pattern of melt is inconsistent with the pattern of NIR albedo change. Both low and high elevations in central and NE Greenland show NIR albedo increases while only low elevations experience melt (and the duration is in fact increasing, not shown). Additionally, the boundary between NIR albedo positive and negative trends falls in the middle of the area with little melt but closely tracks the summit ridge of the ice sheet, suggesting it is accumulation related rather than melt related.

declining but are accompanied by rising B5 NIR albedo. We speculate that this signature could be evidence for substantial surface exposure and accumulation of mineral impurities in this region of GrIS. The spectra of most minerals exhibit higher NIR reflectance than bare ice, leading to the potential for a trend toward increased NIR albedo while visible albedo drops with high mineral content (Adams and Filice, 1967; Painter et al., 2003; Bøggild et al., 2010; Casey et al., 2012; Tedesco et al., 2013). The SE GrIS margin shows decreasing NIR albedo and only isolated change in visible wavelengths at the lowest elevations. We interpret this similarly to NW Greenland, only with much higher accumulation in this region. NIR impacts of increased wet snow and ice presence dominate. Higher accumulation drives the equilibrium line to a lower elevation, and melt accumulation of light-absorbing impurities plays a substantial role in lowering visible albedo below this elevation. Continuing around the periphery of the northern GrIS, the remainder of the ice sheet margin from approximately Scoresbysund to Humboldt Glacier, we see positive trends in NIR albedo and no significant trends in short wavelength visible albedo. In this region of low annual accumulation and long surface exposure times, this signature appears to indicate increasing snowfall, which

is confirmed by our separate preliminary analysis of MAR data in northeastern Greenland showing an increase in accumulation patterns starting in 2013 as well as GrIS surface mass balance climate model results (e.g., Noël et al., 2015). A very small addition of snowfall would cause more rapid surface burial and result in a lower age of surface grains and higher NIR albedo. Ultimately, each of these interpretations only clarifies what physical mechanisms would be consistent with the spectral signature changes observed. These hypotheses should be considered provisional and tested by in situ observation of snow and ice properties, which may be guided by satellite-identified signals.

5.2 Future use of MODIS data

The MODIS record is a powerful tool for assessing surface reflectance and albedo changes remotely. Our results indicate that future investigations should use the latest data recalibration (currently C6) data, and investigators should be aware of the limitations of the sensors, which we here attempt to restate plainly for the community's benefit:

1. Absolute trends in reflectance and albedo on the order of $0.01\,\text{decade}^{-1}$ are near the limits of the sensor calibration accuracy. Though statistically significant albedo trends of 0.01 or less TOA reflectance may exist over some surfaces with particularly stable albedo, trends at this level should be considered provisional and evaluated with great care, as they might not reflect actual physical processes.

2. Calibration drift is band dependent. Small band-dependent calibration degradations can be magnified in band ratio products, such as those used to detect dust mineralogy or aerosols, indicating spurious trends.

3. Data limitations are greatest in recently collected data. Vicarious C6 calibration, based on pseudo-invariant Earth sites, may not fully capture emerging trends in sensor degradation. Increasing Terra–Aqua discrepancies appear since 2014 in C6 data.

4. When the sensors disagree in ways not explained by overpass time, MODIS Aqua is likely to provide more stable data as Terra's calibration is expected to continue to degrade in ways that will make it challenging to characterize. Since C6 calibration is now vicarious (based on observation of pseudo-invariant desert sites), it is likely that emerging trends in sensor behavior will take some time to manifest in a statistically significant way for calibration revisions.

6 Conclusions

MODIS C6 calibration revisions result in substantial modification of decadal albedo trends on the GrIS reported by prior authors based on C5 data (e.g., Box et al., 2012; Stroeve et al., 2013; He et al., 2013; Polashenski et al., 2015). Declining C5 surface reflectance trends that were particularly pronounced in Terra's shortest wavelength bands are smaller or absent in C6 data. MODIS C6 surface reflectance and albedo data over dry snow areas of the GrIS feature mostly small, non-statistically significant trends in visible and NIR wavelengths. (The exceptions are the marginally statistically significant B1 decline in BSA and WSA dry snow albedo and the statistically significant B4 decline in BSA and WSA dry snow albedo.) These findings are consistent with the recent study of Polashenski et al. (2015), which suggested the dry snow albedo decline in C5 data would disappear in C6 after finding no enhancement in light-absorbing impurity concentrations on the interior Greenland Ice Sheet. The declining trends in wet snow and ice surface reflectance and albedo, independently supported by evidence of increased melt activity (Nghiem et al., 2012; Fausto et al., 2016b), remain statistically significant in C6 data, though at lower magnitude.

An examination of spatial, wavelength-specific variability in C6 albedo trends indicates several interesting attributes of GrIS albedo decline that may motivate future work to better understand the mechanisms controlling albedo feedbacks on the ice sheet. At higher elevations, patterns of NIR albedo change, including increasing reflectance in NE Greenland and declining reflectance in southern and western Greenland, highlight possible regional changes in metamorphism, precipitation and surface constituents. In the ablation zone, ratios of visible and NIR albedo trends suggest that differences in snowfall frequency, melt duration and surface exposure of light-absorbing impurities control recent albedo trends – with the net impact of these mechanisms being complex and likely dependent on the location and the seasonal timeframe chosen. Though the majority of albedo reduction occurs on the GrIS in melt-impacted areas, these results may support a crucial role for snow grain metamorphism in initiating (or preventing) feedbacks in dry snow.

Melt-related albedo reductions continue to have the potential to trigger significant ice–albedo feedbacks and accelerate melt and surface mass loss, and, as a result, melt initiation remains a critical process. Our findings, particularly those illustrating the regional complexity in spectral albedo trends, highlight the need for future work on GrIS albedo to define and differentiate the role of processes that control albedo decline.

The implications of this study extend beyond Greenland. The substantial revision from C5 to C6 MODIS products impacts a broad array of investigations. Conclusions based upon trends from C5 data, particularly shorter wavelength band Terra data, should be re-examined for robustness with C6 products. Future investigators should also note the limitations of MODIS products. Investigators should use great caution in evaluating trends of ~ 0.01 decade^{-1} or smaller and note that C6 corrections may not fully capture recent

and emerging trends in sensor degradation, particularly on the challenging-to-characterize Terra sensor.

Competing interests. The authors declare that they have no conflict of interest.

Acknowledgements. We acknowledge funding from NSF grants ARC-1204145 and 1304807 and from NASA grants NNX14AE72G, NNX14AD98G and NNX16AO75G. We thank NASA EOS, LP DAAC and NSIDC for providing MODIS data and Crystal Schaaf and Qingsong Sun for providing an executable file to grid recently released un-gridded MCD43 C5 data. We thank Marie Dumont for her service as editor as well as Jason Box and an anonymous reviewer for constructive comments which improved this manuscript.

Edited by: Marie Dumont

References

Adams, J. B. and Filice, A. L.: Spectral reflectance 0.4 to 2.0 microns of silicate rock powders, J. Geophys. Res., 72, 5705–5715, https://doi.org/10.1029/JZ072i022p05705, 1967.

Alexander, P. M., Tedesco, M., Fettweis, X., van de Wal, R. S. W., Smeets, C. J. P. P., and van den Broeke, M. R.: Assessing spatio-temporal variability and trends in modelled and measured Greenland Ice Sheet albedo (2000–2013), The Cryosphere, 8, 2293–2312, https://doi.org/10.5194/tc-8-2293-2014, 2014.

Barnes, W. L., Pagano, T. S., and Salomonson, V. V.: Prelaunch characteristics of the Moderate Resolution Imaging Spectroradiometer (MODIS) on EOS-AM1, IEEE T. Geosci. Remote, 36, 4, 1088–1100, https://doi.org/10.1109/36.700993, 1998.

Bøggild, C. E., Brandt, R. E., Brown, K. J., and Warren, S. G.: The ablation zone in northeast Greenland: ice types, albedos and impurities, J. Glaciol., 56, 101–113, 2010.

Box, J. E., Fettweis, X., Stroeve, J. C., Tedesco, M., Hall, D. K., and Steffen, K.: Greenland ice sheet albedo feedback: thermodynamics and atmospheric drivers, The Cryosphere, 6, 821–839, https://doi.org/10.5194/tc-6-821-2012, 2012.

Box, J. E.: Greenland Ice Sheet Mass Balance Reconstruction, Part II: Surface Mass Balance (1840–2010), J. Climate, 26, 6974–6989, 2013.

Box, J. E., van As, D., and Steffen, K.: Greenland, Canadian and Icelandic land ice albedo grids (2000–2016), Geological Survey of Denmark and Greenland Bulletin, 38, 69–72, 2017.

Casey, K. A., Kääb, A., and Benn, D. I.: Geochemical characterization of supraglacial debris via in situ and optical remote sensing methods: a case study in Khumbu Himalaya, Nepal, The Cryosphere, 6, 85–100, https://doi.org/10.5194/tc-6-85-2012, 2012.

Casey, K. A., Kaspari, S. D., Skiles, S. M., Kreutz, K., and Handley, M. J.: The spectral and chemical measurement of pollutants on snow near South Pole, Antarctica, J. Geophys. Res.-Atmos., 122, https://doi.org/10.1002/2016JD026418, 2017.

Dumont, M., Brun, E., Picard, G., Michou, M., Libois, Q., Petit, J. R., Geyer, M., Morin, S., and Josse, B.: Contribution of light-absorbing impurities in snow to Greenland's darkening since 2009, Nat. Geosci., 7, 509–512, https://doi.org/10.1038/ngeo2180, 2014.

Enderlin, E. M., Howat, I. M., Jeong, S., Noh, M. J., van Angelen, J. H., and van den Broeke, M. R.: An improved mass budget for the Greenland ice sheet, Geophys. Res. Lett., 41, 866–872, https://doi.org/10.1002/2013GL059010, 2014.

Fausto, R. S., van As, D., Box, J. E., Colgan, W., and Langen, P. L.: Quantifying the surface energy fluxes in South Greenland during the 2012 high melt episodes using in-situ observations, Front. Earth Sci., 4, 82, https://doi.org/10.3389/feart.2016.00082, 2016a.

Fausto, R. S., van As, D., Box, J. E., Colgan, W., Langen, P. L., and Mottram, R. H.: The implication of nonradiative energy fluxes dominating Greenland ice sheet exceptional ablation area surface melt in 2012, Geophys. Res. Lett., 43, 2649–2658, https://doi.org/10.1002/2016GL067720, 2016b.

Fettweis, X.: Reconstruction of the 1979–2006 Greenland ice sheet surface mass balance using the regional climate model MAR, The Cryosphere, 1, 21–40, https://doi.org/10.5194/tc-1-21-2007, 2007.

Fettweis, X., Tedesco, M., van den Broeke, M., and Ettema, J.: Melting trends over the Greenland ice sheet (1958–2009) from spaceborne microwave data and regional climate models, The Cryosphere, 5, 359–375, https://doi.org/10.5194/tc-5-359-2011, 2011.

Flannigan, M., Cantin, A. S., de Groot, W. J., Wotton, M., Newbery, A., and Gowman, L. M.: Global wildland fire season severity in the 21st century, Forest Ecol. Manag., 294, 54–61, https://doi.org/10.1016/j.foreco.2012.10.022, 2013.

Franz, B. A., Kwiatkowska, E. J., Meister, G., and McClain, C. R.: Moderate Resolution Imaging Spectroradiometer on Terra: limitations for ocean color applications, J. Appl. Remote Sens., 2, 023525, 17 pp., 2008.

Hall, D. K., Salomonson, V. V., and Riggs, G. A.: MODIS Terra Snow Cover Daily L3 Global 500m Grid, Version 5, Boulder, Colorado, USA, NASA National Snow and Ice Data Center Distributed Active Archive Center, https://doi.org/10.5067/63NQASRDPDB0, 2006a.

Hall, D. K., Salomonson, V. V., and Riggs, G. A.: MODIS Aqua Snow Cover Daily L3 Global 500m Grid, Version 5, Boulder, Colorado, USA, NASA National Snow and Ice Data Center Distributed Active Archive Center, https://doi.org/10.5067/ZFAEMQGSR4XD, 2006b.

Hall, D. K., Box, J. E., Casey, K. A., Hook, S. J., Shuman, C. A., and Steffen, K.: Comparison of satellite-derived and in-situ observations of ice and snow surface temperatures over Greenland, Remote Sens. Environ., 112, 3739–3749, https://doi.org/10.1016/j.rse.2008.05.007, 2008.

Hall, D. K. and Riggs, G. A.: MODIS Terra Snow Cover Daily L3 Global 500m Grid, Version 6, Boulder, Colorado, USA, NASA National Snow and Ice Data Center Distributed Active Archive Center, https://doi.org/10.5067/MODIS/MOD10A1.006, 2016a.

Hall, D. K. and Riggs, G. A.: MODIS Aqua Snow Cover Daily L3 Global 500m Grid, Version 6, Boulder, Colorado, USA, NASA National Snow and Ice Data Center Distributed Active Archive Center, https://doi.org/10.5067/MODIS/MYD10A1.006, 2016b.

He, T., Liang, S., Yu, Y., Wang, D., Gao, F., and Liu, Q.: Greenland surface albedo changes in July 1981-2012 from satellite observations, Environ. Res. Lett., 8, 044043, https://doi.org/10.1088/1748-9326/8/4/044043, 2013.

Hodson, A. J., Nowak, A., Cook, J., Sabacka, M., Wharfe, E. S., Pearce, D. A., Convey, P., and Vieira, G.: Microbes influence the biogeochemical and optical properties of maritime Antarctic snow, J. Geophys. Res.-Biogeo., 122, 1456–1470, https://doi.org/10.1002/2016JG003694, 2017.

Howat, I. M., Negrete, A., and Smith, B. E.: The Greenland Ice Mapping Project (GIMP) land classification and surface elevation data sets, The Cryosphere, 8, 1509–1518, https://doi.org/10.5194/tc-8-1509-2014, 2014.

Kohshima, S., Seko, K., and Yoshimura, Y.: Biotic acceleration of glacier melting in Yala glacier, Langtang region, Nepal Himalaya, Proceedings of the Kathmandu Symposium, November 1992, IAHS Publication 218, 1993.

Krabill, W., Abdalati, W., Frederick, E., Manizade, S., Martin, C., Sonntag, J., Swift, R., Thomas, R., Wright, W., and Yungel, J.: Greenland Ice Sheet: High Elevation Balance and Peripheral Thinning, Science, 289, 428–430, 2000.

Kwiatkowska, E. J., Franz, B. A., Meister, G., McClain, C. R., and Xiong, X.: Cross calibration of ocean-color bands from Moderate-Resolution Imaging Spectroradiometer on Terra platform, Appl. Opt., 47, 6796–6810, 2008.

LaChapelle, E. R.: Field Guide to Snow Crystals, University of Washington Press, Seattle, WA, USA, 112 pp., 1969.

Levy, R. C., Remer, L. A., Kleidman, R. G., Mattoo, S., Ichoku, C., Kahn, R., and Eck, T. F.: Global evaluation of the Collection 5 MODIS dark-target aerosol products over land, Atmos. Chem. Phys., 10, 10399–10420, https://doi.org/10.5194/acp-10-10399-2010, 2010.

Levy, R. C., Mattoo, S., Munchak, L. A., Remer, L. A., Sayer, A. M., Patadia, F., and Hsu, N. C.: The Collection 6 MODIS aerosol products over land and ocean, Atmos. Meas. Tech., 6, 2989–3034, https://doi.org/10.5194/amt-6-2989-2013, 2013.

Lyapustin, A., Wang, Y., Xiong, X., Meister, G., Platnick, S., Levy, R., Franz, B., Korkin, S., Hilker, T., Tucker, J., Hall, F., Sellers, P., Wu, A., and Angal, A.: Scientific impact of MODIS C5 calibration degradation and C6+ improvements, Atmos. Meas. Tech., 7, 4353–4365, https://doi.org/10.5194/amt-7-4353-2014, 2014.

Meister, G. and Franz, B. A.: Corrections to the MODIS Aqua calibration derived from MODIS Aqua ocean color products, IEEE T. Geosci. Remote, 52, 6534–6541, https://doi.org/10.1109/TGRS.2013.2297233, 2014.

Moody, E. G., King, M. D., Schaaf, C. B., Hall, D. K., and Platnick, S.: Northern Hemisphere five-year average (2000–2004) spectral albedos of surfaces in the presence of snow: statistics computed from Terra MODIS land products, Remote Sens. Environ., 111, 337–345, 2007.

Nghiem, S. V., Hall, D. K., Mote, T. L., Tedesco, M., Albert, M. R., Keegan, K., Shuman, C. A., DiGirolamo, N. E., and Neumann, G.: The extreme melt across the Greenland ice sheet in 2012, Geophys. Res. Lett., 39, L20502, https://doi.org/10.1029/2012GL053611, 2012.

Noël, B., van de Berg, W. J., van Meijgaard, E., Kuipers Munneke, P., van de Wal, R. S. W., and van den Broeke, M. R.: Evaluation of the updated regional climate model RACMO2.3: summer snowfall impact on the Greenland Ice Sheet, The Cryosphere, 9, 1831–1844, https://doi.org/10.5194/tc-9-1831-2015, 2015.

Painter, T. H., Duval, B., Thomas, W. H., Mendez, M., Heintzelman, S., and Dozier, J.: Detection and quantification of snow algae with an airborne imaging spectrometer, Appl. Environ. Microbiol., 67, 5267–5272, 2001.

Painter, T. H., Dozier, J., Roberts, D. A., Davis, R. E., and Green, R. O.: Retrieval of subpixel snow-covered area and grain size from imaging spectrometer data, Remote Sens. Environ., 85, 64–77, 2003.

Painter, T. H., Barrett, A. P., Landry, C. L., Neff, J. C., Cassidy, M. P., Lawrence, C. R., McBride, K. E., and Farmer, G. L.: Impact of disturbed desert soils on duration of mountain snow cover, Geophys. Res. Lett., 34, L12502, https://doi.org/10.1029/2007GL030284, 2007.

Platnick, S., King, M., Meyer, K. G., Wind, G., Amarasinghe, N., Marchant, B., Arnold, G. T., Zhang, Z., Hubanks, P. A., Ridgway, B., and Riedi, J.: MODIS Cloud Optical Properties: User Guide for the Collection 6 Level-2 MOD06/MYD06 Product and Associated Level-3 Datasets, Version 1.0, 145 pp., 2015.

Polashenski, C. M., Dibb, J. E., Flanner, M. G., Chen, J. Y., Courville, Z. R., Lai, A. M., Schauer, J. J., Shafer, M. M., and Bergin, M.: Neither dust nor black carbon causing apparent albedo decline in Greenland's dry snow zone: Implications for MODIS C5 surface reflectance, Geophys. Res. Lett., 42, 9319–9327, https://doi.org/10.1002/2015GL065912, 2015.

Rignot, E., Velicogna, I., van den Broeke, M. R., Monaghan, A., and Lenaerts, J. T. M.: Acceleration of the contribution of the Greenland and Antarctic ice sheets to sea level rise, Geophys. Res. Lett., 38, L05503, https://doi.org/10.1029/2011GL046583, 2011.

Ryan, J. C., Hubbard, A., Irvine-Flynn, T. D., Doyle, S. H., Cook, J. M., Stibal, M., and Box, J. E.: How robust are in-situ observations for validating satellite-derived albedo over the dark zone of the Greenland Ice Sheet?, Geophys. Res. Lett., 44, 6218–6225, https://doi.org/10.1002/2017GL073661, 2017.

Sayer, A. M., Hsu, N. C., Bettenhausen, C., Jeong, M. J., and Meister, G.: Effect of MODIS Terra radiometric calibration improvements on Collection 6 Deep Blue aerosol products: Validation and Terra/Aqua consistency, J. Geophys. Res.-Atmos., 120, 12157–12174, 2015.

Schaaf, C. and Wang, Z.: MCD43A3 MODIS/Terra+Aqua BRDF/Albedo Quality Daily L3 Global – 500 m V006, NASA EOSDIS Land Processes Distributed Active Archive Center (LP DAAC), USGS/Earth Resources Observation and Science (EROS) Center, Sioux Falls, South Dakota, USA, https://doi.org/10.5067/MODIS/MCD43A3.006, 2015.

Soja, A. J., Tchebakova, N. M., French, N. H. F., Flannigan, M. D., Shugart, H. H., Stocks, B. J., Sukhinin, A. I., Parfenova, E. I., Chapin III, F. S., and Stackhouse Jr., P. W.: Climate-induced boreal forest change: Predictions versus current observations, Global Planet. Change, 56, 274–296, 2007.

Stroeve, J. C., Box, J. E., Gao, F., Liang, S., Nolin, A., and Schaaf, C.: Accuracy assessment of the MODIS 16-day albedo product for snow: comparisons with Greenland in situ measurements, Remote Sens. Environ., 94, 46–60, 2005.

Stroeve, J. C., Box, J. E., and Haran, T.: Evaluation of the MODIS (MOD10A1) daily snow albedo product over the Greenland ice sheet, Remote Sens. Environ., 105, 155–171, 2006.

Stroeve, J., Box, J. E., Wang, Z., Schaaf, C., and Barrett, A.: Re-evaluation of MODIS MCD43 Greenland albedo accuracy and trends, Remote Sens. Environ., 138, 199–214, 2013.

Sun, J., Angal, A., Xiong, X., Chen, H., Geng, X., Wu, A., Choi, T., and Chu, M.: MODIS RSB calibration improvements in Collection 6, Proc. SPIE 8528, Earth Observing Missions and Sensors: Development, Implementation and Characterization II, 85280N, https://doi.org/10.1117/12.979733, 2012.

Takeuchi, N.: Temporal and spatial variations in spectral reflectance and characteristics of surface dust on Gulkana Glacier, Alaska range, J. Glaciol., 55, 701–709, 2009.

Tedesco, M.: Greenland Daily Surface Melt 25km EASE-Grid, (2001–2015), New York, NY, USA: City College of New York, The City University of New York, Digital media, 2014.

Tedesco, M., Fettweis, X., van den Broeke, M. R., van de Wal, R. S. W., Smeets, C. J. P. P., van de Berg, W. J., Serreze, M. C., and Box, J. E.: The role of albedo and accumulation in the 2010 melting record in Greenland, Environ. Res. Lett., 6, 014005, 6 pp., 2011.

Tedesco, M., Foreman, C. M., Anton, J., Steiner, N., and Schwartzman, T.: Comparative analysis of morphological, mineralogical and spectral properties of cryoconite in Jakobshavn Isbræ, Greenland, and Canada Glacier, Antarctica, Ann. Glaciol., 54, 147–157, 2013.

Tedesco, M., Doherty, S., Warren, W., Tranter, M., Stroeve, J., Fettweis, X., and Alexander, P.: What darkens the Greenland Ice Sheet?, EOS, 96, https://doi.org/10.1029/2015EO035773, 2015.

Tedesco, M., Doherty, S., Fettweis, X., Alexander, P., Jeyaratnam, J., and Stroeve, J.: The darkening of the Greenland ice sheet: trends, drivers, and projections (1981–2100), The Cryosphere, 10, 477–496, https://doi.org/10.5194/tc-10-477-2016, 2016.

Thomas J. L., Polashenski, C. M., Soja, A. J., Marelle, L., Casey, K. A., Choi, H. D., Raut, J.-C., Wiedinmyer, C., Emmons, L. K., Fast, J., Pelon, J., Law, K. S., Flanner, M. G., and Dibb, J. E.: Quantifying black carbon deposition over the Greenland ice sheet from forest fires in Canada, Geophys. Res. Lett., 44, https://doi.org/10.1002/2017GL073701, 2017.

Toller, G., Xiong, X., Sun, J., Wenny, B. N., Geng, X., Kuyper, J., Angal, A., Chen, H., Madhavan, S., and Wu, A.: Terra and Aqua moderate-resolution imaging spectroradiometer collection 6 level 1B algorithm, J. Appl. Remote Sens., 7, 073557, https://doi.org/10.1117/1.JRS.7.073557, 2013.

van As, D. and Fausto, R. S.: Programme for monitoring of the Greenland ice sheet (PROMICE): first temperature and ablation records, Geological Survey of Denmark and Greenland, 23, 73–76, 2011.

van den Broeke, M., van As, D., Reijmer, C., and van de Wal, R.: Assessing and improving the quality of unattended radiation observations in Antarctica, J. Atmos. Ocean. Tech., 21, 1417–1431, 2004.

van den Broeke, M., Bamber, J., Ettema, J., Rignot, E., Schrama, E., van de Berg, W. J., van Meijgaard, E., Velicogna, I., and Wouters, B.: Partitioning recent Greenland mass loss, Science, 326, 984–986, https://doi.org/10.1126/science.1178176, 2009.

van den Broeke, M. R., Enderlin, E. M., Howat, I. M., Kuipers Munneke, P., Noël, B. P. Y., van de Berg, W. J., van Meijgaard, E., and Wouters, B.: On the recent contribution of the Greenland ice sheet to sea level change, The Cryosphere, 10, 1933–1946, https://doi.org/10.5194/tc-10-1933-2016, 2016.

Velicogna, I. and Wahr, J.: Time-variable gravity observations of ice sheet mass balance: Precision and limitations of the GRACE satellite data, Geophys. Res. Lett., 40, 3055–3063, https://doi.org/10.1002/grl.50527, 2013.

Vermote, E.: MODIS/Surface Reflectance 8-Day L3 Global 500 m SIN Grid V005, NASA EOSDIS Land Processes Distributed Active Archive Center (LP DAAC), USGS/Earth Resources Observation and Science (EROS) Center, Sioux Falls, South Dakota, USA, 2007.

Vermote, E.: MODIS Terra Surface Reflectance 8-Day L3 Global 500m SIN Grid V006, Sioux Falls, South Dakota, USA, NASA EOSDIS US Geological Survey Land Processes Distributed Active Archive Center (LP DAAC), https://doi.org/10.5067/MODIS/MOD09A1.006, 2015a.

Vermote, E.: MODIS Aqua Surface Reflectance 8-Day L3 Global 500m SIN Grid V006, Sioux Falls, South Dakota, USA, NASA EOSDIS US Geological Survey Land Processes Distributed Active Archive Center (LP DAAC), https://doi.org/10.5067/MODIS/MYD09A1.006, 2015b.

Wang, D., Morton, D., Masek, J., Wu, A., Nagol, J., Xiong, X., Levy, R., Vermote, E., and Wolfe, R.: Impact of sensor degradation on the MODIS NDVI time series, Remote Sens. Environ., 119, 55–61, 2012.

Wang, W., Zender, C. S., van As, D., Smeets, P. C. J. P., and van den Broeke, M. R.: A Retrospective, Iterative, Geometry-Based (RIGB) tilt-correction method for radiation observed by automatic weather stations on snow-covered surfaces: application to Greenland, The Cryosphere, 10, 727–741, https://doi.org/10.5194/tc-10-727-2016, 2016.

Warren, S. G. and Wiscombe, W. J.: A Model for the Spectral Albedo of Snow. II: Snow Containing Atmospheric Aerosols, J. Atmos. Sci., 37, 2734–2745, 1980.

Wiscombe, W. J. and Warren, S. G.: A model for the spectral albedo of snow. I: Pure snow, J. Atmos. Sci., 37, 2712–2733, 1980.

Wright, P., Bergin, M., Dibb, J., Lefer, B., Domine, F., Carman, T., Carmagnola, C., Dumont, M., Courville, Z., and Schaaf, C.: Comparing MODIS daily snow albedo to spectral albedo field measurements in Central Greenland, Remote Sens. Environ., 140, 118-129, 2014.

Xiong, X. and Barnes, W.: An overview of MODIS radiometric calibration and characterization, Adv. Atmos. Sci., 23, 69–79, 2009.

Xiong, X., Wu, A., Wenny, B. N., Madhavan, S., Wang, Z., Li, Y., Chen, N., Barnes, W. L., and Salomonson, V. V.: Terra and Aqua MODIS Thermal Emissive Bands On-Orbit Calibration and Performance, IEEE T. Geosci. Remote, 53, 5709–5721, https://doi.org/10.1109/TGRS.2015.2428198, 2015.

Zennaro, P., Kehrwald, N., McConnell, J. R., Schüpbach, S., Maselli, O. J., Marlon, J., Vallelonga, P., Leuenberger, D., Zangrando, R., Spolaor, A., Borrotti, M., Barbaro, E., Gambaro, A., and Barbante, C.: Fire in ice: two millennia of boreal forest fire history from the Greenland NEEM ice core, Clim. Past, 10, 1905–1924, https://doi.org/10.5194/cp-10-1905-2014, 2014.

Zhan, Y. and Davies, R.: Intercalibration of CERES, MODIS and MISR reflected solar radiation and its application to albedo trends, J. Geophys. Res.-Atmos., 121, 6273–6283, 2016.

Permissions

List of Contributors

Laurence Gray and Luke Copland
Department of Geography, Environment and Geomatics, University of Ottawa, Ottawa, ON K1N 6N5, Canada

David Burgess
Geological Survey of Canada, Natural Resources Canada, Ottawa, ON K1A 0E8, Canada

Thorben Dunse
Department of Geosciences, University of Oslo, 0316 Oslo, Norway

Kirsty Langley
Asiaq, Greenland Survey, 3900 Nuuk, Greenland

Geir Moholdt
Norwegian Polar Institute, 9296 Tromso, Norway

Pirmin Philipp Ebner and Martin Schneebeli
WSL Institute for Snow and Avalanche Research SLF, 7260 Davos Dorf, Switzerland

Hans Christian Steen-Larsen
LSCE Laboratoire des Sciences du Climat et de l'Environnement, Gif-Sur-Yvette CEDEX, France
Center for Ice and Climate, Niels Bohr Institute, University of Copenhagen, Copenhagen, Denmark

Barbara Stenni
Department of Environmental Sciences, Informatics and Statistics, University Ca' Foscari of Venice, Venice, Italy

Aldo Steinfeld
Department of Mechanical and Process Engineering, ETH Zurich, 8092 Zurich, Switzerland

J. E. Jack Reeves Eyre and Xubin Zeng
Department of Hydrology and Atmospheric Sciences, University of Arizona, Tucson, 85721, USA

Louise Steffensen Schmidt, Guðfinna Aðalgeirsdóttir, Finnur Pálsson and Helgi Björnsson
University of Iceland, Institute of Earth Sciences, Reykjavik, Iceland

Sverrir Guðmundsson
University of Iceland, Institute of Earth Sciences, Reykjavik, Iceland
Keilir Institute of Technology, Reykjanesbær, Iceland

Peter L. Langen and Ruth Mottram
Danish Meteorological Institute, Copenhagen, Denmark

Simon Gascoin
Centre d'Etudes Spatiales de la Biosphère, Université de Toulouse, CNES/CNRS/IRD/UPS, Toulouse, France

Alexis Berne
Environmental Remote Sensing Laboratory (LTE), École Polytechnique Fédérale de Lausanne (EPFL), Lausanne, Switzerland

Jacopo Grazioli
Environmental Remote Sensing Laboratory (LTE), École Polytechnique Fédérale de Lausanne (EPFL), Lausanne, Switzerland
MeteoSwiss, Locarno-Monti, Switzerland

Christophe Genthon, Brice Boudevillain and Claudio Duran-Alarcon
Univ. Grenoble Alpes, CNRS, IGE, 38000 Grenoble, France

Massimo Del Guasta
Istituto nazionale di Ottica, INO-CNR, Italy

Jean-Baptiste Madeleine
Sorbonne Universités, UPMC Univ Paris 06, UMR 8539, Laboratoire de Météorologie Dynamique (IPSL), Paris, France
CNRS, UMR 8539, Laboratoire de Météorologie Dynamique (LMD), IPSL Climate Modeling Center, Paris, France

John Faulkner Burkhart
Department of Geosciences, University of Oslo, Oslo, Norway
University of California, Merced, CA, USA

Arve Kylling
Norwegian Institute for Air Research, Kjeller, Norway

Crystal B. Schaaf
School for the Environment, University of Massachusetts Boston, Boston, MA, USA

Zhuosen Wang
NASA Goddard Space Flight Center, Greenbelt, MD, USA
Earth System Science Interdisciplinary Center, University of Maryland, College Park, MD, USA

Wiley Bogren
U.S. Geological Survey, Flagstaff, AZ, USA

Rune Storvold and Stian Solbø
Norut-Northern Research Institute, Tromsø, Norway

Christina A. Pedersen and Sebastian Gerland
Norwegian Polar Institute, Fram Centre, Tromsø, Norway

J. Rachel Carr
School of Geography, Politics and Sociology, Newcastle University, Newcastle-upon-Tyne, NE1 7RU, UK

Heather Bell
Department of Geography, Durham University, Durham, DH13TQ, UK

Rebecca Killick
Department of Mathematics & Statistics, Lancaster University, Lancaster, LA1 4YF, UK

Tom Holt
Centre for Glaciology, Department of Geography and Earth Sciences, Aberystwyth University, Aberystwyth, SY23 4RQ, UK

Emmy E. Stigter and Walter W. Immerzeel
Department of Physical Geography, Utrecht University, Utrecht, the Netherlands

Niko Wanders
Department of Civil and Environmental Engineering, Princeton University, Princeton, NJ, USA

Tuomo M. Saloranta
Norwegian Water Resources and Energy Directorate (NVE), Oslo, Norway

Joseph M. Shea
International Centre for Integrated Mountain Development, Kathmandu, Nepal
Centre for Hydrology, University of Saskatchewan, Saskatchewan, Canada

Marc F. P. Bierkens
Department of Physical Geography, Utrecht University, Utrecht, the Netherlands
Deltares, Utrecht, the Netherlands

Christian Katlein and Stefan Hendricks
Alfred-Wegener-Institut Helmholtz-Zentrum für Polar- und Meeresforschung, 27570 Bremerhaven, Germany

Jeffrey Key
Center for Satellite Applications and Research, NOAA/NESDIS, Madison, Wisconsin, USA

Robert Ricker
Alfred Wegener Institute, Helmholtz Centre for Polar and Marine Research, Bremerhaven, Bussestrasse 24, 27570 Bremerhaven, Germany
Univ. Brest, CNRS, IRD, Ifremer, Laboratoire d'Oceanographie Physique et Spatiale (LOPS), IUEM, 29280 Brest, France

Stefan Hendricks
Alfred Wegener Institute, Helmholtz Centre for Polar and Marine Research, Bremerhaven, Bussestrasse 24, 27570 Bremerhaven, Germany

Lars Kaleschke and Xiangshan Tian-Kunze
Institute of Oceanography, University of Hamburg, Bundesstrasse 53, 20146 Hamburg, Germany

Jennifer King
Norwegian Polar Institute, Tromsø, Norway

Christian Haas
Alfred Wegener Institute, Helmholtz Centre for Polar and Marine Research, Bremerhaven, Bussestrasse 24, 27570 Bremerhaven, Germany
Department of Earth and Space Sciences and Engineering, York University, Toronto, ON, Canada

Joseph Graly, Joel Harrington and Neil Humphrey
Department of Geology and Geophysics, University of Wyoming, 1000 E. University Ave. Laramie, WY 82071, USA

Andrew K. Hamilton
Department of Civil Engineering, University of British Columbia, Vancouver, British Columbia, Canada
Department of Geography and Environmental Studies, Carleton University, Ottawa, Ontario, Canada

Bernard E. Lava
Department of Civil Engineering, University of British Columbia, Vancouver, British Columbia, Canada

Derek R. Mueller
Department of Geography and Environmental Studies, Carleton University, Ottawa, Ontario, Canada

Warwick F. Vincent
Department of Biology and Centre for Northern Studies (CEN), Université Laval, Quebec City, Quebec, Canada

Luke Copland
Department of Geography, Environment, and Geomatics, University of Ottawa, Ottawa, Ontario, Canada

Kimberly A. Casey
Thayer School of Engineering, Dartmouth College, Hanover, NH 03755, USA
Cryospheric Sciences Lab, NASA Goddard Space Flight Center, Greenbelt, MD 20771, USA

Chris M. Polashenski
Thayer School of Engineering, Dartmouth College, Hanover, NH 03755, USA
Cold Regions Research and Engineering Laboratory, Alaska Projects Office, US Army Corps of Engineers, Fairbanks, AK 99709, USA

Justin Chen
Department of Computer Science, Stanford University, Stanford, CA 94305, USA

Marco Tedesco
Lamont–Doherty Earth Observatory, Columbia University, NY 10964, USA
NASA Goddard Institute for Space Studies, New York, NY 10025, USA

Index